SURVEY OF THE UNIVERSE

SURVEY OF THE UNIVERSE

DONALD H. MENZEL

Paine Professor of Practical Astronomy, Harvard University
Physicist, Smithsonian Astrophysical Observatory
Cambridge, Massachusetts

FRED L. WHIPPLE

Director, Smithsonian Astrophysical Observatory
Professer of Astronomy, Harvard University
Cambridge, Massachusetts

GERARD De VAUCOULEURS

Professor, Dept. of Physics and Astronomy
University of Texas
Austin, Texas

PRENTICE-HALL, INC., ENGLEWOOD CLIFFS, NEW JERSEY

Current printing (last digit):

10 9 8 7 6 5 4 3 2 1

13–879163–5
Library of Congress Catalog Card Number: 79–149820
Printed in the United States of America

PRENTICE-HALL INTERNATIONAL, INC., London
PRENTICE-HALL OF AUSTRALIA, PTY. LTD., Sydney
PRENTICE-HALL OF CANADA LTD., Toronto
PRENTICE-HALL OF INDIA PRIVATE LTD., New Delhi
PRENTICE-HALL OF JAPAN, INC., Tokyo

Preface

His great facility for communication and for the rapid accumulation of knowledge distinguishes man from all other living organisms. Knowledge to be appreciated esthetically or to be used practically, must be communicated. This book aims to present the basic knowledge of astronomy to readers having no training in mathematics or science beyond a modest secondary school level. Although knowledge in the physical sciences can be concisely expressed in mathematical form, most of it can be well transmitted and appreciated without mathematical formulation or, in some cases, through simple formulae. A lack of mathematical training or aptitude need turn no one from the profitable reading of this book.

We desire to present the reader with an overview of the rapidly expanding knowledge of the physical universe. We hope to impart some of the excitement and enthusiasm for astronomy that has drawn the three of us into life works in research and teaching. The structure of this knowledge is neither rigid nor static but continually evolves and expands. Every statement, every conclusion, every generality is suspect, subject to review, change, and modification as observation and theory mutually restructure and expand our understanding of the universe. Even observations are not sacrosanct because they can be improved and reinterpreted. Theories are fallible under the brutal scrutiny of new techniques of observation. On the other hand, our continually restructured body of knowledge represents a closer and closer approximation to the universe as we can observe it. Only the exciting borderline of exploration need be ill-defined.

We attempt to differentiate clearly the well-founded basic knowledge from speculative ideas, even though the latter are essential to progress near the limits of the known. Because scientific knowledge is the handiwork of individual human beings and consists only of what they have individually contributed, we attempt to show how the major contributors have developed the structure historically.

A frail body limits man's mobility and provides him with little strength and a weak defense against the hazards of his environment. His sensory organs, although surprisingly acute and diversified, are distinctly limited, particularly with regard to the huge bands of radiation outside the visual range. His natural communication range is small. His remarkably compact, effective, and efficient brain can store and recall only a trivial fraction of the enormous input from his sensory equipment. A short life limits his experience to a few brief decades in time.

Knowledge applied via technology extends man's mobility, strength, defense, and communicative powers by almost unlimited amounts. Similarly, it expands his sensory equipment to types of radiation input not dreamed of in ancient times and possibly to energy sources yet to be discovered. Libraries and computers store knowledge for ready access by the human brain to an almost inconceivable potential. Computers correspondingly expand the numerical and, in some senses, the logical powers of the mind. By means of this knowledge, man's lifespan effectively lengthens over the historical period. Astronomy and imagination lead man in time and space through the observable universe.

Now that man, by knowledge and application, has developed comparatively god-like powers he needs more than ever to "find himself" and to use these powers of technology and science for the greatest possible benefit to the individual and to the human race. Although science cannot identify the path that man should choose among an increasing multitude of possible paths, it can help the individual to discover where he stands, in much the same fashion that primitive man's senses and mind enabled him to evaluate his position in a new terrain. The three authors of this book believe that his accumulated knowledge has always been relevant to the decisions that man must make in his potential for action and accomplishment, today even more than ever before. We believe that, among the many branches of science, astronomy is particularly useful in aiding the individual search for his own proper niche with respect to the physical universe. While astronomy cannot answer the basic questions regarding human goals and standards, it can broaden an individual's view of his environment and, perhaps, of himself.

The authors are particularly indebted to Dr. Joseph Ashbrook and Dr. Cecilia Payne-Gaposhkin who contributed to the writing of early versions of much of this text. Mrs. Antoinette de Vaucouleurs contributed in a major fashion to many aspects of exposition, factual data, and tabular material and especially to the index. Dr. Paul Griboval receives our deep appreciation for producing most of the diagrams as does Mrs. Lyle Boyd for her editorial contribution. Dr. Thornton L. Page and Dr. Andrew Young have aided materially through their careful reading and criticism of an early version of this book. We also thank Dr. Bernard Davis and Dr. George Pasteur for advice on Chapter 18, Extraterrestrial Life, which was based in turn on an article by Dr. Menzel, published in The Graduate Journal (of the University of Texas, Volume VII, 195, 1965). For devoted and patient secretarial assistance, we wish to thank Miss Doris Beaudoin. For illustrative material, we are especially indebted to the Lick Observatory, the Mount Wilson and Palomar Observatories, Sky and Telescope Magazine, and the U. S. National Aeronautics and Space Administration.

The Authors

Contents

8   Radiation and Atomic Structure 120

9   The Sun and its Radiations 144

SURVEY OF THE UNIVERSE

Astronomy and the Origins of Science

1-1 Astronomy and Modern Life. Some great thinkers have maintained that astronomy is the mother of all sciences. Sir Arthur Eddington once suggested that, on a completely cloudbound planet, man could sit down with pencil and paper and deduce the stars. We question this possibility and prefer the view of the French physicist Jean Perrin: "It is a feeble light that comes to us from the stars, but without it what would be the present condition of Man's mind?"

There can be no question that the scientific method together with astronomical studies have profoundly influenced philosophy, art, and literature as well as religion (see Fig. 1.1). Certainly the philosophies of recent centuries were affected by astronomical and other scientific developments, and philosophy today is struggling to accommodate itself to scientific results.

Astronomy is no longer an "ivory tower," visited by a few specialists whose minds are remote from people and current events. Man has transcended earth's gravity, sent scientific probes hundreds of millions of miles into space, and has begun to explore the moon and planets directly. Space science and the vast technology of today's scientific explosion provide astronomers with tools and methods of almost unbelievable power, precision, and variety. The atmospheric barrier with its optical unsteadiness and its opacity to most of the radiation from the universe is now an opening window.

FIGURE 1-1 *Mayan Calendar Stone, or Stone of the Sun. (Courtesy Museo Nacional de Antropologia, Mexico City.)*

But how does man fit into the universe? Is he a divine creation, a cosmic accident, or a typical result of universal chemical and physical laws? Even though astronomy may not be able to answer these basic questions and is not expected to deal with human values, it provides new and significant fundamental knowledge, which the individual can incorporate into his philosophy of life to increase both his depth and his breadth of understanding.

Today, as never before in history, an understanding of the scientific method and an appreciation of its power in areas outside science itself are critical to full and successful living. Astronomy—among the first of the sciences—illustrates the development of the scientific method and its potential for the future.

1-2   Early Science.   Almost from the beginning, when man first began to notice the world around him, the things that he saw called for an explanation. What was a rock and why did it fall? What was fire and why did it rise? And water? Why was it sometimes liquid, sometimes hard like a rock, and why did it, when left alone, vanish into nothingness?

The ancients, noting the sequence of night and day and the recurrence of the seasons, certainly wondered about the sun. How and why did it move? Where did it go at night? And how did it manage to get from the west, where it set, around to the east, ready to rise the next morning? Various cultures devised explanations for this phenomenon. Certain South American Indians believed that a savage tribe captured the sun when it reached the western horizon, cut it up into little pieces, fried them in a pan and put them in the sky where they became the stars. A new sun, born during the night, was ready to undertake the journey the following day.

Another idea, widely accepted in early times, regarded the sun as a god who travelled across the sky in a chariot or other conveyance. One culture believed that a race of giants trapped the setting sun in a net, bound it, put it in a boat, and rowed it rapidly out of sight beyond the northern horizon or through vast subterranean caverns, to release it in the east the next morning.

Such nature myths existed in great abundance. Each was devised to explain a single phenomenon. The echo, for example, was a mischievous nymph who pined away until only her voice was left, ready to mimic any call. A giant, chained underground, caused earthquakes as he tried to break his bonds and escape. And so on.

Almost inevitably many cultures came to worship the sun as a god. The sun clearly brought the warm summer and beneficent weather. The chill and dreaded winter came as the days shortened and the sun stayed closer to the horizon.

The fertile valleys of the Euphrates, the Tigris, the Indus, and the Nile fostered early civilizations perhaps 10,000 years ago. The records from those days are fragmentary, but they are enough to show the nature of man's attitude as he began to develop agriculture. The weather seemed as variable as human emotions: love, hate, anger. Hence man invested his gods with human qualities and imagined them controlling the various forces of nature according to whim. He visualized gods who blew the winds with lungs like enormous bellows. Gods of storm, when angry, could unleash the rain, snow, thunder, lightning, hurricanes, and floods.

If an emotional god could arbitrarily and mercilessly impose his will upon the inhabitants of this world, perhaps man could curry favor with that god by bringing gifts, making sacrifices, or performing some ritual. Thus primitive people began to design ceremonies for special occasions.

Drought, for example, is a recurring affliction of nature. The ancients would have noted that frogs croak when it rains. Confusing cause and effect, they concluded that the frogs brought the rain. Hence man painted himself green and hopped around croaking like a frog in an effort to induce rain. And if rain chanced to fall, what further proof could one need of the efficacy of such a rite? If rain failed to come, perhaps the rain god was still angry.

For efficient cultivation and irrigation of fields, man required calendars to keep track of the seasons and seasonal weather (Fig. 1-1). For this purpose he used the three natural divisions of time: the day, the month (the interval between successive full moons), and the year. We shall discuss the problems of calendars later on (Chapter 5). For the moment, let us note that any attempt to set up a calendar required observations of the sun and the moon, and the charting of their slow motions eastward with respect to the stars. Before man mapped the stars, he formed them into constellations—objects, animals, or persons—convenient patterns for ready recognition. The wandering planets joined the sun and moon as major gods. Many of the lesser gods found a place in the various star groups, often illustrating the myths about these gods and their interaction with mankind.

FIGURE 1-2   *Stonehenge, South England, an observatory erected in about 19th century B.C. (Courtesy Robert Citron.)*

The early "astronomers" who observed and began to keep records of the sky surrounded themselves with an aura of mysticism and religion (Fig. 1-2). Many were actually high priests. In this fashion *astrology*, the cult of predicting human events from the positions of heavenly bodies, and *astronomy*, the science of these bodies, developed from a common stem of human need, curiosity, observation, and invention. Even through the middle ages, the distinction between astrology and astronomy remained vague.

1-3   The Scientific Method.   What is the scientific method? How did it come about? Although myths satisfied the psychological needs of ancient man, he had yet to learn to subject his ideas to experimental test. Even today many people are quite happy with a convincingly authoritative answer to a question, regardless of merit. But experimental evidence alone does not necessarily prove or disprove a theory.

Consider the man who made an offering to the god of crops by inserting a dead fish into the ground along with the seed. He soon had visible proof of the god's response, for his corn crop enormously exceeded that of the man who offered no such sacrifice. Here we have a fundamental confidence or faith in the uniformity of nature, an act of observation or experiment, and finally the test of a theory. But these three fundamental ingredients of science do not in themselves constitute the scientific method, even when the theory is augmented into an extensive, embracive logical system.

The failure of logic alone to develop science is beautifully illustrated by early mathematics. The properties of triangles, circles, spheres, cones, and so on follow from logical reasoning alone. One need not resort to experiment or even to measurement in order to prove the Pythagorean theorem: the sum of the squares of two sides of a right triangle is equal to the square of the hypotenuse. The esthetic goal of perfection inherent in early geometrical progress led to a dead end in science. Since the universe seemed to possess a geometrical structure, man sought to study it by applying the same abstract logic that had been successful in mathematics. The only control on the imagination was man's first firm belief in the "perfection" of the universe. Thus Plato speculated that the distances of the heavenly bodies from the earth might follow some simple numerical formula, and all heavenly bodies were considered to be perfect spheres until well after Galileo (1564–1642) showed with his telescope that the Moon was indeed rough and mountainous. Even this observation was almost universally questioned for many years because of the tacit assumption concerning the perfection of heavenly bodies.

Even the multitudinous systematic observations in ancient astronomy and the physical world generally, coupled with the application of reason, were slow to produce the full expression of the scientific method. To see what is lacking, let us return to the primitive man who buried a fish along with a seed. He had indeed proved an important scientific fact of real value to his society: that dead fish improve the crop. But his theory was inadequately developed. Extending it, he should and perhaps did try planting other valuables with his seed, such as rings, pots, or other precious things. Possibly, after experiment, he concluded that gods prefer fish to pots. But scientific progress stopped there, because both the philosophy and methodology of the scientific method were lacking.

The scientific method requires that *every theory be developed as far as possible* and then that *all predictions be checked by observation or experiment* to *expand or narrow the range of application of the theory.*

Thus the complete scientific method stems from all of the following components:

Confidence or faith in the uniformity of nature.

Recognition of the value of systematic observations or experiments.

The exercise of imagination in developing theories that interrelate as many observations or experiments as possible.

The confirmation or validation of each theory by the prediction of critical observations or experiments.

The continuous challenge to every theory by further observations or experiments.

The complete scientific method is an endless spiral of observations, theory, prediction, observation or experiment, theory, prediction, and so on ad infinitum. At no point on the spiral do we find or expect to find perfection. But each circuit leads us to greater knowledge of the universe and more general statements of our knowledge.

Thus simple relations, such as Newton's universal law of gravitation, become only approximations to more general statements, such as Einstein's theory of relativity. Application of the scientific method leads to new physical, chemical, biological, or other relationships that in turn lead to new technologies, new devices, and new concepts. These improve our technological society and provide new fuels for further advance in the field of science. All parts of science are interrelated; indeed, many of our greatest forward steps occur when results in one field are applied in a completely different field. Thus the study of the atomic nucleus leads, on the one hand, to methods of medical and biological progress and, on the other, to methods of unraveling the ancient history of man on Earth, visualizing the origin of the solar system, and describing the nature of stars.

The scientific method, representing the best use of man's reasoning power, his imagination, his ingenuity, and his skill in observation and manipulation, has been used almost exclusively in the material world—primarily in the study of matter and the observable material universe. Although this constitutes what is known as *science* today, it by no means represents the ultimate use of the scientific method. A beginning has been made in applying systematic processes to man's social problems, and the authors believe the scientific method holds enormous potential for solving problems of broader human concern. Thus the humanitarian should join the scientist in a mutual endeavor, applying the scientific method wherever mankind encounters difficulty. A more scientific description of the nature of man's emotions, his aspirations, and his physiological reactions could well form a basis for new social and political structures, as much improved over today's as man's physical situation is over that of recent centuries.

The next three chapters concern the history of astronomy, showing briefly how its scientific structure developed.

CHAPTER *2*

Ancient Astronomy

2-1   Prehistory.   The beginnings of astronomical lore antedate the oldest historical records. Early inscriptions on stones, rocks, bones and antlers, and on the walls of caves that once sheltered the men of the Stone Age—all show man's perennial interest in the phenomena of the skies.

These prehistoric relics are widespread, from the carved rocks used for boundary markers in ancient Babylonia, the cave paintings of southern France, the ox-bone inscriptions of China, to the Mayan temples of Central America. Our ancestors knew the stars and planets, measured their changing positions, and sometimes even worshipped them.

One can hardly look at these ancient records without awe and respect. Our ancestors, thousands of years ago, managed to piece together, in true scientific fashion, many fundamental observational facts about the universe. They were handicapped by ignorance, by primitive ways of handling numbers, and by the lack of scientific instruments, but nevertheless they were able to collect many significant observations, to analyze them, and to draw reasonable conclusions. Even though many of the early theories have been disproved, they had an important influence on intellectual growth.

It is easy to understand why astronomical science developed so early, and why our ancestors were so much interested in the universe. Man is a curious animal, and if curiosity did not exist, science (including astronomy) and possibly man himself would never have developed.

When we study the ancient records, we are forcibly struck with the changelessness of the universe of stars. Although some sort of human life has existed on earth for perhaps four million years, astronomical records go back less than 10,000 years, and the stars move and change so slowly that the constellations have scarcely altered during the last 50,000 years.

2-2 **Time and the Seasons.** One of the first practical aspects of primitive astronomy dealt with the notion of time and its passing. The ancients found it necessary to divide the day into parts, so that they could satisfactorily organize the routine of daily existence. Clocks and watches were unknown, but the sun and the shadow cast by a pointer made good substitutes. The sundial was the earliest and for many ages the only effective form of clock. Lost in antiquity are the reasons for subdividing the day into conventional twenty-four hours.

The changing position of the sun through the year, as it swung south in the winter and north in the summer, added complications that the ancients could resolve only by knowing more about the sun and its apparent path through the heavens.

Study of the sun, in its daily motion, led naturally to an understanding of the seasons and their regular recurrence. This knowledge, too, had its practical application. As men grew more civilized, planting crops instead of searching the forests for food, the need for recording the passage of seasons became urgent. Certain seasons brought special types of weather activity, such as the monsoon, heavy snows, or the flooding of the Nile or Euphrates river valleys. Primitive man needed to know when to expect the different kinds of weather. Hence, each tribe had experts on the sun, the stars, and the weather—men who often became the most influential individuals in the tribe.

Since so much of human life centered about the sun, with its kindly warmth and beneficent effect on crops, worship of the sun, perhaps motivated by a desire to control or influence so powerful a force, became the dominant religion in many places. The Inca Indians of Peru, for example, developed a solar religion of great complexity.

The more that men studied the heavens, however, the more they noted the regularities in the motions of the sun, moon, and stars. They found the "wandering" stars or planets and gave them names. The ensuing emphasis on regularity induced men to fear any unusual event. Eclipses and comets became the most fearful of apparitions. The early Greeks were able to recognize the mechanism of eclipses, but comets plagued mankind with superstitious fears until the eighteenth century, when even these phenomena became part of the law and order characteristic of astronomical science.

FIGURE
2-1

Restoration of Greek Amphora, dating from 582 B.C. Representation of the constellations. Hercules (the Kneeler) shooting an arrow (Sagitta) toward the Eagle (Aquila). The second figure is Prometheus. (The National Museum, Athens. Photo by Menzel.)

2-3   The Constellations.   Constellations meant far more to the ancients than they do to us. In our civilized and sophisticated society, many of our recreations take place indoors: the theater, books, motion pictures, television, to mention but a few. The ancients had none of these, except possibly a primitive version of the theater, and most of their activity took place in the open. The earliest and perhaps the major source of entertainment was the travelling minstrel or wandering storyteller, the man who drifted, gypsylike, from one tribe to another, telling tall tales and otherwise amusing the tribesmen in return for food and bed.

FIGURE
2-2

The constellation of Ursa Major, the Great Bear.

Fragments of their stories have descended to us—tales of daring and adventure handed down through the ages to give us the basic myths of many lands: tales of the Greek, Roman, or Norse gods (Figs. 2-1, 2-2). The stories of different races have much in common, and the great majority of them are woven about the various constellations. For we can readily imagine that the primitive storyteller, perhaps talking to a group of shepherds in the hills of Chaldea, would use the heavens as a picture book and illustrate his stories by the pictures in the sky.

Thus, we have inherited our constellations, with the legendary creatures or personalities whose names they bear. Some astronomer invented constellations to fill in the gaps where the stars were too faint to suggest a celestial figure or where the stars were too far south to be visible to the races who contributed the standard legends.

A complete list of constellations appears on pages 12 and 13. The astronomer still finds them useful in fixing the approximate locations of certain stars or heavenly objects. In this respect, the boundaries of the figures are something like the boundaries of the individual states. To say, for example, that a certain city lies in the south-central area of Michigan fixes its location accurately enough for many purposes. For greater precision we have to specify the exact latitude and longitude. We shall find, later on, that the astronomer uses much the same device, attaching to the sky a coordinate system similar to the geographic one used on the surface of the earth.

2-4   Primitive Concepts of the Universe.   In ancient times, man's universe was almost as limited as his horizons. Travel was difficult and slow and even the nomads covered very little territory. Men accepted without question what their limited experience suggested, namely that the earth was essentially flat, though somewhat wrinkled by hills and valleys. Those few who tried to interpret the universe still further were not particularly successful, because of limited experience and imperfect observations.

Various systems set the earth and its oceans on pillars, or on the back of an elephant. These supports, in turn, required some additional support, such as a table or the broad back of a super-turtle. Somehow, the fact that any given number of supports would still be finite seems never to have concerned the ancients very seriously.

Thus, for many centuries, the idea of a flat earth was quite adequate for man's philosophy. The various models differ in nonessential details, usually of a philosophical nature. Some scientists, such as Thales, floated the flat earth on an ocean of water. Others added a mountain range around the edge, like the rim of a piepan, to keep the ocean from spilling over into the space beneath.

TABLE 2-1 *List of constellations*

Latin Nominative	Latin Genitive	Abbreviation	English
(a) Constellations North of the Zodiac			
Andromeda	Andromedae	And	Andromeda
Aquila	Aquilae	Aql	The Eagle
Auriga	Aurigae	Aur	The Charioteer
Bootes	Bootis	Boo	The Herdsman
Camelopardalis	Camelopardalis	Cam	The Giraffe
Canes Venatici	Canum Venaticorum	CVn	The Hunting Dogs
Cassiopeia	Cassiopeiae	Cas	Cassiopeia
Cepheus	Cephei	Cep	Cepheus
Coma Berenices	Comae Berenices	Com	Berenice's Hair
Corona Borealis	Coronae Borealis	CrB	The Northern Crown
Cygnus	Cygni	Cyg	The Swan
Delphinus	Delphini	Del	The Dolphin
Draco	Draconis	Dra	The Dragon
Equuleus	Equulei	Equ	The Little Horse
Hercules	Herculis	Her	Hercules
Lacerta	Lacertae	Lac	The Lizard
Leo Minor	Leonis Minoris	LMi	The Little Lion
Lynx	Lyncis	Lyn	The Lynx
Lyra	Lyrae	Lyr	The Lyre
Ophiuchus	Ophiuchi	Oph	The Serpent Bearer
Pegasus	Pegasi	Peg	Pegasus
Perseus	Persei	Per	Perseus
Sagitta	Sagittae	Sge	The Arrow
Scutum	Scuti	Sct	Sobieski's Shield
Serpens	Serpentis	Ser	The Serpent
Triangulum	Trianguli	Tri	The Triangle
Ursa Major	Ursae Majoris	UMa	The Great Bear
Ursa Minor	Ursae Minoris	UMi	The Little Bear
Vulpecula	Vulpeculae	Vul	The Fox
(b) Constellations along the Zodiac			
Aries	Arietis	Ari	The Ram
Taurus	Tauri	Tau	The Bull
Gemini	Geminorum	Gem	The Twins
Cancer	Cancri	Cnc	The Crab
Leo	Leonis	Leo	The Lion
Virgo	Virginis	Vir	The Virgin
Libra	Librae	Lib	The Scales
Scorpius	Scorpii	Sco	The Scorpion
Sagittarius	Sagittarii	Sgr	The Archer
Capricornus	Capricorni	Cap	The Goat
Aquarius	Aquarii	Aqr	The Water Carrier
Pisces	Piscium	Psc	The Fishes

Latin Nominative	Latin Genitive	Abbreviation	English

(c) Constellations South of the Zodiac

Latin Nominative	Latin Genitive	Abbreviation	English
Antlia	Antliae	Ant	The Pneumatic Pump
Apus	Apodis	Aps	The Bird of Paradise
Ara	Arae	Ara	The Altar
Caelum	Caeli	Cae	The Graving Tool
Canis Major	Canis Majoris	CMa	The Great Dog
Canis Minor	Canis Minoris	CMi	The Little Dog
Carina	Carinae	Car	The Keel (of the ship)
Cetus	Ceti	Cet	The Whale
Chamaeleon	Chamaeleontis	Cha	The Chameleon
Circinus	Circini	Cir	The Pair of Compasses
Columba	Columbae	Col	The Dove
Corona Australis	Coronae Australis	CrA	The Southern Crown
Corvus	Corvi	Crv	The Crow
Crater	Crateris	Crt	The Goblet
Crux	Crucis	Cru	The Southern Cross
Dorado	Doradus	Dor	The Swordfish
Eridanus	Eridani	Eri	The River Eridanus
Fornax	Fornacis	For	The Furnace
Grus	Gruis	Gru	The Crane
Horologium	Horologii	Hor	The Clock
Hydra	Hydrae	Hya	The Sea-Serpent
Hydrus	Hydri	Hyi	The Water-Snake
Indus	Indi	Ind	The Indian Bird
Lepus	Leporis	Lep	The Hare
Lupus	Lupi	Lup	The Wolf
Mensa	Mensae	Men	The Table
Microscopium	Microscopii	Mic	The Microscope
Monoceros	Monocerotis	Mon	The Unicorn
Musca	Muscae	Mus	The Fly
Norma	Normae	Nor	The Ruler
Octans	Octantis	Oct	The Octant
Orion	Orionis	Ori	Orion
Pavo	Pavonis	Pav	The Peacock
Phoenix	Phoenicis	Phe	The Phoenix
Pictor	Pictoris	Pic	The Easel
Piscis Austrinus	Piscis Austrini	PsA	The Southern Fish
Puppis	Puppis	Pup	The Prow (of the ship)
Pyxis	Pyxidis	Pyx	The Mariner's Compass
Reticulum	Reticuli	Ret	The Net
Sculptor	Sculptoris	Scl	The Sculptor
Sextans	Sextantis	Sex	The Sextant
Telescopium	Telescopii	Tel	The Telescope
Triangulum Australe	Trianguli Australis	TrA	The Southern Triangle
Tucana	Tucanae	Tuc	The Toucan
Vela	Velorum	Vel	The Sails (of the ship)
Volans	Volantis	Vol	The Flying Fish

2-5   Astrology and Astronomy.   Until the eighteenth century astrology and astronomy were so completely intermeshed that it is quite impossible to separate them. Shakespeare expressed a widespread astrological belief when he wrote:

> There is a tide in the affairs of men,
> which, taken at the flood, leads on to fortune.*

Though few educated people today believe in astrology, our vocabularies still reflect the astrological uses of certain words. When we say that a man is "saturnine," we are harking back to the astrological belief that a man born under the planet Saturn possessed a sluggish, gloomy temperament, produced at birth by the malign influence of that particularly malevolent body. Similarly, persons born under the planets Jupiter, Mars, or Mercury were supposed to be jovial, martial, or mercurial. If under the sun, they were possessed of a sunny disposition. The moon (Latin : Luna) could drive people insane (loony) or give them a moony disposition.

We inherit the names of the days of the week from astrology. Since we derive the names of these days from more than one source, we need to know the planets or corresponding gods and goddesses in both the Latin and Saxon forms.

Table 2-2 lists the days of the week in various languages, and the astronomical objects that provided their names.

**TABLE
2-2**   *Days of the week*

Object	English	French	Spanish	Old English	German
Sun	Sunday	dimanche	domingo	Sunnandaeg	Sonntag
Moon	Monday	lundi	lunes	Monandaeg	Montag
Mars	Tuesday	mardi	martes	Tiwesdaeg	Dienstag
Mercury	Wednesday	mercredi	miercoles	Wodnesdaeg	Mittwoch
Jupiter	Thursday	jeudi	jueves	Thunresdaeg	Donnerstag
Venus	Friday	vendredi	viernes	Frigedaeg	Freitag
Saturn	Saturday	samedi	sabado	Saeterndaeg	Samstag

The English names derive from Germanic rather than Latin mythology, with Tiw or Tyr for Mars, Woden or Wotan for Mercury, Thonar (thunder) or Thor for Jupiter, and Fria or Freya for Venus.

The ancients recognized the planets in the order of the times taken to complete a circumference of the heavens. Saturn required the longest. Then, in turn, we have Jupiter, Mars, the Sun, Venus, Mercury and the Moon. This order of the planets is very different from that of their association with the days of the week. To interpret this difference we must again

---

* Julius Caesar, Act IV, Scene 3.

turn to astrology. We have inherited from antiquity, as noted earlier, the division of the day into twenty-four hours. Astrologically, each of the seven "planets" was supposed to rule in turn over the various hours. Thus, if Saturn ruled the first hour, Jupiter controlled the second, and so on. In the same day, Saturn would again hold sway on the eighth, the fifteenth, and the twenty-second hour.

Although the various planets ruled in turn, Saturn would be the dominant planet of the day, because it ruled the first hour. Hence that day would be "Saturn's day," or Saturday. But if Saturn also ruled the twenty-second hour, Jupiter would rule the twenty-third, Mars the twenty-fourth, and the sun would thus take over the first hour of the following day, Sunday. And so on through the week.

Saturn, being the most malignant of all of the planets, was the one that men most feared. Thus, they tended to cease from business activity on that day, the original Sabbath. Our modern use of Sunday as the day of rest started officially in A.D. 321.

2-6 **Greek Astronomy.** Although the Babylonians and other earlier races had contributed to astronomy, the early Greek scientists raised the subject to a new level of understanding.

Of these men, Thales of Miletus [640(?)–546 B.C.], commonly referred to as the founder of Greek astronomy, was an experimentalist of the first order. Although he supposed that the earth was flat, he did understand the principles of geometry and the motions of the stars. He taught the Greeks to navigate by the "Little Dipper," which was the equivalent of the North Star in his day. He successfully predicted an eclipse of the sun and he further reasoned that stars were self-luminous, but that the moon shone by light reflected from the sun. These truths, so simple and so familiar to us today, were important advances.

Pythagoras [582(?)–500(?) B.C.] was the founder of the idealistic school of philosophy. He had good reason to believe in the beauty of mathematics, for he proved one of the most important mathematical theorems in all of geometry, the one that bears his name today: the square erected on the hypotenuse of a right triangle is equal to the sum of the squares erected upon the other two sides. Consider that this result was established without a single measurement, and you may grasp why Pythagoras became convinced that he could prove all of the truths of the universe by similar mathematical reasoning.

One could, of course, resort to experiment. One could even prove the Pythagorean theorem by such means, but the basic truths were mathematical, or so Pythagoras thought. Here is another sample of his reasoning. Ten is a "perfect number," since $1 + 2 + 3 + 4 = 10$. Pythagoras could see but nine heavenly objects—namely the five planets, the sun, the moon, the earth, and the fixed stars. He decided that there must be a tenth somewhere.

This object, he postulated, must be an invisible "counter earth" or "twin earth," which in common with the other nine bodies revolved about some central fire other than the sun.

The planets in their circling were supposed to produce a heavenly harmony, the music of the spheres, a symphony for the gods on Mount Olympus.

In some of his conclusions, particularly those related to the shape of the earth, which he regarded as spherical, and the motions of the planets, Pythagoras came closer to the truth than many of his idealistic followers.

Aristotle (384–322 B.C.) still ranks as one of the greatest philosophers the world has ever known. Not a single science has completely escaped his influence—for better or worse. In astronomy and geology he was less successful than he was in other fields.

The high regard for Aristotle's authority actually retarded the advance of science for many centuries. In some respects he was a mystic. Though he regarded the earth as a sphere, he taught that the sun and stars were pure *aether*, with as much emphasis on purity as on the aether. Aristotle defined this mysterious aether as an absolutely weightless substance, which naturally moved in circles around the earth. It thus differed from matter, whose natural motion, according to Aristotle, was up or down. Inferring that material objects could continue in motion only as long as they were pushed, he endowed the moon and planets with animate souls, whose primary responsibilities were to push and steer the bodies in their courses around the sky.

Aristarchus of Samos (310–230 B.C.), viewed against the background of his time, proves to be one of the greatest astronomers of history. He tried to determine the distance of the sun using ingenious observational techniques and deductions from them that were fundamentally sound. If some of his conclusions were wrong, the error arose from the crudeness of the observational equipment available in his day. Even so, he succeeded in showing that the sun was much more distant and therefore much larger than the moon, a significant fact in itself. As for the moon, he made a rough determination of its distance too.

But Aristarchus was a "modern" in one major respect. He argued for the superiority and simplicity of a solar system in which the planets, including the earth, circled about the sun. He was so far ahead of his time, however, that nearly two thousand years had to elapse before astronomical techniques advanced to the point where direct measurement could prove the truth of his ideas.

Eratosthenes (276–195 B.C.) deserves special mention, because he was the first person to measure the size of the earth. He noted that, on the twenty-first of June, the sun was precisely in the zenith at Syene, Egypt, where it shone straight to the bottom of wells. The city of Alexandria lay some four hundred eighty miles due north of Syene. There, on another June twenty-first, Eratosthenes noted that the sun was seven degrees south

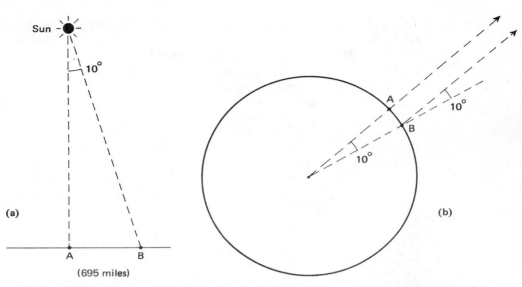

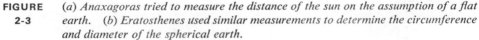

(a)

(b)

FIGURE   (a) *Anaxagoras tried to measure the distance of the sun on the assumption of a flat*
2-3   *earth.*   (b) *Eratosthenes used similar measurements to determine the circumference
               and diameter of the spherical earth.*

of the zenith. Attributing this difference to the earth's sphericity, he
reasoned that four hundred eighty miles must be $\frac{7}{360}$ of the earth's
circumference. And thus he calculated that the earth was some 24,000
miles in circumference—a value within a few per cent of the modern
figures, although the closeness of the agreement was largely accidental. Of
course, Eratosthenes did not use the modern mile as his unit. Instead he
used the "stadium," expressed in terms of the distance an army can walk in
a day.

Hipparchus [160(?)–125 B.C.] has been widely acknowledged as the
"father of astronomy." He was a careful and indefatigable observer. He
built new and more precise instruments for measurement. He analyzed his
own records and compared them with older records available to him. He
mapped the path of the sun, determined the distance of the moon to within
10 percent, charted the entire sky, and estimated both the positions and
brightnesses of stars. He invented and used the science we call "trigonometry," which has special applications to the solution of triangles or the
determination of distances.

Hipparchus also was the first to detect the "precession of the equinoxes"
—the slow motion of the intersection of the ecliptic (the plane of the
earth's orbit) with the celestial equator reflected by the drift of the North
Pole relative to the stars. We know now that this motion is caused by the
wobble of the earth's axis in a period of about twenty-six thousand years
(see Sec. 5.3).

Although Hipparchus was probably aware of the theories of his brilliant predecessor, Aristarchus, he found no need to postulate the motion of the earth about the sun, for he could represent all of his observations quite as well by the more conventional model of the universe, wherein the earth is stationary, and the planets circle it. Hipparchus, like his contemporaries, regarded the circle as a sacred figure. But his observations clearly were inconsistent with the uniform motion of a planet in a precise circle.

To retain the circles, Hipparchus invented a remarkable mechanical system to account for the motions of the planets. He first tried to account for the observed motions by moving the center of the supposedly circular orbit some distance away from the center of the earth. He postulated that the planet moved uniformly along this *eccentric* orbit. And, when this simple displacement of the center still could not account for the observations, Hipparchus further postulated that the planet moved uniformly in a small circle, an *epicycle*, whose center progressed uniformly along the circumference of the greater circle, called the *deferent*.

Claudius Ptolemy (*c.* A.D. 90–168) continued and vastly extended the work of Hipparchus, measuring with remarkable accuracy the positions of stars and planets (Fig. 2-3). He wrote the *Megale Syntaxis*, transmitted to us through its Arabic translation, the *Almagest*, signifying "the greatest."

FIGURE
2-4

The Ptolemaic Model of the Solar System.

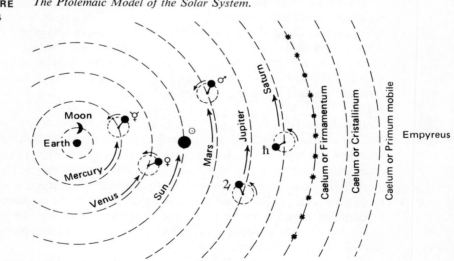

2-9 Medieval Astronomy. In many respects, the *Almagest* was the link that bound the early phases of astronomy with the modern. For, during the thirteen hundred or so years following Ptolemy, the lamp of science all but went out in medieval Europe. The Mohammedans and Tartars kept alive the flickering fires, recopying, revising, and slightly extending the *Almagest*. The Tartar ruler, Ulugh Beigh (1394–1449), stimulated the development of astronomy and built a large observatory in Samarkand.

The great Leonardo da Vinci (1452–1519), painter, writer, inventor, and scientist, touched upon phases of astronomy and recognized, among other things, the earth's rotundity, its rotation, and particularly the fact that the faint glow within the horns of the crescent moon derives from light reflected from the earth.

During these ages men had other interests: war, art, letters, philosophy, and religion. And since the basic philosophy in vogue was that of Aristotle, experimental science and the school of realism received little consideration from the few educated men in those dark times.

3

The Copernican Revolution

With the Renaissance came an awakening of scientific interest in Western Europe. The invention of printing in 1453 enormously assisted the diffusion of knowledge. John Muller (1436–1476)—better known as Regiomontanus (from his native town of Königsberg)—established an observatory equipped with a workshop and a printing press at Nuremberg in 1471, and became the first astronomical publisher, printing a Latin translation of the *Almagest*. The Belgian Cardinal Nicholas of Cusa (1401–1464) was one of the first to discuss and favor Aristarchus' hypothesis of the motion of the earth, but he was unable to support his conviction with decisive arguments. This task fell to the quiet genius of a Polish priest, Nicolaus Koppernigk, better known by his latinized name, Copernicus.

3-1   Copernicus.   Nicholas Copernicus (1473–1543) spent a lifetime studying and writing on theories of the solar system. He revived the old hypothesis of Aristarchus, which set the sun as the primary orb about which the planets revolved. He was able to improve on the earlier model simply because he had better observations. However, increased accuracy made the representation of planetary motions more difficult. In spite of the simplifying introduction of the sun as the center, Copernicus still employed the epicycles of Hipparchus and Ptolemy to account for the departures of the planets from uniform motion along circular paths (Fig. 3-1).

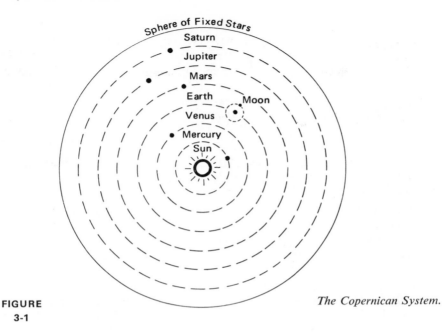

FIGURE
3-1
The Copernican System.

   Many people regarded Copernicus' theory as an interesting or amusing paradox rather than a representation of reality. If the earth moved around the sun, one would expect to see a small displacement of the nearer stars relative to the more distant ones, an effect known as *parallax*. Failure to observe such apparent displacements probably influenced many of the earlier astronomers and philosophers, such as Hipparchus, Ptolemy, and Aristotle, to reject the heliocentric hypothesis. Actually the parallactic motions are far too small to have been detected with the crude instruments used by the ancients. Nevertheless, the absence of parallax was used as an argument against Copernicus. However, the major opposition to the heliocentric system came from the supporters of established philosophy and religion whose beliefs, based on the authority of Aristotle and the Scriptures, were threatened by advocacy of a world system that denied man his central, favored position.

   Despite Copernicus' caution in the presentation of his book *De Revolutionibus Orbium Celestium* (which he prudently dedicated to the Pope, as a work of mathematics written only for mathematicians, "Mathemata Mathematici Scribuntur"), the revolutionary implications of the theory soon became clear to all, while its scientific power was progressively recognized and strengthened by the work of the great astronomers of the sixteenth and seventeenth centuries. Even in 1616, when Copernicus' work was placed on the "Index" of forbidden books by the Church of Rome, it was already too late to prevent the triumph of the modern system of the world.

The fiery Italian priest Giordano Bruno (1548–1600), taking an extremely modern view, declared that the universe was infinite and filled with stars throughout the infinity of space. He advocated the heretical doctrine of the plurality of inhabited worlds. Bruno, like many who have held ideas in advance of their times, suffered greatly for his unorthodox views and eventually met death—burning at the stake—when the Roman Inquisition caught up with him.

3-2   Brahe.   Tycho Brahe (1546–1601) was the greatest of all the observational astronomers who lived prior to the era of the telescope. He built two large observatories on the island of Hven, near Copenhagen. He devised and constructed the most precise of pretelescopic pointers (see Sec. 6-2) and designed a clock more accurate than any previously available. He observed the planets and stars regularly, compiling extensive catalogs and numerous records of all kinds of celestial objects. He proved that comets were celestial bodies, more distant than the moon, and not "exhalations of the earth," as most of his contemporaries maintained.

Tycho also made a model of the solar system. He, likewise, rejected the heliocentric models of Aristarchus and Copernicus because his observations did not disclose a stellar parallax (Sec. 3-1). We shall see later that his observations, fine as they were, still fell far short of capacity to detect the minute shifts of the stars caused by the earth's motion. The celestial system that Tycho advocated, however, was a compromise between the Copernican system and the Ptolemaic. He set the sun moving about the stationary earth, but he made all the planets, except the moon, revolve about the sun. Tycho's system, like that of Ptolemy, made the stars rotate in circles around the earth. It could account for the motions of the sun, moon, and planets quite as well as the system proposed by Copernicus.

3-3   Kepler.   Johann Kepler (1571–1630) inherited the observations of his teacher, Tycho, at the time of Tycho's death. From the philosophical standpoint, Kepler was somewhat of a mystic. He himself was not an observer, but he had great faith in the accuracy of Tycho's observations. From them, he painstakingly sought the laws that governed the motion of planets in the solar system. Years of effort went into his analysis.

Kepler started with two assumptions: (1) the solar system was heliocentric, and (2) the orbit of the earth was a circle. His method was basically simple. He could calculate the revolution period of each planet. Mars, for example, had a year of 687 days. He found, from Tycho's abundant observations of Mars, two observations separated by 687 days, when Mars would be at the same point in its orbit.

Kepler drew a circle to represent the earth's orbit and placed two dots on it to indicate the positions of the earth for the first observation and the

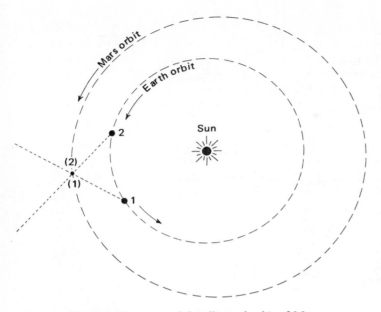

FIGURE   *Kepler's discovery of the elliptical orbit of Mars.*
3-2

measure 687 days later. Knowing the directions of Mars from the earth on those two days, he could draw straight lines, whose intersection gave the position of Mars on that date (Fig. 3-2). From hundreds of such observation pairs Kepler plotted out the orbit of Mars, connecting the points to form the smooth orbit curve shown in Fig. 3-2. The curve was not a circle by any stretch of the imagination. Nor could he represent it with the conventional eccentric and epicycles (Sec. 2-5). Deciding finally that it was an ellipse, he went on to announce the three laws known by his name. (In one respect Kepler was fortunate: the orbit of the earth is, indeed, much more nearly a circle than that of Mars.)

To draw an ellipse, merely place two pins through a piece of paper, with a continuous loop of string around them as shown in Fig. 3-3. Then, with a pencil in the loop, keep the string taut and describe the ellipse. The two pins mark the two foci. The line from one focus to a point on the ellipse is the *radius vector*.

Kepler's three laws are as follows:

1. Each planet moves in an orbit that is an ellipse, with the sun at one of the foci (the other focus is empty).

2. The speed of the moving planet changes with distance from the sun, so that the radius vector sweeps equal areas in equal intervals of time (see Fig. 3-4).

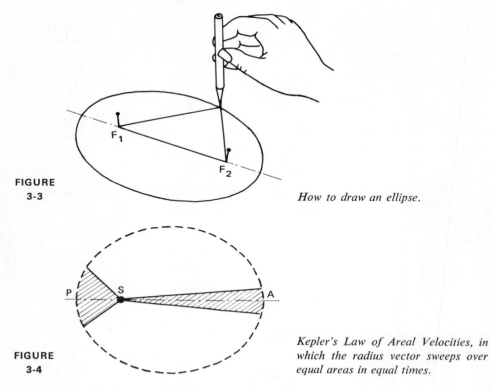

FIGURE
3-3

How to draw an ellipse.

FIGURE
3-4

Kepler's Law of Areal Velocities, in which the radius vector sweeps over equal areas in equal times.

3. The squares of the times that it takes the planet to go round the sun (the periods of revolution) are proportional to the cubes of their mean distances from the sun.

The "mean distance" in the third law is half the major axis of the ellipse. Kepler announced this law in 1619, ten years after he had given the first two.

To illustrate the operation of the third law, suppose that a planet exists whose mean distance is four times that of the earth from the sun. The cube of four is sixty-four, which should be equal to the square of the period of that planet. Hence, taking the square root of sixty-four, we get eight years as the time of revolution.

Kepler's discovery of the ellipticity of the orbit of Mars highlights the importance to science of a simple increase in the accuracy of measurements. The deviation between the ellipse and the previously assumed circle causes a difference of no more than eight minutes of arc in the longitude of the planet; however, in Kepler's own words, "These eight minutes alone have led to a complete renovation of astronomy." Eight minutes corresponds to about one-fourth the Moon's apparent diameter as seen from Earth.

Kepler also wrote a story in which the hero visited the moon. In this early work of science fiction, Kepler considered such modern problems as the possible effects of lowered gravity on man. He further noted—long before Newton announced the law of gravitation—that travel would become easier as the vehicle left the "attractive" power of the earth behind. His picture of the lunar mountains was so accurate and realistic that one must conclude he had learned of Galileo's telescopic observations.

3-4   Philosophical Impact of Kepler's Laws.   The philosophical significance of Kepler's studies was not recognized immediately. The basic facts that many scientists, from the ancients to Tycho Brahe, had tried to explain were the detailed observations of planetary motions. For some time four theories had vied for recognition. Still at the head of the list, by virtue of its long-established authority, was Aristotle's explanation, that the heavenly bodies were pure aether, guided supernaturally along circular paths about the earth as center. Next came the geocentric theory of Ptolemy, which could accurately represent the motions of the planets. The Copernican theory was third, and last came the compromise theory of Tycho.

Indeed, the fact that several independent theories could describe the observations, within the limits of error of contemporary measurements, posed a basic philosophical question. How are we to decide among multiple explanations for a given phenomenon? What sort of rule must the scientist follow? How does he decide which theories are wrong and which are right?

The answers to some questions seem obvious. We cannot decide intuitively about the theories of science. If theories differ fundamentally, some must be wrong, others may be more nearly right. The modern scientist has a good rule for deciding among various theories. He chooses the simplest one consistent with the observational data.

The student should not conclude, however, that all theories of science are of necessity simple. "Simplicity" is relative. What is simple to one person may seem complicated to another. We are implying that theories should be no more complicated than necessary to represent the observations or experiments.

And yet the decisions between alternatives is not always easy. For example, are the circles that constitute the cycles and epicycles of Copernicus simpler or more complex than Kepler's ellipses? Although any one of the three mathematical systems will suffice for the purposes of computation alone, calculations are simplest for the ellipse. But Kepler's contemporaries could not appreciate the tremendous change that Kepler's laws would eventually introduce into scientific thinking.

3-5 Galileo. Galileo Galilei (1564–1642), who changed our mode of philosophical thinking almost more than any other scientist, established a modern school of both experimental and theoretical science.

Galileo was impressed by the apparent logic and simplicity of the Copernican theory, and became an early convert to it. At that time the Copernican doctrines were being taught, if at all, only as paradoxes— heretical notions worthy of some attention as *curiosa*, but clearly wrong because they sought to dethrone the earth from its primary place at the center of the universe. Galileo was probably the first to grasp the fundamental principle of science discussed in the previous section, when he declared in favor of the Copernican theory because it was philosophically *simpler* than any other.

Galileo was also perhaps the first and certainly the most eloquent advocate of the experimental method: the recourse to observation and experience as the supreme test of ideas and theories. In opposition to the traditional scholastic school's recourse to authority and antiquity for decisive argument, he adhered to and proclaimed the modern scientist's view. Let the facts decide! Let Nature itself answer your questions rather than argue endlessly about them in ignorance and superstition.

3-6 The Telescope. The invention of the telescope loosened the bonds of superstition that had constrained astronomical science for so long. However, seeing was not always believing, and many learned men turned away from the sights revealed by the telescope because they thought they were being hoaxed, deluded, or even bewitched into seeing things that their prejudices forbade them to recognize.

In 1609 Galileo heard that a Dutch optician, Hans Lippershey, had invented a device that seemed to bring distant objects closer. Apparently some effort was made to keep this knowledge secret because of its potential military applications. But Galileo grasped the principles involved, readily figured out the proper combination of lenses, and produced the first astronomical telescope. Its magnifying power was low, perhaps only five or six, but nevertheless it opened up new vistas of the universe. Eventually Galileo made a larger one that magnified about thirty times in diameter (Fig. 3-5).

A solar system in miniature was one of the first sights Galileo observed with the newly invented telescope. Turning it on Jupiter, he saw near the planet four bright points of light, which at first he took to be stars. But when he examined their positions on the following night, he noted that the configuration had changed. A careful record made on successive nights established beyond doubt that Jupiter itself was a primary body and had four moons circling about it.

Here was clear-cut evidence of revolution about a body other than the earth. The very existence of this independent system argued against the

FIGURE 3-5 *Marie B. Righini carrying the Galileo telescope and an antique globe to safety after the devastating 1966 flood in Florence, Italy.*

currently accepted doctrine that all celestial bodies must revolve about the earth as center.

The critics of Galileo tried to combat the evidence of the senses by mystic reasoning. "There are," pointed out one of his antagonists, "seven openings in the head, two eyes, two nostrils, two ears, and a mouth. Thus there can be only seven heavenly bodies, as there are indeed seven days in the week. Thus the false stars seen by Galileo surrounding the planet Jupiter simply cannot exist."

But Galileo, flushed with the excitement of discovery, continued his observing. He studied the planet Venus and noted that it showed phases like the moon, ranging from crescent through first quarter to gibbous. Since, in the Ptolemaic system, Venus must always lie between the earth and the sun, never moving far from the sun in the sky, it should always appear as a crescent. The fact that it was sometimes gibbous showed that Venus moved far beyond the orbit prescribed by Ptolemy (see Figs. 2-3 and 3-1). Here was even more concrete evidence in support of the Copernican system, which correctly predicted the observed changes in shape and apparent size of this planet.

The expanded horizons afforded by the telescope produced many other important facts, many of them highly disturbing to the disciples of Aristotle. The sun had spots, dark blemishes that seemed to be attached to its surface and to rotate with it. The sun, then, could not be pure aether. For if it were pure, how could such blemishes exist?

The moon proved even more startling. The telescope clearly revealed mountains on its surface, as well as great plains. The moon, therefore, was a separate world, an independent body not unlike the earth.

Galileo's telescope resolved the hazy glow of the Milky Way into a multitude of stars. And many luminous patches proved to be areas where stars clustered like fireflies. Some objects, like the Pleiades, which showed six or seven stars to the unaided eye, revealed dozens of fainter ones in the telescope.

Galileo eagerly expounded these discoveries and many others. The telescope had opened up a world far more beautiful and immeasurably vaster than anything his predecessors could have imagined.

3-7　Galileo's Contributions to Mechanics. Galileo was the first to appreciate the significance of two basic properties of matter: friction and inertia. Aristotle, who had developed a system of mechanics, took for granted what seemed to be an obvious principle of nature: any object, such as a ball rolling on the floor, will come to a stop unless force is continuously applied or reapplied. He concluded that uniform motion requires the continuous application of a force.

Galileo recognized friction as a property of the object and the surface on which it rolled or moved. He devised a series of experiments in which various weights were allowed to slide down inclined planes, and concluded that, if friction could be reduced to zero, a body would continue to move indefinitely. The planets do not require a continuously acting force to keep them in motion. Once in motion, they continue to move because there is no friction in interplanetary space.

Galileo's experiments on falling bodies were ingeniously devised to prove Aristotle's argument fallacious. One tenet of the older school of thought was that heavy bodies would fall more quickly than light ones. Indeed, our own experiments makes this seem likely, since a scrap of paper or a feather will fall much more slowly than a hammer or a brick. Galileo, however, realized the part that air resistance plays in this phenomenon.

Stevin and de Groot discovered in about 1590 that light and heavy objects experience the same acceleration. Galileo probably repeated these experiments. We have no details of his technique, except for his statement that a cannon ball falls no faster than a musket ball. When Galileo allowed for the slowing effect of the air, he found that all objects, dropped simultaneously from the same height, would strike the ground at the same time, thus proving Aristotle wrong.

Although Galileo continued throughout his life to make observations, to draw sound scientific deductions from them, and to carry out brilliantly conceived experiments, he experienced the fate suffered by many a pioneer. For he was upsetting an established order. Most of the great revolutions of thought have come about gradually. The one precipitated by Galileo was far too sudden for many of his contemporaries to accept.

After 1616, when Church authorities forbade him to teach or defend the "absurd and false" doctrine of the motion of the earth, Galileo continued to think about such problems. Finally, in 1632, the selection of a new and presumably liberal Pope encouraged him to publish his *Dialogue on the Two Principal Systems of the World*, those of Ptolemy and Copernicus. This work made his preference too obvious, and Galileo was once again denounced by his enemies to the Roman Inquisition and condemned in 1633 to "abjure, curse, and detest" his "errors and heresies" concerning the motion of the earth. After spending some time under surveillance in Rome, Galileo, then an old man, broken in health and spirit, was allowed to retire to Arcetri, near Florence. There he died in 1642, the same year that Isaac Newton was born in England—Isaac Newton, the man who was to employ the scientific foundations laid by Galileo and build the structure of modern science upon them.

4

The Law of Gravitation

The discoveries of Copernicus, Kepler, and Galileo clarified the basic plan of the modern "system of the world," and the laws of mechanics became known to scientists of the second half of the seventeenth century. A great English scientist, Isaac Newton, unified and extended this new knowledge by expressing the laws of motion of celestial bodies in mathematical form.

4-1 Newton. Isaac Newton (1642–1727) began the study of astronomy at an early age. Quickly mastering the mathematics and basic natural sciences of his time, he went on to original work in many fields, ranging from pure mathematics to experimental physics. His greatest achievement in mathematics was the invention of the calculus, a new branch of the subject that he called "fluxions." He applied this mathematical technique primarily to analyze the motions of the planets.

In experimental physics, he carried out researches in optics, including investigations of the nature of light and color. Recognizing the disadvantage of the ordinary refracting (lens) telescope, especially the limitations imposed by chromatic aberration (see Sec. 6-5), Newton developed and constructed a reflecting telescope, in which a concave mirror replaced the convex lens (Fig. 4-1).

FIGURE 4-1 *Isaac Newton's reflector. (© Royal Society of London.)*

Newton exerted a profound influence on mechanics and engineering. His interpretation of the nature of force fields and of motions under the action of impressed forces, as well as his mathematical techniques for studying such phenomena, are still basic for modern engineering. Others, however, would have invented the calculus had Newton not done so. In fact Leibnitz, one of his German contemporaries, discovered this technique independently. This mathematical discipline is fundamental to practically all of the physical sciences, both pure and applied.

4-2 Gravitation. Newton is popularly credited with the discovery of gravitation, but at the time he developed the concept, the basic principle was by no means new. Many scientists, long before Newton, had noted the similarity between the tendency of a heavy object to fall to the earth and the tug or pull that someone could exert on an object. The term "gravitas" applied to a body signified simply that the body had "weight," the result of a tugging force between the earth and a falling object.

However, Newton's expression of the law contained many novel features. No one had explicitly connected the tendency of bodies to fall to the ground with the motion of the moon about the earth or of the planets about the sun. It was Newton who realized that gravitational attraction could extend all the way to infinity and that even the moon and the planets were subject to it. If there is any truth in the tale of "Newton and the apple," the important part is the analogy between the apple and the moon, rather than the actual fall of the apple.

Newton supposed that the moon, in its circular orbit, was continually "falling" toward the earth. However, the moon also possessed a motion perpendicular to the radius vector to the earth. In effect, therefore, the moon "fell around" the earth in a closed orbit. The *centripetal force* of gravitation caused the moon to depart from a rectilinear path and move in a curve.

4-3 **Newton's Principia.** Newton's *Philosophiae Naturalis Principia Mathematica*, or *The Mathematical Principles of Natural Philosophy*, often called "the *Principia*," is still regarded as one of the great scientific classics. In it, Newton summarized the basic principles of celestial mechanics, developed the necessary mathematical procedures, simplified the methods, and demonstrated the nature of motions in the universe.

Newton broke away from the pseudoscientific reasoning of many of his predecessors and founded his theories on what we should today call mathematical physics. He made clear statements of the three basic laws of motion, which to a large extent were either refinements or restatements of the earlier results of Galileo. These laws of motion are still valid today.

First Law. A body at rest will remain at rest, and a body in motion will maintain a constant speed in a fixed direction, unless an unbalanced external force acts upon the body.

Second Law. If an unbalanced external force acts upon the body, the body will accelerate. This acceleration will occur in the direction in which the force is acting and will be proportional to the magnitude of the force but inversely proportional to the mass of the body.

Third Law. For every force acting on one body, an equal but opposite force must act on a second body.

Mathematically, Newton's second law relates the acceleration, a, the impressed force, F, and the mass, m:

$$a = \frac{F}{m}, \qquad F = ma, \quad \text{or} \quad m = \frac{F}{a}. \qquad (4\text{-}1)$$

The third form of the equation, in reality, defines the so-called inertial mass, which somehow measures the total amount of matter the body contains.

We emphasize here the distinction between mass and weight. The *mass* of an object remains constant, whether it is on earth, the moon, or freely orbiting in space. *Weight* is the force, F, that the object will exert on spring scales in the specified environment. The acceleration of gravity on the moon is about one-sixth that on the earth. Hence a body weighing 60 kilograms on earth will weigh only 10 kilograms on the moon.

Equation (4-1) implies that, if no external forces are acting, the acceleration will be zero; the velocity will then remain constant in magnitude and direction, as prescribed by the first law of motion.

Newton's third law means that if two bodies act on one another, the forces that rise from the interaction must completely balance. Two perfectly elastic spheres, having identical masses and equal but oppositely directed speeds, colliding head on, will accordingly bounce back with equal and opposite speeds.

4-4   Gravitation, Kepler's Laws, and Angular Momentum. Newton's great contribution to celestial mechanics lay not so much in his recognizing the phenomenon of gravitation as in his defining quantitatively and mathematically exactly how it worked. The law is basically simple. It states that *the gravitational force between any two particles is proportional to the product of their masses and inversely proportional to the square of the distance between their centers of gravity.* Thus, if m_1 and m_2 are the masses of two bodies separated by a distance r, then the gravitational force is

$$F = G \frac{m_1 m_2}{r^2}, \tag{4-2}$$

where G is the "constant of gravitation." The value of G is completely independent of the size of the body, its mass, its chemical composition, and its location in the universe. In fact, G is a "universal constant," whose value can be determined experimentally. Its magnitude depends only on the units employed for mass and length. When we express the masses in grams and the distances in centimeters,

$$G = 6.67 \times 10^{-8}. \tag{4-3}$$

Combining this law for the gravitational force with the three basic laws of motion previously stated, Newton employed his newly invented calculus to show exactly what the motions of the heavenly bodies should be. He found that he could predict all three of Kepler's laws for the specific case where a very massive body controlled the motions of particles or of planets whose masses were negligible compared with that of the primary body. This demonstration was one of his greatest triumphs. Newton further showed that a spherical body, such as the sun or a planet, behaves as if all the mass were concentrated at the center.

Newton went a great deal further than Kepler, for he proved that parabolas and hyperbolas were orbits quite as acceptable as ellipses. However, since these are not closed, an object moving in such a path would come close to the sun only once. Newton then went on to show that some comets were pursuing very nearly parabolic or perhaps even hyperbolic orbits. Since circles, ellipses, parabolas, and hyperbolas can all be described as curves formed by the intersection of a plane with a cone, as shown in Fig. 4-2, these orbits are generally known as *conic sections*.

The philosophical significance of Newton's approach to the problem of planetary motions is quite as important as his mathematical analysis. Almost all of the ancients, from Pythagoras and Aristarchus to Kepler, had tried to describe the orbits of the heavenly bodies in terms of mechanical or geometrical principles, as if the solar system were a giant machine composed of wheels, chains, and gears. The primary motion was usually circular until Kepler introduced the ellipse. Newton showed that none of these curves, not even the ellipse, was sacred.

The elementary theory, which predicted that a conic section would be a permissible orbit for a planet revolving about the sun, neglected the interaction of the planets. However, since every body in the universe attracts every other body, the planets must attract one another, causing some departure from the motions in Keplerian ellipses with the sun at one focus. Indeed, the apparent simplicity of planetary motions derives from

FIGURE 4-2   *Conic Sections, circle* oa, *ellipse* ob, *parabola* occ', *and hyperbola* odd'.

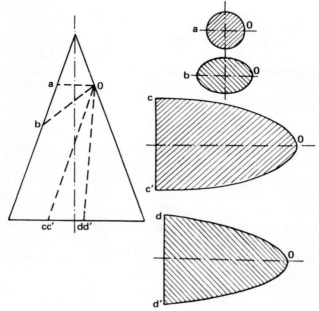

the fact that the sun is much more massive than even the greatest of the planets, and therefore exerts the primary control throughout the solar system.

From Newton's second law we can prove that a body, falling free in a gravitational field, will acquire a velocity proportional to the time. The acceleration of gravity, g, at the earth's surface is about 32 feet per second per second. We neglect the effect of air resistance (see Sec. 3-8). Acceleration means that a body, falling under gravity, adds a velocity of 32 feet per second, every second. Thus any body, light or heavy, when falling from rest, will acquire a speed of 32 feet per second within one second after its release. Since it started from rest, its average speed during the first second would be half this amount, or 16 feet per second. Thus the distance traversed in one second should be 16 feet.

At the end of two seconds, the speed will have doubled, to 64 feet per second. Again, the average speed will be just half this figure, or 32 feet per second. Multiplying this average speed by 2, the duration of the fall in seconds, we obtain the distance traversed in two seconds as $32 \times 2 = 64$ feet.

Let us generalize this reasoning. We say that the velocity, v, is proportional to the time, t, or

$$v = gt. \tag{4-4}$$

As we put $t = 1, 2, 3, \ldots$, we get velocities of 32, 64, 96, $\ldots$ feet per second. This result is a direct consequence of the law of gravitation.

The average velocity of the body, falling freely from rest ($v = 0$) and attaining the speed $v = gt$ after t seconds, is $(0 + gt)/2 = gt/2$. The distance s, that a body will fall in t seconds is equal to the average velocity multiplied by t, the time of fall, or

$$s = \frac{gt}{2} \times t = \tfrac{1}{2}gt^2. \tag{4-5}$$

Measurements on falling bodies give for the value of the gravitational acceleration at the earth's surface,

$$g = 32 \text{ feet per second per second,}$$

or
$$\tag{4-6}$$

$$g = 980 \text{ centimeters per second per second.}$$

In the metric, or cgs system, which employs the centimeter, the gram, and the second as measures of length, mass, and time, respectively, the values of v and s after an interval of 4 seconds are

$$v = 980 \times 4 = 3920 \text{ cm/sec,} \tag{4-7}$$

$$s = \tfrac{1}{2} \times 980 \times 16 = 7840 \text{ cm} = 78.4 \text{ m.} \tag{4-8}$$

These formulas apply, of course, only to regions close to the earth's surface, where we can assume that g is constant. Note from equation (4-1) that the force of gravity equals the acceleration g times the mass of the body acted upon. Further, from (4-2), we find that

$$g = \frac{GM}{r^2},$$

(4-9)

where M is the mass of the earth and r its radius.

Kepler's law of areas is a specific example of a far more general law, known as *conservation of angular momentum*. Figure 4.3 represents a

FIGURE 4-3

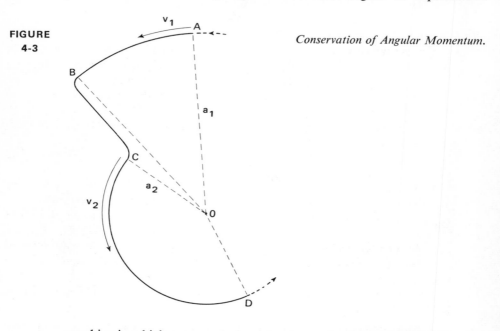

Conservation of Angular Momentum.

machine in which a mass revolves about an axis, O. By mechanical means one can alter the distance of the mass from the center. Suppose that the mass moves initially in a circle of radius a_1. Let P_1 be the time for the body to make a complete circuit. The linear velocity will be the circumference divided by the time P_1, or

$$v_1 = \frac{2\pi a_1}{P_1}.$$

(4-10)

The areal velocity, α_1, will be the area of the circle divided by P_1, or

$$\alpha_1 = \frac{\pi a_1^2}{P_1} = \frac{a_1 v_1}{2}.$$

(4-11)

Now, when the body is at B, suppose that some force, acting from the center, pulls or cranks it into a new circular path, CD, of radius a_2, velocity v_2, and circuit time P_2. The equations for the linear and areal velocities are identical with (4-10) and (4-11), except that the letters will all carry the subscript 2.

Kepler's second law applies strictly only to a single planet orbiting around the sun. For it neglects the mutual gravitational pull of the planets on one another, which can change or distort the original elliptical form of the orbit.

Actually, the law of areas will still apply to the nongravitational system of Fig. 4-3, as long as the forces to move the body come from the center. Hence we will still have

$$\alpha_1 = \alpha_2, \tag{4-12}$$

from which we conclude that

$$a_1 v_1 = a_2 v_2. \tag{4-13}$$

In other words, the smaller the circle, the faster the body must move.

One should not confuse this equation with Kepler's third law, which states that

$$\frac{a_1^{\,3}}{P_1^{\,2}} = \frac{a_2^{\,3}}{P_2^{\,2}}, \tag{4-14}$$

not

$$\frac{a_1^{\,2}}{P_1} = \frac{a_2^{\,2}}{P_2}, \tag{4-15}$$

as required by (4-13). In equation (4-14) the subscripts refer to *different* bodies, moving under gravity, whereas (4-13) and (4-15) refer to the same body, moving in different orbits.

We define angular momentum, L, as the mass of the body times twice its areal velocity about the center, or

$$L = 2M\alpha = Mav. \tag{4-16}$$

If the motion is not circular, v refers to that component of velocity perpendicular to the radius. We disregard the in-and-out motion, the velocity component along the radius.

As long as we are dealing with a single body, equation (4-16) tells us no more than Kepler's law, (4-11). But if we are dealing with a number of bodies, such as the solar system, we can calculate the angular momentum of each planet separately. And to the orbital angular momentum, we can add the angular momentum of rotation, if we wish. The law of conservation of angular momentum states that the sum of all these individual angular momenta must remain constant with time. If gravitational pull causes one body to accelerate, increasing its angular momentum, the other bodies of the system will have their angular momenta correspondingly diminished.

4-5 Proofs of the Earth's Rotation. Before the time of Newton, no direct proof of the earth's rotation existed. The arguments for such rotation were largely philosophical, based on the apparent simplicity of the picture rather than on experimental facts.

Newton showed that the centrifugal force produced by rotation must distort the earth's shape, so that it must bulge at the equator and be flattened at the poles. Newton used this conclusion to account for the "precession of the equinoxes," the slow wobble of the earth's axis, discovered by Hipparchus (Sec. 2-8).

As time went on, other direct evidences of the earth's rotation were sought. Perhaps the simplest depends upon the deflection that falling bodies experience when they drop from a great height because of conservation of angular momentum. For heights of even a few hundred feet the deflections are not inappreciable, though winds and air currents tend to mask the phenomenon (see Fig. 4-4). An object at the top of the Empire

FIGURE 4-4

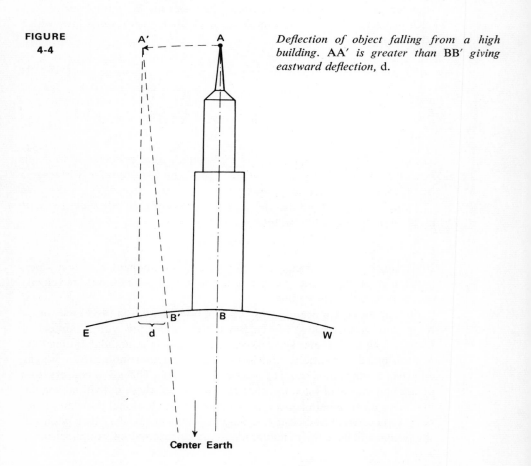

Deflection of object falling from a high building. AA′ *is greater than* BB′ *giving eastward deflection,* d.

State Building, for example, is moving eastward, because of the earth's rotation, more rapidly than an object at the foot of the building, which is nearer to the earth's center. Hence, theoretically, a body dropped from the top of this building would strike the ground about 7 cm or 2.3 inches to the east of the point vertically below the initial point. The formula for the deflection, d, of a body falling vertically is

$$d = \frac{gt^3}{6}\frac{v}{r} = \frac{2\pi}{3}\left(\frac{2}{g}\right)^{1/2}\frac{s^{3/2}}{T}\frac{a}{r} = \frac{2\pi}{3}\left(\frac{2}{g}\right)^{1/2}\frac{s^{3/2}}{T}\cos\phi \qquad (4\text{-}17)$$

or, in cgs units,

$$d = 1.10 \times 10^{-6}s^{3/2}\cos\phi,$$

where S is the height attained (in centimeters), t the time of flight, v the velocity of the earth's rotation at ground level, r the radius of the earth, g the acceleration of gravity, T the length of the day, a the distance of the point from the axis of rotation, and ϕ the latitude.*

In a modern era of satellites and long-range missiles we must reckon with large deflections. The world's first multiple-stage rocket, the "Wac Corporal" (a secondary rocket shot off from a V-2 at its highest point), attained a height of 250 miles (400 km) and experienced a deflection of about one and three-quarter miles to the west. However, the rocket would regain the distance in falling back to the earth, and theoretically should strike at the point from which it was originally launched. Wind currents, air resistance, and the inevitable inaccuracies of the firing would cause considerable departure from this ideal position, of course.

The earth's rotation also influences the direction of winds on the surface, and the deflection of rockets or projectiles shot to the north or south. To consider the nature of the problem, suppose that we have a long-range rocket upon the earth's equator set to point at a target on the same longitude, but at a latitude 45° due north of the point on the equator. Now, if we fire such a rocket, will it strike at the point toward which we have sighted?

The essential argument is as follows. Again conservation of angular momentum applies. The launching point of the rocket is moving eastward with a velocity of rotation at the equator of about a thousand miles an hour; the rocket shares this velocity component toward the east as it is launched. The target, however, has a velocity of only about seven hundred miles an hour, because it is closer to the earth's axis of rotation. When we fire the rocket, there is nothing to slow down the eastward motion that the projectile possessed before the firing. Hence the point struck will be not the target, but a spot east of the target. The deviation is to the right as we face the target (Fig. 4-5). For a rocket moving with constant speed of 0.62 mile

---

* Cos ϕ and sin ϕ are trigonometric functions whose values depend on the latitude. For details see Mathematical Appendix, 6.

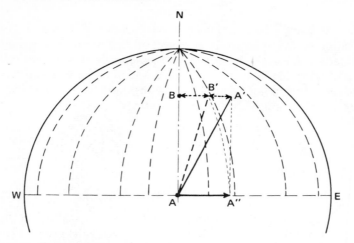

**FIGURE
4-5** *Deviation of projectiles from the force of Coriolis. Projectile at* A *on the equator, aimed northward at* B *along a meridian, possesses greater eastward velocity* AA′ *than that of target* BB′. *The projectile, therefore, strikes eastward (to the right) of the target by the angle* B′AA′.

per second (1 km/sec) the spot struck will deviate somewhat more than 300 miles from the north-south line.

Suppose now we fire a similar rocket from latitude 45° directly toward the equator. The missile, when fired, has an eastward velocity of only seven hundred miles an hour, while the target on the equator has an eastward velocity of a thousand miles an hour. The rocket will strike a point some 300 miles west of the target. Again the deviation is to the right. A projectile shot in the southern hemisphere would swerve to the left.

Analogous deflections occur for winds that blow over long distances. The persistent tendency of winds to deviate to the right, for example, when they approach a low-pressure area somewhere in the northern hemisphere, causes the air to circulate in a counterclockwise direction around such an area. Cyclones, tornadoes, and hurricanes all exhibit this counterclockwise rotation in the northern hemisphere. The direction is reversed in the southern hemisphere.

4-6   The Foucault Pendulum.   The most spectacular demonstration of the earth's rotation ever devised, the Foucault pendulum, in its simplest form, consists of a massive weight hung by a long, thin cord, and allowed to swing back and forth. The French physicist Foucault first performed the experiment in 1851 with a pendulum suspended beneath the highest point of the dome of the Pantheon in Paris. Perhaps the most remarkable feature of this relatively simple but striking demonstration was that no one had thought of it earlier.

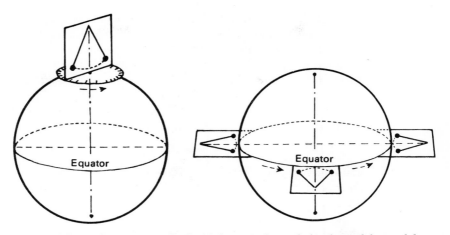

FIGURE
4-6

Foucault's pendulum. At the poles the earth rotates beneath the plane of the pendulum oscillation. At the equator the force of gravity is perpendicular to the axis of rotation; hence the pendulum does not appear to rotate.

To understand how the Foucault pendulum operates, imagine the experiment set up, under the simplest possible cirumstances, over the earth's north pole (Fig. 4-6). We draw the pendulum to one side, releasing it just at the moment when its swing is directed toward a star seen on the horizon. If the point of support is completely frictionless, there will be nothing to make the freely swinging pendulum follow the earth's rotation. The heavy weight will continue to be directed toward the star, while the earth rotates beneath it. If we compare the direction of the pendulum, not with the direction of the star, but with the surface features of the turning earth, we shall see it apparently rotating in a clockwise direction, opposite to the direction of the earth's rotation. If our pendulum were set up over the south pole of the earth, it would appear to turn in a counterclockwise direction.

At lower latitudes, the turning of the Foucault pendulum depends on the angle through which the ground beneath the pendulum "skews around" as the earth rotates. At latitude 40° the pendulum would make a complete turn in slightly more than 37 hours, whereas it would turn once in 24 hours at either pole. The farther from the poles the pendulum lies, the slower its turning rate. The general relation between angular velocity of the plane of the pendulum oscillation ω and latitude ϕ is

$$\omega = \omega_0 \sin \phi. \qquad (4\text{-}18)$$

At the equator, the pendulum will not turn at all (sin $\phi = 0$, hence $\omega = 0$). The ground does not "skew around" at all because the surface of the earth is parallel to the axis of rotation.

A further experimental demonstration of the earth's rotation deserves mention on account of its many practical applications. The axis of a delicately balanced and rapidly spinning top tends to maintain a constant direction in space. A spinning drum or flywheel will thus indicate the direction of the earth's rotation. This principle of the gyroscope applies to the so-called gyroscopic compass, which is standard equipment on many modern ships. Gyro control is fundamental in most rocket guidance systems.

4-7   Masses of the Planets. We have already noted how gravitational forces deflect the paths of planets from simple Keplerian ellipses. Indeed, the observed departures from the elliptical motion can be used to determine the forces that produce the deflections. Since these forces are proportional to the masses of the bodies that produce them, they permit us to calculate the masses of the planets themselves.

We have a check on the accuracy of the calculations for planets that possess one or more satellites. In most of these cases the satellite is so close to the planet that even the gravitational pull of the sun is of secondary

FIGURE 4-7   *Nomogram relating the period,* P, *and semimajor axis* a *of a satellite circling about a central body of mass,* M, *by Kepler's Third Law,* $MP^2 = a^3$.

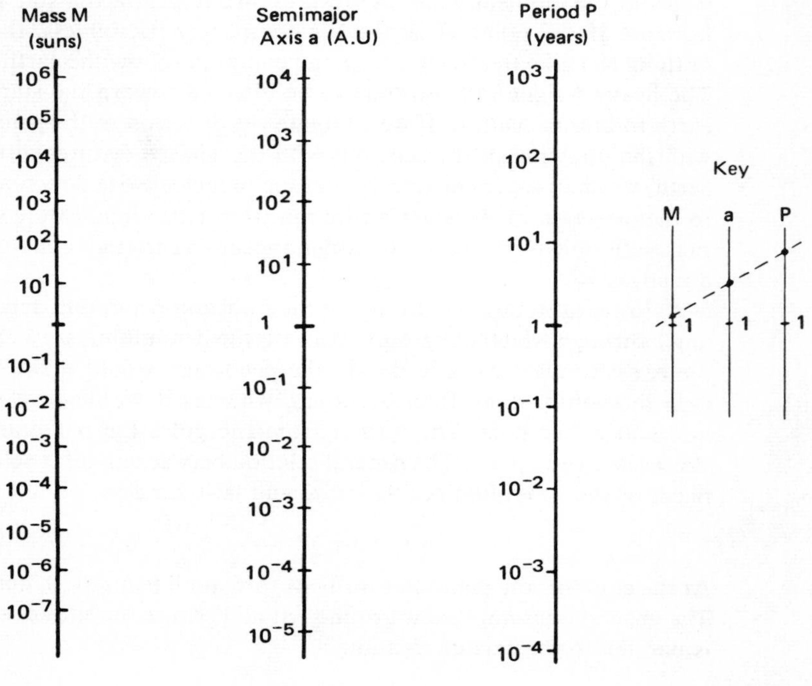

importance. Thus we can apply Kepler's third law to calculate the mass of the planet directly, since we can measure both the periodic time and the mean distance.

Figure 4-7 is a special type of diagram, a *nomogram*, which shows the relationship between the mass M (in terms of the sun's mass), the period P (expressed in years), and a, the semidiameter of the orbit (in astronomical units, the distance from the sun to the earth). The nomogram illustrates the equation

$$MP^2 = a^3. \tag{4-19}$$

To use the nomogram, we assume that two of the three quantities M, P, and a are known. Lay a ruler across the diagram to connect the two known points; then read off the values of the third at the point of intersection. Thus, given P and a we can determine M. If we use $M = 1$, we can reproduce Kepler's third law for the planets of the solar system.

4-8   Celestial Mechanics and the Discovery of Neptune.   Newton's basic scientific methods, applied and developed by outstanding scientists such as Gauss, Laplace, Leverrier, Tisserand, Poincaré, and many others, have produced the science of celestial mechanics. The application of this science has completely clarified the motions of planets in the solar system. It further enables us to predict the future motions of the planets, so that prediction and observation can be compared. The correspondence has been almost perfect. Only the motion of Mercury has shown a slight discrepancy (see Chapter 12), later explained by Einstein's theory (Chapter 34).

Celestial mechanicians discovered Neptune because they had faith in their ability to predict the future positions of the planets with high accuracy. Many years had elapsed since the work of Newton. The British astronomer William Herschel discovered the planet Uranus in 1781. Astronomers added the new planet to the map of the solar system, observed it, and calculated what its path should be. By 1840, the mathematical astronomers had decided that something was definitely wrong: Uranus was not following the expected path. Until 1822 it ran ahead of prediction; then it began to lag behind. Many were the speculations as to the causes of the deviation. Some astronomers suggested that the law of gravitation, which had proved so accurate for the planets out to Saturn, might begin to fail at still larger distances from the sun. Others, however, preferred the suggestion that some planet as yet unknown was perturbing Uranus, pulling it from its predicted path.

Leverrier in France and Adams in England attacked the difficult problem of using the observed deflections to deduce the position of the disturbing planet. Unknown to one another they solved it almost simultaneously in 1846. Adams sent his results to the Astronomer Royal at Greenwich who, though skeptical, passed the information on to the Cambridge Observatory where the astronomers began to make laborious measures of the positions of the stars near the place predicted, in the hope of detecting the planet because of its motion. Leverrier sent his predictions to the German astronomer Galle at Berlin, who fortunately possessed a star map of the region. Thus, by direct comparison with his chart, instead of by laborious measurements, he found the planet within the first few hours of observation on September 25, 1846.

Although Galle was the first to record the new planet, and the honor of the discovery fell principally to Leverrier, the position given by Adams was almost as accurate as that predicted by the French astronomer.

The discovery of this planet, which received the name of Neptune, was the culminating triumph of celestial mechanics—proof of the correctness of Newton's laws, and a demonstration of the universal character of the law of gravitation.

CHAPTER 5

The Motions
of the Earth

The triumphs of Kepler, Galileo, and Newton in mechanics and gravitational theory inspired a host of other discoveries and advances in astronomy. In this chapter we shall describe a few of these scientific milestones and present some of the dynamical and geometric concepts used in astronomy that do not depend on modern physical theory or instrumentation.

5-1   The Velocity of Light.   In 1676 the Danish astronomer O. Roemer, working at the Paris Observatory under the direction of J. D. Cassini, was studying the motions of Jupiter's bright satellites as they moved into the shadow of Jupiter, in order to improve Cassini's tables of their eclipses. He soon found that when the earth was on the side of the sun away from Jupiter (E_2 and J_2 in Fig. 5-1), the eclipses occurred later by nearly twelve hundred seconds than when the earth was closest to Jupiter (E_1 and J_1). He concluded correctly that the eclipses were late at E_2 because the light had to traverse the extra distance equal to the diameter of the earth's orbit.

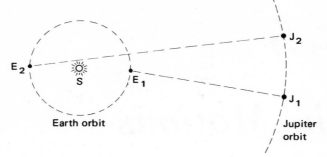

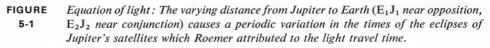

FIGURE
5-1
Equation of light: The varying distance from Jupiter to Earth (E_1J_1 *near opposition,* E_2J_2 *near conjunction) causes a periodic variation in the times of the eclipses of Jupiter's satellites which Roemer attributed to the light travel time.*

From present-day determinations of the velocity of light [299,793 kilometers per second (186,283 miles/sec)] and of the astronomical unit, the mean distance from the earth to the sun [149,598,000 km (92,956,000 miles)], we find that the time required by light to cross the earth's orbit is only 998 seconds, not the 1200 seconds found by Roemer.

Today the best measures of the velocity of light depend on radar studies, being derived from the time a radio signal takes to reach and return from a target at a carefully measured distance. No measurable difference exists between the velocities of light and of much longer radio waves, although the wavelengths of the latter may be millions of times greater.

Roemer has certainly earned the credit for proving that light moves with a finite velocity—a half century before an English astronomer, James Bradley, confirmed this conclusion through independent observations.

5-2 **The Aberration of Light.** Bradley's discovery of the aberration or "wandering" of light is an outstanding illustration of a careful and thorough scientific project that led to a completely unexpected result, far different from the original objective.

For two hundred years after the invention of the telescope, astronomers made sporadic efforts to demonstrate the orbital motion of the earth by the apparent annual displacement of star positions resulting from this motion. James Bradley, in 1725, set himself the task of measuring these small stellar displacements by the most precise technique then available.

Bradley pointed his telescope towards the zenith and fixed it rigidly in a chimney so that there could be no possibility of its being displaced over a period of a year. He then measured the relative position of the star γ Draconis as it passed through the field of his telescope from night to night, expecting the star to move as in Fig. 5-2, since the orbital motion of the earth might cause the star to shift its position, an effect known as *parallax*.

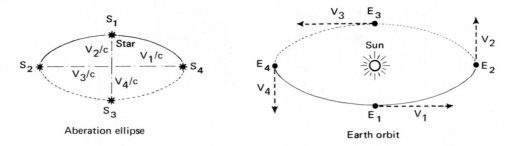

FIGURE   *Aberration ellipse: The apparent position of a star varies according to the changing*
5-2   *direction of the Earth's velocity on its orbit. See text for detailed explanation.*

When the earth moves in its orbital plane, through positions E_1, E_2, E_3, and E_4, a star at right angles to the plane of the earth's orbit should move in a very small orbit through the corresponding sequence S_1, S_2, S_3, and S_4. A star in the plane of the earth's orbit should move back and forth along that plane.

After Bradley had continued his observations a number of months, he discovered that the star indeed moved systematically, but not in phase with his expectations. The position of the earth at E_1 corresponded with that of the star at S_4; E_2 with S_1, E_3 with S_2, and E_4 with S_3. He checked his observations with other stars and other equipment and, for some years, pondered this surprising result. Finally he found the solution. The earth is moving rapidly in its orbit. If light does not travel instantly through space, but with a finite velocity, the combined motions of light and earth cause a displacement of the apparent position of the star from its true geometric position. The effect is called the aberration of light. A simple illustration of this effect is shown by Fig. 5-3, the inclination of the paths of raindrops on the window of a moving automobile or railway carriage. The inclination angle depends on the ratio of the horizontal velocity of the car to the vertical velocity of the drops. The same effect applies to light and the moving earth.

Light entering a telescope at a point O (Fig. 5-4) travels to point B, while the eyepiece of the telescope moves from A to B, so that the eyepiece receives the beam of light as it arrives at the observer's eye. The observer, looking through the eyepiece, sees the star apparently in the direction A-O, rather than in its true direction B-O. Hence, as shown in Fig. 5-3, the star always appears ahead of its true position in the direction of the earth's motion. The radius of the aberration circle in Fig. 5-3, or half the major axis of the aberration ellipse in other positions on the sky, amounts to $20.5''$, the constant of aberration, a large angle in terms of present-day measuring accuracy. The parallax angle, sought by Bradley, amounts to less than one second for the nearest stars and defied measurement for more than a hundred years after his great discovery of aberration (see Sec. 19-2).

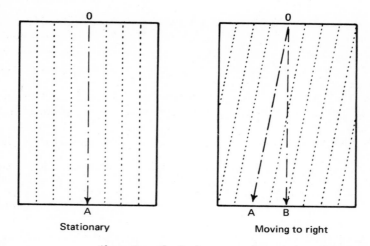

Stationary

Moving to right

**FIGURE
5-3**

Aberration of rain drops traced on window of moving railway carriage moving in direction of arrow.

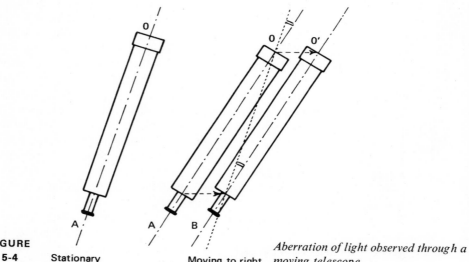

**FIGURE
5-4**

Stationary

Moving to right

Aberration of light observed through a moving telescope.

5-3 Precession of the Equinoxes. The *ecliptic* is the great circle where the plane of the earth's orbit cuts the celestial sphere. The *celestial equator* is another great circle where the plane of the earth's equator cuts the celestial sphere. These two circles are inclined to one another by 23°.5. The two points of intersection are known as the *equinoctial points*, because when the sun reaches them, about March 21 and September 21, days and nights are equal. As early as about 120 B.C. Hipparchus found that the equinoctial points did not agree with measures made by earlier astronomers, and thus discovered that the celestial pole is slowly moving among the stars (see Sec. 2-8).

We know today that earth's pole moves along a small circle of radius 23°.5 in a period of some 26,000 years. When the great pyramids of Egypt were built, the pole star was α Draconis, which is at present some 25 degrees from the pole. In about 13,000 years, Vega will be a very conspicuous but rather inaccurate pole star [Figs. 5-5 and 5-6(b)].

This phenomenon, known as the precession of the equinoxes, causes the equinoctial points to move westward (clockwise when one looks down from the north) along the ecliptic at a rate of some 50.2 seconds of arc per year. The explanation for this motion of the earth's pole had to wait some 1800 years until Newton had fully developed his law of gravitation and had applied it in detail to the motions of the earth and moon.

FIGURE 5-5   *Motion of celestial pole among the stars from 1600 to 2200. The "Pole Star," α Ursae Minoris will be nearest to pole in the year 2105.*

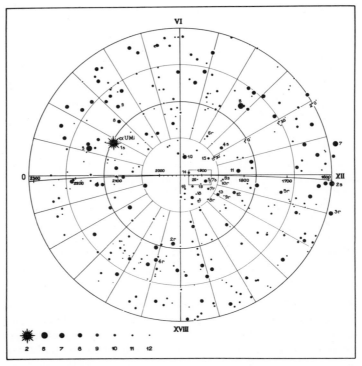

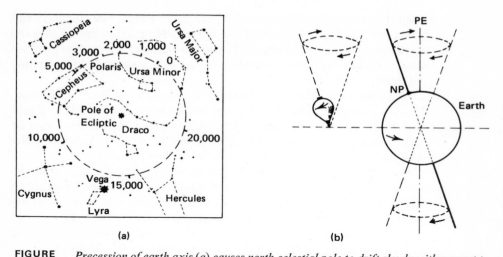

(a)   (b)

FIGURE 5-6   *Precession of earth axis (a) causes north celestial pole to drift slowly with respect to the starry heavens approximately along a small circle of 47° aperture in 26,000 years. (b) Similar effect in motion of top where restoring forces are opposite in direction.*

Why should the earth's axis turn in space like the axis of a top? Newton approached the problem from the point of view that the earth *actually* spins like a top [Fig. 5-6(a)]. Why, then, does the axis of a top precess? The answer is that the top, if not set spinning with its axis absolutely vertical, tends continuously to fall over. Because the top spins, the overturning force produces a conical motion of the axis. Such a motion in a spinning body is generally known as gyroscopic. A qualitative explanation follows in the next paragraphs.

We readily discern the analogy between the earth and a top, in that both are spinning, and we also recognize that the force of gravity acts to tip over a top. But what force tends to tilt the axis of the earth? To answer this question, we must first recall that the earth is flattened by its rotation. If the earth were a perfect sphere, the attraction of the sun and moon could be considered as acting on the earth's center, and there would be no force tending to tilt the earth's axis. But the rotation produces an equatorial bulge, and the plane of the earth's equator is tipped some 23.5 degrees to the plane of the earth's orbit, the ecliptic.

Consider now the moon's attraction on the earth. The spherical central section will be attracted as though it were concentrated in a point mass, but the remaining portion of the earth will act much as though it were concentrated in two equatorial bulges, *A* and *B* of Fig. 5-7. The schematic drawing exaggerates the distortion, which actually amounts to only 27 miles in the earth's diameter. The moon (and sun) will attract the point *A* with a force directed in part down toward the ecliptic, and also the point *B* with a force directed in part up toward the ecliptic. The upward and downward forces on the central spherical mass will, of course, exactly compensate because of symmetry. Thus the attractions of the sun and moon on the bulge act to turn the earth's pole towards the pole of the ecliptic.

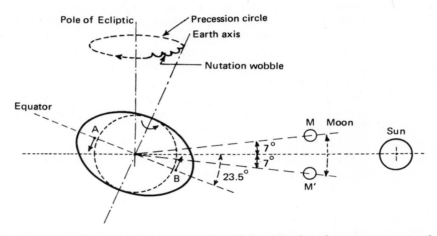

FIGURE
5-7   *Attraction of moon (and sun) on equatorial bulge* AB *of earth causes precessional motion of earth axis. Precession of inclined moon orbit* (MM') *in 18.6 years causes nutation wobble.*

The earth's pole would wobble back and forth like the mast of a boat in an ocean swell if the earth were not rotating. But as we look at the mass *A* in the diagram, it is moving directly toward us, while the mass *B* is moving away from us. These masses respond to the lunar and solar forces in the sense that both *A* and *B* tend to move toward the ecliptic in addition to their normal motion of rotation. Hence *A* moves out of the paper and down, while *B* moves into the paper and up. The effect of these two motions on the pole is in the same direction—that is, the pole tends to move out of the plane of the diagram. Consequently the earth's pole rotates about the pole of the ecliptic in a clockwise direction as we look down on the earth from the north. The line of the equinoxes, where the earth's equator intersects the ecliptic, must move in the same direction, opposite to the sense of the earth's rotation and revolution.

We can predict the rate of precessional motion, when we introduce the measured values of the masses, sizes, and motions of the earth, moon, and sun into the complicated equations for precession. The calculated and observed rates agree within the accuracy of the observations. The whole process involves a straightforward application of Newton's law of gravitation.

5-4 **Nutation.** As Bradley continued to search for the parallactic displacement of the stars, he finally discovered another small shift in star positions, called *nutation*. This second great and unanticipated discovery also resulted, like aberration, from his zealous search for a displacement that was too small to be measured with his equipment.

The earth's pole wobbles about 9.2 seconds of arc on either side of its mean position in a period of 18.6 years as it pursues its precessional circle around the pole of the ecliptic. Thus this circle is really a scalloped or wavy curve, and the wobbly motion is called nutation, a nodding of the pole. Nutation arises from a variation in the moon's precessional force, caused by the 5-degree tilt of the moon's orbital plane from the ecliptic. The earth's equatorial bulge and the sun disturb the moon's motion in a precessionlike fashion, to swing the pole of the moon's orbit clockwise about the pole of the ecliptic in a period of 18.6 years. The precessional force acting on the earth varies with the same period and produces the small nutation.

For artificial satellites precession is a major orbital correction, sometimes amounting to several degrees per day.

5-5 **Tidal Forces and Ocean Tides.** Except for the revolution of the earth about the sun, the only effects of external gravitational forces that we can detect on the earth itself are the tides. Newton was the first to explain quantitatively how the attracting forces of the moon and earth could produce the oceanic tides.

The shape of the earth is distorted because the moon and sun exert a greater attraction on the near side of the earth than on the far side, a consequence of the inverse-square relation in gravitation. The earth moves in its orbit in response to the average of all the forces acting upon it. The average motion and the average force apply to the center of the earth; hence, the side closer to the moon is drawn away from the center of the earth, while the farther side of the earth is less strongly attracted than the center. Thus the center is relatively pulled away from the far side.

All of these forces stretch the earth in both directions along the line to the moon, and contract it in the perpendicular directions. The earth is therefore stretched and squeezed into an elongated form, like a symmetrical egg, with a circular cross-section perpendicular to the moon's direction. Figure 5-8 illustrates the effect (with the scale, as usual, grossly exaggerated to make the tide-raising forces more evident). The sun exerts less than half the tide-raising force of the moon but still plays an important role in producing the tides.

If the solid body of the earth had no rigidity, it would respond instantly to the tide-raising forces of the moon and sun and we should see no tides in ocean water. On the other hand, if the earth were absolutely rigid, the oceans would show even greater tides than those we now observe. The

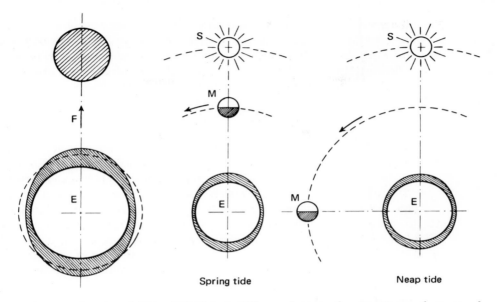

Spring tide   Neap tide

FIGURE   *Tides: the tidal force F (left) is the difference between the attraction on the near and*
5-8   *far sides of the ocean. The combined actions of the sun and moon (center) cause*
             higher tides near new or full moon than at quadrature (right).

earth, in fact, yields partially to these distorting forces, about as much as a
steel ball of similar size would (see Sec. 10-7), so that the oceans display
appreciable tides, with an average range of several feet along their shores.
As the earth rotates, the two tidal bulges, on opposite sides of the earth
along a line to the moon, move westward over the oceans. Only a foot or
two high in midocean, the tidal waters pile up around the irregular sea-
coast. In large funnel-like bays, such as the Bay of Fundy, they can grow to
60-foot tides, which come roaring in like a major flood.

Since the piling up of tidal waters requires time, the actual tide may lag
more than 12 hours behind the peak of the tidal bulge. Predictions of the
tides include a correction for the lag of the tides, called the *establishment of
the port*, which varies greatly from point to point along a coast.

If the moon revolved in the plane of the earth's equator, we should
expect successive high tides with almost equal amplitudes at intervals of
about 12.4 hours. Because the moon moves along the ecliptic instead of
along the equator, however, one tidal bulge may pass over a station while
the opposite bulge may miss it by a large distance. Thus successive tides
may differ greatly in height.

The sun, being more distant but much more massive, exerts about $\frac{5}{11}$ of
the moon's tide-raising force. The two bodies combine their action when
they are nearly in line, at new moon or full moon, to produce a total tidal
force over twice ($\frac{16}{6}$) that when their forces are opposed, at first or last

quarter. *Spring tides* (high high-tides and low low-tides) occur at new and full moon, while *neap tides* (high low-tides and low high-tides) occur at first and last quarters. The greatest tides of all occur when the moon is nearest to the earth (perigee) at the time of new or full moon. A comparison of local tide tables with lunar phases will demonstrate how Newton's theory of the tides applies in practice.

5-6   Geometry of the Seasons.   As the spinning earth moves around the sun (in the plane of the ecliptic), its equator remains tilted 23.5 to the ecliptic, while the direction of the earth's axis remains practically fixed in space, apart from the slow precession. Thus the sun crosses the plane of the equator twice during the year, first at the *vernal equinox*, about March 21, from the south side to the north side (see Fig. 5-9). Then six months later it crosses back to the south side of the equator at the *autumnal equinox*, about September 21. During these six months the north pole of the earth is tilted toward the sun, reaching its maximum inclination at the *summer solstice*, about June 21. Correspondingly, the south pole tilts towards the sun during the other six months, with a maximum at the *winter solstice*,

FIGURE   *The seasons are caused by the inclination of the plane of the equator to the plane of*
5-9   *the ecliptic. They are identified here for the northern hemisphere.*

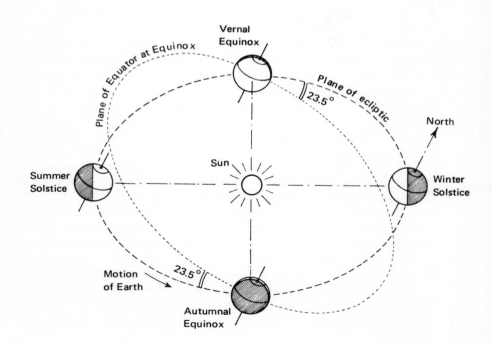

about December 21. The terms equinox and solstice can be used both for the points on the ecliptic and for the time when the apparent position of the sun is at these points.

Thus geometrical circumstances, properly interpreted, tell the story of the seasons. At the equinoxes, for example, the sun lies in the plane of the equator to illuminate the earth [Fig. 5-10(a)]. Thus, if we neglect atmospheric refraction (see Sec. 6-13), the sun appears just on the horizon as viewed from either the north or south poles. All parts of the earth are in sunlight for exactly half the 24 hours, so that day equals night. From this phenomenon comes the word equinox, *equi-nox* or equal night. As we shall see later, our calendar, unlike most older calendars, is so arranged that the equinoxes occur at nearly fixed dates in the spring and fall.

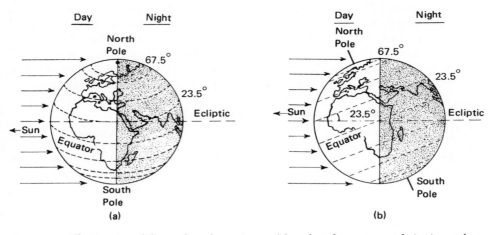

FIGURE   *Illumination of the earth at the equinoxes (a) and at the summer solstice in northern*
5-10   *hemisphere (b).*

At summer solstice [Fig. 5-10(b)] the sun shines 23.5 beyond the north pole. Thus, north of latitude 66.5 on earth the sun shines for 24 hours on June 21, when we can observe the midnight sun. Indeed, the north pole itself enjoys six months of continuous sunshine. Correspondingly, the south pole is in darkness for the same months of summer in the northern hemisphere.

Since the observer spins around with the earth at a fixed latitude, Fig. 5-10(b) shows that if he is located in the northern hemisphere he will spend more time in sunlight than in darkness during the six months from March 21 to September 21 Thus the days are longer than the nights, and the difference increases at increasing latitudes. On the equator all days and nights are geometrically equal. In the southern hemisphere, the seasons are the reverse of those in the northern hemisphere.

Note that the sun can pass directly overhead (through the zenith) at noon sometimes during the year at any latitude between 23°.5 south and 23°.5 north. It can sometimes shine all day or remain below the horizon all day north of latitude 66°.5 N and south of 66°.5 S. Hence these special latitudes have been given names: 66°.5 S: Antarctic circle; 23°.5 S: Tropic of Capricorn; 23°.5 N: Tropic of Cancer; 66°.5 N: Arctic circle. They define the boundaries of the polar, temperate, and equatorial zones.

5-7 Physics of the Seasons. Experience demonstrates that the longer the sun shines on a given spot the warmer that spot is apt to become. Even the north pole warms up considerably in summer. Hence the longer days and shorter nights in summer contribute to the warmer weather of that season. During the summer also the sun's rays tend to fall more nearly perpendicular to the average ground surface than during the winter. Since a given bundle of sun's rays provides a given amount of heat over a given area, [1.4 kilowatts per square meter (about 1170 watts per square yard)], this heat will be more concentrated if the beam strikes a surface at a right angle (Fig. 5-11). When the rays fall obliquely, they cover more surface and therefore deliver less heat per unit of area. The limit is zero, of course, when the sun is on or below the horizon.

FIGURE 5-11

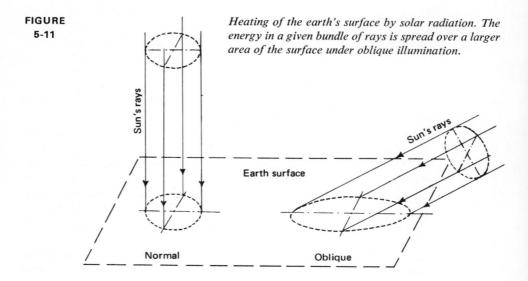

Heating of the earth's surface by solar radiation. The energy in a given bundle of rays is spread over a larger area of the surface under oblique illumination.

We can also understand this effect better if we see, as in Fig. 5-10(b), that during our summer a larger area of the northern hemisphere is exposed to the sun's rays and therefore receives more insolation (solar radiation) than during the winter (more than twice as much at the summer solstice).

The earth, of course, does not freeze immediately when the sun sets. The solid surface, the atmosphere, and especially the oceans, having stored the heat during the daytime, slowly radiate it away into space. The surface of the earth is truly a great heat reservoir that warms slowly in the spring and cools slowly in the fall. Thus March 21 is generally cooler in the northern hemisphere than September 21, even though the average radiation received from the sun is equal at the two equinoxes. The maximum summer temperature may lag as much as six weeks, from June 21 to as late as early August. This lag of the seasons may push the bitterest winter weather from December 21 into late January or early February, according to local circumstances.

All of these seasonal effects occur in the southern hemisphere six months later (or earlier) than those in the northern hemisphere. But the seasonal temperature range is generally less at a given latitude in the southern hemisphere than in the northern, even though the earth is nearest the sun about the first of January, so that one might expect northern winters to be relatively milder and the summers cooler. Apparently the large expanses of ocean in the southern hemisphere act as temperature regulators on the land areas to moderate the seasonal differences.

5-8   Problems of Calendar Making.   A satisfactory calendar must enable the average person to predict recurrent events, such as the seasons, without undue mental exertion. The higher the civilization, the greater are the demands upon its calendar, imposed by business arrangements and the planning of social and religious events. Preferably a calendar should repeat itself, or almost repeat itself, with the seasons—that is, once a year. The year refers to one revolution of the earth about the sun, but we must decide what reference point to adopt as the measure of a complete revolution.

For astronomical uses, perhaps the most fundamental year is the *sidereal year*, the time required for the earth to complete one revolution about the sun as measured with respect to the fixed stars. Its length is $365^d \, 6^h \, 9^m \, 9\overset{s}{.}5$ of ordinary mean solar time. This sidereal year, however, does not keep in step with the seasons because of the precession of the equinoxes (Sec. 5-3). A more practical year for everyday use is the *tropical year*, or year of the seasons, the time required for the sun to return to the vernal equinox (Sec. 5-6). Since the equinox moves westward among the stars, the tropical year is some 20 minutes shorter than the sidereal year, lasting $365^d \, 5^h \, 48^m \, 46\overset{s}{.}0$ in mean solar time. We note immediately that neither the sidereal nor the tropical year is convenient for calendar making because neither is equal to a whole number of days, with the remainder a simple fraction of a day.

The ancient Egyptians avoided this complication by arbitrarily choosing a year of exactly 365 days. As a consequence, the seasons kept moving ahead in their calendar by nearly a quarter of a day every year, completing a full circuit in some 1508 of our years. Thus every year the Egyptians had to subtract approximately one-quarter of a day from the previous year's reckoning to make the best prediction for the Nile's rising. This peculiarity of the Egyptian calendar has proved useful in dating their past history, since their calendar date of some seasonal event, such as the rising of the Nile, closely specifies the year, when other historical evidence selects the proper 1508-year interval.

The tropical year also does not contain an integral number of months, or intervals between successive New Moons. The average duration of the lunar month is 29^d 12^h 44^m 2^s78. Twelve of these lunar months add up to some eleven days less than the tropical year (see Table 5-1). A number of Islamic nations still use a lunar calendar with twelve lunar months per year of alternately 354 and 355 days. This calendar gains about one year in 33 on the tropical year, which is the basis of our present-day calendar.

TABLE
5-1
Lengths of various years

Year	Length
Sidereal	365^d25636
Tropical	365.24220
Egyptian	365.00000
Julian	365.25000
Gregorian	365.24250
Synodic (12 mo.)	354.36708

5-9   The Julian and Gregorian calendars. In 45 B.C., Julius Caesar asked the Alexandrian astronomer Sosigenes to help untangle the hopelessly confused Roman calendar. Sosigenes immediately abandoned the tradition of keeping a whole number of lunar months within the year and arbitrarily adopted a year of length 365^d ¼. He introduced a year of twelve months, instead of the customary ten plus extra months interposed at arbitrary intervals to keep the lunar and solar calendars on a roughly even basis. By adding an extra day every fourth year, the leap years, to a normal year of 365 days, he made the average exactly 365^d25. As a further innovation, he started the year on January 1 instead of in March. Caesar took advantage of the revision to rename the previous fifth month, July, after himself. His successor, Augustus Caesar, not be outdone, similarly commandeered the sixth month. Hence only the last four months of our present calendar year bear names based on the original Roman numbering (VII to X).

The Julian calendar was a definite improvement over earlier calendars, but the year was longer than the tropical year by 0^d00780 which amounts to one day in every 128 years. In A.D. 325, at the time of the Church Council of Nicea, the vernal equinox occurred on March 21. By A.D. 1582, the equinox had fallen behind by eleven days to March 10. Pope Gregory XIII, disturbed by the increasing confusion, asked the astronomer Clavius (after whom the largest crater on the near side of the moon was later named), to work out some method to correct the calendar. At the suggestion of Clavius, the Pope ordered that ten days be dropped in 1582, so that October 15 followed October 4. The *Gregorian calendar* does not consider as leap years those century years not divisible by 400. The year 2000 will be a leap year, while the years 2100, 2200, 2300 will not be leap years. The length of the Gregorian year then is shorter than the Julian year by 3 days in 400 years or $0\overset{d}{.}0075$ per year. The remaining departure of our average present calendar year from the tropical year thus amounts to only one day in 3300 years, a minute discrepancy that need not worry our calendar makers for some time to come.

Corrections to a calendar, incidentally, rarely occur without considerable opposition by the public. When the English Parliament finally adopted the Gregorian calendar in 1752, dropping eleven days immediately following September 2, the public complained bitterly. Although parliament had very carefully selected the period of the change to avoid unfairness in financial transactions, many people felt that they had been unjustly robbed of eleven days—even though everyone lived eleven days longer by the new calendar.

5-10   Calendar Reform.   Even though the Gregorian year errs by only one day in 3300 years, calendar reform is needed for other reasons. Men waste countless hours in excursions to the calendar because the days of the week do not fall on the same days of the month, either from month to month or from year to year. Each year brings a new pattern of holidays, festivals, and business dates, falling on different days of the week or month.

The most generally approved calendar modification is that recommended by the Association for World Calendar Reform. The year contains exactly 52 weeks, plus New Year's day, which does not bear a week-day name ($52 \times 7 + 1 = 365$). Leap day, every four years, is an extra day set between weeks between June and July. The year then divides simply into repeating quarters of three months, each containing thirteen weeks, as in Table 5-2.

A competing reform calendar of thirteen months, each of exactly four weeks, with a similar treatment of New Year's and leap days, has also been advanced, though it departs more radically from the present calendar. It has the distinct disadvantage that the year does not readily subdivide into halves or quarters.

Both of these reform calendars have suffered strong opposition from certain religious sects because the sequential repetition of the seven days of the week would be broken by the insertion of New Year's and leap days. Major changes appear unlikely in the foreseeable future.

TABLE
5-2
Proposed reform calendar

31^d January April July October							30^d February May August November							30^d March June September December						
S	M	T	W	T	F	S	S	M	T	W	T	F	S	S	M	T	W	T	F	S
1	2	3	4	5	6	7					1	2	3	4					1	2
8	9	10	11	12	13	14	5	6	7	8	9	10	11	3	4	5	6	7	8	9
15	16	17	18	19	20	21	12	13	14	15	16	17	18	10	11	12	13	14	15	16
22	23	24	25	26	27	28	19	20	21	22	23	24	25	17	18	19	20	21	22	23
29	30	31					26	27	28	29	30			24	25	26	27	28	29	30

5-11 Determination of Time. A clock is essential to the determination of time and even to its definition. The clock mechanism may be a mechanical, regular motion, a chemical reaction, an electrical phenomenon, an atomic vibration, a nuclear disintegration process, the periodic motion of a body in space, or any of many other types of processes. Astronomical time most commonly utilizes the rotating earth as a clock mechanism. The sky could be the dial of the clock and a plumb bob its hand, but we find it easier to think of the earth as fixed and the sky moving.

FIGURE
5-12
The celestial sphere. North Pole: P. Zenith: Z. Vernal Equinox: Y. Star: S. Horizon: NESW. Meridian: NPZ. Celestial Equator: EAYW. Hour Circle: ZSB. Hour Angle: ZPS.

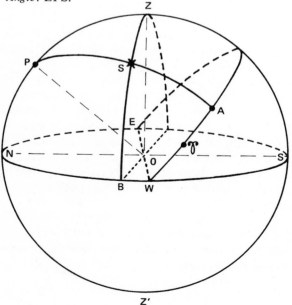

To see how the astronomer's earth clock actually gives us the time we must first define our system of directions carefully. We shall consider the sky as the *celestial sphere*, of infinite radius for our present purposes, with the observer exactly at its center. The north and south poles of the earth project as points on the celestial sphere (Fig. 5-12) to define the *celestial poles*, with the *celestial equator* as a great circle 90° from the poles. Then the direction of the observer's plumb line punctures the sphere above in his *zenith* and below at his *nadir*. His horizon lies midway between, with north and south points defined by the *meridian*, a great circle passing through the poles and the zenith. The east and west points occur where the equator and horizon circles intersect.

An *hour circle* is an important great circle drawn from the pole through a star, the sun, or some other point on the sky and perpendicular to the celestial equator. The meridian is a special hour circle passing through the observer's zenith. The *hour angle* is the angle between the meridian (south direction) and the hour circle of the object, measured positively towards the west.

Let us now look down on the earth from a northerly position (Fig. 5-13). A line from the north celestial pole makes a suitable hour hand. When the sun crosses over the meridian the *apparent solar time* will be 12 o'clock noon. The line to the sun defines the hour circle of the sun. The apparent solar time is the hour angle of the actual sun plus 12 hours.

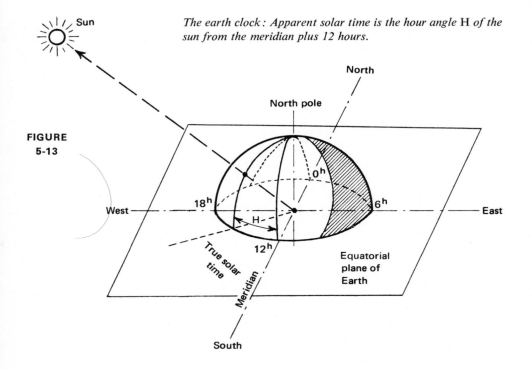

The earth clock: Apparent solar time is the hour angle H *of the sun from the meridian plus 12 hours.*

FIGURE 5-13

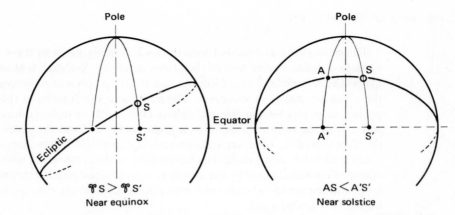

Pole

Pole

A   S

Equator

S

S'

A'   S'

Ecliptic

♈S > ♈S'

AS < A'S'

Near equinox

Near solstice

**FIGURE
5-14**
Inclination of ecliptic to equator is the major contributor to the equation of time.

5-12   The Equation of Time.   The sun makes a poor hour hand for our clock because its hour circle does not progress uniformly with respect to the stars. Near January 3 the earth is nearest to the sun, at *perihelion*, so that the sun appears to move more rapidly among the stars than it does near July 3, when the earth is farthest away, at *aphelion*. But even more important, the real sun moves in the ecliptic, whereas we measure time by a projection on the equator. Thus even though the sun were to move uniformly along the ecliptic, its hour circle would turn among the stars more slowly near the equinoxes than near the solstices, as shown in Fig. 5-14.

To avoid these irregularities in apparent solar time (which are conspicuous even on a sun dial) astronomers measure solar time by the hour angle of an imaginary sun that moves uniformly along the equator. The hour angle of this *mean sun* measures *mean solar time* and agrees with the apparent solar time about April 16, June 14, September 2, and December 25. The difference between apparent solar time and mean solar time is

**FIGURE
5-15**
Equation of time: Mean solar time minus apparent solar time.

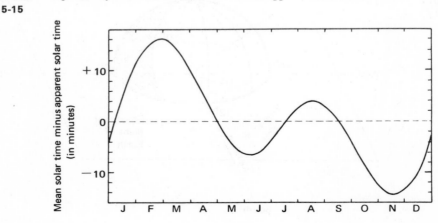

Mean solar time minus apparent solar time (in minutes)

+ 10

0

− 10

J F M A M J J A S O N D

called the *equation of time*. It reaches a maximum value of over sixteen minutes in early November, when apparent solar time is ahead. Mean solar time takes its greatest lead by some 14 minutes in the middle of February (Fig. 5-15). The maximum contribution to the equation of time by the earth's orbital eccentricity is less than eight minutes, and by the inclination of the ecliptic to the equator nearly ten minutes.

5-13 Civil and Standard Times. *Civil time*, which is identical with local mean solar time (Sec. 5-12), begins more conveniently at midnight than at noon, twelve hours earlier than mean solar time. Hence the hour angle of the mean sun plus twelve hours defines civil time. Until 1925 the astronomer's day, appropriately enough, actually began at noon rather than midnight. The practice was finally dropped to avoid confusion.

Even civil time is not quite applicable for everyday use because it is defined in the terms of the observer's meridian. Each observer has his own meridian and hence his own private civil time—an obvious practical difficulty. In the United States, for example, the local times of two points one hundred feet apart in east-west direction differ by about one-tenth of a second. Each meridian is determined by the observer's astronomical longitude, measured westward from the meridian passing through the axis of the Airy transit instrument at the old Greenwich Observatory, England, taken as origin of longitudes. Figure 5-16 shows that the difference in

FIGURE 5-16 *Local time and longitude: The difference of local times in O_1 and O_2 is equal to the longitude difference.*

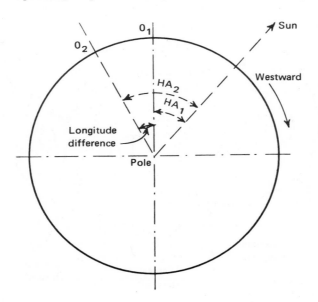

longitude between two meridians (O_1 and O_2) equals the difference in hour angle for any point S in the sky. Thus all varieties of time are greater toward the east by an amount exactly equal to the longitude difference. Since 360° of angle equal 24 hours of time, 15° of angle equal one hour, and so on. The difference in time equals the difference in longitude. For example, when it is noon on the west coast of the United States, it is later, 3 p.m., on the east coast, with a difference in longitude of approximately 45°.

Standard time is the civil time of a chosen meridian. The population in any area of the earth chooses a nearby standard meridian of longitude and adopts its civil time for everyday use, thus avoiding the longitude difficulty of local civil time. An integral hour of longitude is the favorite choice for a standard meridian, and the communities located within a longitude difference of roughly 30 minutes of time, or 7° 30′, east or west, use its standard time throughout the time zone so defined. Thus Eastern Standard Time is the civil time of the 75th meridian, 5 hours west from Greenwich, Central Standard Time of the 90th, Mountain Standard Time of the 105th, and Pacific Standard Time of the 120th. The irregularities of the adopted time zones in North America are quite conspicuous (Fig. 5-17). *Universal Time* used for astronomical observations is the Standard Time of the Greenwich Observatory meridian.

Daylight Saving Time is simply one hour later than the appropriate standard time, or the standard time of the adjacent time zone to the east.

**FIGURE
5-17**   *Time zones in the United States.*

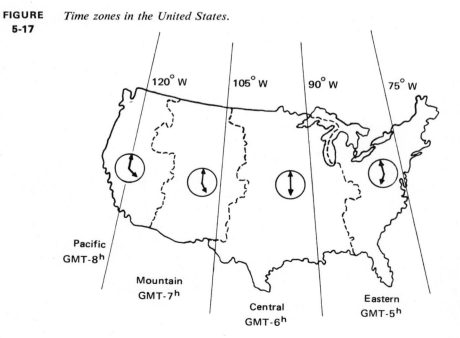

The use of standard time applying to a zone somewhat to the east of the actual longitude is quite common over the world.

Let us trace the time zones eastward, as if we were in an extremely fast jet plane capable of circumnavigating the earth in a minute or so. As we move eastward from the Pacific zone, say, the time becomes progressively later, by three hours in the eastern zone and eight hours at Greenwich. Somewhere we shall pass midnight and change to the next day, say from Monday the 18th to Tuesday the 19th. But as we complete our circuit around the earth, we come back a day late! By our supposition we have been away for only a minute and hence the day in the Pacific zone is still Monday. To solve this dilemma, we arbitrarily change the date when we cross the *International Date Line*, near longitude 12^h or $180°$ from Greenwich, somewhere in the middle of the Pacific. In crossing to the west, the traveler adds a day. Monday the 18th becomes Tuesday the 19th. Crossing to the east, he subtracts a day, independent of possible changes in watch time. Figure 5-18 illustrates a typical distribution of standard times and dates around the world.

Crossing the International Date Line by boat or aircraft is always a surprising experience, as one apparently gains or loses a day in his lifetime. If, of course, the traveler returns, or continues on around the world to his point of departure, he neither gains nor loses in his reckoning. A person

FIGURE 5-18   *Time zones around the world.*

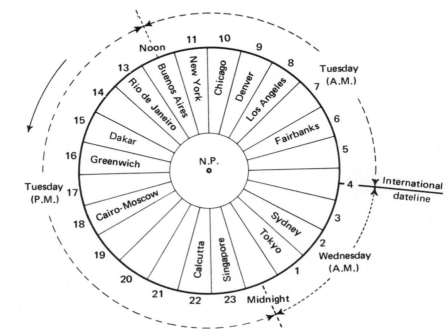

who makes a permanent change in residence across the International Date Line may gain or lose almost an entire day. Those who are especially fond of Sundays may enjoy as many as ten Sundays in February by judiciously crossing and recrossing the Date Line. To do so, choose a leap year in which February 1 is a Sunday and enjoy it on the west side of the International Date Line until nearly midnight. Cross the date line to the east at midnight and enjoy Sunday February 1 again. Go west again during the week and rest there on Sunday February 8. Repeat process for this each double Sunday through February 29th.

5-14 **Navigation.** Since the development of a portable chronometer to provide accurate time at sea, or the more recent development of radio time signals, a navigator has been able to determine his position accurately at sea or in the air by astronomical observations. The detailed techniques of celestial navigation are too lengthy and mathematical for convenient presentation here, but the basic principles are simple.

The navigator easily measures the apparent altitude of a heavenly body above the sea horizon or above an artificial horizon established by gravity, such as provided by a bubble sextant, even though he cannot readily determine his meridian. Such a sextant has a built-in "level," with a bubble of air whose position will indicate to the observer when he is holding the instrument precisely horizontal. After he corrects the observation for instrumental errors, the bending of light (refraction) passing through the atmosphere, and his elevation above the sea, he obtains an accurate measure of the *zenith distance* of the body, or the angle between its direction

FIGURE 5-19   *Sumner circle: Constant zenith distance of sun.*

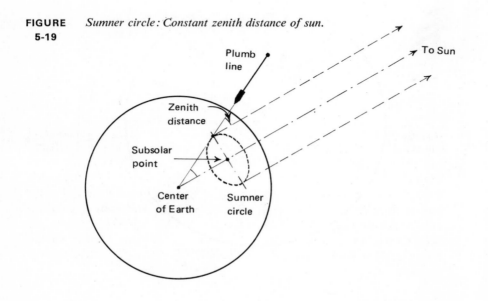

and the direction of a plumb line pointing upward. This observation places him somewhere on small circle, or *Sumner circle*, on the earth (see Fig. 5-19). Suppose that the navigator observes the sun. The center of the Sumner circle is the *subsolar* point, where a line from the center of the earth to the sun cuts the surface. The small circle as seen from the center of the earth has an angular radius equal to the zenith distance of the sun.

Thus the navigator must observe a second heavenly body, or make a later observation of the sun, to locate himself completely. Suppose he immediately observes the moon and thus establishes a second small circle on the earth. The two circles will intersect in two points, and unless the navigator is indeed badly lost he will know which of the two points gives his correct position (see Fig. 5-20). In practice the Sumner circles are so large that the navigator plots only their tangents near his approximate position on a navigational chart. These tangents are known as *Sumner lines*, and their intersection fixes the navigator's position. The accuracy on shipboard is often within a mile of the true position and from aircraft about a dozen miles.

FIGURE 5-20 *Use of moon and sun Sumner lines for navigation.*

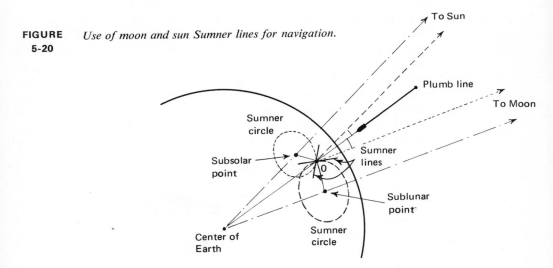

5-15 Ephemeris Time. Up to the beginning of this century astronomers had no reason to believe that the rate of rotation of our earth clock was not perfectly regular. However, as we have noted, there was no reason other than convenience to choose the earth's rotation rather than any other regular motion as our standard of reference. We could very well have chosen the motion of the earth—or indeed of any other planet—around the sun, or the motion of the moon around the earth as our primary clock. As long as our various clocks agree, there is no problem. If they should disagree, we must decide which is more nearly correct and why the other is in error.

Observations of the positions of the sun and moon through the centuries showed persistent departures from the positions predicted by celestial mechanics. These departures can be considered as the sum of two components, a regular and systematic trend called the secular acceleration (of the moon or sun), and irregular, erratic changes. After much searching, E. W. Brown in the United States and H. Spencer Jones in England recognized in 1925–26 that these vagaries of the sun and moon's motions—and indeed of the planets in proportion to their respective angular velocities—must arise from irregularities of the rotation of the earth. If all celestial bodies seem to speed up or slow down in unison, it is because our earth clock is not a perfect timekeeper. Hence the "secular accelerations" of the sun and moon reflect in fact a slowing down of the earth's rotation by tidal friction (both in the oceans at the surface and in the plastic layers inside the earth).

The remaining irregularities reflect fluctuations in the earth's rate of rotation caused by seasonal changes in mass distribution and by erratic mass motions in the earth. This example affords another application of the law of conservation of angular momentum (see Sec. 4-4).

In order to free astronomical computations from the effect of the irregularities of the earth's rotation, a new definition of time based on idealized motions of the sun and moon (presumed completely uniform and obeying Newton's laws) was adopted and introduced in the astronomical ephemerides on January 1, 1960. This presumed idealized Newtonian time is called *Ephemeris Time*. One second of Ephemeris Time is a specified fraction of the tropical year 1900*; the difference between Ephemeris Time and Greenwich Standard Time (Universal Time), plotted in Fig. 5-21 for the past 150 years, is derived for each year from a comparison of the

* Tropical year 1900 = 31,556,925.9747 seconds of Ephemeris Time.

FIGURE 5-21 *Irregularity of earth rotation, shown by variable difference between Ephemeris Time and Universal Time.*

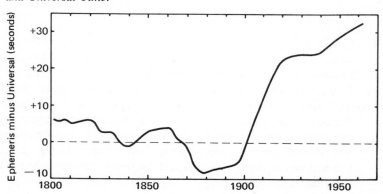

observed and computed positions of the moon. The theory of the moon's motions is now believed to be so perfect as to leave no room for any significant error other than the apparent one attributed to faulty time-keeping by the earth clock. There is no way of predicting the future trend of fluctuations in the earth's rotation, although a short extrapolation of the curve in Fig. 5-21 usually permits a reasonable guess for a few years ahead.

In recent years quartz clocks and "atomic" (or molecular) clocks relying on the very stable frequency of vibration of piezoelectric crystals or of simple molecules have given us means of interpolating and extrapolating Ephemeris Time to a still higher degree of precision.

5-16 **Celestial Coordinates and Sidereal Time.** In Sec. 5-11 we defined the celestial sphere, which employs the celestial equator as a fundamental circle. On this circle we shall choose the vernal equinox as a convenient zero point. Then the angle between the hour circle to the vernal equinox and the hour circle to the star, measured eastward along the equator, gives one coordinate of the celestial object, known as *right ascension*, usually stated in units of time instead of angle. The second coordinate, the *declination*, is the angle to the star, measured along the hour circle, north or south from the equator. Thus, as compared with the terrestrial system, right ascension is analogous to longitude and declination to latitude.

Star catalogues list right ascensions and declinations of the various objects. Since the equinox is moving, because of precession, celestial coordinates change with time; hence the catalogue must also indicate the year (or equinox) for which the coordinates apply. Simple tables and formulas enable the astronomer to correct the position for any specific date.

Star time or *sidereal time* is simply the hour angle of the vernal equinox —that is, the number of sidereal hours, minutes, and seconds that have elapsed since the vernal equinox was on the meridian. Alternatively, the sidereal time (S.T.) equals the right ascension (R.A.) of an object on the meridian. Then the hour angle (H.A.) of a body, measured westward from the meridian is

$$\text{H.A.} = \text{S.T.} - \text{R.A.} \tag{5-1}$$

If the H.A. is negative, the object lies east of the meridian.

Sidereal time and mean solar time coincide only at the autumnal equinox, about September 21. The sun moves apparently eastward among the stars so that, exactly a year or 365.2422 solar days later, the sun has made one complete revolution with respect to the stars. Hence the stars will have completed 366.2422 revolutions, one more than the sun. A sidereal day is about four minutes (actually $3^m56{.}^s56$ sidereal time) shorter than the mean solar day.

CHAPTER 6

Some Tools of Astronomy

The advance of science depends on tools, both instrumental and conceptual. Mathematics is one of the most important conceptual tools of astronomy. In this chapter, however, we shall limit our discussion to instruments, beginning with the human eye, the earliest astronomical tool.

6-1 **The Eye.** The human eye is a truly remarkable instrument. Even the best modern television cameras can reproduce its performance only crudely. At the distant beginnings of astronomy, and for thousands of years thereafter, the eye was practically the only astronomical instrument, and it was directly responsible for man's interest in the heavens. The eye's quick response to extremely faint light sources still sets a standard for the designer of sensitive optical equipment. In the perception of small differences of light intensity it has few rivals. Its capacity to perceive shape or configuration is a remarkable asset. However, its remarkable adaptability to a wide range of intensities limits its usefulness as an instrument for measuring absolute luminosity.

Apart from the complicated physiological and chemical properties of the retina, and the psychological aspects of nerve communication to the brain, the eye is a relatively simple optical device. It possesses a lens that projects an image on the retina by means of *refraction*, the bending of light.

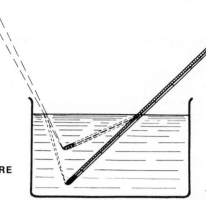

FIGURE 6-1

Refraction of light. A thermometer placed in a jar of water appears bent.

When a light beam, as in Fig. 6-1, meets at an angle the flat surface of a transparent solid or liquid, refraction changes its direction. The beam travels nearer the perpendicular to the surface within the denser medium. This property of light is exploited in the eye and in all other lenses. Suppose light is coming from a remote point source such as a star and meets a curved surface separating air from a transparent denser medium, as in Fig. 6-2(a). In being bent towards the perpendiculars, *P*, at the surface, the rays at *A* and *B* are bent towards the *focus F*, where they meet the central ray passing through *C*.

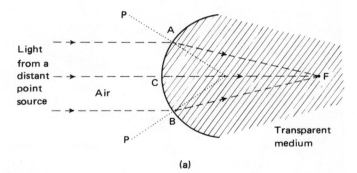

(a)

FIGURE 6-2

(a) Light rays from a distant point source come to a focus in a simplified model of the eye. (b) Two point sources produce images on the retina of the eye.

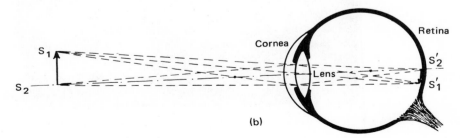

(b)

Most of the focussing in the eye is produced by the outer surface of the cornea, the lens proper being used mostly to improve the image quality and to change the focus. In Fig. 6-2(b) two light sources S and S' focus on the retina at points F and F', so that an extended object is imaged with nearly its proper shape, upside down on the retina (the brain interprets the nerve signals so that we see the object right side up).

**FIGURE
6-3**

Mural quadrant of Tycho, 1587, Uraniburg Observatory. Courtesy of the Bettman Archive.

6-2   Pointers. A gunsight used as a means of determining direction represents the first astronomical instrument of the pretelescopic era. The general principle of sighting at distant objects through fixed marks must have been applied long before the date of our earliest records. It was the practice of ancient astronomers. Many prehistoric buildings were oriented so that the sun would be seen to rise in line with certain parts of the structure. The tradition lived on until medieval times, and survives today in the east-and-west orientation of many churches.

Tycho Brahe carried the principle of a gunsight on a graduated circle to its highest precision (Fig. 6-3). When Kepler used Tycho's observations of Mars to determine the fundamental laws of elliptic motion, he had such confidence in their accuracy that he believed the errors could not reach eight minutes of arc. Note that an eight-minute error corresponds to a quarter of the moon's apparent diameter. We know today that Tycho's average error was only two minutes of arc, which corresponds to a really good shot.

Even though pretelescopic astronomy was crude and inaccurate by modern standards, we should realize that perhaps the greatest single advance in astronomy, Newton's formulation of the law of gravitation, was based essentially on nontelescopic observations. Newton probably would have made his discovery even if the telescope had not been invented. But we may question whether it would have been accepted without the telescopic evidence of facts supporting the Copernican theory, such as the phases of Venus, the mountains on the moon, or the satellites of Jupiter. Thus Galileo's observations with the new tool, the telescope, paved the way to an understanding of the fact that the earth was itself one of the heavenly bodies.

6-3   The Telescope. A telescope consists chiefly of two optical parts: an *objective*, or large focussing device, such as a lens or mirror, and an *eyepiece*, or small focussing device, usually a lens. The two are held together by a tube or framework that can be pointed in the desired direction. Since all astronomical objects lie at great distances, the light rays from them come in practically parallel as they fall upon the objective of the telescope. The objective O then brings them to a focus in the image, located in the *focal plane F* (Fig. 6-4); OF is the focal length of the objective, FY that of the eyepiece. The eyepiece is placed so that its distance behind the focal plane is equal to its own focal length. Thus, the diverging light rays from the focal plane pass through the eyepiece and again become parallel as they enter the observer's eye.

Although the eye can adjust its focal length for various distances, it is ideally most relaxed for parallel light coming from a distant object. Hence the observer sets the eyepiece so that all rays of light from a given part of

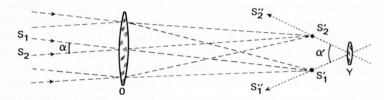

FIGURE
6-4 *Refracting telescope. Objective O receives light from two neighboring stars, S_1 and S_2, and images them at S_1' and S_2' in the focal plane, F. The eyepiece at Y magnifies the separation, so that the eye views the apparent images at S_1'' and S_2''.*

the original object emerge parallel or nearly parallel to one another. That the objective forms *real* images of distant objects in the focal plane becomes evident when we put a piece of paper at this position, or use a magnifying glass to focus the sun's image and burn a hole in a piece of paper. Since there are real images in the focal plane, the eye sees two star images separated by the angle α' in Fig. 6-4. Without the telescope in the line of sight, the eye would see them separated by the angle α at which they entered the objective. Hence the telescope has increased the separation of the images in the ratio α'/α, which we define as the *magnification, M,* of the telescope. Since the angles α and α' are both small, and the corresponding triangles have a common base in the focal plane, the ratio α'/α very nearly equals F/f, where F is the focal length of the objective, f that of the eyepiece. Hence the magnification of a telescope is given by the equation:

$$M = \frac{F}{f}. \tag{6-1}$$

The magnification is equal to the ratio of the focal length of the objective to the focal length of the eyepiece. For example, the combination of a telescope objective of 100 centimeters focal length with an eyepiece of focal length 2 centimeters yields a magnification of 50 times. The moon, one-half degree in angular diameter, would appear to be about 25 degrees across with such a telescope. Thus the astronomer can readily vary the magnification of a telescope. He may select, from among a number of eyepieces of different focal lengths, one that provides the magnification best suited to a given kind of observation.

6-4 Light-Gathering Power and Resolving Power. These two properties of a telescope are very valuable in addition to the magnifying power. A telescope can collect much more light than the eye, to make visible objects fainter than the naked eye could see. Also, a telescope enables the observer to distinguish objects that would otherwise be too close together for the eye to separate.

The area of the objective determines the ability of a telescope to collect light. Since the area varies as the square of the linear aperture (diameter) A, a telescope 1 meter in aperture, compared with one of 5 centimeters, can show objects whose apparent brightness is less by a factor of $(100/5)^2$ or 400 times. Since the aperture of a night-adapted eye is about 0.8 cm ($\frac{1}{3}$ inch), we can say that the light-gathering power of a telescope, relative to the eye, is roughly $1.5\,A^2$, where the aperture A is measured in centimeters ($9A^2$ if A is in inches). A telescope of 25 cm aperture, therefore, will reveal stars nearly 900 times fainter than those visible to the naked eye.

The brightness of point images such as stars follows this A^2 law. The speed of a telescope in photographing extended areas, such as the moon or Milky Way, depends, however, on the *focal ratio*, $F/A =$ (focal length)/ (aperture). As in ordinary photography the exposure time is proportional to the square of the "F-number" or "F-value," or to $(F/A)^2$. The total light collected varies as A^2, but it is concentrated on the photographic emulsion inversely as the square of the focal length. In everyday camera practice the symbol $F/6.3$ means that $F/A = 6.3$. A *fast* system has a small F-number, perhaps $F/A = 1.5$. In astronomical photography, then, the speed of a telescope for extended objects is proportional to $(A/F)^2$, while for point images such as stars the light collected is proportional to A^2, and the actual exposure time depends on the image quality or size.

The important ability of a telescope to separate small angular objects such as close pairs of stars, or to delineate fine angular structures such as craters on the moon, is called its *resolving power*, R. Strictly defined, R is the smallest angle between point images that can be separated by a telescope. The practical resolving power naturally depends upon the quality of the optical surfaces, but even for ideal surfaces it is limited by the wave nature of light. Because light acts like a wave—like a ripple expanding from a stone dropped into still water—it can turn corners as can the water ripple. This property of light, *diffraction*, prevents an ideal telescope from producing a perfect point image even from a perfect point source of light.

The wavelength of light, measured from crest to crest of the wave in vacuum, is very small for visual light, only 5×10^{-5} cm (1/50,000 inch) in the green where the eye is very sensitive. Since the aperture of a telescope is many wavelengths wide, the light waves are well contained, so that an ideal point image is only slightly smeared out by diffraction. In fact the diameter of the *diffraction disk* is inversely proportional to the aperture, A, and directly proportional to the wavelength, having a diameter of the order of λ/A in radian measure. Since there are about 2×10^5 seconds of arc in a radian, we have for this diameter in seconds of arc:

$$D = 2.06 \times 10^5\, \frac{\lambda}{A}. \qquad (6\text{-}2)$$

To be seen easily as two separate images in a perfect telescope, two equal stars must be separated by slightly more than the diameter of the diffraction disk. Hence the minimum resolution limit, R (in seconds of arc), in visual yellow light is generally adopted as

$$R = \frac{11.4}{A \text{ (cm)}} = \frac{4.5}{A \text{ (inches)}}. \tag{6-3}$$

The equation implies that an observer with a telescope of 11.4 cm (4.5 inches) aperture should be able to resolve two stars as close together as one second of arc. A telescope with an aperture just over a meter (45 inches) would theoretically resolve a pair of stars only 0.1 second apart. Since a night-adapted eye can separate stars only about three minutes of arc apart, a large telescope is able to reveal not only much fainter objects, but also far greater detail, than does the naked eye.* In practice, however, atmospheric turbulence, which causes variable *seeing conditions*, often reduces the performance of a large telescope far below its theoretical value (see Sec. 6-15).

Equation (6-3) holds for a visual light at wavelength 5×10^{-5} cm. In terms of a more general wavelength, λ, the resolution limit becomes

$$R = 2.3 \times 10^5 \frac{\lambda}{A} \qquad \text{(in sec of arc)} \tag{6-4a}$$

or

$$R = 64 \frac{\lambda}{A} \qquad \text{(in degrees),} \tag{6-4b}$$

where λ and A measured in the same units.

Outside the atmosphere, where space telescopes can collect the shorter waves of ultraviolet light, a telescope of a given aperture could give, in principle, finer images than in visual light because of the shorter wavelengths of ultraviolet rays. For instance, if $A = 100$ cm and $\lambda = 0.2$ microns $= 2 \times 10^{-5}$ cm, equation (6-4a) gives $R = 0.''046$.

Equation (6-4b) serves even for radio telescopes (see Sec. 6-9), where λ may be several meters. Note that the smaller the resolution limit, the finer the detail that can be observed. Clearly, a radio telescope will have a resolution limit numerically much greater than a visual telescope of the same aperture and will be able to register correspondingly less detail. For instance, if $A = 1830$ cm $= 60$ ft and $\lambda = 21$ cm, equation (6-4b) gives $R = 0.73$ degree.

Having seen how a telescope performs its three primary functions—magnifying, collecting light, and resolving fine detail—we shall now discuss the two main types of telescopes: refractors, fitted with a lens for an objective, and reflectors, fitted with a mirror.

---

* Note that equation (6–2) does not apply to the naked eye, whose resolution limit is determined not by diffraction but by optical aberrations (defects) and by the cell structure of the retina.

**FIGURE
6-5**   *Dispersion by a prism. Short-wave violet light, V, undergoes greater
bending than green light, G, of intermediate wavelength or the long-
wave red light, R.*

6-5   Aberrations of Refractors.   Some optical properties of the simple
lens are a nuisance for astronomical observations and must be corrected
to improve the performance of refracting telescopes. As we have noted,
when a ray of light passes from air into glass through a smooth surface, it
is bent in direction by refraction. It also spreads out into the colors of the
spectrum, because refraction generally decreases with increasing wavelength, producing *dispersion.* The flashing prismatic colors of a diamond
are produced by dispersion. Figure 6-5 shows parallel light coming from
the left and undergoing dispersion by the successive action of two surfaces
of a prism. The rays of blue light of smaller wavelengths undergo greater
deviation than those of red light. They separate from the latter and emerge
in a slightly different direction. We can now readily visualize the effect of
dispersion for a simple lens if we consider the lens to consist of two small-
angled prisms with their bases in contact. Here the dispersion will bring
blue light to a shorter focus than red light (Fig. 6-6). Such an effect, known
as *chromatic aberration,* is undesirable in a telescope, because all colors
should come to the same focus to form a sharp image.

**FIGURE
6-6**   *Refraction of a simple lens. The focus for violet light, V, is closer to the
lens than for green light, G, or red light, R. As a result, a star image is
never sharp, but possesses a colored (chromatic) ring.*

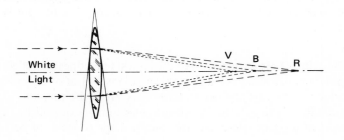

In practice, we correct a simple lens for chromatic aberration by adding another lens made of glass of different dispersive power. A *positive lens* is thicker in the center than at the edges, and produces a real image, while a *negative lens*, thicker at the edges than at the center, produces the opposite effect, spreading out the incident beam of light. To correct a positive lens for chromatic aberration, we combine it with a negative lens made of glass with higher dispersive power, but with less refractive power than the positive lens. The added negative lens removes the chromatic aberration and also partially reduces the focussing power of the positive lens. Such a lens combination, shown in Fig. 6-7, is called an *achromatic lens*—that is, a lens without color effect.

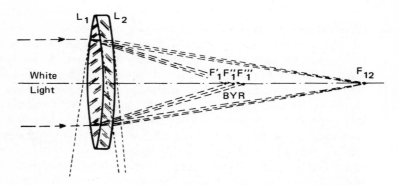
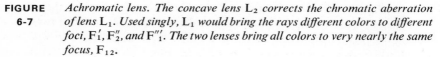

FIGURE 6-7 *Achromatic lens. The concave lens L_2 corrects the chromatic aberration of lens L_1. Used singly, L_1 would bring the rays different colors to different foci, F'_1, F''_2, and F'''_1. The two lenses bring all colors to very nearly the same focus, F_{12}.*

Although no lens has been made perfectly achromatic, many are nearly free of chromatic aberration and serve admirably over a wide color range. The largest refracting telescope, at the Yerkes Observatory of the University of Chicago (Fig. 6-8), has an achromatic objective, 40 in. in diameter. Possibly no one will build much larger refractors, because the sagging of such a huge objective under its own weight tends to spoil its performance, and its great thickness leads to excessive absorption of light in the glass.

6-6 Reflecting Telescopes. All the largest telescopes are reflectors, which employ a concave mirror as an objective, instead of a lens, to bring light to a focus. The early reflecting telescopes, from the time of Newton until the middle of the nineteenth century, had mirrors of polished metal. The mirrors of modern telescopes are usually made of glass, or better of fused

FIGURE   *The 36-inch refractor of the Lick Observatory of the University of*
6-8   *California.*

silica, with a thin coating of highly reflective aluminum evaporated on the concave face. Metal mirrors, however, are again being used in astronomy.

Most reflectors make use of the geometrical principle that parallel rays, striking a parabolic surface parallel to its axis, are reflected to converge at a point (F in Fig. 6-9). In a small reflecting telescope the observer cannot put the eyepiece at the *prime focus* (Fig. 6-10a, b) to which the mirror converges the rays, because his head would block the incoming light. Instead, the converging light beam is reflected to one side by means of a small flat mirror placed within the telescope, to bring it to the *Newtonian focus* (so called because Newton used this device in his original reflector). Or the beam is reflected back along the axis by a small convex mirror, passing

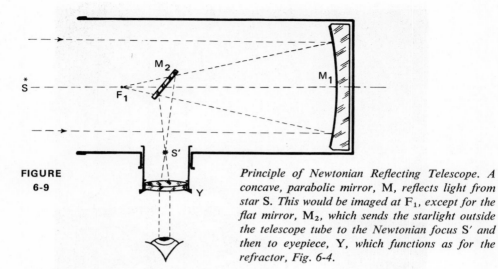

FIGURE 6-9

Principle of Newtonian Reflecting Telescope. A concave, parabolic mirror, M, reflects light from star S. This would be imaged at F₁, except for the flat mirror, M₂, which sends the starlight outside the telescope tube to the Newtonian focus S' and then to eyepiece, Y, which functions as for the refractor, Fig. 6-4.

FIGURE 6-10(a)

Diagram of the Lick Observatory 120-inch Reflector, showing the various focal points available: prime, newtonian, cassegrainian and coudé (sky and telescope).

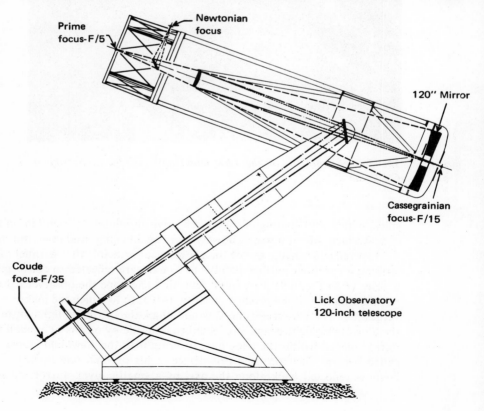

Prime
focus-F/5

Newtonian
focus

120" Mirror

Cassegrainian
focus-F/15

Coude
focus-F/35

Lick Observatory
120-inch telescope

FIGURE   *The Lick Observatory 120-inch Reflector.*
6-10(b)

through a central hole in the main mirror to converge at the so-called
Cassegrainian focus (after a seventeenth-century French optician, Cassegrain). In using such an arrangement, the observer stands at the lower end
of the telescope, as he would with a refractor. The Cassegrainian focus can
be used to increase the effective focal length of the telescope if such is desired for some instrument attached at the focus.

Very large telescopes use a series of flat mirrors to reflect the light to any
desired focus—for example, down the mounting axis to the basement
below, a *coudé focus* (*coudé* is French for elbow). The 200-inch reflector on
Mount Palomar has so large a mirror that the observer can sit in a cage at
the prime focus without blocking off too much of the incoming light (Fig.
6-11).

A reflecting telescope has the advantage that all incoming light rays are reflected in the same way regardless of color; thus it is completely achromatic. Furthermore, good mirrors can be made much larger than lenses because they use only the upper surface. The light does not pass through the mirror, hence small imperfections within the glass have a negligible effect on the optical performance. Theoretically, however, a parabolic reflector will produce a perfect image only on its axis. Star images show increasing deformation with distance from the axis, developing a fuzzy tail known as *coma*. Special lenses can improve the image quality. A complicated mechanical system of multiple supports carries the weight of the mirror and prevents it from sagging.

Since a reflecting telescope is achromatic, it can easily be constructed with a short *focal ratio* (focal length/aperture). The 200-inch telescope, for example, has a focal length of 660 inches or a focal ratio of only 3.3. Refractors, typically, have been built with focal ratios of 6 to 15, because of

FIGURE 6-11 *The prime focus of the 200-inch reflector of Mount Palomar. (Mount Wilson and Palomar Observatories.)*

the difficulties in constructing the steep curves of achromatic lenses. Careful lens design, however, results generally in much larger fields of good focus for refractors than for reflectors. Shortening the focal ratio greatly reduces the length of the supporting tube structure, and hence the weight and cost of the tube, with corresponding reductions for the driving mechanism and the protecting dome structure.

The main disadvantage of a parabolic reflector is its very small field of good focus. The 200-inch reflector, for example, shows its best image quality over an area only two minutes of arc across, or $\frac{1}{15}$ of the moon's diameter. Many photographic refractors, by way of contrast, have good photographic image quality over areas of five degrees or more.

Thus, while the parabolic reflector can be made extremely large to concentrate an enormous amount of light from objects in a small area of the sky, it cannot compete with the refractor for survey problems in which a single photograph of a relatively large area of the sky is important. In recent years, however, a camera invented by the German optician B. Schmidt has combined some of the best features of the reflector and the refractor into a single instrument.

6-7 **The Schmidt Camera.**   The Schmidt camera combines a spherical mirror with a correcting lens, as shown in Fig. 6-12. Used by itself, the spherical mirror focusses equally well all parallel beams of light passing through its center of curvature, C, but a spherical surface does not bring a parallel beam to a sharp focus. The rays striking an outer zone of the

FIGURE
6-12

Principle of the Schmidt Camera. Starlight from S_1 *and* S_2 *enters through the glass corrector plate,* C*, whose inner surface has a complex curve. After reflection from the concave spherical mirror,* M*, light converges to the curved focal surface,* S_1' *and* S_2'*. The photographic plate,* P*, bent under pressure, conforms to the focal surface.*

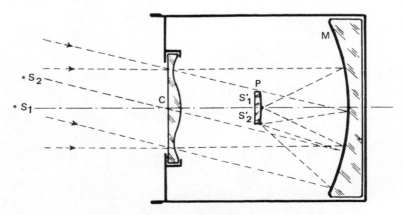

spherical surface come to a shorter focus than those striking an inner zone. This effect, known as *spherical aberration*, is gross for a hemispheric mirror. The specially curved surface of the lens, or *correcting plate*, compensates for the spherical aberration so long as a relatively small fraction of a sphere is encompassed by the mirror. Hence the Schmidt telescope yields excellent image quality over a wide angular field, greatly surpassing both reflectors and refractors in this respect. Schmidt systems also can be made with large aperture and short focal length, providing a small focal ratio, F/A, or fast system. Thus Schmidt systems can be efficient for photographing faint, extended nebulae.

The photographic plateholder at the focus blocks only a small fraction of incoming light. The focal surface (Fig. 6-12) is not plane, but curved, being a portion of a sphere with the same center, C, as the mirror. This minor inconvenience is overcome by the use of thin glass plates bent to the proper curvature without breaking. Some special Schmidt telescopes employ molded photographic film.

The great Schmidt telescope at Mount Palomar Observatory in California has an aperture of 48 inches, a mirror of diameter 72 inches, and a focal ratio $F/A = 3$—not particularly fast but with beautiful image quality over a field more than seven degrees in diameter. It has surveyed the entire sky visible from latitude North 33°.4, both in blue and in red light. These

FIGURE 6-13 *Baker-Nunn Schmidt Camera for satellite tracking. A = F = 20 inches. Field = 30°. (Photograph at Curacao, West Indies by Smithsonian Astrophysical Observatory.)*

magnificent reference photographs provide astronomers with a wealth of information and are used for comparison purposes whenever new or interesting objects are discovered by optical, radio, or other means.

James G. Baker, of Harvard Observatory, has designed modified Schmidt cameras in which an extra mirror provides a flat focal surface. This system has worked admirably in the 32-inch Baker-Schmidt telescope of the Boyden Observatory, Bloemfontein, South Africa. In the Baker "super-Schmidt" cameras used for meteor photography, an aperture of 12 inches and a focal length of 8 inches (focal ratio $F/A = 0.65!$) give a usable field of 55 degrees diameter—ten million times larger in area than a parabolic reflector of similar aperture ratio (see Fig. 15-12). Schmidt-type systems have been very effective in tracking artificial earth satellites (Fig. 6-13).

6-8   Telescope Mountings.   Telescope mountings for professional use usually permit the telescope to point at any part of the sky, and to follow the slow daily motion of the stars that is caused by the rotation of the earth. An *equatorial mounting* has two axes around which the telescope may move. The *polar axis* points toward the celestial poles parallel to the earth's axis of rotation. A clock or motor slowly rotates this axis to turn the telescope westward, at the exact rate to compensate for the earth's rotation. The result is that the telescope, always pointing toward the same star, is fixed with respect to the celestial sphere, while the earth is turning under it. A second axis, perpendicular to the polar axis, is called the *declination axis*.

Accurately divided circles on these axes allow the observer to set the telescope on any selected position of the sky, so that one need know only the celestial coordinates of an object to be able to find it. The 200-inch Mount Palomar telescope can be pointed with an accuracy of one second of arc.

Several types of equatorial mountings appear in the photographs of Figs. 6-8, 6-10, and 6-13. The Yerkes refractor has a so-called German mounting, much used for this type of telescope. The 200-inch reflector employs an unusual yoke mounting, which supports the telescope in a large cradle. This cradle rotates about the polar axis on large oil-pad bearings at its upper (north) end.

There are many other types of telescope mountings for specialized uses. A modern trend for very large telescopes returns to the practice of Sir William Herschel, who used an *alt-azimuth* instead of an equatorial mounting. Here the entire system turns in azimuth around the vertical while the second axis is horizontal, providing motion in altitude or elevation. Relatively simple computers control the driving motors to follow the apparent motions of celestial objects. The great U.S.S.R. 6-meter (240-inch) aperture reflector is supported in an alt-azimuth mounting. The solar tower telescope (Fig. 9-15) of Sacramento Peak Observatory also employs such a system. The entire telescope, which weighs some 6000 tons, floats on a

FIGURE 6-14 *The Dome of the 107-inch Reflector of the MacDonald Observatory.*

huge pool of mercury. It slowly turns on a vertical axis to compensate for the rotation of the solar image.

Optical telescopes must be protected from the weather when not in use and usually from stray light, both natural and artificial, when in use. Thus, they are most commonly enclosed in a dome that is closed during the day and in bad weather. A slit in the dome can be opened for observations and the entire dome turned as required so that the telescope can follow the diurnal rotation of the sky (Fig. 6-14).

6-9   Radio Telescopes. Radio waves from the sun and other celestial sources can be collected by special arrays of receiving antennas or by scaled-up versions of reflecting telescopes. Because radio waves have much greater wavelengths than light waves, the aperture A must be much greater to achieve a reasonable resolving power (see Sec. 6-4). For example, to match the visual resolution limit $R = 4\overset{''}{.}5$ of an optical telescope having an aperture of only 1 inch [equation (6-2)], a radio telescope used at wavelength λ would require [equation (6-4a)] an aperture of $\lambda \times (2.3 \times 10^5)/4.5 = 51,000\lambda$ (cm)—or 10 km (6.4 mi) for $\lambda = 21$ cm and proportionately for longer wavelengths. However, radio astronomers, rather surprisingly, have overcome this difficulty by placing receivers thousands of miles apart. In such *radio interferometers*, the stations are connected by radio transmission links or by highly precise clocks, so that the waves of the incoming wave trains can be counted and compared as the earth turns. Thus the interferometers make radio competitive with optics in separating nearby point sources or in measuring small diameters for sources. Eventually the base lines can exceed the earth's diameter as radio telescopes go into orbit.

Even though smaller resolving powers of radio "dishes" give useful information on the radio emission of celestial bodies, radio astronomy since its beginnings in the 1930's has been a race for ever larger receiving apertures. Larger telescopes are necessary, not only to reduce the angle of the incoming beam of radiation, but also to improve the reception of faint signals. The energy that we receive from the universe in the form of radio waves is indeed exceedingly small—much smaller than in the light waves—and only the surprising sensitivity of radio receivers enables us to detect them at all. Among the most sensitive amplifying systems developed we note only the MASER (short for Microwave Amplification by Stimulated Emission of Radiation), which uses some special molecular properties of crystal, such as ruby, at very low temperatures. In principle energy is stored in the crystal by optical or other means and then released by a triggering action from the radiation being studied.

No one "looks through" a radio telescope. The faint signals from radio sources, usually *noise*, after great amplification, activate pen recorders or accumulate in counters or computers to be read out as numbers. The background radio noise, measured in nearby regions of the sky, must always be subtracted from the radiation measured in the source.

The largest dish-type radio telescope, built at Arecibo, Puerto Rico, by Cornell University, has an overall diameter of 1000 feet (Fig. 6-15). The

FIGURE 6-15   *Radio Telescope at Arecibo, Puerto Rico. Diameter: 1000 feet. (Arecibo Ionospheric Observatory, Cornell University.)*

FIGURE 6-16   *Radio Telescope at Parkes, N.S.W., Australia. Diameter: 210 feet (C.I.S.R.O.)*

dish itself is fixed in position, pointing toward the zenith. The sensitive receiver, suspended on cables far above the dish, can move so as to compensate partially for the object's apparent motion caused by the earth's rotation. The National Radio Astronomy Observatory 300-foot antenna at Greenbank, West Virginia, rocks back and forth along the meridian, but it cannot track objects east or west. The largest fully steerable dish, located at Jodrell Bank, Manchester, England, has a diameter of 250 feet. The Parkes antenna, in Australia, is 210 feet across (see Fig. 6-16). Plans for new developments include a 400-foot antenna for Jodrell Bank and a 100-meter (328-ft) dish for Germany.

Much larger antenna systems have been designed that are cheaper to build and do not raise such formidable engineering problems as the giant movable reflectors of England and Australia. The Australian radio astronomer, B. Mills, built a highly successful antenna in the form of a giant cross whose arms are 1500 feet long precisely oriented in the east-west and north-south directions. Such an antenna has approximately the same resolving power as a square reflector having sides equal to the lengths of the arms, but a much smaller collecting area, hence the reduced cost of construction. For instance, the original 1500-ft Mills cross ($A = 450$ meters) operating at a wavelength $\lambda = 3.5$ meters (or a frequency $v = c/\lambda = 85$ megacycles per second) had a resolution limit $R = 3.5/450 = 0.078$ radian $= 45$ minutes of arc (see Fig. 6-17a, b).

FIGURE 6-17 *The One-Mile Mills Cross Radio Antenna, Australia. (a) Extended view. (b) Close-up of section. (Molonglo Radio Observatory.)*

In order to achieve equivalent results at a smaller cost and with more practical antennas, British radio astronomers at the Cavendish Laboratory in Cambridge have developed *aperture synthesis*, a technique that employs one arm of a Mills cross and a small movable antenna that *successively* occupies the positions of the elements of the second arm of the cross. A large electronic computer records the signal intensities at various positions of the second antenna and synthesizes the reading of the corresponding complete antenna. It takes longer to make the observations (often many months), but the cost of construction and maintenance is minimized.

When a higher resolving power is needed to make a detailed study of small radio sources, such as the sun, or to measure their positions with precision, radio astronomers apply the same interference techniques used by optical astronomers to measure close double stars or stellar disks (see Chapter 22). Two radio telescopes are used as an *interferometer* (see Sec. 22-3) and "look" simultaneously at the same source.

6-10 Astronomical Photography. The modern photographic emulsion was perfected in the 1880's. This emulsion consists of a suspension of photosensitive microcrystals of silver halide, usually bromide, in a thin gelatin layer on glass or acetate film. Astronomers mainly use glass plates, which give better geometric stability to the fragile gelatin layer and are easier to handle. Light activates the grains chemically so that the *developer* changes the silver halide to silver. Then the *hypo* (sodium thiosulfate), or fixing agent, dissolves away the undeveloped silver halide, leaving black silver grains, forming a negative image of the original bright image. Washing finally removes the hypo and its solute.

Photography has almost completely replaced the human eye for direct celestial observations, although it must, in turn, be replaced by television-type sensors. Modern telescopes are essentially huge cameras that form images of the sky on photographic plates or on other receivers. Let us consider photography as typical of such sensing devices.

The photographic emulsion has many advantages over the human eye. First of all, a photographic record is objective and permanent. It can be stored and studied at leisure. Second, a photographic plate can record simultaneously images of thousands—in principle millions—of star images or picture elements, while the eye can study one (or only a few) at a time. Third, and most important, the photosensitive emulsion has "integration power." In other words, the action of light is cumulative. Increasing the length of exposure to a faint source of light will build up a detectable image even if the object is invisible to the eye in the same telescope. The eye lacks almost completely this integrating power. What it cannot see in a few tenths of a second (or perhaps a few seconds for exceedingly faint sources), it will never see, no matter how long and hard we look.

Thanks to photographic records accumulated since the late nineteenth century in many observatories, astronomers can determine the positions, motions, luminosities, colors, and forms of all manner of celestial objects and follow their changes in time, if any. The Harvard Observatory collection of nearly half a million photographs of the sky provides a detailed record of celestial activities and changes between about 1885 and 1955, while the Palomar Atlas gives a double record, in blue and in red light, to a much fainter limit.

The positions of the stars are measured on photographic plates by means of a coordinate measuring machine in which precision screws can move the plate carrier or stage in two mutually perpendicular directions, x and y. Graduated drums or scales give the x,y coordinates to an accuracy of 0.001 millimeter (1 micron = 0.4 ten-thousandth of an inch). The operator, looking through a microscope, brings each star image in turn under a fine reticle engraved in glass, or under cross-wires made of a selected spider cocoon strand. In modern instruments the reticle is replaced by a pinhole in which the star image is centered automatically with extreme precision by an electronic-mechanical system governed by a photoelectric sensor (see Sec. 6-11). Punched cards or tape automatically record the readings, replacing the human operator and his pad and pencil. Computing machines then read the cards, make the necessary calculations, and even print the final results.

The motions of stars can, of course, be found from a comparison of celestial coordinates measured at different epochs. However, astronomers, who are often more concerned with motions—that is, position changes—than with positions per se, use a special instrument to compare directly two photographs of a given star field taken at two different epochs, say a few years or decades apart. This instrument, called a "blink comparator," projects on a screen in rapid alternation and exact superposition the images of two photographs of the same field. If a star has perceptibly moved in the interval of time between the two photographs, its image appears to jump back and forth while the others stay fixed. Thus the discovery of stars with unusually large "proper motions" (Chapter 19) is made easier. The large photographic-plate collections were systematically surveyed to search for fast-moving stars.

The luminosities of stars, nebulae, planets, and so on can be measured on photographs. The brighter a star, the bigger its image on a photograph because of light scattering in the photographic emulsion. Chromatic aberrations of the lens in refractors, atmospheric unsteadiness (see Sec. 1.14), and guiding errors contribute to the image size. In spite of undesirable effects, astronomers correlate image size with brightness and so measure approximately the photographic luminosities of the stars (see Sec. 6-11). Similarly, the greater the surface brightness of an extended object, the darker its negative image on a photographic plate. Measurements of the opacity of the silver deposit in different parts of an image make it possible to map the distribution of luminosity in the extended objects (see Sec. 6-11).

The notion of "color" refers to a subjective impression that depends on the distribution of energy in the various wavelengths of light from a source of radiation (see Chapter 8). The natural photographic emulsion is sensitive mainly to blue, violet, and ultraviolet light. However, impregnation with special dyes renders silver bromide crystals sensitive to other wavelengths. Thus the sensitivity of the emulsions can be extended to green and yellow light (orthochromatic plates), to orange and red light (panchromatic plates), and even to infrared light (infrared plates), up to wavelengths nearly twice the wavelength of the greenish-yellow light to which the eye is most sensitive. Thus, the apparent brightness of a star or other light source can be measured on photographs taken through various color filters. Comparison of such photographs permits an objective and quantitative determination of color (see Sec. 6-11). The ability of photographic emulsions to record radiations over a much wider spectral range than the human eye has played a vital role in the progress of astrophysics over the past one hundred years.

Finally we should mention that the use of moving pictures—either speeded up or slowed down by means of time-lapse photograph—has not only produced vivid demonstrations of known motions and changes, but has also disclosed new phenomena in the atmosphere of the sun (see Chapter 9).

6-11   Photometers.   Apart from some gravitational effects (Chapter 4), most of what we have learned about the universe has come through a study of the light received from celestial bodies. Hence the importance of photometers—devices that measure the intensity of light.

The earliest photometers were, naturally enough, visual instruments for comparing and evaluating the apparent luminosities or "magnitudes" of stars with one another or with a standard candle. By definition the apparent magnitudes of two stars differ by 5 units if their apparent luminosities bear a ratio of 1 to 100 (see Chapter 19). The planet Venus at maximum brightness has a visual magnitude of -4, while the faintest stars just visible to the naked eye are of magnitude about $+6$; the difference of 10 magnitudes corresponds to a luminosity ratio of $(100)^2 = 10,000$. More generally, the magnitude m is defined by the equation

$$m - m_0 = -2.5 \log \frac{\mathscr{F}}{\mathscr{F}_0},$$ (6-5)

where $\mathscr{F}$ refers to the "illumination" or flux density produced by the light source on a screen perpendicular to the light rays, and the subscript zero indicates some comparison source used for reference.

The zero point of the magnitude scale has been arbitrarily set from observations of actual stars, the average magnitude of the 20 brightest stars being approximately $m = 1$. The sun, visually of magnitude -26.8, is thus 27.8 magnitudes brighter than the average first-magnitude star, or, from equation (6-5), $m - m_0 = -27.8$ and the light ratio or flux-density ratio is $10^{11.12}$. It is now the custom to correct magnitudes for the dimming effect of our atmosphere (see Chapter 19) and to reduce all data to "outside the atmosphere." Although astronomers must compare stellar sources of light with terrestrial sources to calibrate their system in laboratory energy or flux units, their photometric studies usually involve comparison of stars with one another, sometimes through an intermediate comparison source.

The photographic plate has been used extensively in astronomical photometry. At Harvard Observatory, for instance, in the 1910's and 1920's, E. S. King exposed photographic plates at various distances *beyond* the focus of a telescope where the sharp image of a bright star was formed.

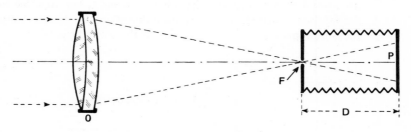

FIGURE   *Principle of King method of out-of-focus photometry.*
6-18

The illumination of the plate is inversely proportional to the square of the distance D of the plate from the point source of intensity I (Fig. 6-18); that is,

$$\mathscr{F} = \frac{I}{D^2}. \tag{6-6}$$

This is the fundamental law of photometry. Hence if two stars S_1, S_2 produced the same opacity in the plate when exposed for the same length of time at extrafocal distances D_1, D_2, their intensities were in the ratio $\mathscr{F}_1/\mathscr{F}_2 = (D_2/D_1)^2$, from which relation the difference of magnitudes was computed through equation (6-5).

By first measuring photographic magnitudes m_{pg} on ordinary blue-sensitive plates, then so-called "photovisual" magnitudes m_{pv} on yellow-sensitive (orthochromatic) plates exposed through a yellow filter (so as to match roughly the color sensitivity of the eye), we can obtain a measure of color or "color index":

$$CI = m_{pg} - m_{pv}. \qquad (6\text{-}7)$$

Through the work of many astronomers, especially F. H. Seares at Mount Wilson Observatory, an international list of standard stars of known photographic and photovisual magnitudes, hence also color indices, was set up for a number of stars near the North Celestial Pole. This North Polar Sequence, officially adopted in 1922 by the International Astronomical Union, defined the scale of stellar magnitudes for nearly thirty years.

6-12   **Photocells and Image Tubes.** In recent years more precise techniques have largely replaced photographic methods of stellar photometery. The photoelectric cell (Fig. 6-19) is basically a photosensitive layer or photocathode deposited on the inner surface of an evacuated glass or quartz tube. The photocathode material is highly specialized for efficiency in ejecting electrons when exposed to light. The negative photoelectrons are attracted to a positive electrode (or *anode*) raised to some suitable potential by a battery. The current induced in the circuit, amplified and measured, provides a determination of the star's brightness.

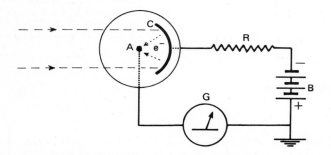
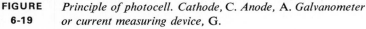

FIGURE
6-19   *Principle of photocell. Cathode,* C. *Anode,* A. *Galvanometer or current measuring device,* G.

The great advantage of the photocell over the eye or the photographic plate is that the photocurrent is, in principle, rigorously proportional to the illumination of the photocathode. Hence, if one star causes a current twice as intense as that from another star, the first star is twice as bright as the second. In other words, in equation (6-5) we can set the current, i, proportional to the light flux $\mathscr{F}$. We express this property by saying that the photocell is a *linear* receiver.

In modern photoelectric photometers the simple photocell is often replaced by the photomultiplier, a device yielding a built-in internal amplification of several million times or more prior to further external amplification by electronic techniques. The photoelectrons, accelerated by electric fields, successively collide against a number (often a dozen) of photoelectric metal targets, which at each stage release more electrons than strike them. This snowballing or avalanche effect has led to devices capable of counting individual photons from celestial sources so faint as to require the ultimate sensitivity imposed by the discrete nature of light (see Chapter 8).

Figure 6-20 is a schematic diagram of a modern photoelectric photometer used to measure starlight at the Cassegrain focus of a large reflecting telescope, with the amplified current measured by a pen recorder. Readings are made alternately on the star image passed through a small aperture in the focal plane of the telescope, then a nearby "blank" region of the sky.

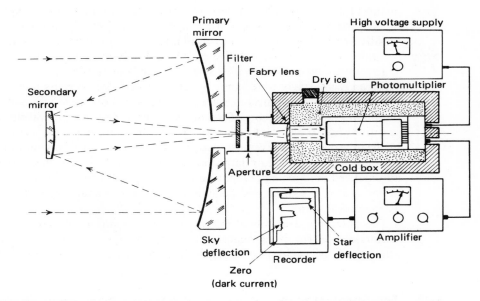

FIGURE 6-20 *Photoelectric photometer. Note chart recorder shows readings of star, of nearby sky, and of closed aperture.*

The difference of the deflections measures the star brightness. Measurements through two or more filters define photoelectric color indices—for instance, blue minus visual (that is, yellow), noted $B - V$; or ultraviolet minus blue, noted $U - B$; or again visual minus red, $V - R$; or red minus infrared, $R - I$. Such photoelectric color indices have proved to be powerful tools for the study of stars, star clusters, and galaxies (see Chapters 19, 21, 26, 30, 31, 32).

The photocell, although enormously precise as a radiation-measuring device, suffers from an inherent weakness: it measures only one light source at a time. The photographic plate can literally measure millions. Developments in electronics and in television techniques now provide methods for measuring multiple light sources both simultaneously and precisely. Note that the need occurs for spectra (see next section) and for extended emission areas on the sky as well as for star fields.

So many electronic imaging devices are being developed, tested, and used that no detailed description is feasible here. In principle a sensitive irradiated surface can be made to produce charged areas in wavelength bands ranging from the far ultraviolet to the infrared, depending upon the surface materials. These areas can be scanned as in television, or the electrons accelerated in controlled electrical fields to focus on photographic emulsion or on a second area that can be scanned. The scanning process by a narrow electron beam not only permits the production of a television picture; more importantly, it permits storing the data in magnetic-tape or other devices now extensively used in digital computing machines. The sensitivities and exposure times available by *image tube* now compete favorably with the most sensitive photographic emulsions.

The advantages of image-tube systems over photography are threefold. First, an increasingly wide range of sensitive materials permits more efficient operation, sometimes in wavelength regions not easily accessible to the photographic emulsions. Second, both the wide range in wavelength sensitivity and the ease of data transmission by radio make the use of an image tube desirable for remote instruments in balloons, rockets, and space vehicles. Third, the ease of data storage in memory devices makes possible automatized analysis by computing machines. Such procedures remove a major source of tedium and delay in reduction and analysis of observed astronomical data.

The observing demands of the space age have produced many special sensing devices for measurements of X rays, γ rays, the far-infrared and ultra-shortwave radio spectra, magnetic fields, particles in space, and so on.

6-13   The Spectrograph.   Among the astronomer's many observing tools, perhaps the most fundamental is the spectrograph. Various chemical substances, when heated to incandescence, radiate special characteristic colors, as explained in Chapter 8. Sodium, vaporized in a flame, emits two unique shades of yellow. Hydrogen, excited in an electric discharge, emits a series of colors unique for that element. Every substance—atom or molecule— has its own colors, enabling the scientist to identify the presence of that substance in the radiation source—whether in the laboratory, the atmosphere of a star, or a gaseous nebula.

To facilitate precise identification of color, the astronomer employs a glass prism to spread the light into a band of colors, like the rainbow, from red to violet. Such a band was named the *spectrum* by Newton, who first performed the experiment. The instrument used to disperse the colors is the spectroscope, first developed by the German optician, Fraunhofer, in 1815. When adapted for photographic recording the device is a *spectrograph*.

The simplest spectrograph of all consists of a prism placed in front of the lens of a camera designed to photograph the stars. The prism refracts the starlight at different angles, blue light more than the red. Hence the point image of a star spreads out into a streak or spectrum from red to violet. Let the star drift perpendicular to the spectrum to widen it, and the various special colors spread out to form lines. Such lines, bright lines in emission or dark ones in absorption, are characteristic of the elements present in the source and provide a means for chemical analysis. This simple device is called the *objective-prism spectrograph* (Fig. 6-21a).

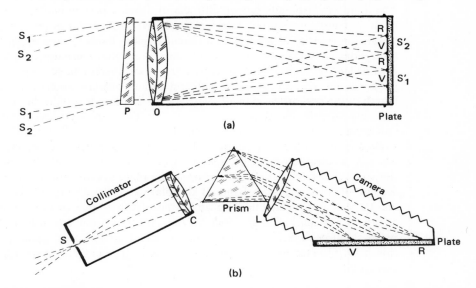

FIGURE   *The Spectrograph. (a) Objective-prism spectrograph used without slit for stellar*
6-21   *point sources,* S_1 *and* S_2 *produce dispersed images at* S_1' *and* S_2'. *(b) Slit spectro-*
          graph to be used with slit, S, *at the focus of a telescope.*

In a more elaborate instrument (Fig. 6-21b) the telescope focusses the image on a narrow slit (S). Another lens, called the collimator (C), renders the rays parallel as they pass through a prism or series of prisms, which disperse the colors as before. Still another lens, the camera lens (L), focusses the light on the photographic plate (P) to record the spectrum.

This type of spectrograph can be employed to impress the comparison spectrum from some known laboratory source. Using the known wavelengths of the spectral lines from the laboratory source, the astronomer can calculate the exact wavelengths of the lines in the star or other astronomical source. This method provides information for precise identification of substances in the star, or for determining shifts caused by the relative motion of the source toward or away from the observer (see Sec. 8-14).

Diffraction gratings are now frequently used in place of prisms to disperse the light. These gratings usually consist of fine parallel rulings on a plane glass surface upon which aluminum is vaporized to increase the reflectivity. The rulings, scratches formed by a diamond cutter, have a regular but arbitrary spacing, usually ranging from about 300 up to 1200 lines or more per millimeter. The grooves are usually cut with saw-tooth cross section to increase the intensity of reflection in a given direction; such a grating is said to be "blazed." The diffraction grating might also be called an interference grating; its function depends on the wave nature of light. The light wave spreads out or diffracts in a plane perpendicular to each narrow-spaced groove but interferes with the waves from other grooves, producing a series of spectra at different angles. These spectra appear on both sides of the faint, noncolored, direct reflection, usually called the zero-order spectrum. The successive orders of spectra have dispersions proportional to the order. The different orders overlap, so that a wavelength of the first order coincides with one of half that length in the second and one-third in the third, and so on. The blaze of a grating tends to concentrate the rays in a given range of angles.

As one moves the spectrograph slit across an extended source, such as the sun, a selected line in the spectrum records bright and dark areas of the source. If one moves the photographic plate synchronously across a second slit placed just in front of the plate to expose only the selected spectrum line, he can build up an image of the source in the specially selected wavelength. The *spectroheliograph*, invented independently in 1893 by Deslandres in France and by Hale in the United States, operates on this principle (see Sec. 9-7). In effect, this instrument is a narrowband color filter.

6-14 Atmospheric Refraction. The earth's atmosphere, a mixture of gases in various proportions, extends several hundred miles above the surface. It corresponds, in weight, to a layer of water 10.3 meters (34 feet) thick. This atmosphere profoundly affects stellar and planetary light rays that pass through it. Clouds, haze, dust, smoke, and even pure molecules absorb and scatter light from the heavenly bodies. Even a clear, uniform atmosphere deviates the direction of the incoming rays. Light rays from a star are bent toward the vertical. Figure 6-22 illustrates the character of this *atmospheric refraction*.

As a consequence of refraction, heavenly bodies appear higher in the sky than they actually are, by amounts that vary from about half a degree at the horizon to zero overhead. Thus when the sun or moon seems to be just rising, it is in fact geometrically below the horizon. Refraction increases so rapidly near the horizon that the sun and moon appear to flatten as they rise or set, because the lower edge experiences greater elevation than the upper. The amount of refraction diminishes rapidly

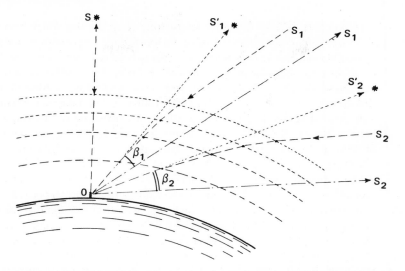

FIGURE 6-22 *Atmospheric refraction makes stars appear higher in the sky than their geometrical direction. Thus S_1 appears to be in direction S_1', etc. The refraction angle, β, increases towards the horizon as in β_1 and β_2.*

with elevation above the horizon. At an altitude of 30° the refraction amounts to only 100 seconds of arc; at 60° it is only 30 seconds. All precise observations of astronomical position require correction for this effect. The navigator who neglects to apply the refraction correction may err in his geographical position by many miles.

The atmosphere also exerts a sort of prismatic effect on the light rays. The blue rays are more refracted than the red ones. As a consequence, the light from a star, especially when near the horizon, is drawn out into a spectrum, a phenomenon usually called *atmospheric dispersion* or *differential* refraction. A bright star or planet, just above the horizon, appears red below and green or blue above. At sunset, the last starlike point of the disappearing sun sometimes looks like a tiny green or blue bead. This phenomenon, known as the *green flash*, appears only when the horizon is extremely clear.

6-15 Atmospheric Turbulence. Twinkling of a star is an atmospheric effect. Irregularities of temperature in the turbulent lower layers cause the light from a star to reach the eye over more than one path. Changes of apparent light intensity, frequently caused by constructive or destructive interference (see Sec. 22-3) between the different beams, result from these irregularities. Each color tends to twinkle independently. Hence stars appear to change in color as well as intensity.

Planets have less tendency to twinkle than do the stars, because they are not point sources and each part of their appreciable surface area twinkles somewhat independently, not precisely in unison.

Just a moment prior to total solar eclipse, when only a small segment of the sun's disk remains uncovered, a series of alternate bright and dark shadow bands often appear, moving rapidly across the ground. These, also, are related to twinkling, being produced by traveling waves and irregularities in the earth's atmosphere. The bands are analogous to the pattern of bright and dark flecks visible at the bottom of an undulating swimming pool on a sunlit day.

The dancing or blurring of the image of a star or planet, caused by atmospheric turbulence, the astronomer refers to as *bad seeing*. The unstable air above a hot stove or radiator furnishes a highly magnified example of this phenomenon. The distortion of stellar images, associated with twinkling and bad seeing, greatly reduces the effectiveness of a large telescope. In fact, atmospheric *seeing* rather than theoretical diffraction usually limits both the resolving power of a large telescope and the brightness of stars that can be photographed with it. Optically, the 200-inch Palomar reflector could resolve stars separated by only 0.02 second of arc, whereas the actual resolution limit is rarely better than 0.2 second [see equation (6-2)]. When observing conditions are poor, the "seeing disk" of a star may easily increase to a diameter of several seconds. In such cases a telescope having an aperture of only a few inches could have about the same resolving power as a far larger instrument, though the image would be far less brilliant. Poor seeing enlarges photographic images, spreads out the light of stars over larger areas of the photographic emulsion, and thus reduces the effectiveness of a telescope.

6-16 Atmospheric Transparency. The earth's atmosphere both scatters and absorbs light rays passing through it. The molecules of air and the solid particles much smaller than the wavelength of light selectively scatter the short more than the longer wavelengths. We term this phenomenon Rayleigh scattering, after the British physicist who used it to explain the blue, daytime sky. This type of scattering takes place in all directions, so that a pure "Rayleigh sky" is deep blue right up to the edge of the sun.

Larger particles, such as smoke, dust, fog, haze, or smog, tend to scatter all colors of light about equally. Usually this scattering is greater near the forward direction. As a consequence, a bright, white glare surrounds the sun, and in many locations a whitish haze dilutes the brilliant blue of a clear Rayleigh sky. This phenomenon, often called Mie (pronounced "me") scattering, we shall meet again in the scattering of light by cosmic dust clouds deep in interstellar space (see Sec. 29-4).

Atmospheric gases absorb as well as scatter light. Each gas has its own characteristic absorption lines and bands, which imprint its own characteristics on the spectra of the various heavenly bodies. We refer to these absorptions as the *telluric* lines, caused by components of the terrestrial atmosphere. Molecular oxygen, for example, takes two enormous bites from the deep red, the so-called A and B bands. Ozone, a molecule consisting of three oxygen atoms, is concentrated in a double layer extending from about 15 to 30 kilometers in altitude and sharply cuts out the ultraviolet beyond about 2900 Å. Water vapor and carbon dioxide absorb heavily in the infrared.

These and still other molecules absorb various radio wavelengths. Free electrons in the ionospheric layers, from 100 to 250 km above the surface, reflect the longer radio wavelengths, preventing them from entering. This effect is most pronounced during the daytime, when absorption of the sun's far ultraviolet tears away the electrons (ionizes the atoms and molecules). Absorption and twinkling effects, similar to those impressed by the lower atmosphere on optical radiation, occur for radio sources (see Chapter 33).

Efforts to minimize the difficulties caused by poor seeing and low atmospheric transparency have led astronomers to build observatories on sites at high altitudes, carefully selected after long studies of sky conditions and weather. The chosen sites often lie far from large cities, not only in order to avoid man-made smoke and dust, but also to reduce troublesome artificial illumination of the night sky. The street lights and advertising signs of a large city can easily reduce photographic efficiency for the observation of faint stellar objects by a factor of five or ten. The night sky over the great Mount Wilson Observatory now always reveals the red lines of neon gas and the green line of mercury from the lights of nearby Los Angeles.

By sending balloon-borne telescopes into the stratosphere, to altitudes of about 30 kilometers, astronomers can avoid much of the absorption and irregular refraction that occur mainly below these levels, obtaining high-resolution photographs of the sun or planets from such altitudes. The low sky brightness is also advantageous for other programs. The telescopes are usually unmanned and remote-controlled.

Balloons, however, are of little avail for observing the ultraviolet spectrum between 2900 Å and about 2400 Å. The ozone responsible for this absorption is about as opaque as a concrete roof. However, an atmospheric "window" exists in the range from about 1800 to 2400 Å, so that solar spectra have been obtained from balloons. Rockets and satellites are important vehicles for study of the ultraviolet. Also, because they operated outside the earth's turbulent atmosphere, the relatively small cameras of the U.S. "Lunar Orbiter" space vehicles have recorded details of the surface of the moon with far greater resolution than could the world's greatest earthbound telescope (see Chapters 11 and 16). Since these vehicles did not return to earth, the photographs were "scanned" electronically and sent back to earth by a sort of television technique. This process produced the striped appearance of some of the records. The photographs returned by the Apollo astronauts, of course, did not suffer from this defect.

6-17 Computing Devices. Long of service to the astronomer, logarithm tables, slide rules, and desk calculators are still useful means of simplifying the long and tedious calculations of astronomy. However, in recent years the high-speed, electronic, digital computer has enormously reduced the labor of computations. Such a device can, in only seconds, carry out calculations that would have taken a full man-year by the older methods.

We have already referred to the use of machine methods for the analysis of stored observational data. Many special machines now automatically measure and record the positions of stars on photographic films or plates. Others may determine brightness of stars, planets, or other astronomical objects. Still other machines examine and measure the details of celestial spectra. The data are stored on punched cards, magnetic tape or drums, or in electronic devices of various kinds.

In the various control centers, machines analyze radar and other data from artificial satellites and space probes, tracking the orbits and determining the need for special midcourse corrections. The basic techniques are available for improving our knowledge of the orbits of planets, satellites, and comets.

Radar measurements of echoes received from planets and satellites require data processing in advanced computers, to furnish new information about the scale of the solar system and planetary detail.

Lunar Orbiters and the Mars Mariner contain devices for automatic processing of photographs, sending back the information point by point, encoded according to the intensity of the photographic record. Again digital computers process the data, even correcting the observations if necessary, and drawing a map of the surface so exactly that the detail almost equals that of the original photograph. Some of the lunar and Mars photographs (see Chapters 11 and 12) were taken in this manner.

Automation and computers now play a basic role in almost all phases of astronomy. Their use is rapidly increasing in theoretical astrophysics, as well as for rapid analysis of observations Computers calculate theoretical intensities of spectral lines of atoms and molecules. They enormously simplify the calculation of stellar models, from the deep interior to the outer atmosphere, taking proper account of energy generation by nuclear fusion (Chapter 25). In brief, high-speed computers have revolutionized the science of astronomy and will undoubtedly play an increasingly important role in the future.

CHAPTER *7*

Measurements in the Solar System

7-1   On Measurements.   Measurement is the foundation of all physical science. Theory without measurement degenerates into speculation; it cannot explore the natural universe, and it even lacks the power of prediction. On the other hand, measurement without theory tends to be a random process, lacking direction. Measurements furnish the "proof of the pudding." When compared with the predictions of theory they suggest improvements to the theory; the improvements, in turn, lead us to new critical measurements.

Even in the era of rockets and satellites, astronomers can conduct few direct experiments. Our measurements and observations must therefore be planned to answer specific questions. Without knowledge of such fundamental data as the sizes and masses of the astronomical bodies we observe, we could make little progress in understanding the universe. This chapter will outline observations necessary to measure the distances, dimensions, and masses of the planets and the sun and will describe the physical principles applied. From the information thus obtained we can construct physical models of these bodies to learn about their internal constitutions, densities, and physical properties. These models in their turn furnish the necessary foundation for theories of the origin of the solar system and of its significance in the universe (see Chapter 17).

The earth is our natural starting point. Here we must establish the base line necessary for the measurement of distance, and the physical constants needed for the determination of mass.

7-2 The Size of the Earth. How can we measure the size and shape of
our own planet? We cannot conveniently wrap a tape measure around the
circumference of the earth, nor can we roll a marked wheel around it and
count its revolutions, as short distances were often measured in ancient
times. For large-scale surveying, instead, we may use light rays, which
travel in straight lines. We shall employ optical instruments to adapt and
extend the methods that a surveyor uses to measure the distance across
a river without crossing it.

Suppose, in Fig. 7-1, that we wish to measure the direction and distance
from the point A to the point F, which cannot be seen from A. We select a
point B visible from A over fairly smooth terrain, and a point C visible
from both A and B. Since we can determine by geometry all the angles and
sides of a triangle if we can measure two angles and the included side, we
can begin our process of *triangulation* by measuring the distance between
A and B with an extremely precise tape. With a surveyor's theodolite we
next measure the angles a and b, and solve for the distance AC. Again, if a
point D is visible from point B and point C, but not from A, we can measure
the angles b' and c, to determine the length CD. We can continue the
process until we have solved all the necessary triangles leading from A to F.
Finally, by somewhat more complicated trigonometry, we solve for the
distance AF and its direction with respect to the meridian that passes
through A.

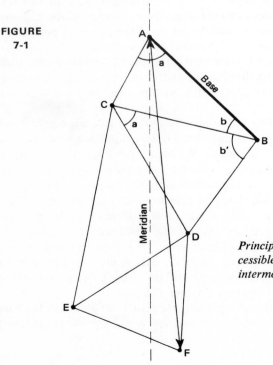

Principle of triangulation from point A *to inaccessible point* F *by means of base line* AB *and
intermediate points* C, D, *and* E.

The result of lengthy triangulations along a meridian, or great circle of longitude, gives the circumference of the earth. The length of the meridian passing through the poles is about 40,000 kilometers or 25,000 miles.

Such measurements as those just described can be carried out with a precision of about one part in a million over a large distance on land. The results can be checked by measurements of other survey networks that also involve the points *A* and *F*. The actual triangulations are, of course, complicated by the rotundity of the earth and by the differences of elevation of the survey points. But the oceans are the true barriers to the triangulation of the earth's surface. Whereas the errors can be kept down to about 15 feet over 3000 miles of land, they are at least ten times as great over a similar range of ocean. The uncertainty in the distance between two specific points in Europe and America was several hundred feet until measurements with satellites reduced the error to about 30 feet.

7-3 **The Shape of the Earth.** In previous chapters we have already anticipated the results of the many arduous triangulations that have been made over the surface of the earth. The length of a degree of latitude can be measured, by triangulation, between points on the earth's surface whose latitudes and longitudes have been accurately determined by astronomical measurements. A degree of latitude is 110 km (68.7 miles) long at the equator, and 111 km (69.4 miles) at the poles. This difference is the direct result of the shape of the globe: the earth is *oblate*, with a cross section

FIGURE 7-2 *Oblateness of the earth. The radius of curvature at the surface is greater at the poles (R_P) than at the equator (R_e).*

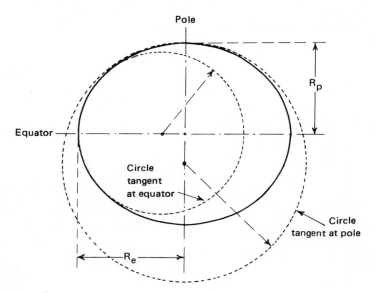

through the poles that is shown schematically in Fig. 7-2. A circle with the same curvature as the earth's cross section at the equator would be smaller than a circle having the curvature of the earth at the poles. Hence a degree of latitude, which is defined by the curvature of the surface of the earth, has a greater length at the poles than at the equator.

The *oblateness* of the earth is defined as the difference between the equatorial and polar radii divided by the equatorial radius, $(R_e - R_p)/R_e$, as in Fig. 7-2. The oblateness derived from satellite data is $1/298.25$. The equatorial radius exceeds the polar radius of 6378.16 km (3950 miles) by 21.4 km (13.4 miles). This difference is by no means trivial. The Mississippi river, for example, empties into the Gulf of Mexico some four miles farther from the center of the earth than its source. The river flows uphill, insofar as distances from the center of the earth are concerned; but still it does not defy the law of gravitation.

The centrifugal force produced by the earth's rotation, as mentioned in Chapter 4, distorts the earth into an oblate spheroid such that, at every point of the surface, the combined gravitational and centrifugal forces are always exactly perpendicular to the average level surface. Hence, although the oceans of the earth are distorted from a spherical form, they do not tend to flow steadily north or south in the two hemispheres. In the same way, rivers do not tend to flow away from the equator, but move downhill with respect to the average surface of the earth spheroid.

7-4   The Distance to the Moon. The principle used in determining the distance from the center of the earth to the center of the moon is the same as that used by a surveyor in measuring the distance across a lake. To simplify our argument, consider two observers in the same geographic longitude, who observe the moon simultaneously when it crosses their common meridian. In Fig. 7-3 the observers at A and B measure the angles

FIGURE 7-3

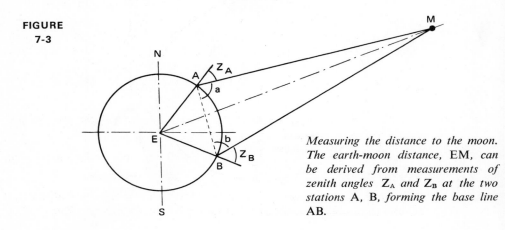

Measuring the distance to the moon. The earth-moon distance, EM, *can be derived from measurements of zenith angles* Z_A *and* Z_B *at the two stations* A, B, *forming the base line* AB.

Z_A and Z_B from the center of the moon to their respective zeniths. They can calculate all parts of the triangle ABE from their known positions on the earth, and the angles a and b from these quantities and the measured values of Z_A and Z_B. They can thus solve the triangle ABM completely. A further application of trigonometry yields the distance EM from the center of the earth to the center of the moon.

The mean distance between the centers of the earth and the moon, or the semimajor axis of the moon's average orbit, is 384,401 km (about 238,900 miles), now confirmed by radar. Because of the eccentricity of the moon's orbit, and the slight variations in the shape of the orbit caused by the sun's attraction, the distance ranges between 406,718 km and 356,428 km (about 252,800 and 221,500 miles). The average velocity of the moon in its orbit turns out to be 1 km per second, or approximately 2300 miles per hour. An observer on the moon would see the earth as a nearly spherical body with an average apparent diameter of $1°54'$, almost four times as large as the moon looks to us.

7-5 Elements of Planetary Orbits. Astronomers can predict the directions of astronomical bodies in the solar system by gravitational methods with surprising accuracy, without needing to know a single distance in miles. Kepler's third law relates the revolution periods in years to the distances of the planets expressed in astronomical units. The *astronomical unit* is the mean of the maximum and minimum distances from earth to sun —that is, the semimajor axis of the earth's orbit. When we express all of the distances in the solar system in astronomical units, observations of the periods of the planets, in conjunction with the theory of gravitation, will give the relative positions in space to an extremely high degree of accuracy, of the order of one part in a million. Even though the moon goes around the earth as both go around the sun, we can still predict its position as seen from the earth without knowing its actual distance in miles.

Let us consider the problem of predicting planetary positions. We require only six quantities to specify completely the elliptical orbit of a planet around the sun, and to determine the planets position at any time, the distance always being specified in astronomical units.

Two quantities give us the size and shape of the orbit (Fig. 7-4):

1. *Semimajor axis, a,* in astronomical units.
2. *Eccentricity, e,* the ratio of the separation FF' of the foci of the ellipse to the major axis $2a$.

Two angles locate the plane of the orbit with reference to the plane of the ecliptic:

3. *Node, Ω,* the angle from the vernal equinox eastward along the ecliptic to the point on the celestial sphere where the orbital plane crosses

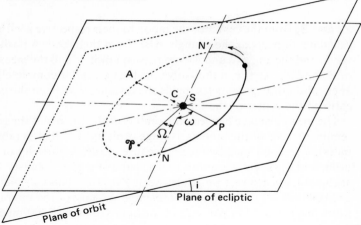

FIGURE
7-4 *Definition of orbital elements where* A = *Aphelion.* P = *Perihelion.* C = *Center of ellipse.* S = *Sun* = *Focus of ellipse.* p = *Planet.* NN¹ = *Line of nodes.* γ = *Vernal equinox on plane of ecliptic. Semimajor axis,* a = $\overline{AP}/2$. *Eccentricity,* e = $\overline{CS}/\overline{CA}$.

the ecliptic from south to north. This crossing point is also called the *ascending* node.

4. *Inclination, i,* the angle between the orbital plane and the ecliptic, which is acute, betweeen 0° and 90°, for direct motion like the earth's, and obtuse, between 90° and 180°, for retrograde motion, clockwise as seen from the north pole of the ecliptic.

A further angle tells us how the orbit is oriented in its plane:

5. *Argument of the perihelion, ω,* the angle at the sun from the ascending node of the orbit to perihelion, measured along the direction of the planet's motion.

Finally, we must know where the planet was at a given time, so we need to specify:

6. *Perihelion time, T,* any one of the instants at which the planet was at its perihelion.

Since by Kepler's third law the semimajor axis, *a,* determines the period of revolution, we now have all the data necessary to predict the position of the planet at any time in the future. To predict its direction on the celestial sphere as seen from the earth we must also know the earth's precise position.

In practice, such predictions require mathematical calculations of considerable complexity, and small corrections must be introduced on account of the mutual attractions that the planets exert on each other. These forces, minute as they are, cause slow changes of the orbital elements, known as *perturbations*. These orbital changes have been the subject of enormous amounts of calculation. As an example of perturbations, the eccentricity of the earth's orbit, now 0.01673, is decreasing at a rate of 0.00004 per century. After 23,800 years it will reach a minimum of 0.0033, and then it will begin to increase again, reaching a maximum of 0.0211 in 70,000 years.

The semimajor axes and eccentricities of the planetary orbits cannot change sufficiently to produce marked alterations of the periods, even though the nodes and perihelion directions may in time swing completely around.

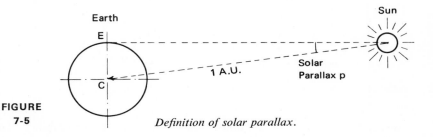

FIGURE
7-5

Definition of solar parallax.

7-6   The Solar Parallax. Small differences may still exist between the predicted and observed directions of the planets, as described in the last section, because the positions are referred to the center of the earth, while, of course, we observe from its surface. A displacement equal to the earth's radius causes a change of perspective that is not negligible. We must therefore correct practically all observations for the difference, known as the *parallax* correction.

In Fig. 7-5 we define the *solar parallax* as half the angular equatorial diameter of the earth as seen from the sun. We define the *horizontal parallax* of any other object in the solar system in an analogous fashion. The solar parallax amounts to only about 8.8 seconds of arc, and consequently it is rather difficult to measure with high precision.

The relationship between the parallax and the distance appears in Fig. 7-6. Let R be the radius of the earth, and p the parallax in seconds of arc of the point p, a very small angle. On a circle of radius D, equal to the distance of the object, we take an arc of length equal to D; it subtends an angle of one radian, equal to $57°296$, or $206,265''$, when seen from the center P. Thus we have

$$\frac{p}{206,265} = \frac{R}{D} \tag{7-1}$$

or

$$D = R \times \frac{206,265}{p}. \tag{7-2}$$

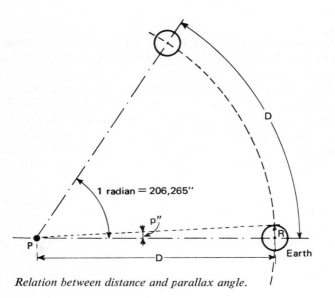

**FIGURE
7-6**

Relation between distance and parallax angle.

Hence, the distance to an object varies inversely as its parallax. The smaller the parallax angle, the larger the distance.

It would not be practical to observe the sun itself from two different points on earth to measure the solar parallax (as was done for the lunar parallax), because the angle is too small and the sun's edge not sharp enough. Instead, we take advantage of our precise knowledge of the mutual distances of the planets through Kepler's laws. If we measure the parallax of a planet close to the earth, the angle is larger, and the solar parallax follows by a simple proportion. The oldest methods introduced in the seventeenth and eighteenth centuries used Mars or Venus, the former near opposition when its position can be measured rather precisely with respect to stars, the latter during its transits in front of the sun (see Chapter 12), when it follows different paths when seen from different points on earth. However, the large disks of these planets and various optical effects seriously limit the precision attainable.

A much more precise method, introduced toward the end of the nineteenth century, measures the parallax of asteroids, some of which, such as Eros, pass much closer to earth than any major planet. The stellar appearance of the asteroids makes the measurement of their positions, relative to stars, much easier. The minor planet Eros, discovered in 1898, sometimes comes within 16 million miles of the earth, less than one-fifth the distance of the sun, and its parallactic displacement as measured between observatories in the northern and southern hemispheres exceeds one minute of arc, a quantity that can be measured to better than one part in a thousand.

Extensive observations of Eros, made by international cooperation in 1901 and 1931, led to solar parallax values of 8″.806 and 8″.790, respectively. Although a high accuracy of 1 part in 10,000 was claimed for the 1931 value, more recent studies indicate that it is affected by systematic errors. The mean value 8".798 is, in fact, in good agreement with more modern determinations.

Other precise methods of measuring the sun's distance depend on radial-velocity measurements of stars (see Chapter 21), which give in effect the constant of aberration (Chapter 5)—that is, the ratio of the earth's orbital velocity to the velocity of light—and on gravitational effects, especially in the moon's motion, which compare the attractions of the earth and of the sun, and thus involve the sun's distance.

The newest and most accurate method of measuring the astronomical unit uses a radar technique. The delay between the emission of a powerful radar pulse aimed at Venus near inferior conjunction (see Chapter 12), when it is nearest to us, and the return of the weak echo reflected by the planet can be measured with extreme precision by electronic techniques. If the time of travel of the radio waves to Venus and back is $2T$, and D is the distance of the planet, then obviously $D = T/c$, where c is the velocity of electromagnetic (radio and light) waves in empty space ($c = 299{,}793$ km/sec.) Measurements made by powerful radars for the first time during the inferior conjunction of April, 1961, and confirmed by later observations give a solar mean distance of 149,598,000 km. This value, corresponding to a solar parallax of 8".7943, is in good agreement with the classical determinations, but has superior accuracy.

7-7   Diameter Measurements.   With the astronomical unit evaluated in kilometers or miles, we are in a position to calculate the actual distance to any of the planets, at any time, directly from the orbital elements. We can also calculate the linear diameters of the planets from their angular diameters, using the geometrical methods that we have already applied. Since the sizes of all the objects in the solar system are small compared to their distance from us, Fig. 7-6 or equation (7-3) gives the linear diameter, d (in km or miles), in terms of the angular diameter, δ (in seconds of arc), and the distance, D (in km or miles). Thus

$$d = \frac{D \times \delta}{206{,}265}. \tag{7-3}$$

For example, the sun's diameter subtends an angle of just over 0°.5, or 1919".3, at a distance of one astronomical unit. Hence, by equation (7-3), its true diameter is 1,392,000 km (865,400 miles). The moon, with an apparent diameter nearly equal to the sun's, has a diameter of only 3,476 km (2,160 miles).

Some observers have made very precise measurements of diameters by matching the apparent image of a planet or asteroid with small disks of various sizes, introduced into the optical system. Even the best determined diameters can rarely be trusted to an accuracy better than 0".1, or some 50 miles at the sun's distance. Atmospheric turbulence and other optical effects complicate measurements of diameter, sometimes because of the earth's atmosphere, sometimes because of the planet's. For instance, the true diameter of Venus cannot be measured optically because of the clouds that hide its surface (see Chapter 12). On the other hand, radar has measured Venus' true diameter by the difference in pulse delay between the limb and the center and by the displacement of the near edge from the orbital motion of the center. The radar method proves to be accurate for the moon and has also been applied to Mars and Mercury.

7-8   The Mass of the Earth and the Constant of Gravitation. We weigh an object on the surface of the earth by measuring the force of the earth's attraction on it. When an object has been weighed, we can determine its mass, because, as Newton showed, the weight of a body is proportional to its mass.

The weight, W, of a body of mass m equals the Newtonian force exerted on it by the earth, as follows:

$$W = G\frac{mE}{R^2} = gm, \tag{7-4}$$

where E is the mass of the earth, R the radius of the earth, and G the Newtonian constant of gravitation (see Sec. 4-3). The equation is a simple one because, as Newton showed, a body attracts distant bodies as though all its mass were concentrated at one point within it, its *center of gravity*. If the body is a sphere, with the material symmetrically arranged around its center, the center of gravity and attraction lies at the geometric center of the sphere. The equation depends on the spherical form of the earth, and it neglects the corrections that result from the earth's oblateness. The distance between the body to be weighed and the center of gravity of the earth is therefore equal to the earth's radius.

Equation (7-4) enables us to determine the mass of the earth. We take a small body of known mass and measure its weight; then we compare this weight—that is, the force of attraction exerted by the earth—with the attraction exerted by another body of known weight, placed at a known distance from the first body.

The experiment is a delicate one, for this second force is very small. It was first performed in 1797 by the English physicist Henry Cavendish, who measured the force of attraction between two large fixed lead balls and two small balls by observing the torsion, or twisting, of the wires from which the small balls were suspended. Since this force of attraction was equal to the product of G, the constant of gravitation, and the two masses, divided by the square of the distance between the m, Cavendish obtained a direct measure of G, which in fact was not improved upon for nearly a century. The mass of the earth could then be determined from equation (7-4).

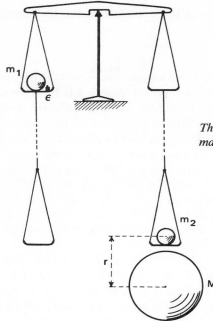

The von Jolly balance for measuring the mass of the earth.

FIGURE
7-7

In 1881, the mass of the earth was determined more precisely by von Jolly by means of the balance shown in Fig. 7-7. Here two glass globes, filled with mercury and equal in mass, m_1 and m_2, are supported in sets of identical balance pans. The lower pans hung by long wires are separated by 21 meters (69 feet). Alternate pans carry identical glass globes, sealed to compensate for differential air pressure. Small correction masses are added to allow for the fact that the lower mass, m_2, is nearer the center of the earth than the upper one, m_1, since the scales must be balanced at the start of the experiment, and though the masses of m_1 and m_2 are equal, their weights, on account of their difference of position, are not. After the scales are balanced, a large lead ball, of known mass M_1, is placed in the position shown in the diagram. Since M_1 attracts m_2, the balance is now disturbed, and a small mass, ε, found by trial, must be added to just restore the balance. Thus the force, F, exerted by the earth's mass, M_e, on the mass, ε at distance R, the earth radius, is equal to the force exerted on m_2 by M_1, at the distance r, between their centers of gravity:

$$F = G\frac{\varepsilon M_e}{R^2} = G\frac{m_2 M_1}{r^2}, \qquad (7\text{-}5)$$

which gives the mass of the earth:

$$M_e = \frac{m_2 M_1}{\varepsilon} \times \frac{R^2}{r^2}. \qquad (7\text{-}6)$$

etween F, G, and the known masses in the experiment is

$$G = \frac{Fr^2}{m_2 \, M_1}.$$ (7-7)

etermine G from the known masses and the distance r, when
xerted by the earth on the mass ε has been measured. We
's second law (Chapter 4): a force is measured by the acceler-
ces. If the acceleration produced at the earth's surface, or
surface gravity (see Sec. 7-13), by the earth's gravitational force is g, the
weight of a body is the product mg, where m is the mass of the body. But
the weight of ε is the force exerted on it by the earth's gravitational field,
and is equal to εg. Thus, if we write εg for F in equation (7-7), we have an
expression for G that involves measurable quantities—the masses M_1, m_2,
and ε, the distance r, and the acceleration g. The last can be determined in
various ways, from falling bodies, bodies rolling down inclined planes, or
experiments with a pendulum.

Von Jolly used a large lead ball of mass nearly 7 tons and small mercury
masses of 5.775 kg, or about 11 pounds. With a distance of 56.86 cm (22
inches) between the centers of the large and small balls, he found that a
small additional mass of only 0.00059 gram compensated for the attraction
of the large ball. Von Jolly's value for the earth's mass differed only slightly
from the best modern value of 6.6×10^{21} metric tons.

Earlier attempts to measure M_e and G depended on the attraction of a
mountain on a plumb bob, or the change of gravity deep within a mine
shaft, on a mountain summit, or even atop an Eygptian pyramid. In spite of
many experiments, the value of G is still known only within about 0.1 percent. The accepted value, in metric units, is 6.67×10^{-8} cm^3 gm^{-1} sec^{-2}.

The force of gravity is small compared with other forces exerted between
particles of matter at small distances. Although we can formulate its
behavior exactly, its cause is as mysterious today as it was in Newton's
time. Einstein's theory of relativity (Chapter 34) has added some extremely
small corrections to Newton's law, but it has not made gravitation any
easier to understand. The powerful forces that are exerted between
particles of matter at distances comparable to atomic dimensions are
electrical rather than gravitational. As yet we have no satisfactory theory
relating gravitational and electrical forces.

7-9 Masses of the Sun and of Planets with Satellites. Masses
of the sun and certain planets can now be readily calculated from observations of relative motions. Consider the sun, of mass M, with a relatively
small planet, of mass m, moving about it in a circular orbit of radius a. If v
is the velocity of the planet in its orbit, then an inward force is required to
keep the planet from flying away in a straight line tangent to the orbit.
This force is mv^2/a, measured in the laboratory by rotation experiments. It

must precisely equal the force of gravity if the planet is to remain in its circular orbit. Hence,

$$\frac{GMm}{a^2} = \frac{mv^2}{a}.$$ (7-8)

If the period of one revolution of the planet is P, then the velocity v is equal to the orbital circumference divided by the period, or $2\pi a/P$. By solving the equation for M with this value of the velocity, we find that

$$M = \frac{4\pi^2}{G} \times \frac{a^3}{P^2}.$$ (7-9)

Now that we know the value of the constant of gravitation G, we can immediately calculate the mass of the sun, using the earth as the planet in equation (7-9). In this case, a is the length of the astronomical unit (1.495×10^{13} cm) already found, and P is one sidereal year (3.16×10^7 seconds). The mass of the sun comes out to be $M = 2.0 \times 10^{33}$ grams.

Whatever planet we may choose in the derivation, we must, of course, obtain the same mass for the sun. Therefore the ratio a^3/P^2 is a constant, the same for all planets, as stated by Kepler's third law (see Sec. 3-3).

We can also use equation (7-9) to compute the mass of a planet with satellites, if we insert the proper value for the period and semimajor axis of one of its satellites. Again, all satellites of the same planet must give the same value for the planet's mass. The masses of Mars, Jupiter, Saturn, Uranus, and Neptune can be computed in this way. In fact, however, the orbits of satellites are very difficult to observe precisely, so that better values of these planetary masses are derived from their mutual gravitational effects and Jupiter's effect on the motion of asteroids.

One property of equation (7-9) limits our ability to determine masses. The mass m of the smaller body has cancelled out of the equations (7-8) and (7-9). As long as the mass of a satellite is very small compared with that of the larger body about which it revolves, our present procedure gives us no means of determining the satellite's mass.

Kepler's law takes a slightly more complicated form when the mass of the smaller body in a system bound by gravitation is comparable to that of the larger. If their masses are M_1 and M_2 (expressed with the sun's mass as a unit), the more general form of Kepler's law is

$$M_1 + M_2 = \frac{a^3}{P^2},$$ (7-10)

when the period is measured in years and the semimajor axis in astronomical units. This equation will be extremely valuable when we encounter the problem of determining the masses of double stars (Chapter 21). To determine M_1 and M_2 separately it is necessary to derive not only $M_1 + M_2$, but also M_1/M_2 by some other, independent method. A similar problem arises in measuring the mass of the moon.

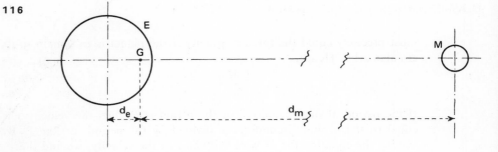

FIGURE
7-8   *The center of gravity of the earth-moon system lies at point G.*

7-10   The Mass of the Moon. The methods described in the preceding section cannot be used to determine the moon's mass, which is not negligible in comparison with the mass of the earth. As a consequence the earth and moon revolve about their common center of gravity, a point where an imaginary rigid lever connecting the two masses would exactly balance if placed in an external gravitation field such as the sun's. This point lies inside the earth globe, some 2900 miles from the center. It is this center of gravity of the earth-moon system, and not the center of the earth, that moves around the sun in a Keplerian elliptic orbit in one year.

The problem of determining this center of gravity turns out to be very similar in principle to that of determining the parallax of the sun or of a nearby planet. During the course of the month, the earth's revolution about the center of gravity causes a displacement like that produced by its rotation (diurnal parallax). Hence, the observed positions of nearby objects in the solar system deviate from the positions they would have if viewed from this center of gravity. The corresponding parallactic effect gives not the value of the earth radius, but the distance d_e of the center of gravity from the center of the earth.

If d_e and d_m in Fig. 7-8 represent the distances of the centers of the earth and moon from the center of gravity of the system, we have the following relation:

$$M_e \times d_e = M_m \times d_m. \tag{7-11}$$

The mass of the moon so derived—for example from observations of Eros and other asteroids—is $1/81.4$ of the mass of the earth, or 8×10^{19} tons. An improved value of $1/81.30$ has been provided by radar measures of Venus.

7-11   Masses of Isolated Planets. The masses of isolated planets, or of most satellites, are particularly difficult to determine. In fact, measures of their mass must depend upon the gravitational disturbances, or perturbations, that they produce on other bodies in the solar system. Venus disturbs in varying degrees the orbits of the earth, Mercury, and Eros and

especially of space probes passing close to it, so that we ca
mass with fair accuracy. Mercury, on the other hand, is so sm
can evaluate its mass to within only a few parts in a thousand
perturbations it causes in the motion of Eros and space probes.
presents the most difficult case of all (see Chapter 12). We do not bel
that the small and poorly determined perturbations on Uranus an
Neptune ascribed to Pluto are real—but we have no other method of
measuring its mass.

The problem of mass determination among the satellites in the solar
system is even harder than for the unaccompanied planets. We can
determine the masses of the four Galilean satellites of Jupiter and of the
larger satellites of Saturn by the mutual perturbation of the satellites upon
each other's orbits. In most cases we shall never know the masses of the
satellites with high accuracy unless, by chance, some wandering comet
should happen to come very close to one, or until space travel to these
distant planets becomes possible or radar systems become powerful
enough to reach them from earth.

The situation is even more hopeless with regard to the masses of comets
and asteroids, although in a few rare instances upper limits to the masses
may be established by analysis of extremely close approaches of comets to
planets or satellites. Brooks' comet, for example, which in 1866 came
within the orbits of Jupiter's inner satellites without perturbing their
motions in a measurable way, was estimated to have a mass less than a
ten-thousandth of the earth's mass.

7-12 **Planet Densities.** The densities of planets follow directly from
measures of planetary masses and diameters. The density, ρ, for a spherical
planet of radius r, is the mass M divided by the volume, $4\pi r^3/3$, or

$$\rho = \frac{3\,M}{4\,\pi r^3}. \qquad (7\text{-}12)$$

We derive the somewhat surprising result that the sun's mean density
is only 1.42 times that of water, while Saturn is so tenuous that, if it had a
solid shell, it would actually float in a hypothetical pool of water, for its
density is only 0.71. The other giant planets, Jupiter, Uranus and Neptune,
have densities comparable to that of the sun, while the terrestrial planets,
Mercury, Venus, Earth, Mars, and the moon, are more like very heavy
rocks with densities in the range 3.3 to 5.5 times that of water. The earth
and Mercury are the densest sizable bodies we know of in the solar system,
5.5 times the density of water; only small metallic fragments of meteorites
have higher densities (see Chapter 15).

The peculiar densities of the planets present an intriguing problem
whose solution will help toward a better understanding of the nature,
origin, and evolution of our solar system (see Chapter 17).

ies. On the surface of a planet the weight of a unit gram, measures the force of gravity. By the straightfor- f Newton's law of gravitation, and remembering that for e mass acts as if concentrated at the center, we can find tion of gravity, g, of a planet. If the planet has a mass both expressed in terms of the earth's mass and radius, celeration is given by the simple equation

$$g = \frac{M}{R^2} g_E, \tag{7-13}$$

where g_E is the earth's surface gravity. We learned in Sec. 7-8 how to determine the value of g_E.

From this equation we find that the surface gravity on the moon ($R = 0.27$, $M = 1/81.3$) is only about $\frac{1}{6}$ that of the earth, while on the sun it is nearly 28 times as great. Thus a man weighing 180 pounds on the earth would weigh only about 30 pounds on the moon, but some 2.5 tons on the surface of the sun (if he could survive). On Venus, Saturn, Uranus, or Neptune he would hardly notice the difference from his earthly weight, but on Jupiter he would register nearly 500 pounds. On Mars he would weigh only 70 pounds and on the satellites of Mars less than half an ounce.

7-14   Velocities of Escape. The critical velocity that would be just suffici-ent to carry a body from the surface of the object to infinity, completely out of its gravitational attraction, is known as the velocity of escape. For a sphere of mass M and radius R, the velocity of escape, V_e, is given by the equation

$$V_e = \sqrt{\frac{2GM}{R}}, \tag{7-14}$$

where G is again the Newtonian constant of gravitation.

The velocity of escape for the earth is 11.2 km/sec (7.0 miles/sec); for the moon, 2.4 km/sec (1.5 miles/sec); for Mars, 5.0 km/sec (3.1 miles/sec); for the sun, the colossal value of 617 km/sec (383 miles/sec). These numbers also represent the minimum velocities with which matter from space can fall upon the surface of planets, satellites, or the sun. Thus meteors enter the earth's atmosphere with geocentric velocities greater than 11.2 km/sec. The square of the velocity of escape equals twice the minimum energy (per unit mass) that a rocket must possess if it is to escape from a planet and embark on space flight.

A satellite circling just above the surface of a planet will move at a speed of $1/\sqrt{2}$ or 0.707 times the velocity of escape. Thus an artificial satellite travelling in an orbit around the earth just above the atmosphere moves at about 7.9 km/sec (17,800 mph), and the period of revolution is approxi-mately 90 minutes.

Equation (7.14) shows that the velocity of escape from a body decreases inversely as the square root of the distance from it. Thus it can be used to calculate the velocity of escape at any distance R from any astronomical body. At the earth's distance from the sun, a body could escape completely from the solar system with an initial velocity of 42.1 km/sec (26.2 miles/sec). Thus a space ship designed to leave the solar system would require a velocity nearly four times as great as that needed to leave the earth itself. Well away from earth, however, and moving at the earth's velocity, the space ship, if sent forward along the tangent to the orbit, needs only an additional 12 km/sec (8 miles/sec) to attain parabolic velocity, the velocity that a body moving about the sun in a parabolic orbit possesses when it crosses the earth's orbit. Comets and meteors moving in extremely elongated, almost parabolic orbits possess very nearly this heliocentric velocity when they encounter the earth (see Chapters 14, 15).

8

Radiation and
Atomic Structure

8-1 **Nature of Radiation.** The nature of radiation has long been one of the fundamental problems of science, a problem still not completely solved. Radiation can assume many forms, including gamma rays, X rays, light, and radio waves.

We feel that we know more about radio waves than about other waves, because we can produce them in the laboratory with a radio transmitter and thus control their characteristics. We can, for example, send out radio waves for which the distance between successive wave maxima (the wavelength) ranges all the way from many kilometers to less than a millimeter.

We know, moreover, that heat waves and light waves have much in common with radio waves. In fact when the English physicist J. C. Maxwell proposed the theory that light waves are electromagnetic in character, back in 1860, he actually discussed radio waves, although such waves had never been produced or even considered up to that time. By now, we possess ample evidence that radio, infrared, light, ultraviolet, and even X and γ rays are all basically similar. All are electromagnetic energy, characterized by a specific wavelength and a definite speed of travel. In a vacuum, all electromagnetic waves travel with the same speed, 299,793 km/sec (about 186,000 miles/sec).

The wavelengths of light waves are so small that they are most conveniently expressed in very small units. These units are defined in terms of the metric system. For the longest radio waves, indeed, we may use kilometers. For intermediate and short radio waves we use meters, centimeters, or even millimeters. But for heat and light waves a smaller unit is needed.

The micron, which is a thousandth of a millimeter, and is designated by the symbol μ, is a useful unit for expressing the wavelengths of heat or infrared waves. But for light we usually express wavelengths in Ångström units (Å); one Ångström is 10^{-8} centimeter, or 10^{-4} μ. To get an idea of the wavelength of light, we may note that green light has about 20,000 waves per centimeter, or 50,000 to the inch. The wavelength is 5000 Å $= 0.5$ $\mu = 5 \times 10^{-5}$ cm.

The electromagnetic spectrum covers an enormous range in wavelengths, and the subdivisions are somewhat arbitrary. Figure 8-1 indicates the basic subdivisions and the primary uses of radiation of different wavelengths.

The human eye is a remarkable instrument, but it is sensitive to only a very small part of the complete electromagnetic spectrum. If our ears were equally limited, we should be able to detect only a single musical octave, and should be as deaf to the rest of the notes as we are blind to the rest of the radiation spectrum. Perhaps this blindness has some advantages. Without it, we should continually be conscious of the heat radiations and radio waves that now travel by us, and even through us, completely unnoticed.

The fact that we commonly speak of radio or light waves implies that radiation possesses some wave characteristics. This view was held by many scientists as long as three centuries ago. However, Newton subscribed to an alternative idea: he thought of light as "corpuscular," consisting of tiny projectiles rather than waves. There was much to be said in favor of both views. But J. C. Maxwell's theories were supported by many experiments carried out during the latter part of the nineteenth century, which seemed to establish the wave character of radiation beyond question.

Just at the turn of the century, however, the German physicist Max Planck found some properties of radiation that he could not readily explain on any hypothesis that strictly involved waves. He therefore revived the long-dead corpuscular theory and suggested that, even though light has properties that suggest wave motion, it actually occurs in little bundles of energy, which he termed *quanta*. He suggested that each quantum, or bundle or photon possesses an amount of energy proportional to the *frequency* of the light's vibration.

If λ is the wavelength of the light in centimeters, than the number of waves per centimeter is $1/\lambda$. If the light travels with velocity c centimeters per second, then the number of wave crests that flash by in a second is equal to c/λ, the frequency of the light vibration. Thus the relation between frequency ν, wavelength λ, and the velocity of light c is given by the simple equation

$$\nu = \frac{c}{\lambda}. \tag{8-1}$$

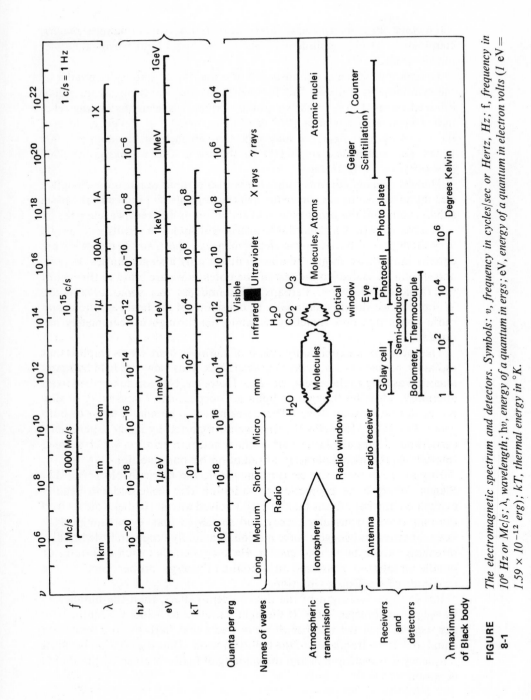

FIGURE 8-1

The electromagnetic spectrum and detectors. Symbols: ν, frequency in cycles/sec or Hertz, Hz; f, frequency in 10^6 Hz or Mc/s; λ, wavelength; $h\nu$, energy of a quantum in ergs; eV, energy of a quantum in electron volts (1 eV $= 1.59 \times 10^{-12}$ erg); kT, thermal energy in °K.

The relation given by Planck between the energy ε of a quantum and the frequency ν of the light is:

$$\varepsilon = h\nu, \tag{8-2}$$

where h is a universal constant, known as Planck's constant; its numerical value is $h = 6.62 \times 10^{-27}$ gm cm^2 sec^{-1}, in the centimeter-gram-second or cgs system.

8-2 Properties of Radiation.   The basic properties of radiation fall into four main classes. First, we may say that the radiation has *direction*. We can state where the light is coming from. Second, radiation has *intensity*. An ordinary object, such as a cloud or a distant tree, has a definite extent, and when we look at it we see that the brightness or intensity of the light varies over its surface, depending on the precise direction in which we look. A distant star may be so small that it has no apparent extension at all. But for most ordinary objects the direction and intensity of the light define the basic appearance (shape and brightness) that the object would have, for example, on a black-and-white photograph.

A third property of radiation is *color*, which is uniquely defined by the wavelength. Visible light covers only a small range of wavelengths. Hence we shall not ordinarily use the restrictive word "color," which implies the reaction of the eye, but its more general equivalent, wavelength, to specify this property of radiation.

Finally, we note that radiation can be "polarized." The vibrations associated with light waves are perpendicular to the direction in which the wave is traveling. In this sense, light waves are like ripples on a pond. The amplitude of the wave determines its intensity, and the spacing between successive crests gives the wavelength. But we must also specify the direction of the ripple motions with respect to the direction in which the wave is moving. A light wave moving through space can vibrate in a complicated fashion in any direction perpendicular to the direction of propagation. We say that only *transverse* vibrations are possible, in contrast with longitudinal vibrations such as in sound waves. According to Maxwell's theory the light wave consists of a coupled electrical vibration and a magnetic vibration perpendicular to one another.

The total energy in a beam of radiation is proportional not to the amplitude, but to the square of the amplitude of the vibration. If the direction of the vibrations remains constant along the light beam, so that the vibrations are all in the same plane, we say that the light is *plane polarized*. But if the direction of the electrical maximum is rotating uniformly in space, tracing out a helical path, something like the thread of a screw, we say the light is *circularly polarized*. If the motion is spiral, but the intensity varies during the cycle, we speak of the polarization as *elliptical* (Fig. 8-2).

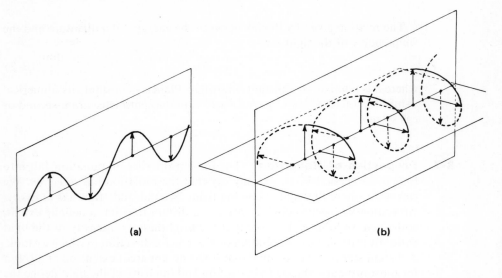

**FIGURE
8-2**
*Plane (a) and circular polarization (b). In (a) the vibration lies in the vertical plane;
in (b) it rotates around the direction of propagation to describe a helix.*

Formerly, the polarization of light was of less interest to the astronomer than its other properties. But currently we are finding polarization increasingly important in the interpretation of the physical condition of planets, stars, nebulae, and even interstellar space.

8-3 The Detection and Measurement of Radiation. The means and methods used to detect and measure radiation differ greatly according to the wavelength of the radiation being observed, as illustrated by Fig. 8-1, which indicates the ranges over which various types of radiation receivers or detectors are sensitive. The lengths of some of these lines are somewhat arbitrary. For example, although ordinary *photographic plates* have little, if any, sensitivity to wavelengths longer than that of green light, a plate sensitized with organic dyes will respond to red and infrared light (see Sec. 6-10). The ranges shown in Fig. 8-1 are generally extreme ranges.

The *thermometer* is one of the earliest known detectors of radiation. W. Herschel employed it in 1800 to explore the solar spectrum and discovered the existence of infrared radiation from the sun. It is now seldom used by astronomers because of its low sensitivity. The *thermocouple* consists of a junction of two dissimilar metals, such as antimony and bismuth. Radiation of any kind absorbed by the junction heats it, causing an electrical current to flow through wires connected to the couple. The strength of the current can be used to measure the intensity of the radiation.

The *bolometer*, which also is used especially for infrared radiation, contains a fine wire whose resistance changes with temperature. We determine the change of resistance by measuring the amount of current flowing along the wire in a simple electric circuit. The *radiometer* is a device for detecting the pressure exerted by molecules heated by absorption of radiation. It may be likened to a set of miniature scales, with the "pan" made of a fly's wing, or some other very light material. The deflection of this pan, when radiation falls on it, measures the amount of energy. Another type of radiometer, responsive to radio frequencies, is used in radio telescopes.

The more recently invented *Golay cell* is, in principle, a tiny aneroid barometer, whose internal pressure increases when heated by radiation. The sensitivity of this device depends on its speed or response. Rapid interruptions of the incoming beam of radiation, by means of a rotating sector, produce pulsations that can be turned into an alternating current and amplified electronically. This detector has been used for very short radio waves, with wavelengths of the order of 0.1 millimeter. *Photosensitive transistors*—miniature substitutes for vacuum tubes and photocells—thermistors, and other "solid-state" detectors or radiation are also coming into use.

One of the most valuable tools of the astronomer is the *photoelectric cell* (see Sec. 6-12), in which radiation creates an electric current. Different substances are sensitive to different spectral ranges, from ultraviolet to infrared. Various compounds of the alkali metals sodium, potassium, rubidium, and cesium are in wide use. Their sensitivities to infrared light increase in the order given. For use in the far infrared, photoconductive substances such as lead sulfide or lead telluride are more responsive.

Finally, as described in Sec. 6-9, radio waves from space are collected by radio telescopes and antennas of various kinds. The very weak energy flux is amplified by extremely sensitive receivers.

The foregoing detection devices are those in widest use, but the list is not complete. With the growth of electronics, new methods of detection continually appear, and astronomers are alert to apply them (see Sec. 6-12).

8-4   How Radiation Is Produced.   The production of radiation is simplest to understand for radio waves. A radio transmitter has two essential parts: an oscillator, which sets electrical charges in oscillation, and an antenna, in which these electric charges move alternatively up and down.

We know from elementary physics (Oersted's experiment) that a magnetic field surrounds every wire in which an electric current is flowing. Even though the radio antenna does not represent a closed electrical circuit, it can nevertheless carry a current as the electrons within it surge up and down. When the electrons of the wire oscillate on some particular frequency, the associated electromagnetic field will spread out into space in the form of electromagnetic waves of that frequency. Transmission efficiency is greatest when the wavelength of these waves is close to the length of the antenna.

We know also that atoms contain electric charges, and that under some circumstances the electrons in an atom will surge back and forth in much the same way as they do in the radio antenna. During such oscillations, the atom radiates energy into space. in the form of waves of light (see Sec. 12).

8-5 The Kirchhoff-Draper Laws. Three laws with many applications in astrophysics govern the emission and absorption of radiation by atoms under various physical conditions. They are often attributed to the German physicist G. P. Kirchhoff, who published them in 1859, but the American astronomer W. Draper was their first discoverer, as Kirchhoff himself acknowledged.

1. An incandescent solid, liquid, or gas under high pressure radiates a continuous spectrum, containing all colors or a wide range of colors.

2. An incandescent gas under low pressure radiates a discontinuous or bright-line spectrum—that is, it emits light of a limited number of distinct colors, characteristic of the gas.

3. When white light traverses a cool gas, the gas absorbs those colors that it would itself emit, if it were incandescent. Such a gas will therefore produce an absorption, or dark-line, spectrum.

These are the three basic laws of spectroscopy, and we use them all in our study of the spectra of the heavenly bodies. The sun and most of the stars send out a background of continuous radiation. Hence we can infer that the source of such light must be a gas under high pressure, since the temperatures of the sun and stars are too high for any of their material to be solid or even liquid. We shall find later that the pressures need not even be extremely high.

The spectra of the sun and most of the other stars also show dark absorption lines, characteristic of the chemical elements that compose their atmospheres. The hot interiors of these stars are surrounded by cooler gaseous atmospheres, which absorb from the continuous background the colors characteristic of the atoms composing the gas.

Finally, we observe emission, or bright-line, spectra for a great variety of objects—comets, certain stars, and the great luminous gas clouds in space called emission nebulae. We can infer that they are composed of, or contain, luminescent gas at a low pressure.

8-6 Radiation and Temperature. That temperature can have an influence on radiation is a familiar and significant fact. The hotter a fire, the more intense is its output of heat and light. But we must define temperature precisely before we can proceed to a quantitative discussion of radiation.

The word temperature is often loosely used. As we shall see, it can be defined in a number of different ways.

The most familiar device for the measurement of temperature is a thermometer, a capillary tube with a reservoir of liquid mercury or alcohol at one end. Heat causes the liquid to expand and to rise in the capillary tube; cold causes it to contract and to fall in the tube. A scale attached to the tube enables us to tell what the temperature is. The scale itself has been graduated by readings made at certain known temperatures.

The Fahrenheit thermometric scale is the one in common everyday use in the United States. On this scale, water freezes at 32° F and boils at 212° F. The zero point of the Fahrenheit scale was arbitrarily chosen by the inventor as the lowest reading encountered during his weather observations. When the temperature is below the zero of the Fahrenheit scale, we have to express it by such terms as "15 degrees below zero" or as −15° F.

The Celsius thermometric scale was constructed by this eighteenth-century Swedish physicist on a more rational basis. Its scale was defined by the freezing and boiling points of distilled water. Ice melts at 0° C and water boils (at sea level under standard conditions) at 100° C. Because of this 100-degree range the scale is often referred to as centigrade.

For many astronomical purposes, we use a modified centigrade scale introduced in the nineteenth century by the British physicist, Lord Kelvin. This scale is based on the concept of absolute zero, the lowest possible limit of physical temperature. On this scale, the absolute zero of temperature is 0° K; water freezes at 273° K and boils at 373° K. Simple relations relate the scales of temperature to one another:

$$T_F = \tfrac{9}{5}T_C + 32, \tag{8-3}$$

$$T_C = \tfrac{5}{9}(T_F - 32), \tag{8-4}$$

$$T_K = T_C + 273. \tag{8-5}$$

But what do we actually measure with such a thermometer? Certainly not radiation, for it gives the same reading in the dark as in the light. However, if we take a thermometer out into direct sunlight, we know that it will usually give a reading far higher than the air temperature, because absorbed radiation will warm the liquid in the bulb. Temperatures read "in the sun" usually exceed those measured "in the shade."

The temperature that we determine with a thermometer is really a measure of the degree of agitation of the molecules of the liquid in the bulb. The temperature of the air is a measure of the degree of agitation of the air molecules. If we could reduce the temperature of air to 0° K, the individual atoms or molecules would be essentially motionless with reference to one another. Indeed, at this temperature air would no longer be gaseous, because it is the relative motion of the particles that keeps them apart, differentiating between the solid, liquid, and gaseous phases of a substance. Temperature, then, describes the average energies of the particles. We can define these average energies most easily for a gas, in which the particles are separated and moving about freely. But we can extend the same concept to a liquid or even a solid, and we know that (except where certain physical changes of state are involved) most liquids and solids expand when heated.

Thus we have one definition of temperature, which we term the *gas-kinetic temperature* because it measures the kinetic energy, or energy of motion, of the atoms or molecules of a gas.

8-7 The " Black-Body " Radiation.

We could also define temperature in terms of radiation. Let us suppose that we place our gas inside a sealed, completely insulated enclosure, like a large oven, leaving one tiny window through which the radiation can escape so that we can measure it. Such an enclosure with almost no window is, appropriately enough, called a *black body*. The escaping radiation is such a small fraction of the total penned up inside the oven that its loss does not have any sensible effect on conditions inside the enclosure.

The black-body radiation escaping from such an enclosure possesses some very special properties, which depend only on the gas-kinetic temperature of the enclosure and are independent of the composition of the gas. Let us mention some of these properties. As we raise the temperature until some of the radiation falls within the visible range, we first note a deep red glow that changes successively to cherry red, orange, yellow,

FIGURE 8-3 *Wien's law relates the peak wavelength of black-body emission to absolute temperature T ($\lambda_m T = 2900$).*

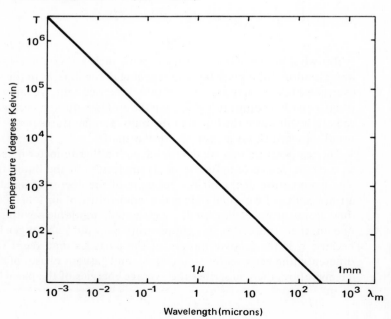

straw-color, white, and finally a dazzling blue-white. These colors furnish a rough measure of the quality of the radiation. Precise measurements would show that, as the temperature increases, the wavelength λ_m at which the radiation is most intense decreases in such a way that the product $\lambda_m \times T$ is constant (Fig. 8-3). When the wavelength is expressed in microns and the temperature in degrees Kelvin, the quantitative expression, known as "Wien's law", is:

$$\lambda_m \times T = 2900. \tag{8-6}$$

Experiments prove that the total rate of radiation escaping through the opening in the oven is proportional to the fourth power of the absolute temperature. This law, discovered experimentally by Stefan, was proved theoretically by Boltzmann. It is known as the "Stefan-Boltzmann law." The equation that expresses this law is

$$E = \sigma T^4, \tag{8-7}$$

where E is the escaping energy, in ergs per square centimeter per second, T is the absolute temperature on the Kelvin scale, and σ is a constant of proportionality whose value is 5.67×10^{-5} in cgs units (erg cm^{-2} sec^{-1} deg^{-4}).

FIGURE 8-4   *Planck's curves illustrate the spectral energy distribution of black-body radiation at different temperatures.*

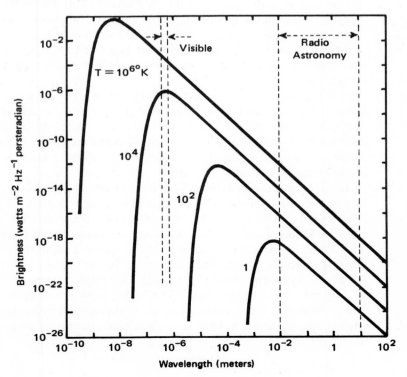

The Wien law and the Stefan-Boltzmann law are actually only parts of a more general relationship, known as Planck's radiation law. Planck's law defines the precise values of the radiation intensity at all wavelengths, not merely at the wavelength of maximum energy. The total energy is equal to the area underneath the black-body radiation curve. Several representative curves for selected temperatures appear in Fig. 8-4. Note the extreme steepness of the fall toward the shortwave or ultraviolet part of the spectrum.

If the opening in the "oven" is large, or if there is no enclosure at all (as is the case for the surface of a star), the properties of the escaping radiation can depart markedly from the theoretical predictions of the radiation laws. Hence the gas-kinetic temperature may sometimes differ markedly from the radiation temperature.

8-8   Electrons and Atomic Structure.   The many peculiarities of the observed spectra of the chemical elements presented difficult problems to physicists in their early attempts to understand atomic structure. The experiments of Rutherford had shown that most of the mass of an atom, and all of its positive electrical charge must lie in a very limited region at the atom's center. This concentrated region, which occupies usually less than 10^{-12} of the whole volume of the atom, is the *nucleus*. In a later chapter we shall find that the nucleus itself has an internal structure that plays an important part in the generation of solar and stellar energy. For the present we shall regard it as merely a small region that contains practically the whole of the atomic mass, and a specific number of units of positive electric charge, 1, 2,..., 100,.... The proton, the simplest unit of positive charge, has a charge equal and opposite to that of the negative electron. The mass of the proton, however, is 1836 times greater than that of the electron.

A neutral atom has zero electric charge. Hence, if a nucleus contains Z units of positive charge, there must also be Z negative electrons arranged in some manner around the nucleus. The atom as a whole will then be electrically neutral. The atomic scientist faces the major problem of finding some arrangement of the electrons that will account for both the chemical and physical properties of the atom.

In 1913 the Danish physicist Niels Bohr proposed a model for hydrogen, the simplest of all atoms, which consisted of a positive proton with $Z = 1$, and a single electron revolving around it like a planet moving round the sun. He followed with models like miniature solar systems for more complex atoms, in which Z negative electrons revolved about the positive nucleus. But Bohr's model, which accounted well for many of the properties of atomic hydrogen, failed significantly for complex atoms.

A decade later, the invention of "wave mechanics" or "quantum mechanics" furnished a new approach to the problem of electron arrangement. It was developed and applied to the external structure of atoms by L. de

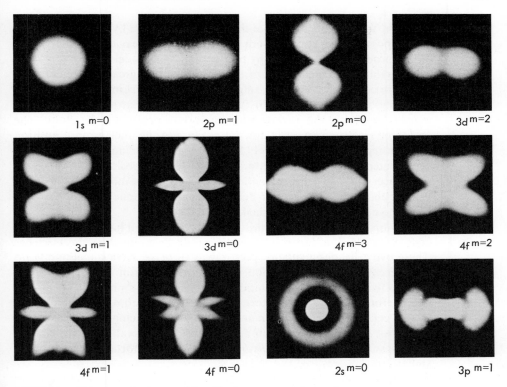

1s m=0	2p m=1	2p m=0	3d m=2
3d m=1	3d m=0	4f m=3	4f m=2
4f m=1	4f m=0	2s m=0	3p m=1

**FIGURE
8-5** *Wave mechanics models of hydrogen atom for various energy states.*

Broglie in France, E. Schrödinger and W. Heisenberg in Germany, and
P. A. M. Dirac in England; many others followed their lead. According to
wave mechanics, electrons close to the nucleus lose their individuality as
particles. They seem to dissolve into a sort of haze, and can arrange themselves in a wide variety of patterns (Fig. 8-5). The point of view of wave
mechanics has been more fruitful than Bohr's simple model in accounting
quantitatively for the properties of complex atoms.

As long as the arrangement of the electrons remains unchanged, the atom
will not radiate energy. But during the interval when the pattern is changing
from a given form to one that is more stable, the pulsing charges of elec-
tricity within the atom behave like those in a radio antenna. Hence the
atom can act like a miniature radio transmitter, sending out electromagnetic waves into space.

8-9   The Hydrogen Spectrum.   The spectrum of hydrogen presents the
simplest problem of all, because the hydrogen nucleus contains only a
single unit of positive charge, and hence the hydrogen atom has a single

electron. We say, therefore, that the atomic number of hydrogen is 1. Some of the possible pulsating patterns of this atom appear in Fig. 8-5; H. E. White produced these figures by photographing a model atom.

The relatively simple atom of hydrogen helps us greatly to understand the more complex atoms. The single electron can exist in any one of many vibrating patterns. We can in fact classify these patterns in ways that vaguely resemble the various electron orbits postulated by Bohr. The hydrogen electron can settle down very close to the nucleus, in a position that we represent by number 1. It can move further out, to the position represented by the number 2, and so on through positions 3, 4, 5, . . . , each successive integer representing a position where most of the negative electric charge lies at a greater distance from the atomic nucleus. We may term these positions shell 1, shell 2, ...

A complication appears at this point. The first shell has, in essence, two cells or "holes," either one of which could accommodate the electron. These two cells, designated respectively as + or −, have something to do with *electron spin*, the direction in which the electron appears to rotate relative to the direction of orbital revolution in Bohr's model.

Sometimes a free electron, torn from some other atom, will fall into this second cell of a hydrogen atom and stay there, producing a hydrogen atom that has two electrons and, in consequence, an overall negative charge. Such atoms, called *negative hydrogen ions*, are important in theories of stellar atmospheres (see Sec. 9-1).

A diagram showing the structure of the hydrogen energy levels appears in Fig. 8-6. An electron in shell 7, for example, can lose energy and spontaneously fall down to any one of the lower shells. The farther it falls, of course, the more energy is released, and hence, by Planck's equation (8-2), the higher the frequency (or the shorter the wavelength) of the radiation emitted.

The transitions 2-1, 3-1, 4-1, . . . , give rise to the Lyman series of spectral lines, labeled as Lα, Lβ, Lγ, ... They lie in the far ultraviolet and have not been directly observed in the spectra of the stars, because the ozone in the earth's atmosphere absorbs the stellar radiations at such short wavelengths. However, scientists have recently succeeded in photographing the early members of the Lyman series in solar and stellar spectra, by means of spectrographs sent up in rockets and satellites. The lines often appear in emission (see Sec. 9-1).

The transitions 3-2, 4-2, 5-2, . . . , are called the Balmer series, and carry the labels Hα, Hβ, Hγ. . . . The respective series terminating in 3, 4, and 5 are called the Paschen, Brackett, and Pfund series. Still other series are known, farther to the infrared, some even in the radio range. The Balmer series is the one best known and first discovered in stellar spectra, since it lies in the range readily observed, between the red and the near ultraviolet. The other series listed all lie in the infrared, and some of them have been detected in the spectra of the sun and stars.

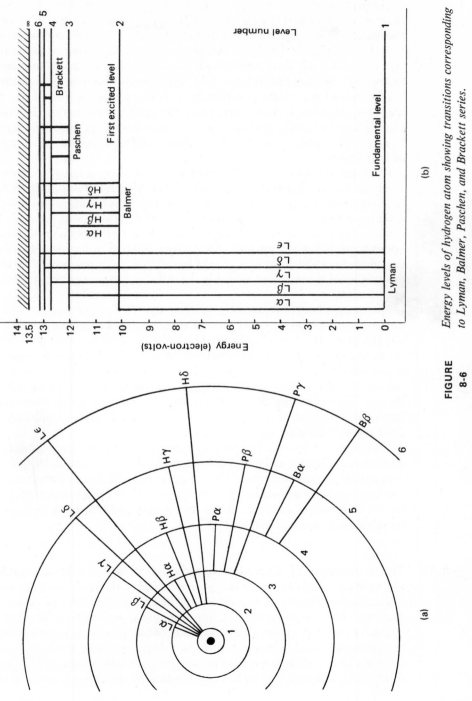

FIGURE 8-6

Energy levels of hydrogen atom showing transitions corresponding to Lyman, Balmer, Paschen, and Brackett series.

133

The high-level transitions in hydrogen, which produce radio emissions, have been detected. They come from regions of hydrogen gas near very hot stars. We still use the Greek letters α, β, γ,... to indicate jumps of 1, 2, 3,... quantum numbers, but we have given up trying to name the series. Instead, we write down the lower of the two quantum numbers, following it by the appropriate Greek letter. In this notation, the Balmer line Hγ would receive the notation 2γ. A few examples of lines found in the radio spectrum include: 90α, 104α, 109α, 158α. Presumably the complete spectrum occurs, though some lines are masked by terrestrial interference. Also, radio receivers are not always available for the entire range of frequencies.

Hydrogen has one other special emission, of a very unusual character. We have already noted that the level of lowest energy has a position for two types of electrons. As we noted earlier, electrons behave as if they are spinning on an axis, like a miniature earth. The nucleus, also, seems to be rotating. The electron is more stable when the direction of its rotation is opposite to that of the nucleus. An electron rotating in the same sense as the nucleus is strictly not in a level of lowest energy; it possesses a certain probability of turning over, emitting the difference in energy as a quantum of radiation.

The energy change is extremely small, so that the frequency of the radiation, by equation (8-2), is very low, 1420 megacycles per second, or 1.42×10^9 Hz, where Hz stands for hertz. The corresponding wavelength, 21 cm, lies in the radio region of the spectrum. Where most ordinary transitions in the atom have very small mean lifetimes, of the order of 10^{-7} or 10^{-8} second, the lifetime for emission of the 21-cm line is about 11 million years. We say that the energy level is *metastable* and refer to the radiation as a *forbidden line*, a term indicating that the emission results from atomic transitions of an improbable character.

The 21-cm radiation, therefore, is extremely weak. However, the predominance of hydrogen in interstellar space and the vastness of the galaxy compensate for the low probability of emission. The 21-cm line, detected and measured by radio telescopes, has proved to be a most important tool for the study of the structure of our own galaxy and of other galactic systems. (See Chapters 29, 31, and 32.)

8-10   The Structure of Complex Atoms.   Consider a neutral complex atom with Z units of positive charge in its nucleus and Z orbiting negative electrons. By experiment we can always determine the value of Z, which fixes the chemical nature of the atom uniquely, no matter what special characteristics the nucleus may have otherwise. Thus $Z = 1$ corresponds to hydrogen, $Z = 2$ to helium, $Z = 3$ to lithium, and so on.

To get neutral helium we must fit two electrons into the atom. The most stable state, however, will be found when both these electrons occur in the lowest level—in the innermost shell or orbit. If we now bombard such a helium atom with energetic electrons, we may cause one or both electrons to jump to a level of higher energy, an *excited* level. As these electrons then cascade spontaneously down to lower levels, they will emit the radiations characteristic of the atom of neutral helium. Figure 8-7 depicts the energy levels of neutral helium. In all of those shown, only one electron is excited, and the other remains in shell 1.

We can continue for lithium, with $Z = 3$. As for helium, the most stable condition will find two of the electrons in the inner shell. But this shell is now complete, and the third electron must go somewhere else. Since the first shell is filled, we must assign the third electron to one of the upper shells, say shell 2.

The electrons can, of course, make complicated jumps from one shell to another, and sometimes even between energy levels of the same shell. Each jump leads to the emission of radiation if the final state is one of lower energy. If, on the other hand, the final state has greater energy than the initial state, the transition requires the absorption of radiation, and results in the formation of an absorption line. As the atomic number Z increases, the spectrum may become extremely complex, and we then find special groups of related lines. Sodium, for example, shows pairs of lines or doublets throughout the spectrum, such as the familiar pair of D lines in the bright yellow region. Magnesium, on the other hand, generally displays either single lines or groups of three lines (triplets). The neutral atoms of even atomic number, such as magnesium ($Z = 12$), have odd *multiplicity* in their spectra. Atoms of odd atomic number, such as sodium ($Z = 11$), have even multiplicity.

Every kind of atom has its own specific set of energy levels. Transitions or jumps of the electron between these levels give rise to the various spectral lines, unique for that substance. Some of the transitions between the lower energy levels have a very low probability of occurring. The 21-cm line of hydrogen is an example of such a transition. We term these lines "forbidden" lines, because they occur in violation of some of the laws governing normal transitions and are very weak under ordinary laboratory conditions. Yet some of them possess enormous astrophysical significance. They are present as strong lines in the spectra of the sun's corona, gaseous nebulae, and even in the earth's aurora. The vastness of the sources more than compensates for the low transition probability.

No physicist has yet found a reasonable explanation for the limitation in the number of electrons that can be assigned to the various shells. From empirical considerations, W. Pauli formulated the *exclusion principle* that bears his name. It defines how electrons are allocated and determines the maximum number that can enter any given shell. This maximum number is $2n^2$, where n is the number of the shell. Thus, the inner shell, $n = 1$, can have two electrons; the next shell, $n = 2$, can have eight electrons, and so on.

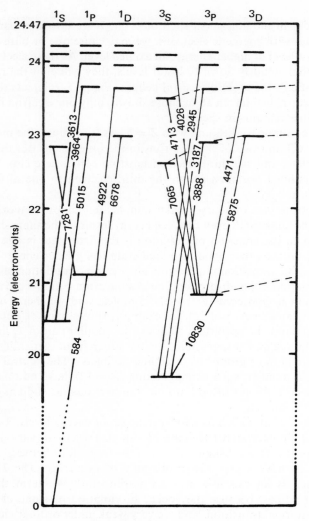

FIGURE 8-7 *Energy levels of neutral helium atom (Grotrian diagram). The wavelengths of the major permitted transitions are given in angströms.*

The Russian chemist D. I. Mendeleev more than a century ago was the first to note certain regularities among the chemical elements. Table 8-2 gives the periodic table in modern form. We now know that all the atoms in a given column of the periodic table possess somewhat similar patterns of outer electrons. It is this structure that determines the chemical behavior of the elements. Thus, atoms in the same column possess similar chemical (and spectroscopic) similarities.

TABLE 8-1

The chemical elements

Z	Name	Abr.	W
1	Hydrogen	H	1.00797
2	Helium	He	4.0026
3	Lithium	Li	6.939
4	Beryllium	Be	9.0122
5	Boron	B	10.811
6	Carbon	C	12.01115
7	Nitrogen	N	14.0067
8	Oxygen	O	15.9994
9	Fluorine	F	18.9984
10	Neon	Ne	20.183
11	Sodium	Na	22.9898
12	Magnesium	Mg	24.312
13	Aluminum	Al	26.9815
14	Silicon	Si	28.086
15	Phosphorus	P	30.9738
16	Sulphur	S	32.064
17	Chlorine	Cl	35.453
18	Argon	Ar	39.948
19	Potassium	K	39.102
20	Calcium	Ca	40.08
21	Scandium	Sc	44.956
22	Titanium	Ti	47.90
23	Vanadium	V	50.942
24	Chromium	Cr	51.996
25	Manganese	Mn	54.9380
26	Iron	Fe	55.847
27	Cobalt	Co	58.9332
28	Nickel	Ni	58.71
29	Copper	Cu	63.54
30	Zinc	Zn	65.37
31	Gallium	Ga	69.72
32	Germanium	Ge	72.59
33	Arsenic	Asz	74.9216
34	Selenium	Se	78.96
35	Bromine	Br	79.909
36	Krypton	Kr	83.80
37	Rubidium	Rb	85.47
38	Strontium	Sr	87.62
39	Yttrium	Y	88.906
40	Zirconium	Zr	91.22
41	Niobium	Nb	92.906
42	Molybdenum	Mo	95.94
43	Technetium	Tc	—
44	Ruthenium	Ru	101.07
45	Rhodium	Rh	102.905
46	Palladium	Pd	106.4
47	Silver	Ag	107.870
48	Cadmium	Cd	112.40
49	Indium	In	114.82
50	Tin	Sn	118.69
51	Antimony	Sb	121.75
52	Tellurium	Te	127.60
53	Iodine	I	126.9044
54	Xenon	Xe	131.30
55	Cesium	Cs	132.905
56	Barium	Ba	137.34
57	Lanthanum	La	138.91
58	Cerium	Ce	140.12
59	Praseodymium	Pr	140.907
60	Neodymium	Nd	144.24
61	Promethium	Pm	—
62	Samarium	Sm	150.35
63	Europium	Eu	151.96
64	Gadolinium	Gd	157.25
65	Terbium	Tb	158.925
66	Dysprosium	Dy	162.50
67	Holmium	Ho	164.930
68	Erbium	Er	167.26
69	Thulium	Tm	168.934
70	Ytterbium	Yb	173.04
71	Lutetium	Lu	174.97
72	Hafnium	Hf	178.49
73	Tantalum	Ta	180.948
74	Tungsten	W	183.85
75	Rhenium	Re	186.2
76	Osmium	Os	190.2
77	Iridium	Ir	192.2
78	Platinum	Pt	195.09
79	Gold	Au	196.967
80	Mercury	Hg	200.59
81	Thallium	Tl	204.37
82	Lead	Pb	207.19
83	Bismuth	Bi	208.980
84	Polonium	Po	—
85	Astatine	At	—
86	Radon	Rn	—
87	Francium	Fr	—
88	Radium	Ra	226.0254
89	Actinium	Ac	—
90	Thorium	Th	232.038
91	Protactinium	Pa	231.0359
92	Uranium	U	238.03
93	Neptunium	Np	237.0480
94	Plutonium	Pu	239.0522
95	Americium	Am	—
96	Curium	Cm	—
97	Berkelium	Bk	—
98	Californium	Cf	—
99	Einsteinium	Es	—
100	Fermium	Fm	—
101	Mendelevium	Md	—
102	Nobelium	No	—
103	Lawrencium	Lw	—

TABLE 8-2

Periodic table of the elements

1	2				Transition Elements						3	4	5	6	7	0
1H																2He
3Li	4Be										5B	6C	7N	8O	9F	10Ne
11Na	12Mg										13Al	14Si	15P	16S	17Cl	18A
19K	20Ca	21Sc 22Ti 23V 24Cr 25Mn 26Fe 27Co 28Ni 29Cu 30Zn									31Ga	32Ge	33As	34Se	35Br	36Kr
37Rb	38Sr	39Y 40Zr 41Nb 42Mo 43Tc 44Ru 45Rh 46Pd 47Ag 48Cd									49In	50Sn	51Sb	52Te	53I	54Xe
55Cs	56Ba	*	71Lu 72Hf 73Ta 74W 75Re 76Os 77Ir 78Pt 79Au 80Hg								81Tl	82Pb	83Bi	84Po	85At	86Rn
87Fr	88Ra	**	103Lw													

Triads: { 26Fe 27Co 28Ni }, { 44Ru 45Rh 46Pd }, { 76Os 77Ir 78Pt }

		Transition Elements												
Lanthanides *	57La	58Ce	59Pr	60Nd	61Pm	62Sm	63Eu	64Gd	65Tb	66Dy	67Ho	68Er	69Tm	70Yb
Actinides **	89Ac	90Th	91Pa	92U	93Np	94Pu	95Am	96Cm	97Bk	98Cf	99Es	100Fm	101Md	102No

8-11   Molecules and Their Spectra.   Molecules and molecular spectra are almost as important in astronomy as atoms and atomic spectra. Many stars are so hot that few molecules, if any, can exist in their atmospheres, but the cooler stars, the planets, and the comets exhibit molecular spectra of many kinds.

Modern physical chemistry interprets the tendency of atoms to stick together (affinity) and form molecules in terms of shared electrons. The concept of atomic structure presented in the previous section suggests that atoms are chemically most stable when their electrons are arranged symmetrically, which happens only when enough electrons are available to fill all the cells of a given kind in a given shell. Hydrogen, for example, is chemically active because it has just one electron, so that the first shell is not filled. Helium, on the other hand, is extremely stable, because its inner shell, shell 1, when unexcited, contains two electrons, so that the inner shell is symmetrical. Thus helium does not readily enter into chemical combinations with other atoms.

Hydrogen, however, can achieve symmetry of pattern in a number of ways. The simplest, perhaps, occurs when two hydrogen atoms come together and "share" their two electrons, thus forming a hydrogen molecule. In this way, both atoms achieve stability.

Consider, now, the atom of fluorine, which has nine electrons, two in shell 1 and seven in shell 2. Fluorine, therefore, needs one more electron to fill shell 2. Hence, if it encounters a wandering atom of hydrogen, the fluorine atom will capture the electron and use it to complete the second shell, forming a molecule of hydrofluoric acid, HF.

Some atoms have more electrons than symmetry requires. Nitrogen, for example, with two electrons in shell 1 and five in shell 2, can "lose" three of the less tightly bound electrons by sharing them with three atoms of hydrogen to form the molecule of ammonia, NH_3. Carbon, in a similar way, can form CH_4 (methane, or marsh gas). We note the interesting fact that both these substances are abundant in the atmospheres of the giant planets. The molecules in a hot stellar atmosphere are not necessarily complete, for we find such molecular fragments (or radicals) as CH in the atmospheres of the sun and stars.

Molecules can possess internal motions of many different kinds that are not meaningful for an atom. A molecule has at least two heavy nuclei, whereas an atom has only one. The nuclei can vibrate with respect to one another, and may rotate around one another. The shared electrons that surround the pair of nuclei can form their vibrating patterns as they did around a single atom, but the extra flexibility of the multiple nucleus makes these patterns more complex. The fully developed molecular spectrum therefore consists of a combination of rotational, vibrational, and electronic transitions, all superposed. The radiation or absorption occurs in *bands*, which are actually groups of closely spaced lines, as shown in Fig. 8-8.

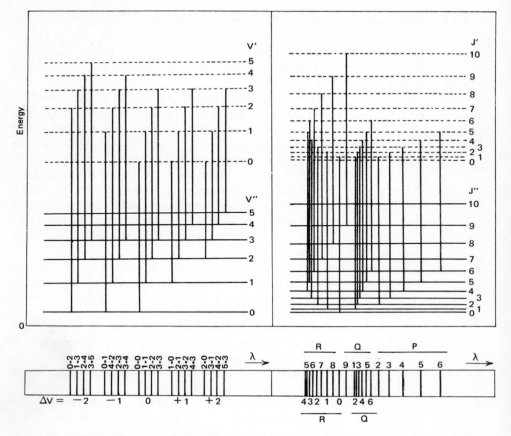

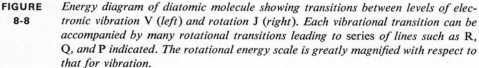

FIGURE 8-8   *Energy diagram of diatomic molecule showing transitions between levels of electronic vibration V (left) and rotation J (right). Each vibrational transition can be accompanied by many rotational transitions leading to series of lines such as R, Q, and P indicated. The rotational energy scale is greatly magnified with respect to that for vibration.*

8-12   The Excitation of Atoms.   At very low temperatures practically all the atoms of a gas are in their lowest neutral energy states. But if the temperature of a gas gradually increases, more and more atoms will have their electrons excited to higher energies. In terms of our atomic model, electrons will be knocked from one of the inner cells into an outer cell of higher energy, from which point they can fall back with the emission of radiation. Here we must refer back to the physical significance of temperature, as discussed in Sec. 8-7. We have noted that the two possible methods of defining temperature, one in terms of the random motions of atoms and the other in terms of the quality and quantity of the radiation, are identical when the gas is completely enclosed in a black body. The outer layers of a star, however, are far from being completely enclosed, and hence the radiation temperature can differ from the kinetic temperature. We now introduce a third concept of temperature, namely *excitation temperature*, on which the distribution of electrons on the various energy levels depends.

Many years ago, L. Boltzmann derived the mathematical law that specifies what proportion of atoms will have their electrons in the different energy levels at a given temperature. The distribution among the various energy levels can therefore be used to specify the excitation temperature.

The atoms in a gas can be excited either by collision with other atoms or particles or by the absorption of radiation. Thus, if the gas-kinetic temperature and the radiation temperature are identical, the excitation temperature will generally be the same also. But if these two temperatures differ, the excitation temperature will usually lie somewhere between the two, and nearer to the one that corresponds to the physical process predominant in the atomic excitation.

At low temperatures, therefore, we can expect that absorption or emission lines will arise only from the lowest levels of the atom. But as the temperature rises, a greater proportion of the atoms will have electrons in cells of higher energy, and lines from these excited levels will begin to appear. We should therefore expect—and later this expectation will be confirmed—that the relative intensities of lines from levels of low and high excitation will provide a measure of the temperature of the gas.

8-13   Ionization.   When the temperature is so high (or the excitation effects so great) that electrons are torn completely from their nuclei and wander freely through the gas, the gas is said to be ionized. The material then consists not of atoms or of molecules, but of *ions*—the new positively charged atomic fragments—and free negative electrons. We should note, however, that a neutral atom or molecule can, under certain conditions, capture a free electron to form a negative ion (see Sec. 8-9).

From our previous discussion of atomic spectra we can infer that each time an atom loses an electron, the whole character of its spectrum must change, since the spectrum of the ionized atom will come from a diminished group of electrons. For example, lines of neutral silicon are prominent in the spectrum of the sun. Singly ionized silicon (with one electron removed) has strong lines in the spectra of certain stars with temperatures about twice that of our sun. In the spectra of still hotter stars we detect lines of doubly and even triply ionized silicon (with two and three electrons removed, respectively). Neutral silicon is denoted by the symbol Si and its spectrum by Si I, singly ionized silicon by Si II, doubly ionized silicon by Si III, and so on for still higher stages.

Ionization occurs more readily in an atmosphere that contains a few free electrons than in one where electrons are numerous. When the *electron pressure* is high, there are many free electrons close by, ready to reunite with any atom that has lost an electron and thus make it neutral again. Conversely, at low electron pressures electrons are scarce, and ions encounter them only rarely. Hence, of two masses of gas having the same chemical composition and the same temperature, the one whose pressure or density is lower will possess the higher percentage of ionized atoms. This important fact enables us to distinguish, from the spectrum alone, between highly distended giant stars and denser dwarf stars at the same temperature (see Chapter 20).

The phenomenon of ionization suggests the definition of still another temperature—the *ionization temperature*. For a gas in an enclosure, ionization temperature will coincide with the temperatures previously defined, but differences may occur under other conditions.

We should further note that the ionization of different kinds of atoms requires different amounts of energy; in other words, their *ionization potentials* differ. Some atoms, such as sodium and potassium, are easy to ionize. Silicon, iron, and magnesium, which are very abundant in stellar atmospheres, require more energy to ionize. Hydrogen is much more difficult, and helium the most difficult of all to deprive of an electron. Hence, a mere look at the spectrum of a star will give some idea of the star's effective ionization temperature, if we note which atoms appear in the ionized state and which in the neutral state. The fundamental formulas that govern the amount of ionization were first given by the great Indian scientist, M. N. Saha.

8-14   The Doppler-Fizeau Effect. We have seen that each chemical element has its own characteristic array of spectral lines of definite wavelengths or frequencies. The frequency, as calculated from equation (8-1), specifies the number of waves that reach the observer's eye per second, on the assumption that the source of light is stationary with respect to the observer. If, however, the source is moving toward or away from the observer, the number of waves received per second will be increased, or decreased.

In 1842 the Austrian physicist C. Doppler noted that the apparent pitch of a sound is altered by the relative motion between the source and the observer, and he suggested that, in a similar fashion, the color of a star might be changed according to its velocity of approach or recession relative to an observer. In 1848 the French physicist H. Fizeau showed that stellar velocities are much too small to cause appreciable color changes, but that the Doppler effect could be detected by very small changes in the wavelengths of individual spectral lines. The difficult experiment was successfully performed for the first time in 1868 by the English astronomer W. Huggins, who was able to detect and measure the minute spectral shift of the hydrogen lines in the spectrum of Sirius corresponding to a line-of-sight velocity of recession of 29 miles per second. If v_0 is the frequency of light for a source at rest with respect to the observer, the observed frequency, v, for a source moving toward or away from the observer with velocity v is

$$v = v_0 \left(1 - \frac{v}{c}\right),$$ (8-8)

and, since $\lambda = c/v$, by equation (8-1),

$$\frac{\lambda - \lambda_0}{\lambda_0} = \frac{v}{c},$$ (8-9)

where c is the velocity of light. According to this formula velocities of approach should be counted as negative and velocities of recession as positive. If, then, a star is approaching, the frequency of the light that reaches the observer will be increased and the wavelength decreased, and its spectral lines will lie to the violet of their normal positions. If, on the other hand, the star is receding, its spectral lines will be displaced toward the red. Hence, by measuring the precise wavelength or frequency, and if the frequency at rest is known, one can determine the velocity of the source with respect to the observer. Spectra showing the Doppler effect appear in Fig. 8-9. The measurements of Doppler shifts in stellar spectra have many important applications in astrophysics. They provide us with a powerful means of determining star and gas motions in our galaxy, and the motions of other galaxies in the whole observable universe (see Chapter 34).

FIGURE 8-9 *Doppler shifted iron lines in stellar spectrum. The iron absorption lines in these two spectra of the constant velocity star Arcturus taken about six months apart show a velocity difference of 50 km/sec entirely due to the orbital velocity of the earth. With respect to the comparison lines the upper spectrum (a) taken on July 1, 1939 shows a velocity shift of +18 km/sec, the second spectrum (b) taken on January 19, 1940 shows a velocity shift of −32 km/sec. (Mount Wilson and Palomar Observatories.)*

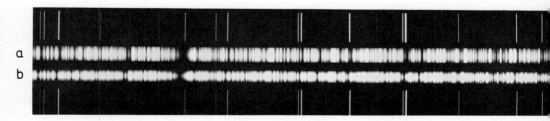

a
b

CHAPTER 9

The Sun and its Radiations

9-1 The Photosphere. The solar photosphere, or "sphere of light," is the astronomer's name for the highly luminous visible solar surface. Precise measurements confirm what the eye, protected by dark glass, infers from direct inspection: that the edge of the sun is sharp and circular. Since the sun rotates, we expect it to bulge at the equator, but the bulge is too slight to be detected except by the most refined techniques. Recent observations, tending to verify the existence of a minute equatorial bulge, need further confirmation.

The best determinations give the distance of the sun from the earth as 149,598,000 km (92,956,000 miles) and its diameter as 1,391,000 km (865,400 miles) (see Secs. 7-6 and 7-7). As seen from the earth, the sun has an apparent diameter of about half a degree.

Careful observation shows that the photosphere is not uniformly bright. Under high magnification, when the earth's atmosphere is extremely steady and the "boiling" of the sun's image at a minimum, the solar surface looks like the mottled skin of a lemon—an appearance usually called *granulation* (Fig. 9-1). Actually, many structural gradations exist, from relatively large-scale mottlings to minute granules visible only under the best optical conditions. These smallest granulations consist of irregular bright patches bordered with a dark edge. Their diameters range between several hundred and a thousand miles. This delicate structure changes quickly, altering its pattern completely within a few minutes. Astronomers associate granulation with turbulence, which originates in the violent convection of the subsurface layers. The larger-scale mottlings persist for hours with only minor changes of form. This semipermanence suggests that the formation is associated with magnetic fields.

The radiation laws already mentioned (Sec. 8-7) enable us to estimate the sun's temperature from observations of the quantity and quality of its radiation. The energy flux or power of solar radiation falling upon the earth (before partial absorption by our atmosphere) is known as the *solar constant*. Its value, measured from the X-15 rocket aircraft at altitudes above 50 km, is 1.952 calories per square centimeter per minute, which amounts to the significant total of 1.5 horsepower per square yard, or approximately one kilowatt per square meter. The "constant" probably changes slightly with variations of solar activity. The greatest fluctuations of solar radiation occur in the far-ultraviolet and soft X-ray regions of the spectrum. This energy cannot pass through the earth's upper atmosphere

FIGURE 9-1   *Solar Granulation and Sunspot. (Courtesy Sacramento Peak Observatory.)*

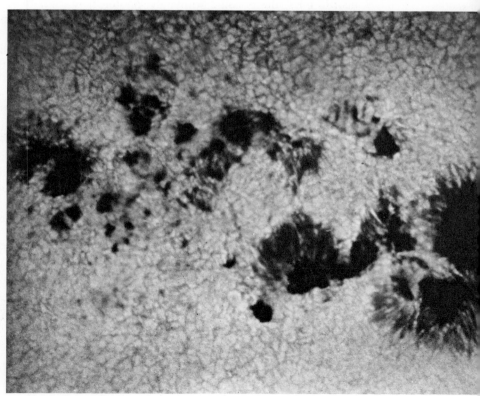

and reach the surface, but we can study it from rockets and artificial satellites. The intensity of solar radiation furnishes a means of determining the effective temperature of the surface of the sun. At the sun's surface, the energy flux is 6.44 kilowatts per square centimeter or 6.44×10^{10} erg/cm^2. Using the Stefan-Boltzmann law [see equation (8-7)], we obtain 5800° K for the value of this temperature.

When observed in visual or photographic light, the sun appears slightly less bright at the edge. This effect, which is known as *limb-darkening*, clearly shows that the solar atmosphere is partially opaque. At the center of the sun's disc, the rays escape vertically and we view the deeper, hotter layers. Near the edge of the sun we look obliquely through the solar atmosphere, hence we see only to a higher level of slightly lower temperatures. If we isolate the more brilliant center of the sun's disc, and apply the Stefan-Boltzmann law, we find that the temperature is higher than the one obtained from the solar constant, but agrees closely with the value of about 6500° K deduced from the distribution of the energy in the spectrum by means of the Wien and Planck laws of radiation.

At such temperatures the sun must be entirely gaseous. Even the element tungsten, the most refractory of all known substances, melts at 3643° K and boils at 6200° K. The gaseous photosphere must therefore be a region of gradually decreasing transparency rather than a sharply bounded surface. Since highly ionized gases are far more opaque to radiation than neutral ones, the pressure within the photosphere must be far less than that at the surface of the earth. The change from almost complete transparency to almost complete opacity takes place in a layer some 200 km thick. This figure is a negligible fraction ($\frac{1}{3500}$) of the solar radius. Hence we are not surprised to find that the sun, as viewed from the earth, appears to have a sharp boundary.

Actually, we see down to slightly different depths in different wavelengths. Studies by R. Wildt and S. Chandrasekhar in the United States have shown that the substance primarily responsible for the opacity of the solar atmosphere is the *negative hydrogen ion*, a hydrogen atom to which an extra electron has become attached, as described in Sec. 8-9. In the far ultraviolet, absorption by atomic hydrogen is important; a broad Lyman-alpha line ($\lambda = 1215$ Å) reduces the intensity of solar radiation over a spectral range of hundreds of angstroms. In the ultraviolet there also occur intense absorption bands of the molecule carbon monoxide. Still further in the ultraviolet strong emission lines appear in the solar spectrum.

Although the sun is gaseous throughout, astronomers generally refer to those portions lying above the photosphere as the sun's atmosphere. This atmosphere, in turn, consists of several well-defined layers or regions, which will receive special attention in other sections of this chapter. These layers, in order outward, are as follows: the *photosphere*, which radiates the bright continuous background; the *reversing layer*, which produces the dark or absorption lines of the solar spectrum; the *chromosphere* or upper reversing layer, which emits a spectrum of emission lines at the onset of totality in a solar eclipse; *prominences*, luminous gas clouds of various sizes and shapes extending upward, like giant flames, from the solar chromosphere; and the *corona*, an outermost envelope of hot solar gases, best seen at total solar eclipse.

9-2   The Spectrum of the Sun.   The radiation of the photosphere gives a continuous spectrum. However, in the sun's atmosphere, the atoms and molecules impress on the continuous spectrum a series of dark absorption lines. The region of the atmosphere where this absorption takes place is called the *reversing layer*. It is somewhat cooler than the radiating surface,

FIGURE   *Photograph of the Solar Spectrum from 3900 Å to 6900 Å. (Courtesy Mount Wilson*
9-2   *Observatory.)*

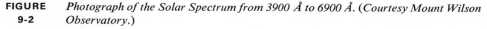

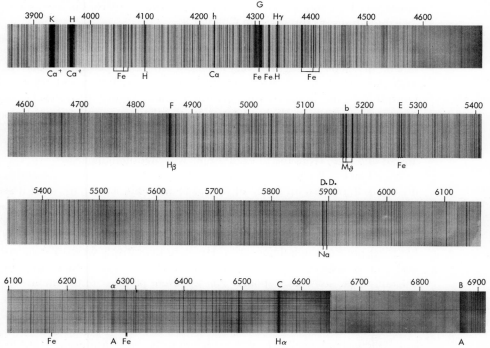

and determinations of its temperature indicate an approximate value of 4500° K. The most intense of the absorption lines are the "Fraunhofer lines," named after the brilliant German optician who, in 1814, first studied them in detail, though the Englishman, Wollaston, had noted their existence a decade earlier. Fraunhofer assigned letters of the alphabet to the strongest lines, from A in the red to K in the violet (see Fig. 9-2).

These absorption lines make possible the determination of the chemical composition of the sun's atmosphere. We can identify some of the elements, those with the strongest lines, by a simple matching of the solar spectrum against the laboratory spectra of known elements. For example, several hundred of the strongest lines of the solar spectrum are iron lines. Even though sodium does not show many lines in the solar spectrum, two outstanding ones in the yellow correspond exactly to the familiar sodium pair and hence confirm its occurrence. Other elements with strong solar lines include magnesium, aluminum, calcium, titanium, chromium, and nickel. As noted in Sec. 6-15, oxygen molecules in the earth's atmosphere produce the telluric bands A and B in the deep red.

For less abundant elements, or for those whose most intense lines fall in the spectral regions absorbed by our atmosphere (see Sec. 10-2), the identification becomes somewhat more difficult. Mere coincidence of a few spectral lines measured in the laboratory with some faint unidentified lines of the sun's atmosphere may not be conclusive. For a more definite identification we have to take into account some of the properties of atoms and their radiations, as discussed in Chapter 8.

We have already noted the effect that temperature can have upon the spectrum of an atom. Certain lines that would be very weak at a temperature of, say, 2000° K might be very intense when the temperature is 4500° K or higher. The effect depends on the degree of excitation and stage of ionization of the atoms. Low temperatures tend to favor the low-level lines of neutral substances. Higher temperatures enhance the intensity of lines from levels of intermediate or high energy. At still higher temperatures the atoms may be mostly ionized; lines of the neutral atom will be weak or absent and lines from the ionized atoms will predominate.

We infer that the intensity of a given line in the spectrum of the sun is not of itself an absolute index of that element's abundance. We must first make due allowance for effects of temperature and pressure. For example, the lines of singly ionized calcium (Fraunhofer's H and K) are the most intense solar lines in the visible spectral region. The red line of atomic hydrogen, one of the Balmer series, $H\alpha$, is comparatively weak. Yet calculation shows that hydrogen is probably 500,000 times more abundant than calcium in the sun's atmosphere. This apparent paradox in the intensities of the spectral lines arises from the fact that at the temperature of the sun's atmosphere only a very small percentage of the hydrogen atoms exist in the excited atomic levels necessary for the absorption of the hydrogen radiation, whereas practically all of the calcium in the solar atmosphere at this temperature contributes to the production of the H and K lines.

However, an analysis based on all of the available evidence about the atoms, including the effects of temperature excitation and ionization, established with some certainty the chemical elements responsible for the observed spectral lines of the solar atmosphere.

TABLE 9-1

Chemical elements in the sun

Classification	Element															No.
Present	H	He	Li	Be	C	N	O	Ne[a]	Na	Mg	Al	Si	P	S	Ar[b]	63
	K	Ca	Sc	Ti	V	Cr	Mn	Fe	Co	Ni	Cu	Zn	Ga	Ge	Rb	
	Sr	Y	Zr	Nb	Mo	Ru	Rh	Pd	Ag	Cd	In	Sn	Sb	Ba	La	
	Ce	Pr	Nd	Sm	Eu	Gd	Dy	Tm	Yb	Lu	Hf	W	Os	Ir	Pt	
	Au	Pb	Th													
Present?	Tb	Er														2
Further study needed	B[c]	As	Ho	Hg												4
Absent	F[d]	Cl	Se	Br	Kr	Tc	Te	I								16
	Xe	Cs	Ta	Re	Tl	Bi	Ra	U								
Not to be expected	Remaining radioactive elements															

[a] Evidence based on discovery of Ne VIII and Ne VII in the ultraviolet spectrum.

[b] Evidence based on discovery of [Ar X] in the coronal spectrum.

[c] BH has been reported as present, but further evidence is needed.

[d] No atomic lines present. MgF has been reported as present, but further evidence is needed.

SOURCE: Courtesy of Charlotte Moore Sitterly, National Bureau of Standards.

At the eclipse of 1868, J. Janssen of France and N. Lockyer of England had observed a brilliant yellow line in the spectrum of the solar chromosphere—a line that was not identified in the laboratory until 1895, when the British chemist, W. Ramsay, found it in the gaseous products of the radioactive decay of uranium. In the meanwhile, the substance responsible for this radiation was called *helium*, from the Greek "helios," the sun. Helium, we should note, emits this yellow line only when highly excited. Its mere presence in the chromosphere and prominences indicated that the temperature in these regions was probably 20,000° K, or higher.

A detailed study of the solar spectrum by H. D. Babcock and C. M. Sitterly has led to a definite identification of about two-thirds of the 92 known natural chemical elements, as listed in Table 9-1. Some elements have their strongest lines in a spectral region inaccessible from the earth's surface (see Sec. 6-16), so that their identification was long in question. However, ultraviolet spectra obtained by means of rockets and satellites removed some of the uncertainties and contributed new identifications.

Thus, looking down the periodic table (Table 8-1), we see clearly that the apparent absence of some of these elements is a selection effect. Some of the rarer elements towards the end of the list are indeed too scarce to contribute to the observed solar spectrum.

Table 9-1 confirms what chemists have known for years, that spectroscopic analysis is an extremely sensitive means of detecting the presence of minute impurities. A substance that is "chemically pure" may reveal, by its spectrum, the presence of many impurity elements.

R. Tousey of the Navel Research Laboratory and H. Hinteregger of the Air Force Cambridge Research Laboratories have employed rockets to extend our knowledge of the far-ultraviolet solar spectrum. Leo Goldberg of the Harvard College Observatory and his associates have developed even more sophisticated equipment carried in the Orbiting Solar Observatory (OSO). OSO-IV, launched into a nearly circular orbit in October, 1967, carried an instrument that functioned either as a spectrograph or spectroheliograph (see Sec. 9-7). It operated over the wavelength range from 300 Å to 130 Å.

9-3 **The Curve of Growth.** The term *curve of growth* denotes the relationship between the intensity of an absorption line (the amount of energy absorbed by it) and the total number of atoms acting to produce it. In other words, the curve of growth indicates how an absorption line will "grow" in intensity as the concentration of a given substance increases. We measure the amount of absorption in a line in terms of *equivalent width*, W, which tells how broad a completely black line would have to be in order to block out as much energy as the actual absorption line. Equivalent widths are therefore expressed in angstroms (or, more conveniently, in milliangstroms).

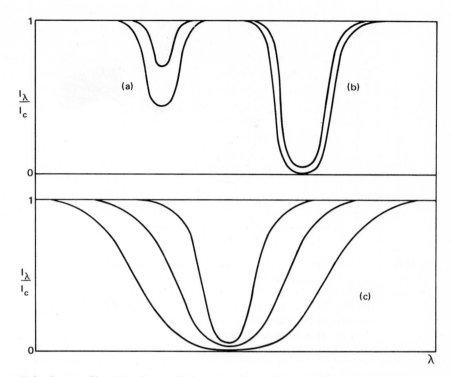

FIGURE
9-3
(a) (b) (c)

Solar line profiles. The shapes of absorption lines change with increasing intensities. Doubling the number of atoms acting for weak lines (a) approximately doubles the amount of absorption (area between the straight line 1, and the curve describing the intensity of the absorption line). For lines of intermediate intensity (b), doubling the theoretical intensity produces only a minor change in the total area. Very strong lines (c) grow mainly in the wings of the line, with similar increases in the intensity.

A line of equivalent width 1 Å is fairly intense, and one of equivalent width 0.01 Å is very weak. Figure 9-3 illustrates the forms of the absorption lines for different numbers of atoms and indicates the equivalent widths. The important point is that for weak lines, the intensity of an absorption line increases in direct proportion to the number of atoms. However, a line that is fairly black at the center has attained practical saturation, and thereafter grows at a very slow rate indeed. We have to add many more atoms before the curve of growth again takes an upward trend (Fig. 9-3e).

Of some significance is the fact that spectral lines have finite widths. In a sense, they behave like radio transmitters or receivers. In tuning a radio receiver, we all know that reception is best at some definite point on the dial, but that the transmission is still perceptible at nearby points. Atoms

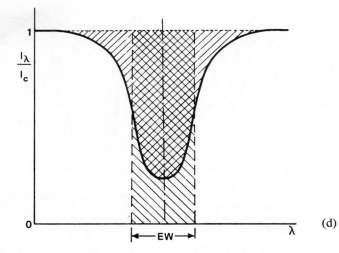

(d)

FIGURE 9-3 (d) (e) *Definition of equivalent width. (d) The rectangular-shaped line cuts out from the continuous spectrum as much energy as does the bell-shaped line profile. (e) The curve of growth plots, W/λ vs. the theoretical intensity of the atomic line. The letters (a), (b), (c), refer to weak, medium, and strong lines.*

(e)

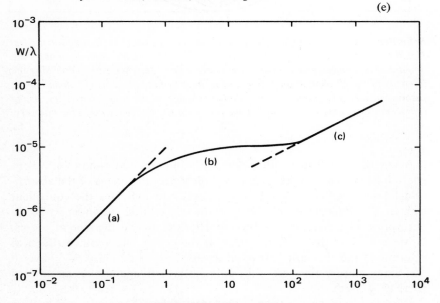

Adjusted theoretical intensity

respond in much the same way to light vibrations coming from the photo-sphere. Each absorbs most strongly on a certain particular frequency, and the ability to absorb the light energy falls off quickly on either side of this peak. Actually, each individual atom could produce sharper absorption lines than the observed widths suggest. The difference comes partly from the Doppler effect (Sec. 8-14), because the atoms are moving rapidly in all directions with respect to one another, on account of the high temperature that prevails in the atmosphere. Atoms moving toward the observer tend to absorb, as a consequence, in the high-frequency or "violet" wing of an absorption line; those moving away absorb in the low-frequency or "red" wing. Thus, the hotter the atmosphere, in general, the broader will be the absorption lines in its spectrum. Lighter atoms also move more rapidly than heavier ones at a given temperature. Thus we should also expect to observe wider lines for the lighter elements in the periodic table, and this expectation is fulfilled.

9-4   Sunspots.   Sunspots have been known from antiquity but were redis-covered through the telescope by Galileo (see Sec. 3-7). Not infrequently they are large enough to be seen with the naked eye when haze sufficiently dims the setting sun for comfortable examination. Sunspots appear as

FIGURE   *Active sunspot photographed in Hα light, partly covered by flare. (Courtesy*
9-4   *Sacramento Peak Observatory.)*

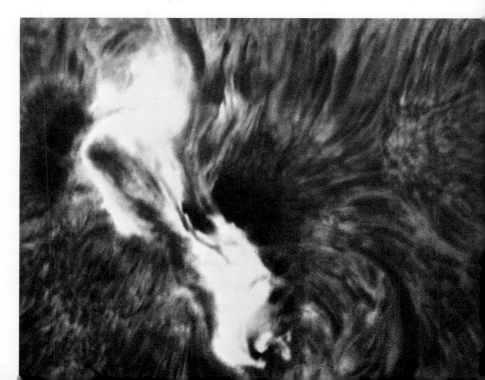

dark areas only because they are surrounded by the very luminous background of the photosphere. Actually, a large spot radiates much more light than the full moon. In all but the smallest spots, the center, called the *umbra*, is surrounded by a less dark ring called the *penumbra*. The spectrum of the dark umbra indicates a temperature of approximately 4000° K, about 2000° K less than that of the photosphere but exceeding that of the carbon electric arc.

A large sunspot is a fairly complicated structure, with multiple dark nuclei and many small pores surrounding them (Figs. 9-1 and 9-4). Spots form and disappear in the sun's atmosphere. Many last only a few days, and a very few endure for more than a month or two.

The regular daily progression of a sunspot across the solar surface proves that the sun rotates about an axis. A complete turn takes about 25.3 days at the equator and 27.5 days at solar latitude 40°. Hence the sun does not rotate like a solid body, If, therefore, a series of spots were lined up along a meridian as shown in Fig. 9-5(a), at the end of one revolution they would appear as shown in Fig. 9-5(b), because of the more rapid rotation of equatorial spots—the so-called *equatorial acceleration*.

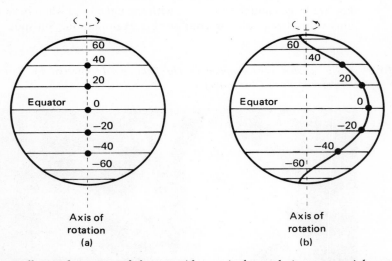

FIGURE 9-5 *Differential rotation of the sun. After a single revolution, equatorial spots gain appreciably over those at higher latitudes.*

The number of sunspots varies from day to day and even more significantly from one year to the next. The spots come and go in marked cycles, with pronounced maxima occurring at intervals of about eleven years. This is the famous *sunspot cycle*, discovered in 1843 by the German astronomer, H. S. Schwabe. This cycle is not completely regular because the

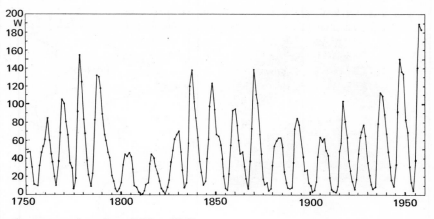

FIGURE
9-6

Sunspot numbers from 1750 to 1965.

observed intervals between successive maxima have been as short as 7.5 and as long as 16 years (Fig. 9-6).

No one knows exactly why the spots vary in this manner. In all probability some deep-seated internal circulation in the sun is responsible. Sunspots, however, are only one manifestation of the solar cycle. Other features that we shall presently discuss, such as the chromosphere, prominences, corona, and the output of ultraviolet radiation, also undergo change. In general, the greater the number of sunspots, the more disturbed the solar atmosphere will be. In the next chapter we shall discuss some of the effects of solar variability on the earth.

Sunspots themselves seem to occur in regions of highly turbulent action, with great jets or geysers of gas rising from their vicinity or, in many cases, forming in clouds and raining back to the surface. These clouds of gas are the *prominences*, which are conspicuous when seen at the edge of the sun; they will be discussed more fully in Sec. 9-9.

Sunspots never occur far from the sun's equator; most of them lie in the belt 35° north and 35° south. Spots that occur at the beginning of a new sunspot cycle, just after the minimum, generally are farther from the equator than those that appear later in the cycle. Figure 9-7 illustrates this variation, known as Spoerer's law. If we note the presence of a spot on a given day, we put a mark at the appropriate latitude. The diagram clearly indicates the tendency of the spots to form closer and closer to the equator as the cycle progresses. The appearance of such a plot has suggested for it the name *butterfly diagram*, first drawn by E. W. Maunder.

9-5 Magnetic Fields of Sunspots. In 1908 the American astronomer G. E. Hale noted that the spectral lines of the spots divided into several components, each of which was distinctively polarized (see Sec. 8-2). The

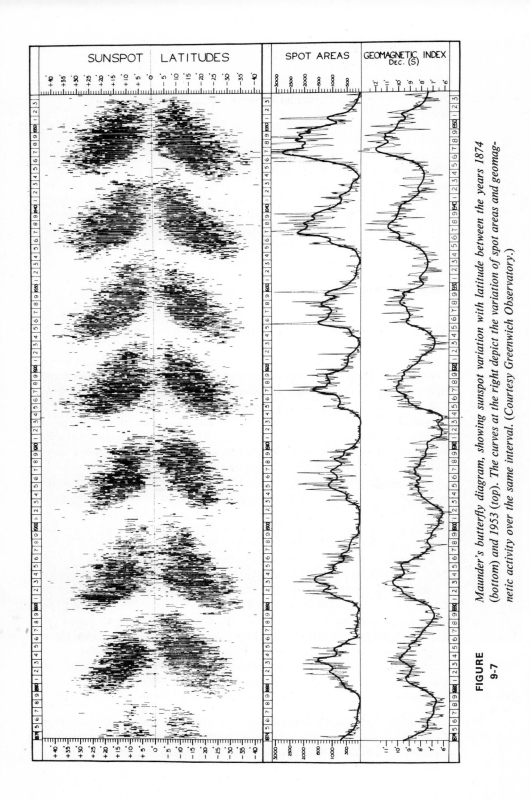

FIGURE 9-7

Maunder's butterfly diagram, showing sunspot variation with latitude between the years 1874 (bottom) and 1953 (top). The curves at the right depict the variation of spot areas and geomagnetic activity over the same interval. (Courtesy Greenwich Observatory.)

Dutch physicist P. Zeeman had proved in 1896 that this kind of splitting occurs when atoms emit light in an intense magnetic field. Thus Hale's observations clearly established that sunspots are powerful magnets. Since a hot gas cannot possibly form a permanent magnet like an iron bar, the existence of such magnetic fields proves that an intense electric current circulates in a ring around the sunspot. The current required to produce the field of a large spot may be as large as 10^{13} (ten million million) amperes.

Studies by T. G. Cowling in England, H. Alfvén in Sweden, and others have shown that such currents with their associated magnetic fields would be difficult indeed to initiate in the solar atmosphere. Hence the magnetic fields displayed by sunspots may be the results of electric currents that have existed from the time of the sun's origin. An alternative possibility is that the sun may be some kind of giant dynamo, whose rotation, internal circulation, and convection produce electric current. The magnetic fields within the larger spots are about as strong as those we can readily produce with powerful electromagnets in terrestrial laboratories—several thousand gauss. Smaller sunspots have weaker fields. The earth's magnetic field, by comparison, is only about one-half gauss (see Fig. 9-8).

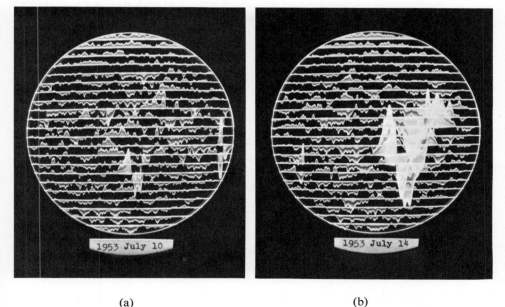

1953 July 10	1953 July 14
(a)	(b)

FIGURE 9-8 *Magnetograms of sun; upward deflections indicate north polarity, downward south polarity, along strips of measurement. (a) 1953 July 10, small activity. (b) 1953 July 14, strong sunspot fields. Note the bipolar fields, marked by upward and downward deflections. Note also reversal of polarities on opposite sides of the equator (central horizontal line). (Photographs by H. Babcock, Mt. Wilson Observatory.)*

Hale discovered another very important property of the magnetic fields of sunspots. Spots tend to occur in pairs—so-called bipolar groups—in terms of their magnetic properties. If the preceding (western) member of a pair is a magnetic north pole, the following (eastern) member will be a south pole. Moreover, if this relation holds in the sun's northern hemisphere, the reverse will be true in the southern hemisphere, where the preceding spot will be a south pole and the following spot a north pole. This type of association persists throughout one sunspot cycle. And then, at the beginning of the next cycle, the polarities reverse in both hemispheres. Thus the length of the true sunspot cycle is not eleven, but twenty-two years.

The presence of magnetic fields in sunspots raises the interesting question whether the sun itself possesses a general magnetic field. Early attempts to observe such a field gave somewhat inconclusive results. However, the behavior of prominences and the structure of the sun's corona so clearly suggested the presence of a magnetic field, that most astronomers believed that one must exist. But very sensitive methods were necessary for its detection.

Horace Babcock at Mount Wilson Observatory finally succeeded in detecting the existence of a weak general field with a strength of perhaps 5 or 10 gauss. The field is complicated and variable. Babcock's study proved, to everyone's surprise, that the sun's general field actually reverses —though not precisely synchronously with the spot cycle. For a short time the sun's two geographic poles had magnetic polarities of the same sign. The sun, therefore, has properties akin to those of magnetic variable stars (see Chapter 26).

9-6 Physical Basis of Sunspots. The physical significance of sunspots has been a controversial topic ever since Galileo. And Galileo, in suggesting that spots were clouds in the solar atmosphere, was certainly more correct than certain of his followers. Sir William Herschel, for example, thought that sunspots were holes in the fiery clouds through which one could see the cool, dark, solid, and presumably habitable surface beneath.

The most popular theory, still advocated in many textbooks, regards the spots as great, vortical storms, like hurricanes or tornadoes. Such storms derive their rotation from the large-scale, horizontal flow of gas into a region of low pressure.

Sunspots, however, appear not to be vortices. Most of the distinctive features of spots result from the presence of powerful magnetic fields and associated electric currents. We can understand this in terms of an established principle of nature: that matter tends towards a position of equilibrium. Rain falls to the ground and courses its way to the sea. Winds blow toward regions of low pressure, as if to fill the void and make the pressure uniform. Heat flows from a region of high to one of lower temperature, tending to equalize the temperature throughout the region.

Application of this principle to a sunspot shows that a static condition can occur only if both the temperature and pressure are low near the spot center, where the magnetic field is strongest. The detailed analysis is somewhat complex, and will not be given here. However, it shows that a true static condition cannot persist, because heat will leak from the hot exterior into the cool interior of the spot. As a result, gas will flow upward and outward, along the magnetic lines of force. Expansion of the gas maintains the low temperature. The sunspot behaves as if it were a pump. In effect the spot is an enormous refrigerator, maintaining vast regions at temperatures 2000° or so lower than their surroundings (Fig. 9-9).

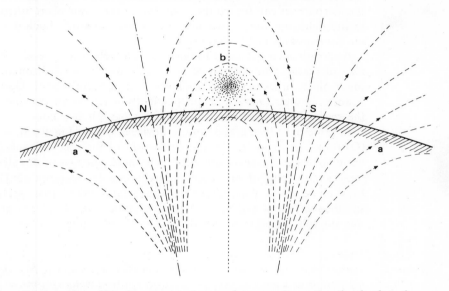

FIGURE   *Cross-section of bipolar sunspot. Arrows show pumping action that leads to forma-*
9-9   *tion of flare filament (b).*

The spot itself, apart from the gentle outward flow and occasional violent flares, is a quiet region compared with the simmering convection of the photosphere. The magnetic fields also tend to restrict or inhibit convection in the spot regions. One must realize, however, that a spot, being dark relative to the surrounding regions, fails to carry its share of the radiation that flows continuously from the solar interior. It acts as a sort of dam. But the energy must escape somehow. In all probability, the diverted energy is in part responsible for the brighter areas around sunspots, the *faculae*, *flocculi*, and *plages* discussed in Sec. 9-8.

9-7 Light Filters and the Spectroheliograph. The use of some special light filters has contributed greatly to our understanding of the solar atmosphere. The photographer who wishes to record terrestrial cloud formations usually employs a red filter. The red glass, by cutting out the blue of the sky, increases the contrast of the white clouds. Most photographic filters, made of glass or of stained gelatin films mounted in glass, transmit rather broad spectral regions—several hundred angstroms or more.

More selective color filters are now available. Some will transmit a spectral band as narrow as a tenth of an angstrom. These narrowband filters can be constructed to transmit some selected spectral line, such as the red Balmer line of hydrogen or the violet K line of calcium. In this way, the astronomer can record the solar disc or the atmosphere surrounding it in the light that these particular atoms are radiating. Such photographs are sometimes called "filtergrams."

Although such specialized filters are relatively recent developments, the idea of taking photographs of the sun in nearly monochromatic light is much older. In 1890, G. E. Hale in the United States and H. Deslandres in France simultaneously and independently invented an instrument called the *spectroheliograph*, which uses a spectrograph as a filter.

In principle, the operation of this device is simple (Fig. 9-10). A telescope (O) forms an image of the sun on the slit of the spectrograph (F_1). The various lines in the spectrum, which are merely images of this slit, are also images of the portion of the sun falling in this slit. A second slit (F_2) at the focal plane of the spectrograph camera isolates any desired spectral line— say the red line of hydrogen. A photographic plate (P) placed just behind this slit will therefore record the image of the segment of the sun in that

FIGURE 9-10   *Optical diagram of spectroheliograph, showing mirrors M_1, M_2 of heliostat, objective 0 of telescope, and the spectrograph (collimator, prisms, and camera) between the entrance and exit slits described in text.*

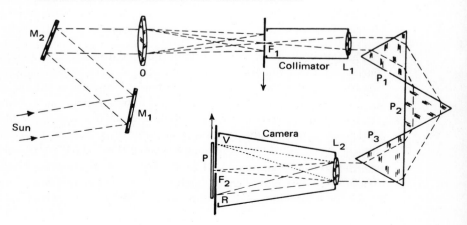

particular light including all the variations of light intensity that may occur along the slit. If, now, we move the image of the sun across slit F_1 and at the same time move the photographic plate perpendicularly to slit F_2, we shall draw out a continuous image of the sun in the light of the isolated line, much as a television tube builds up a complete image from a series of overlapping lines (see Fig. 9-10).

The American astronomer R. R. McMath modernized the spectroheliograph in the 1930's by adding motion-picture recording. His films, made at the McMath-Hulbert Observatory in Michigan, were the first to present in animated form the motions of luminous clouds in the sun's atmosphere. Prior to such records, astronomers had studied prominence motions from measures of "still" pictures taken at intervals of about fifteen minutes. Nowadays movies of the sun's atmosphere in selected spectral lines are made around the clock at observatories in a worldwide network, with Lyot-type filters (see Sec. 9-9).

9-8   The Study of Spectroheliograms.   The photographic records made with a spectroheliograph or a narrowband filter reveal the remarkable structure of the solar surface and the sun's atmosphere. Analysis of such pictures gives the distribution and general excitation of the atoms responsible for the spectral line whose light falls upon the film or plate.

The sun's atmosphere is more transparent outside of the spectral lines than within the lines. Indeed, this *line opacity* in the atmosphere is the cause of the sun's Fraunhofer spectrum. Thus, with a monochromatic filter selecting light from the center of an absorption line, we observe a much higher atmospheric level than on a direct photograph in full light without a filter.

Spectroheliograms taken in the light of the K line of ionized calcium (Fig. 9-11a) display bright patches of calcium emission, especially near sunspots. These are the *calcium plages* (French for "region," pronounced plazh), or *flocculi*, roughly similar to the faint patterns visible in full light, which bear the name of *faculae*. The bright plages are sensitive indicators of disturbed conditions in the sun's atmosphere. They appear before the outbreak of a group of sunspots and may persist for some days after the group itself has vanished. They vary with latitude in the same way as sunspots, progressing toward the equator in the latter part of the sunspot cycle. They seem to be regions where the sun's atmosphere is thicker than elsewhere and possibly somewhat hotter.

Filtergrams taken in hydrogen light (Fig. 9-11b) display mottlings that are much less coarse than those of ionized calcium. Both bright and dark patches occur, but their appearance differs from those on the calcium records; bright flocculi are much less conspicuous near the spot areas.

In both calcium and hydrogen pictures, long fibrous *filaments* sometimes stretch for several hundred thousand miles across the solar disk (Fig. 9-11).

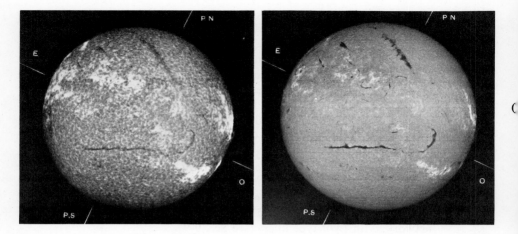

FIGURE 9-11 *Spectroheliograms made on March 25, 1949. (a) In the K line of Calcium. (b) In the Hα line of Hydrogen. (Courtesy Paris Observatory at Meudon.)*

Those at higher latitudes tend to lie roughly parallel to the solar equator. The leading (western) edges of filaments in low latitudes often curve sharply toward the equator and vanish in the neighborhood of a sunspot. Such a filament often resembles a long, tall hedge. These solar hedgerows may be some 50,000 miles in height, 10,000 miles wide, and from 100,000 to 500,000 miles in length along the sun's surface. Though they appear dark by contrast with the surface, as *prominences* beyond the sun's edge, they stand out as luminous clouds against the darker sky. Records show that these dark filaments are more prevalent near sunspot maximum than at minimum.

The Harvard Orbiting Solar Observatory experiment on OSO-IV (see Sec. 9-2) returned to earth encoded pictures of the intensity distribution of various preselected ultraviolet wavelengths over the solar surface. This shortwave radiation emanates from the most highly excited regions of the solar atmosphere. Figure 9-12 shows three such pictures, taken respectively in the light of Lyman continuous spectrum of atomic hydrogen, oxygen VI, and magnesium X, in order of increasing excitation. The records are "negatives" in the photographic sense, in that the areas of most intense emission are darkest. Presumably the lines of higher excitation come from the hotter, more elevated regions of the corona. Hence a complete series of this type, ranging up to iron XVI, with 15 electrons missing, provides a three-dimensional view of the solar atmosphere. The fourth picture is an H-alpha photograph from Sacramento Peak Observatory. Note the close correspondence between the bright areas in all four records. The active areas associated with the northern and southern sunspot zones are well shown.

**FIGURE
9-12**

(a) (b) (c)

Spectroheliograms in the far ultra-violet obtained on October 27, 1967 by Harvard–NASA Orbiting Solar Observatory in light of (a) the Lyman continuum of hydrogen at λ = 897 Å,

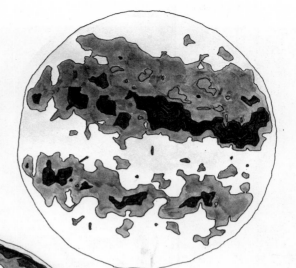

(a)

(b)

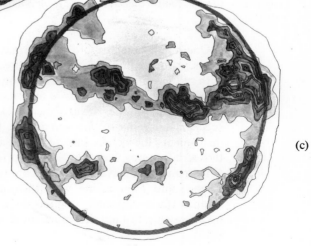

(b) the emission line of 5-times ionized oxygen at λ = 1032 Å,

(c) the emission line of 9-times ionized magnesium at λ = 625 Å. Note increasing extension into the corona for the latter records.

(c)

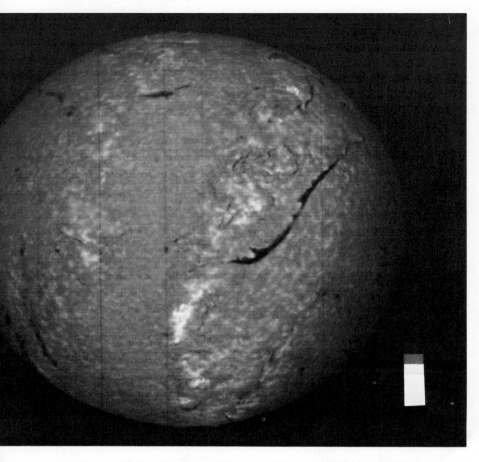

FIGURE 9-12(d) (*d*) *For comparison, Hα record from Sacramento Peak Observatory.*

One major objective of the experiment is the detailed mapping of solar flares, with the hope of improved forecasts of events that could interfere with radio communications on earth or, by their harmful radiation spewed into space, prove dangerous to astronauts in orbit.

R. Tousey of the Naval Research Laboratory has obtained excellent spectroheliograms of the sun in the light of Lyman-α radiation, from instruments in rockets sent to levels above the absorbing layers of the terrestrial atmosphere. In still shorter, even X-ray wavelengths, H. Friedman of the Naval Research Laboratory and L. Giacconi of American Science and Engineering with their associates have photographed the solar surface—especially regions of high excitation. Spectroheliograms taken in this fashion resemble those obtained in the light of Hα (Fig. 9-13).

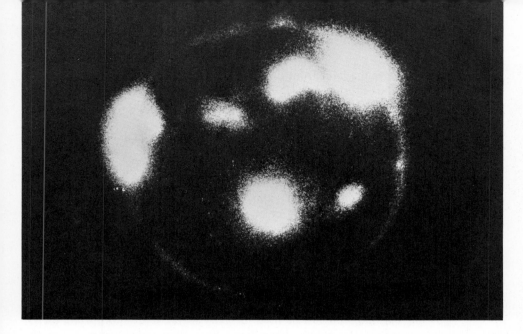

**FIGURE
9-13**
X-ray photograph of sun taken May 20, 1966 with 16-sec exposure from Aerobee rocket through filter transmitting the spectrum intervals 60–44 Å and 16–3 Å. (Courtesy National Aeronautics and Space Administration.)

9-9 The Coronagraph. The outermost extension of the solar atmosphere, known as the corona, is almost a million times fainter than the disk of the sun. For centuries it could be observed only during total solar eclipse. In 1931 the French astronomer B. Lyot invented an ingenious device, the *coronagraph*, for recording the faint light of the corona at any time.

The coronagraph consists essentially of two telescopes in tandem, one behind the other (Fig. 9-14). The first telescope (L_1) forms an image of the sun, which a metal disk (Dk) placed at the focus artificially eclipses. The second lens (L_2) images the first lens (L_1) on the third lens (L_3), which in

**FIGURE
9-14**
The optical system of the coronagraph. The image of the sun's photosphere is covered by the disk, Dk, and the surrounding corona is imaged by L_2, L_3 on the photographic plate P. The diaphragm D_1 and disk D_2 stop light internally reflected in the system.

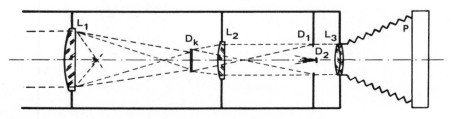

turn images the light from the sun's atmosphere that extends beyond the edge of the eclipsing disk (Dk) upon a photographic film or plate (P). The diaphragms (D_1 and D_2) serve to remove scattered light and spurious faint images from the system.

The unusual feature of the coronagraph is the first objective, which consists of a single lens, so that it is not achromatic. However, the use of a monochromatic filter avoids the blurring of the image that chromatic aberration would otherwise introduce. Also, the lens must have an unusually perfect optical polish, and a complete absence of streaks, scratches, or bubbles that would scatter or reflect light. It must be kept absolutely clean and free from dust. A proper arrangement of baffles, lenses, and screens minimizes the effects of the scattering of light within the instrument. Finally, for optimum operation, the coronagraph must be installed on a high mountain, where the atmosphere is pure and transparent (see Fig. 9-15). The haze that generally occurs at low altitudes scatters sunlight and produces a brilliant white glare, which swamps the faint solar corona.

The coronagraph can be used in many ways. With appropriate filters, which isolate the colors of emission lines that are conspicuous in the spectrum of the corona, it can be used to make photographs or motion pictures of the prominences or of the corona. It can also be employed in conjunction with a spectrograph, and will then give information about the spectra of the sun's atmosphere and the corona. With the occulting disk removed, the coronagraph can photograph the solar surface directly, a

FIGURE 9-15(a) *Large solar instruments at the Sacramento Peak Observatory.* (*a*) *Coronagraph and horizontal telescope.* (*Courtesy Air Force Cambridge Research Laboratories.*)

**FIGURE
9-15(b)**   (*b*) *Solar tower vacuum telescope.* (*Courtesy Air Force Cambridge Research Laboratories.*)

narrowband filter being adjusted to some selected wavelength. The resulting filtergram is equivalent to a spectroheliogram. A special photoelectric attachment can sense the faint polarized coronal light and isolate it from the sky glare. This very sensitive device is called the *coronascope.*

9-10   Prominences.   Prominences are clouds of hot, luminous gas, best seen when they project beyond the edge of the sun's disk (Figs. 9-16, 9-17). Large prominences can appear on spectroheliograms as bright or dark markings. They are structures in the solar atmosphere, elevated above the surface of the photosphere. Since the spectra of prominences show bright emission lines of hydrogen and helium, atoms relatively difficult to excite, requiring temperatures of about 10,000° K and 20,000° K, respectively, for emission, we infer that some of these clouds are much hotter than the solar surface. Nevertheless, they do absorb some energy, so that the long, fibrous filaments appear dark on hydrogen filtergrams.

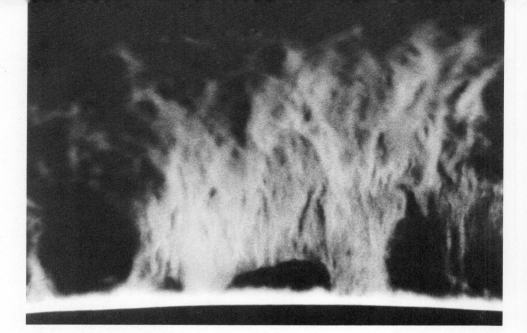

FIGURE 9-16 *Hedgerow prominence, which will show in absorption against disk as a curved, dark filament. See Fig. 9–11. (Courtesy Sacramento Peak Observatory, Air Force Cambridge Research Laboratories.)*

FIGURE 9-17 *Loop prominences, near active sunspot. Note filamentary structure. (Courtesy Sacramento Peak Observatory, Air Force Cambridge Research Laboratories.)*

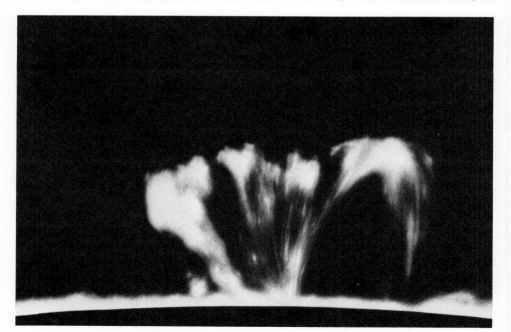

FIGURE
9-18
Nine Frames of Corona in λ = 5303 Å line of Fe XIV over a period of several hours showing changes near active sunspot. (Courtesy Sacramento Peak Observatory, Air Force Cambridge Research Laboratories.)

Motion pictures of prominences show that they are in continuous and often rapid movement (Fig. 9-18). Even though a prominence may appear to keep the same form for hours or even days, only the outline is permanent for material usually flows through it at a fairly rapid rate. Even the prominences that look like vast clouds of condensed steam are neither continuous nor uniform. Under good observing conditions, a prominence consists of a tangled mass of luminous threads, which form on one side, move through the prominence, and disappear at the opposite edge. The threads themselves display knots or condensations, which drift along the length of the threads.

The shapes of prominences led the earlier observers to suppose that most of them, if not all, were some sort of "exhalation" from the sun's surface, rather like displays of fireworks. But the motion-picture records have revealed the astonishing fact that in most prominences the material is flowing downward, like gigantic rainstorms of hot, luminous gas. The flow, however, is rarely vertical, or in straight lines. Usually it sweeps in long, graceful arcs that may even curve upward before the material finally swings in towards the solar surface. Many prominences have fairly sharp upper boundaries, where the falling material first seems to become luminous. The glowing gas then descends slowly through the structure, forming prominences of various types, including the dark "hedgerows" seen in projection against the solar surface (Sec. 9-8). The downward-flowing material appears to be "condensing" out of the solar corona, the sun's outer envelope of tenuous gas.

Prominences show a consistent tendency to move upward over only two types of area on the solar surface. Close to sunspots we often observe vast surges or geysers that burst forth with explosive violence, forming columns of gas from 5000 to 10,000 miles across, and not infrequently rising higher than 200,000 miles above the surface. These surges are among the most spectacular prominences. However, they do not seem numerous enough to

FIGURE 9-19 *The spicule-structure of the solar chromosphere. (Courtesy Sacramento Peak Observatory, Air Force Cambridge Research Laboratories.)*

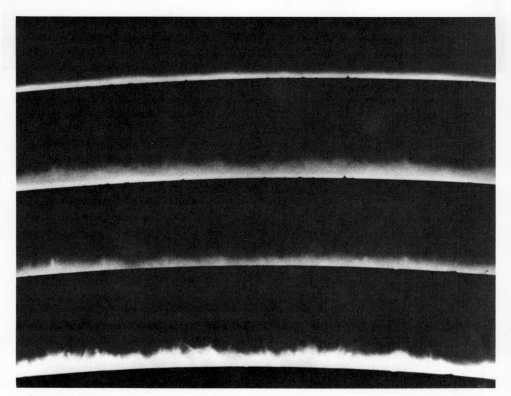

account for all the down-coming material, unless indeed there are some surges so hot and so tenuous as to be completely invisible.

Near the poles of the sun, too, we find a predominance of upward motion. In the polar areas W. O. Roberts observed with the coronagraph of the High-Altitude Observatory at Climax, Colorado thousands of small surges, known as *spicules*, whose motion seems to be directed upward (Fig. 9-19). The complete lifetime of a spicule is generally between one and five minutes. The gas spurts out in a threadlike jet, and may vanish while still moving, perhaps to form a filament in the sun's polar coronal streamers. Although a connection between the spicules, the corona, and the descending material in the larger prominences has never been fully demonstrated, such a relationship seems likely (Fig. 9-20).

The downward motions of the gases in the prominences are generally

FIGURE 9-20   *Very high dispersion spectrum of lines near* $\lambda = 5188$ *Å on the solar disk showing velocity displacements over small distances. (Courtesy Sacramento Peak Observatory, Air Force Cambridge Research Laboratories.)*

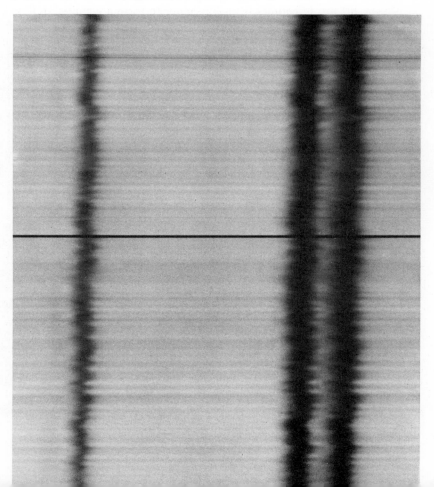

far slower than one would expect from the size of the gravitational acceleration at the sun's surface. Some other force must be opposing gravitation and supporting the prominence. Possibly the sun's corona may serve as a resisting medium. But the most recent interpretations of the forms and motions of prominences indicate that magnetic fields are mainly responsible, the matter flowing parallel to the lines of force.

Near sunspots the flow of material through a prominence is often spectacular, the structure changing rapidly. Some prominences are quite unstable, as though the interplaying forces are not in perfect equilibrium. Oscillations often occur. Occasionally, formations that have appeared stable for some time may explode violently.

The dark filaments, hedgerows seen in silhouette, are of particular interest, since they are among the most stable of all solar features. Sometimes they persist for several months. Their forms remain roughly constant while they undergo thousands of complete changes of material. Then, sometimes within a few hours, a rapid flow of matter begins, more and more of the structure flowing rapidly downward. Finally the whole filament may lift up, sometimes from the center to form a gigantic arch, at other times from one end, like a scarf blowing in the breeze. This phase in the life of certain prominences is known as *eruptive* or *ascending*. The latter term is preferable, for these ascending arches are not nearly as eruptive or explosive as the surges.

Ascending prominences may rise to heights greater than a solar diameter, usually fading as they rise. Most of their material appears to fall back upon the sun's surface, but some escapes into outer space and may later reach the earth.

Prominence activity usually increases with the sunspot cycle, though we often find small prominences at times when sunspots are completely absent.

9-11

Solar Flares. Solar flares are bright flashes of light in the sun's atmosphere, which may last a few minutes or may show remarkable changes even within seconds (Figs. 9-4, 9-21). We generally observe these flashes in the light either of calcium or hydrogen, but extremely bright ones are visible in white light without a filter. The exceptionally intense outbursts or radiation that constitute flares are usually found near sunspots—especially active, growing spots.

A flare usually involves only a small area of the solar disk, which often lies between the two components of a bipolar sunspot, where intensely luminous clouds of gas frequently appear. Motion pictures of such areas show recurring bright flashes among the darker streaks that appear as absorptions. Several intense flares, which seemed to be particularly violent surges, have been recorded near the edge of the sun. Hence one may

FIGURE 9-21 *Solar flare in hydrogen H_α Light, lower right. (Courtesy Sacramento Peak Observatory, Air Force Cambridge Research Laboratories.)*

suppose that flares are some sort of vigorous prominence activity, usually at a low level in the solar atmosphere. Astronomers divide flares into three classes—1, 2, and 3, in order of increasing violence and intensity of the outburst.

In Sec. 9-6 we discussed briefly the probable flow of gases in the neighborhood of a spot. The flow becomes extremely complex in the neighborhood of a bipolar spot, as indicated in Fig. 9-9. The diagram is a schematic cross section of the pair, with a north component on the left and a south component on the right. The lines represent the magnetic fields accompanying the pair. The arrows indicate the direction of gas flow, resulting from the pumping action.

The regions to the far left and far right, indicated as *a* on the chart, give no problem. But the region labeled *b*, where the lines of force of the one spot join with those of the other, does present a problem. As the arrows indicate, both spots are trying to pump simultaneously into the same region, between the spots. Gas pressures and temperatures steadily increase in region *b*, until finally the magnetic field of the spot, no longer able to retain the hot, expanding gas, suddenly tears under the strain. A brilliant filament of gas momentarily appears between the spots. An explosion or flare occurs. A violent shock wave races over the solar surface at a speed of 2000 km/sec or more. After a minute or two the flare usually subsides, though it may repeat the action many times, at unpredictable intervals. This model of a sunspot and associated flares is by no means definitely proved. Some astronomers have attributed flares to a short-circuit of the electric currents and transformation of their energy into heat.

The effects of solar flares on the earth are of special significance. Just at the moment when a bright flare occurs, we note a disturbance in the terrestrial magnetic records. There is a small wobble of the compass needle, accompanied by a fadeout of shortwave radio signals on the sunlit face of the earth. Apparently an intense blast of X-ray radiation accompanying the flare is largely responsible for the sudden disturbance observed in the earth's atmosphere. The flare also seems to eject clouds of electrons, ions, and atoms through the corona into space. These clouds reach the earth a day or two after the event.

The earth's magnetic field, however, acts as a sort of bumper, deflecting most of the atomic cloud of ions and electrons, which in turn compresses the field in front and streams behind the moving earth like a wake. Occasionally the bumper may give way, allowing the charged matter to become entangled with the field, to form the Van Allen belts (see Chapter 10). Jets of the material may escape from the inner edges of these belts. These jets, streaming into the earth's upper atmosphere, especially in the polar regions, will excite the atoms and molecules, causing them to shine. Thus, solar activity directly causes the aurora borealis and aurora australis (see Sec. 10-5). The active, flaring sun also emits bursts of radio waves— like static, to be discussed in Sec. 9-14.

9-12   The Chromosphere.   The chromosphere is a layer of the sun's atmosphere that lies just above the reversing layer. The temperature of the reversing layer is not far from 4500° K. At an altitude of about 1500 kilometers above the sun's surface, the temperature begins to rise until it reaches a value of 10,000° or more. We infer this temperature, as we did for the prominences, from the presence of emission lines of helium, which are strong in the upper chromosphere but apparently quite weak in its lower levels. We even find lines of ionized helium. Relatively simple calculations show that these gases cannot possibly radiate where the temperature is as low as 6000° K; indeed, if there is appreciable emission from ionized helium, the region must be at least as hot as 20,000° K.

The sun's chromosphere is by no means a uniform atmospheric layer. Like the prominences, it has a discontinuous structure. It consists of many fine, luminous threads, including spicules, which appear side by side like blades of grass, and rapidly change in intensity. In many instances these threads consist of material falling downward through the sun's atmosphere. In others, we detect surges that rise and fall.

Some astronomers have suggested that this material moving downward in the sun's atmosphere may have been swept up and captured by the sun as it moves through interstellar space, but this hypothesis is wrong. It cannot account for the enormous amount of material actually observed to fall to the sun's surface in the course of a day. Also, on this theory, the influx should be sensibly constant, independent of the sunspot cycle. Finally, observation clearly shows that material is moving away from, not into the sun, in the form of the solar wind.

The height of the chromosphere is not constant. It varies with the sunspot cycle from about 12,000 to 30,000 kilometers. Even though its upper boundary is not well defined, the chromosphere is not uniformly thick in all solar latitudes.

During a total eclipse of the sun, when the moon has just covered the photosphere, the chromosphere extends beyond the moon's limb and shows as a thin crescent. The hydrogen-alpha light, which is fairly intense, lends a faint pinkish tinge to the radiation, which accounts for the origin of the term chromosphere, or sphere of color. Photographs of the chromosphere at total eclipses are our best source of information concerning the physical constitution of this portion of the sun's atmosphere.

Figure 9-19 shows the structure of the solar chromosphere. Figure 9-22 represents a portion of the flash spectrum, as astronomers refer to the spectrum of the chromosphere, seen just at the beginning or end of totality in a solar eclipse. Each crescent indicates radiation of various chemical substances. Most of the spectral lines seen in the Fraunhofer

**FIGURE
9-22**   *Flash spectrum of solar chromosphere during eclipse of August 31, 1932, by D. H. Menzel, Lick Observatory. The strong emission line at the left is Hβ.*

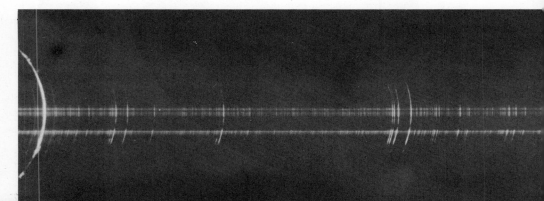

spectrum also appear in the flash spectrum, with the significant difference that lines from ionized atoms are enhanced in the chromosphere. This increased ionization is the result of lower pressure (Sec. 8-13) and higher temperature. Shock waves passing through this region of the solar atmosphere are probably responsible for the greater temperature.

FIGURE 9-23 (*a*) *Corona near sunspot minimum, Sept. 21, 1922. (Courtesy Lick Observatory.)* (*b*) *Corona near sunspot maximum, June 19, 1936. (Photograph by Irvine C. Gardner, © National Geographic Society.)*

(a)

(b)

9-13   The Corona. The sun's corona is the faint halo that comprises the outermost regions of the sun's atmosphere (Fig. 9-23) Its total brightness has been measured as about equal to that of the full moon, or 500,000 times fainter than the sun itself. Thus one can readily appreciate why this faint extension of the sun could be studied only at those rare moments during a total solar eclipse when the moon completely hides the solar photosphere. The invention of the coronagraph (Sec. 9-19) first made possible the daily study of the corona.

A general glow pervades the corona, whose color differs little from that of sunlight. The absorption spectrum is apparently "washed out" in the inner corner by the rapid motions of the atoms or particles that reflect the sunlight, because of the Doppler effect (Sec. 8-14 and 9-3).

The Dutch astronomer H. Van de Hulst has suggested that the shallow lines in the spectrum of the outer corona are scattered by small dust grains in the space between the earth and the sun. As we shall see in Chapter 15, dust particles in space, whose diameter is about equal to the wavelength of light, do scatter that light very effectively. On this hypothesis, some of the light recorded when we photograph the solar corona has been scattered by small particles within the orbit of the earth—the same particles that produce the zodiacal light (see Secs. 6-16 and 15-9).

The inner corona has a fairly intense continuous spectrum, from which absorption lines are absent, Free electrons in the highly ionized inner corona scatter sunlight, and are responsible for the strong continuous spectrum. The electrons have extremely high velocities in this very hot inner corona, and the Doppler effect has blurred out even the most intense lines.

A certain amount of structure is present in the corona, appearing like fine rays and streams. The form of the corona delineated by these outer rays varies from day to day. It also varies with the sunspot cycle. At sunspot minimum the corona exhibits fanlike rays or streamers, which extend to great distances above the solar equator but are almost nonexistent at the poles, so that the eclipsed sun looks like a black disc with extended wings (Figs. 9-23, 9-24). At sunspot maximum, the corona is more uniform. There is much less distinction between the areas over the sun's poles and equator, and the eclipsed sun has been compared to a big dahlia.

The corona tends to show peculiar arch or "helmet" formations over some of the larger prominences. Occasionally the coronal pattern is distorted. The rays may bend or sag, especially near a great hedgerow prominence. Such phenomena are characteristic of sunspot maxima. These appearances contrast with those at sunspot minima, when the coronal pattern suggests the pattern of lines of force in a magnetic field. The correspondence can hardly be a mere coincidence. Magnetic fields must govern the shape of the corona.

Although most of the light of the sun's corona comes from its continuous spectrum, a few well-defined emission lines appear—radiation that greatly

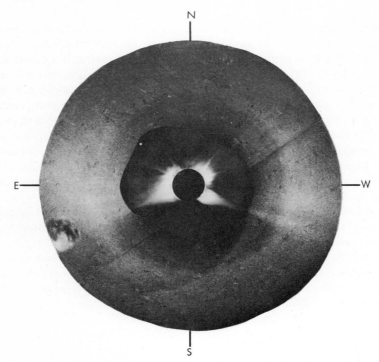

FIGURE 9-24 *Photograph of the outer corona from a rocket. Note the marked radial streamers. This picture was taken on November 12, 1966, shortly after a total eclipse of the sun. The "new moon," its features visible in earth shine, clearly occults the corona. The center of the picture is a reduction to scale of the coronal photograph shown in the frontispiece of this book. (Photograph by R. Tousey, U.S. Naval Research Laboratory.)*

puzzled early astrophysicists. In much the same spirit in which astonomers had labeled the element helium, responsible for the mysterious yellow line (see Sec. 9-2), they dubbed the hypothetical substance responsible for some bright lines in the corona, coronium. Advances in chemistry and physics gradually filled in the gaps of the periodic table (see Sec. 8-10), so that no unknown elements could be called on to account for the coronal radiation.

The Swedish physicist B. Edlén finally, in 1941, resolved the mystery of coronium. He followed up a suggestion of the German spectroscopist W. Grotrian that an intense red coronal line at 6374 Å might perhaps come from the atoms of iron that had lost nine of their normal complement of twenty-six electrons. Edlén continued this analysis, and showed that the most prominent (green) coronal line at 5303 Å was due to iron atoms that had lost thirteen electrons. We designate these ionized atoms respectively as Fe X and Fe XIV. Other lines were also attributed to Fe XI and Fe XIII.

Nickel and calcium, in particular, account for the other important coronal lines. A few radiations in the sun's corona are still unidentified.

Though Edlén's identifications are undoubtedly correct, they introduce many problems. Iron will lose thirteen electrons only under extreme excitation. Calculations show that such high ionization requires a temperature of at least 1,000,000° K. But high as this value may seem, there is other evidence that the outer regions of the sun's atmosphere are extremely hot. Several investigators have measured the widths of coronal lines and have found that they are consistent with such temperatures, occasionally with even higher temperatures, up to 2,000,000°. We believe that these enormous temperatures result from the shock waves developing from solar flares and other active regions well below the chromosphere.

The solar corona is an important part of the sun's atmosphere, for it is the medium that links the active regions of the sun with interplanetary space. The ionized gases, entrapped in the magnetic field of the sun, form a region analogous to the earth's Van Allen belt (see Sec. 9-11 and Chapter 10). The corona appears to be a dynamic rather than a static structure, associated with the solar wind, driven outward by shock waves from lower, active levels, and focussed by magnetic fields.

9-14   Solar Radio Noise.   The detection of solar radio noise is a military contribution to astronomy, which came during World War II. As early as 1900, Sir Oliver Lodge of England had conjectured that such noise might exist and attempted to detect it. The effort failed, largely because the available radio equipment was not sensitive enough, especially at the shorter wavelengths that could penetrate the earth's ionosphere (see Sec. 33-1). Marconi, in 1916, suggested that some of the static disturbances might be of solar or cosmic origin.

Shortly before World War II, a few radio amateurs had surmised that a peculiar form of static occasionally picked up by their receivers might originate in the sun. The first definite record of solar radio emission was made during World War II by British military radar teams (see Sec. 33-1). After the war, when astronomers turned sensitive radar receivers, developed for military purposes, toward the sun, they detected strong radio emissions. Moreover these emissions varied in intensity.

By using Planck's law, we can calculate the amount of solar emission to be expected at radio wavelengths. If the sun radiates at a temperature of 6000° K, the calculated amount is not high, especially in the longer microwave region of astronomical interest. But the observed radiation could be much more intense if the source were in the corona with its million-degree temperature.

Our studies of solar-radio energy are to some extent limited by the fact that the earth's atmosphere is opaque to the longer radio waves. In fact, it is the reflection of such waves from the ionosphere (Chapter 10) that makes long-distance radio transmission possible. The same phenomenon that keeps our own radio signals in also keeps out the radio emissions from the sun. Hence, until we go to wavelengths shorter than 10 or even 5 meters, we cannot expect to receive much of the solar radiation at the surface of the earth. If the sun radiated energy in accordance with a *single* temperature anywhere between 5000° K and 5,000,000° K, the distribution of energy in the radio portion of the spectrum should follow one of the curves shown in Fig. 9-25. Actually, the radiation does not conform at all to any one of these curves, even when the sun is in its least disturbed condition. The intensity may vary considerably from day to day, and even from minute to minute during periods of greatest disturbance. The curve marked *qq* corresponds to the *quiet sun*. It shows how the apparent temperature varies with wavelength.

If we attempted to interpret the solar radio emission in terms of temperature, we should deduce a different temperature for each wavelength. There is a relatively simple explanation for this effect. Our studies of the corona and the chromosphere allow us to determine, at least roughly, the distribution of material and also the gradation of temperature. We have seen that the temperature of the corona increases with height. Compared with the sun itself, the corona sends us practically no visible radiation,

FIGURE 9-25   *Theoretical emission for various temperatures as a function of wavelength. The emission q–q of the quiet sun indicates a different temperature for each wavelength.*

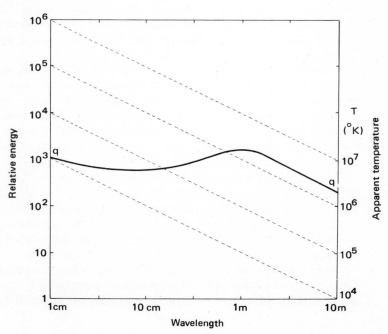

because in this part of the spectrum its gases are extremely transparent. But as we move toward longer wavelengths, we finally reach some for which even the tenuous corona is substantially opaque. Any energy observed in these spectral regions, therefore, must originate in the relatively hot corona and not in the cool photospheric layers. The opacity results primarily from the free electrons in the corona. The longer the wavelength and the greater the number of electrons in a given volume, the more intense will be the absorption. Thus, the energy will correspond approximately to the value predicted by the Planck formula for the temperature of the corona at the region where it becomes opaque. In other words, we see into the sun's envelope to different depths at different wavelengths. The radiation coming from the sun corresponds to the temperature at that particular depth. At the shortest radio wavelengths studied (the highest radio frequencies), we again "see" right through the corona and down into atmospheric layers whose temperature is not far above that of the sun's surface, between 5000° and 6000° K. At such temperatures the radiation is fairly weak. But at longer wavelengths we can "see" into only a more tenuous and hotter layer, where the radiation can be much more intense.

At times of solar disturbance we encounter a very different situation. The sun sends out intense blasts of radiation, which interpreted in terms of temperature would correspond to values of hundreds of millions of degrees. This radiation, however, is generally regarded as being nonthermal in nature. It comes from electrons oscillating in enormous clouds of ionized gas. The most intense emission of this sort occurs during solar flares. Only limited areas of the sun produce this unusual radiation.

Radio astronomers recognize at least four distinct types of bursts of solar radio noise, designated as classes I to IV. Type I bursts are usually part of a long-continuing storm of radio noise. The individual pips are extremely short, lasting from one-tenth of a second to about one second. The most affected frequencies lie in the range of 100 to 300 megacycles per second. The pips cover only a narrow band of frequencies and show no pronounced change during an individual pip. Solar flares often seem to trigger off such events, but the storm may continue for hours or even days, whereas the flare is of relatively short duration. From the fact that the radiation is usually circularly polarized, we deduce that it originates in regions of strong magnetic fields, perhaps near sunspots.

During Type II disturbances, the frequencies of the solar noise change in a radical but characteristic manner. During early stages, the frequencies tend to be high, in the range of 100 megacycles per second or so. Thereafter the frequency slowly drifts to lower values, reaching a level of some 40 megacycles in about four minutes on the average. The distribution of the radiation with frequency is extremely complex, as shown by the radio spectrum analyzer (Fig. 9-26). Many of the records show a second pattern of emission at about double the fundamental frequency. This behavior suggests the presence of harmonics in some kind of large-scale oscillation.

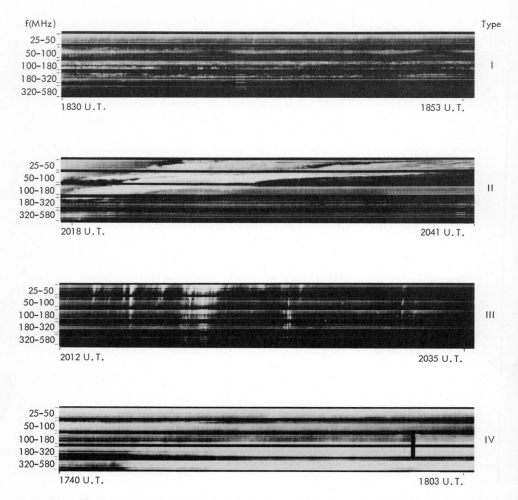

FIGURE 9-26 *The four types of solar radio noise. For each type the frequency increases from top to bottom, 25–580 MHz. Time intervals 23 minutes from left to right. (Courtesy A. Maxwell, Harvard College Observatory, Fort Davis Station.)*

Type III bursts resemble those of Type II, except that their rate of drift is some 400 times faster. The complete shift from high to low frequency takes place in an interval of about a second.

Type IV bursts consist of radiations of much lower frequency, from about 300 megacycles per second down to 30 megacycles per second or lower, according to the limit where the emission can get through the earth's ionosphere. This radiation usually accompanies a flare, especially flares that produce bursts of high-energy protons. These particles, in turn, can enter the earth's ionosphere and enhance the absorption of radio waves, especially in the vicinity of the magnetic poles.

Observations of radio noise at solar eclipses, when the moon acts as an occulting disk, indicate that the noise sources are by no means uniformly spread over the solar disk. We notice abrupt changes in noise emission as the moon covers up an active sunspot area. Moreover, eclipse observations show that at radio frequencies, the limb of the sun is brighter than the edge of the disk. (As we have noted, in visible light the opposite effect occurs, a limb-darkening.) This *limb-brightening* results from the opacity of the corona to radio waves, and from the higher temperature of the upper layers of the corona, which hence radiate more strongly. The solar limb-brightening at radio frequencies is more marked near the sun's equator than near its poles.

10

The Earth as a Planet

Among the billions of human beings on the earth, a few restless souls are dissatisfied until they have explored the unknown beyond the horizon. Man has conquered almost the whole of the earth's surface, even the nearly inaccessible peak of Mount Everest. He has carried his explorations down to the ocean bottom and through the atmosphere, reaching on to interplanetary space. The physical scientist is also an explorer. He seeks the unknown in the universe, from the interior of the earth to the distant galaxies.

In the present chapter we shall explore the earth.

10-1 The Earth Viewed from Space. From space the earth looks like a twin sister of Venus, except that the terrestrial atmosphere produces a bluish haze, and the earth is less brilliant because of its greater distance from the sun and the less complete cloud coverage (Fig. 10-1). We can, in fact, determine the brightness of the earth as seen from a point outside, without actually observing it from a spaceship. We simply measure the earthshine on the moon when the moon is in crescent phase. The dark side of the moon is then illuminated by sunlight reflected from the surface of the earth, as shown in Fig. 10-2. The brightness of the earth varies somewhat, because changing amounts of cloud on the sunlit side of the earth affect the earth's reflectivity. The fraction of sunlight reflected by the earth, *the albedo*, averages about 0.35, ranging from 0.25 to 0.50. It is

FIGURE 10-1 *Earth in crescent phase was first televised August 23, 1966 from the vicinity of the moon by U.S.* Orbiter *space probe. (National Aeronautics and Space Administration.)*

FIGURE 10-2 *Earthshine on the moon faintly illuminates night side of our satellite. (F. Quénisset, Flammarion Observatory, JUVISY, France.)*

consistently greater for blue and violet light than for yellow and red light, so that the earth's light is bluish as the astronauts testify.

From space the continents and waters of the earth are not so clearly outlined as on our maps and globes. Clouds and haze obscure large areas of the oceans and continental masses. An astronaut can discern their outlines only faintly through gaps in the clouds, and he would be able to make a complete map of the surface only after many terrestrial rotations. Figure 10-3 shows the earth from a satellite. Cloud streaks extending for a thousand kilometers or more, great, variable cloud masses, and the brilliant icy polar caps are the most conspicuous characteristics at planetary distances.

FIGURE 10-3   *Earth surface televised August 8, 1967 from U.S. Orbiter Satellite. (Courtesy National Aeronautics and Space Administration.)*

10-2   The Composition of the Atmosphere. The air near the earth's surface changes continually in composition because its water-vapor content depends markedly on local weather conditions. The composition of dry air, however, is remarkably constant, with percentages of the principal gases as given in Table 10-1. Nitrogen molecules, N_2, constitute the bulk of the atmosphere, while the oxygen molecules, O_2, so indispensable to higher living organisms, contribute only about 21 per cent. The inert argon atoms occupy third position, constituting almost all of the remaining 1 per cent. A small amount of carbon dioxide is present and only traces of other gases.

TABLE   *Composition of dry air[a]*
10-1

Gas	Percent by Volume	Gas	Parts per Million
Nitrogen (N_2)	78.084	Neon (Ne)	18.2
Oxygen (O_2)	20.946	Helium (He)	5.2
Carbon dioxide (CO_2)	0.033	Krypton (Kr)	0.1
Argon (A)	0.934	Methane (CH_4)	1.5
Others[b]	0.003	Hydrogen (H_2)	0.5
		Nitrous oxide (N_2O)	0.5

[a] According to E. Glueckauf.

[b] The water-vapor content varies from 0 to 0.2 per cent; ozone, O_3, not listed, also varies.

The small carbon dioxide content of the air results from the very delicate balance on land and sea between the absorption of CO_2 by plants and rocks, and its release by decomposition of organic matter, by weathering of rocks, and from volcanic activity. There is evidence that the percentage of carbon dioxide in the air has been increasing steadily since about 1890 (see Fig. 10-4). It turns out that the rate of increase of carbon dioxide just about matches the production of this gas from combustion in modern industry. Will the increases continue, and possibly produce profound effects on the climate of the earth? The answer is probably no, since the carbon dioxide balance in the atmosphere may be expected to reestablish itself at some slightly higher level of CO_2, at least if the combustion of fossil fuels remains constant.

The carbon dioxide and water vapor in the lower part of the atmosphere help regulate the temperature of the earth. They transmit visible light, but absorb heat (infrared) radiation. The sunlight that passes through them raises the surface temperature of the earth, but the heat absorbed by the earth is radiated out at a much lower temperature in the form of infrared waves. The carbon dioxide and water vapor, which are not transparent at most longer wavelengths, reflect back the earth's radiation and produce the

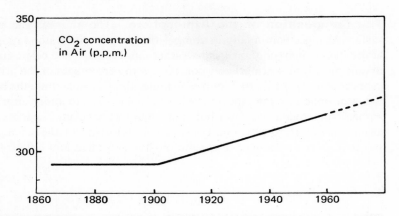

FIGURE 10-4
Slow increase of carbon dioxide content of air (in parts per million) during this century.

so-called *greenhouse effect*, keeping the surface warmer than it would be otherwise. A considerable increase in atmospheric carbon dioxide or water vapor might, therefore, have a marked effect on our climate. Possibly a stronger greenhouse effect caused by water vapor or carbon dioxide was a factor in maintaining a much warmer climate in the humid age of the dinosaurs.

Industry has contributed about one part per million of sulphur dioxide to the atmosphere, as anyone who lives on the leeward side of a large city can readily believe. But traces of industrial carbon monoxide, ammonia, and nitrous oxide have almost no effect on the overall composition of the earth's atmosphere. The ultraviolet radiation of the sun, on the other hand, does transform a certain amount of the atmospheric oxygen into ozone (O_3), especially at higher altitudes, as we shall discuss more fully in Sec. 10-4.

The composition of the earth's atmosphere largely determines the existence and development of terrestrial life. Apart from man-made pollution, it contains only traces of gases antagonistic to organic life, while the oxygen so necessary to our existence is abundant. Undoubtedly we could survive in a much less favorable atmosphere. If we could "improve" our atmosphere by artificial means, we might simply ask for more of it, because it acts as a shield. It absorbs Xrays, gamma rays, and ultraviolet radiation from outer space, and destroys meteors before they reach the earth's surface. It cannot, however, prevent secondary high-energy particles and rays from the more energetic cosmic rays from passing through. Cosmic rays consist of atomic nuclei, mostly hydrogen protons, which move at practically the velocity of light. Because of their intense interaction with the air a secondary particle stabs through each of us about once a minute, and produces a minute amount of ionization among our body molecules. Although biologists have not established precisely to what extent cosmic rays are injurious, some mutations in living cells may have their origin in the basic effect of cosmic rays on cell structure. However, natural radioactivity in surface rocks is probably more effective in causing mutations. The man-made atmospheric contamination that has followed nuclear explosions has added only very little to the natural background radiation on a worldwide basis; so far it has been dangerous only near the explosion sites.

10-3 **The Study of the Upper Atmosphere.** Many devices and methods now bring "the upper atmosphere down to us" for study. Only specialists in the field can keep track of the day-to-day developments. Scientists who lowered instruments into the ocean were said to "sound" the depths. Now, sounding balloons, rockets, satellites, and space probes carry instruments to extreme altitudes while radio waves, magnetism, light, and even sound carry back information about conditions in the upper atmosphere. The much-used sounding balloon, or *radiosonde*, carries a "payload "of a pound or so to a height of about 30 km (100,000 ft), where it bursts. During the ascent it transmits back the air temperature, pressure, and humidity. Larger balloons are sent to even greater altitudes, the maximum being over 40 km (140,000 ft). Radar and radio direction-finders often track the balloons for thousands of miles as winds carry them over the surface of the earth.

Microwave radars "watch" storms in the regions of the atmosphere where weather occurs, below some 10 km (30,000 ft) altitude, while the U.S weather satellites are now revolutionizing the ancient art of weather forecasting. They can transmit pictures of the clouds, cloud motions, temperature, and other data for the entire earth on a daily basis. Before Tiros I, even the great cloud patterns on the earth were not visualized accurately from the scanty data given by sparsely distributed weather stations, ships, and aircraft. The vast amount of weather data needed and now available could never be analyzed effectively, were it not for great electronic computing machines. Weather forecasting at last is becoming a science! And the National Center for Atmospheric Research, in Boulder, Colorado, is the headquarters for a new and modern approach to the problems of understanding and forecasting the weather.

Above about 40 km and up to about 160 km (25 to 100 miles) only rockets and experimental aircraft (such as the X-15) can reach into the atmosphere, while satellites become useful at somewhat greater altitudes. In 1946, scientists in the United States began an extremely ambitious program of rocket research in the upper atmosphere. Early in the program, V-2 rockets, captured in Germany, carried surprisingly heavy loads, a thousand pounds or more, of delicate and complicated equipment to a maximum altitude of 114 km (370,000 ft). Later the Aerobee rocket became the workhorse of the upper atmosphere. Such vehicles, truly flying physical laboratories, carry many types of sensitive measuring equipment—special instruments to measure temperature, pressure, density, or composition of the air; air-sampling bottles and mass spectrographs; radio and radar for study of the ionized layers in the atmosphere; photographic or television cameras for a "top-side" view; photoelectric tubes, phosphors, photographic cameras, and other devices to study the far-ultraviolet light, X rays, and gamma rays from the sun; devices to measure magnetic fields and space charges; cosmic-ray detectors; seeds, mice, monkeys, and fruit flies for biological experiments; and other special devices.

Satellites carry many similar devices and novel ones for special tasks in space. Most of the instruments make their own readings automatically and transmit their measures back to earth via a radio *telemetering* system. This method of remote instrument reading is highly satisfactory unless the experiment demands recovery of the payload, such as mice.

Ordinary parachutes are useless for rocket payloads dropped from very high altitudes, because they either burn up or melt in the frictional heat produced as they enter the lower atmosphere. A fairly successful method of recovering equipment without too serious damage involves an explosion of the rocket to separate the nose cone from the main body on the return leg of the journey. A large rocket, falling from a height of, say 80 to 100 km, lands with a velocity of well over a mile per second, digging a crater some 12 meters (40 ft) deep and equally wide. When blown apart, however, the two aerodynamically unstable parts of the rocket fall much more slowly, so that in a good fraction of the firings, photographic film (in a steel container) and other sturdy equipment fall to earth with little or no damage.

Recovery from satellites requires, first, retrorockets to bring the satellite into the atmosphere at an altitude of perhaps 100 km (60 miles) or less, where atmospheric drag takes over. Then the capsule requires a special shape so that the material of the blunt nose can take up the great heat, with some loss of material by boiling. Finally, parachutes bring the capsule to the ground.

At great altitudes a thermometer is useless for measuring the temperature, because the air is so rarefied that months or years would be needed for the thermometer to register the correct value. In all rocket research we must employ measures of air pressure, air density, or the velocity of sound in place of direct temperature measures, and deduce the temperature from them. The Army Signal Corps exploded grenades at various heights and measured the arrival of the sounds at the ground to determine the velocity of sound. In air, the velocity of sound varies as the square root of the absolute temperature, independently of the pressure or density so long as the composition remains unchanged. Air drag on satellites or falling spheres provides deceleration or nose-pressure measures, which can be interpreted in terms of air density, air pressure, or the velocity of sound.

10-4   Results of Upper-Atmosphere Research. The Upper Atmosphere
Rocket Research Panel, an unofficial American research group, measured
the variation in temperature as one rises in the atmosphere to an altitude of
160 km (100 miles), as shown in Fig. 10-5. The striking changes in temperature with altitude confirm earlier results obtained by several methods,
including the photographic study of meteors conducted at the Harvard
College Observatory, and extensive observations of sound waves from
explosions on the surface of the earth.

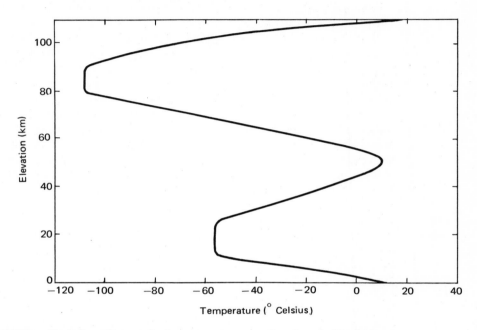

FIGURE   *Variation of atmospheric temperature as a function of altitude,* h.
10-5

The earth's surface is relatively warm because of direct solar heating.
With increasing altitude the temperature decreases, as one would expect in
a gas held by gravity. Gas cools as it expands if no heat is added; hence, air
ascending to a region of lower pressure cools, and vertical circulation is
frequent up to altitudes of 12 to 20 km (7 to 12 miles). At greater altitudes,
however, the temperature ceases to diminish and remains more or less
constant, until it subsequently rises in the higher stratosphere. Thus
vertical circulation is strongly damped at moderate altitudes.

This rise in temperature with altitude undoubtedly results primarily
from the presence of ozone (O_3). In the atmosphere, ozone forms from
molecular oxygen, O_2, by the action of solar ultraviolet radiation. This gas
strongly absorbs the ultraviolet below about 3000 Å and heats the higher
atmosphere by solar radiation. The percentage of ozone remains extremely

small—around one part in 30,000,000 at the height of its maximum concentration, which is usually between 20 and 30 km (12 to 19 miles)—but it varies with time and place.

The temperature continues to rise to a maximum in the region of 40 to 60 km (25 to 35 miles). Near an altitude of 80 km the percentage of ozone falls to an imperceptible value and the temperature drops to a minimum. At still higher altitudes the absorption of the far-ultraviolet sunlight by oxygen molecules and other atmospheric gases raises the temperature.

A high temperature of 500° C or more around 140 km does not mean that a person or material object at that altitude would be burned or baked. The air is so rarefied that, despite their high kinetic temperature, the molecules cannot transfer heat to the surface of the body as fast as the body will radiate away its heat, or be warmed by solar radiation.

The pressure and density of the air fall roughly to $\frac{1}{10}$ of their sea-level values at a height of 16 km (10 miles) to 10^{-6} at about 100 km (62 miles), to about 10^{-9} at 160 km (100 miles) and to about 10^{-11} at 500 km (300 miles), as shown in Table 10-2 and Fig. 10-6. At the same time the *mean free path* of the molecules, the distance that they travel before colliding with others, increases from about 10^{-5} centimeters at sea level to 1 meter at 110 km. The mean free path reaches 1 km at an altitude above 500 km.

TABLE 10-2   *Densities in the atmosphere (water = 1.0)*[a]

Height, km	Density	Height, km	Density
0	1.22×10^{-3}	120	1.5×10^{-11}
10	4.1×10^{-4}	140	3.0×10^{-12}
20	8.9×10^{-5}	160	1.1×10^{-12}
30	1.8×10^{-5}	200	3.6×10^{-13}
40	4.0×10^{-6}	300	4.1×10^{-14}
50	1.1×10^{-6}	400	7×10^{-15}
60	3.5×10^{-7}	500	2×10^{-15}
70	1.0×10^{-7}	600	5×10^{-16}
80	2.1×10^{-8}	700	2×10^{-16}
90	2.8×10^{-9}	1600	3×10^{-18}
100	3.7×10^{-10}	Space	1×10^{-22}

[a] U.S. Air Force Model Atmosphere (1959) to 300 km; after L. G. Jacchia above 300 km.

At still greater heights an upgoing molecule will move in an orbit with little chance of colliding with other molecules until it either escapes the earth or falls back into a denser region of the atmosphere. This region of the atmosphere is known as the *exosphere*. L. Spitzer predicted and L. G.

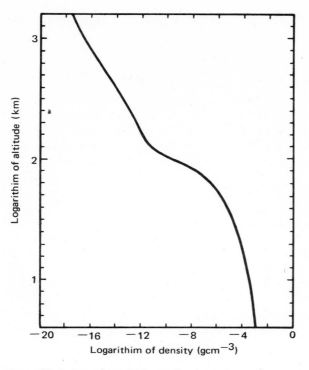

FIGURE *Variation of atmospheric density, ρ (gcm^{-3}), as a*
10-6   *function of altitude,* h.

Jacchia confirmed from satellite data that the temperature in this region is of the order of 1500° K.

Satellites provide the best measures of air density above 200 km (120 miles). The air drag near perigee reduces their periods and the apogee distances until the orbits are fairly circular. Thereafter the orbits contract rapidly. The rate of period change measures the air drag near perigee, from which air density can be calculated. From many satellites, including the 30-m (100-ft) balloon Echo I, L. G. Jacchia discovered that the upper atmosphere is heated and rises in a bulge on the daylight side of the earth (see Fig. 10-7). The rotation of the earth displaces the bulge about 30° east of the subsolar point because of a two-hour lag in the heating process. At great altitudes, to about 1600 km (1000 miles), the entire atmosphere rises with increasing solar activity as measured by solar radio noise at micro-wavelengths. Solar flares bringing magnetic storms to the earth can also heat and expand the upper atmosphere. Typically there is a 24-hour lag between the occurrence of the flare and the resulting atmospheric expansion,

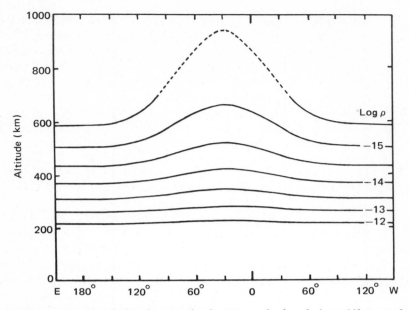

**FIGURE
10-7** *Upper-atmosphere bulge due to solar heating is displaced about 30° east of sun's direction (0°) by the earth rotation. Density, ρ, is in grams per cm³.*

**FIGURE
10-8** *Effect of solar activity on the earth's atmospheric temperature in 1961, compared with magnetic index, Ap, and solar noise, F_{10}, at 10-cm wavelength. (L. G. Jacchia, Smithsonian Astrophysical Observatory.)*

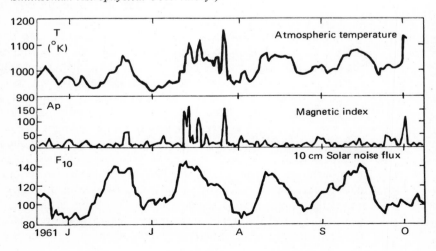

and a peak of about the same duration. Figure 10-8 shows the strong dependency of high-altitude air densities upon solar activity.

Optical observations of satellites by the amateur *Moonwatch* observers and by the photographic tracking stations of the Smithsonian Astrophysical Observatory were mainly responsible for these early discoveries on atmospheric densities and their variations.

From various rocket, satellite, and ionospheric measurements M. Nicolet, of Belgium, concludes that the composition of the upper atmosphere remains constant to about 120 km (75 miles), where separation of the light elements such as hydrogen and helium begins, and the dissociation of the oxygen molecule into atoms becomes appreciable. Above 1000 km (600 miles) helium and hydrogen become the major constituents.

10-5   The Earth's Magnetic Field.   The earth's magnetic field causes all charged particles near the earth to follow twisted orbits, with extremely complicated coils and spirals. It affects all cosmic-ray particles and ionized atoms and electrons bombarding us from the sun or space. That many are trapped in this field was the first great discovery made from earth satellites, by J. A. Van Allen, an American physicist for whom the zones are named. Thus it is important to understand the mechanism of the earth's magnetism and how it affects these charged particles near the earth.

Everywhere on the earth's surface the needle of an ordinary compass is oriented in response to the local magnetic field. In most places the field is directed more or less north and south. It tends to converge on the *north magnetic pole* of the earth, which is now located in the Hudson Bay region, at latitude 73°, nearly due north of Chicago. The earth's north magnetic pole is actually a *south* pole, since it attracts the *north* (seeking) pole of an ordinary magnet. At the time of Columbus the compass pointed nearly northward in Europe, while today it points west of north. The north magnetic pole is now steadily shifting in a clockwise direction, and in a circle of about 17° radius about the pole of rotation (Fig. 10-9).

The earth's magnetic field roughly approximates that of a large bar magnet within the earth (Fig. 10-10). The two poles, however, do not lie on diametrically opposite sides of the earth. A line joining them would pass about 1000 km from the earth's center, and it would make a considerable angle with the earth's axis. The poles that represent the best fit to a symmetrical field are called the *geomagnetic poles*. These poles, in 1945, lay at latitude 79° N, longitude 70° W and 79° S, 110° E. On the other hand, the *dip poles*, where the magnetic field is vertical at the surface, were located in 1945 at about 73° N, 98° W and 68° S, 145° E. The earth's magnetic field is quite irregular over the earth's surface. Its source is not definitely known. The problem is bound up with that of the physical condition of the earth's

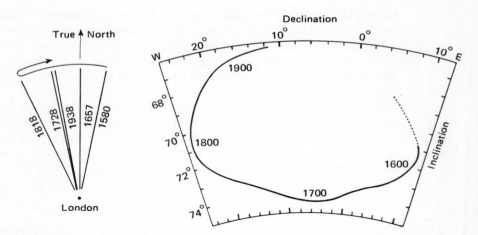

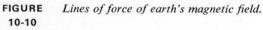

**FIGURE
10-9** *Wandering of magnetic pole is shown by variations of magnetic elements at London since sixteenth century.*

**FIGURE
10-10** *Lines of force of earth's magnetic field.*

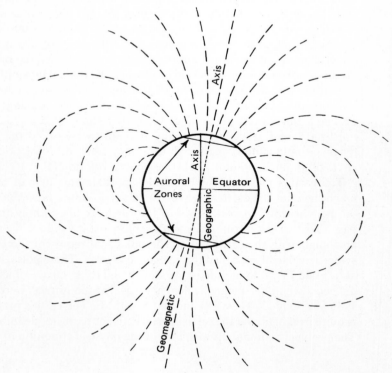

interior (see Sec. 7-8). The interior of the earth is not permanently magne-
tized. The most likely interpretation is the *dynamo theory*: that differential
rotation in the fluid core (or inner shell) of the earth produces electrical
currents and consequently the magnetic fields that we observe. Extremely
slow circulation currents as low as 0.03 cm/sec may be adequate. Magnetism
measured in old Roman pottery and in identical modern replicas shows
that the field was some fifty percent stronger then than now. The magnetic
record in the rocks shows that the magnetic field has reversed sign several
times in geological time.

Fast-moving charged particles headed toward the earth can slide along
the magnetic lines, as shown in Fig. 10-11, but will twist in tight spirals

**FIGURE
10-11**

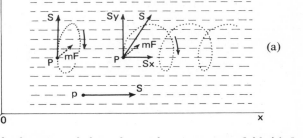

*Helical motion of charged particles in magnetic field. (a) In homogeneous field,
(b) Slow ion from the sun forming increasingly tight helix as it encounters the
stronger field of the earth, (c) Relatively fast solar ion encountering earth's field.
(Our Sun, by D. H. Menzel, Harvard University Press.)*

when they attempt to cross the lines. In consequence, such particles can easily slip toward the north and south magnetic poles of the earth, but they encounter more and more difficulty toward the regions of the earth's magnetic equator. The earth's field acts as a sort of bumper, warding off all but the most energetic particles. Thus only the charged particles with high kinetic energy can penetrate the earth's equatorial regions.

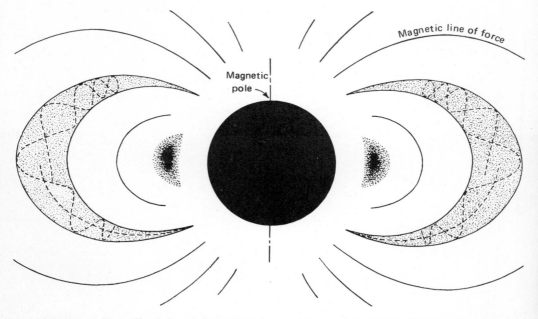

FIGURE 10-12   *Van Allen belts of charged particles trapped in magnetic field of earth. The shaded areas are those of typically higher concentration.*

Not only can the earth's magnetic field change the motions of incoming charged particles, it provides a trap to hold some for considerable periods of time. Figure 10-12 shows a cross section through the earth's magnetic poles of the Van Allen belts discovered in 1958 by the early satellites launched during the International Geophysical Year. They are "doughnut" shaped or *toroidal* volumes around the polar axis. Charged particles moving in the plane of the diagram oscillate nearly parallel to the (magnetic) equator when near it. As they move north or south near the "horns" of the belts, they are reflected back toward the equator by the converging lines of the magnetic field. Those that dip too low toward the higher atmosphere run the risk of being slowed down or freed from the trap by becoming neutralized electrically.

The great surprise in Van Allen's discovery was that so many charged particles are trapped. The energy striking a surface at the maximum of the inner belt can amount to hundreds of ergs per square centimeter per second, a hundred times the corresponding solar radiation in the far-ultraviolet and X-ray region. In the inner belt the energies of the particles, mostly protons (hydrogen nuclei), are greatest. Many of these protons arise from decayed neutrons that came from cosmic-ray encounters with the earth and atmosphere (see Sec. 10-6). The outer belt contains lower-energy particles that appear to be replenished by streams of charged particles shot out from the sun. The extent and symmetry of the outer belt varies considerably with solar activity.

FIGURE   *Photographs of aurora from two stations shows parallactic displacement with*
10-13   *respect to Venus. (C. Störmer.)*

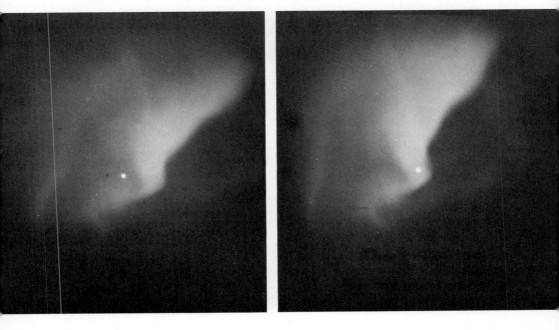

These clouds of ionized solar gas produce the northern lights or *aurora borealis* in the general region of the magnetic pole in the northern hemisphere and the *aurora australis* in the south (Fig. 10-13). The *auroral zones*, or regions of maximum auroral activity, occur not at the poles, but in rings some 17° from the geomagnetic poles, at altitudes generally above 100 km (60 miles).

The spectrum of an aurora, such as that shown in Fig. 10-14, sometimes shows not only atmospheric lines and bands, but also widened hydrogen lines, presumably by the Doppler effect (see Sec. 8-14), which suggests that protons strike the atmosphere with velocities up to 3000 km/sec. At sunspot maximum, when solar storms are stronger than at sunspot minimum, aurorae also are brighter and more frequent. They reach their peak some two years after sunspot maximum because the proton storms from the sun reach their peak at that time.

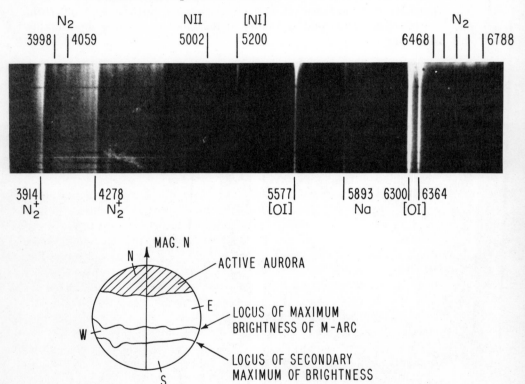

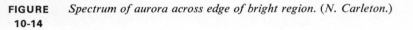

FIGURE 10-14 *Spectrum of aurora across edge of bright region. (N. Carleton.)*

At all latitudes the night sky is illuminated by *air glow*—atmospheric radiations from regions within the height range 70 to 300 km (40 to 200 miles). The air-glow intensity exceeds that of the average background starlight and varies with time of night, latitude, and season but not strongly with sunspot and magnetic activity. The light consists of lines and bands primarily from oxygen atoms, OH molecules, and sodium.

Incoming streams of charged particles encounter and disturb the earth's magnetic field. Thus we find that magnetic storms on the earth are associated with aurorae and various types of solar activity. During such storms, the earth's magnetic field fluctuates slightly in strength and direction, as a sensitive compass needle will show. Very large electric currents are often induced in long-distance power cables, telegraph cables, or telephone lines. Transcontinental teletype machines occasionally type out unintelligible records during an intense magnetic storm. The ionized clouds from the sun are observed to disturb the Van Allen belts. Possibly such disturbances distort the "horns" of the outer belt, causing it to unload high-energy particles into the atmosphere to produce the aurorae. (See also Sec. 15-10 on the solar wind.)

10-6   Cosmic Rays.   Physicists have torn the cloak of mystery from cosmic rays, except for their origin. Protons, hurtling through space with nearly the velocity of light and possessing energies of billions of electron volts, constitute the major fraction of primary cosmic rays. Included also are some nuclei of helium atoms (alpha particles) and a relatively few nuclei of heavier atoms—in much the same abundance ratio to hydrogen as in the observable universe. Finally, there are high-energy electrons and possibly some photons or gamma rays.

Most cosmic rays strike and break up atomic nuclei in the upper atmosphere. Such nuclear debris includes electrons, uncharged neutrons (which have about the same mass as protons), alpha particles, various types of mesons (short-lived, charged particles with masses intermediate between those of electrons and protons), and other elementary particles (see Fig. 10-15 and Chapter 24). Thus the highest cosmic-ray flux, resulting mostly from these secondary particles, occurs at an altitude of some 20 km (12 miles) in the atmosphere. The atmosphere absorbs most of the primary and secondary particles before they reach sea level, but a few of the most energetic penetrate even to the deepest mines and caves.

Physicists measure and count cosmic rays with Geiger counters, cloud or bubble chambers, photographic emulsions, or other devices that can reveal the sudden production of free ions or electrons. The most energetic cosmic ray, if it could deliver its 10^{20} electron volts of energy as forward motion to a marble, could give the marble a speed of some 300 m/sec (330 yd/sec). A marble, in turn, carrying the same energy per unit mass as the cosmic-ray particle, could impart to the entire earth a velocity of 100 cm/sec (1 yd/sec). Fortunately, particles larger than the nuclei of atoms never appear to attain such energies.

Elementary charged particles must have been accelerated to such unbelievably high energies by strong electric or magnetic fields somewhere in the universe—but where? The sun certainly produces some cosmic rays, as S. E. Forbush first suggested and as the early space research abundantly proved. It is difficult to ascertain what fraction of the cosmic rays come from the sun, or to what extent the changing magnetic fields about the sun

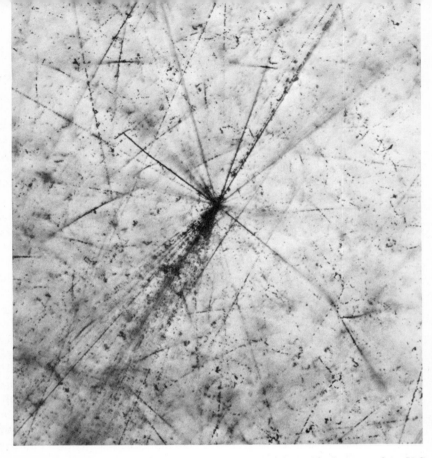

FIGURE
10-15 *Cosmic ray shower of mesons produced in an emulsion block exposed in U.S.
Discovery No. 32. (H. Yagoda, Air Force Cambridge Research Laboratories.)*

and earth cause changes in the cosmic-ray flux that correlate with solar
activity. It appears, however, that the solar contribution to the cosmic-ray
energy spectrum is mainly in the low-energy range, mostly less than 10^9
electron volts (10^3 MeV).

There are enough weak magnetic fields in space so that cosmic rays, even
of very high energies, change their directions markedly as they move over
great distances. Thus we cannot detect any sources in space by the directions from which cosmic rays arrive at the earth. E. Fermi has shown that
fluctuating magnetic fields in interstellar space can accelerate charged particles that have been sprayed into space from unknown sources, perhaps
stars. We do know that nuclear disintegrations cannot produce the rays
directly, because too high energies are needed, nor can cosmic rays have
persisted since the universe began (if it indeed had a beginning) (see Chapter 35). Possibly stellar outbursts of the most violent kind, the supernovae,
of the mysterious quasars, are primary sources (Chapters 27, 34).

10-7 **The Earth's Rigidity and Elasticity.** *Rigidity* measures the degree to which a body actually deforms when we apply a distorting force. Thus steel is quite rigid, while liquids have practically no rigidity at all. Tidal data prove that the earth has some degree of rigidity (see Sec. 5-6).

A concept often confused with rigidity is the *elasticity* of the earth. To understand some of the differences between elasticity and rigidity, imagine that you hold a rubber ball in one hand and a ball of putty in the other. A sudden squeeze may deform both of them about the same amount, in which case they are equally rigid to a quick-acting force. But the rubber ball will immediately become round again, while the putty ball remains deformed. Thus we say that rubber is elastic while putty is viscous, the converse of elastic. Now if we squeeze and continue to squeeze the two balls, at a certain point we find no further change in the rubber one while the putty ball loses its original shape entirely. Thus putty has practically no rigidity to long-sustained forces, while it is fairly rigid to quick-acting forces. "Sili-putty," a silicone compound, is an intermediate type of material; it bounces like rubber, but flows like thick syrup if left standing. We shall see that the earth acts much like Sili-putty—or better, like glass, which also flows in a viscous fashion if one waits long enough, perhaps thousands of years.

An earthquake produces rapidly acting forces on the earth. Earthquake waves are low-frequency sound waves that move swiftly through the earth. Their speeds, 5 to 8 km per second at depths of a few kilometers below the surface and increasing at greater depths, are similar to the speed of sound waves in glass or iron, substances that are very rigid and very elastic. Thus, for extremely quick-acting forces, the earth (except near its center) acts much like a steel ball.

The tide-raising forces are slower than seismic vibrations. To measure the earth's rigidity and elasticity, A. A. Michelson and H. G. Gale in 1913 studied the tides in long pipes buried on the grounds of the Yerkes Observatory in Wisconsin. They buried two 500-foot-long pipes, one north-south, the other east-west, half filled them with water, and observed the tides with delicate optical methods. They found that the tides in the east-west pipe had 70 per cent the amplitude calculated for a completely rigid earth, those in the north-south pipe, only 50 per cent. Thus they concluded that the local rigidity of the earth may vary with direction, but that generally the earth is slightly more rigid than steel.

Michelson and Gale discovered an additional surprising fact: the tides in the pipes did not lag behind the predicted times of maximum and minimum. Thus the earth is also elastic, about as elastic as steel, at least to forces that last for short times.

The steel-like rigidity and elasticity of the earth disappear, however, when the distorting forces persist for several thousands of years. The earth flows "like cold molasses" when a thick enough ice cap forms during a glacial period. The geophysicist B. Gutenberg showed that in one area of

Finland, where the Pleistocene ice sheets were thick, the land is now rising as rapidly as a meter per century, still recovering from the depression caused by the now vanished load of ice. This plasticity of the earth's outermost layer (which is 10^8 times more viscous than ice) is known as *isostasy*, the ability of the surface layer to adjust its level according to the load. Thus the layers beneath the oceans tend to be denser than land masses. They are lower and therefore covered with water, which roughly equalizes the gravitational load of the earth's crust. As a river deposits sediment in its deltas, such as those of the Mississippi River or the Nile, the earth under the delta tends to sink slowly, maintaining shallow water which is available to receive further deposition.

In summary, we find the earth to be a puttylike ball, when considered over astronomical time intervals, but perhaps similar to steel in rigidity and elasticity when subjected to forces that persist less than scores or hundreds of years.

10-8   Earthquakes and the Earth's Interior. Since even our deepest mines and borings extend only a few miles below the earth's surface, we must infer the conditions of the deep interior from indirect evidence. Volcanoes bring forth magma from within the earth, but these may represent special regions, which certainly extend to only a minute fraction of the earth's radius below the surface. Earthquake, or *seismic*, waves travel deeper within the earth and by their velocity and nature provide vital information about the deep interior.

Seismic waves vibrate in many ways, but we may limit ourselves to the two main types, the *pressure*, or P-waves, which move by compression, like sound through air, and the *shear*, or S-waves, which move transversely in the earth, as ocean waves do on the surface of the water (see Fig. 11-16). The P-waves travel faster than the S-waves, up to 8 km per second at moderate depths compared with 5 km per second for the S-waves, and therefore they bring the first news of a distant earthquake to the seismographic station.

FIGURE 10-16 *Seismographic record of P- and S-waves and surface waves (time increases to the right).*

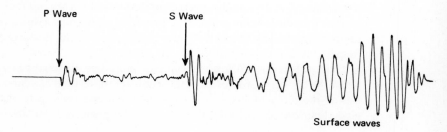

P Wave   S Wave

Surface waves

By measuring the travel time of shocks from distant earthquakes and calculating the paths of the wave fronts through the earth, seismologists deduce the seismic velocities at various depths. For instance, the P-waves speed up to about 13.7 km per second at a depth of 3100 km, and there suddenly slow down to about their surface velocity. They accelerate again more slowly toward the center to some 11 km per second. The S-waves behave similarly except that they cannot penetrate into the core of the earth, being totally reflected at the 2900-km discontinuity (see Fig. 10-17). A second discontinuity occurs very near the center.

These surprising results tell us two facts about the earth's interior. The earth must become denser with depth, because seismic waves generally travel faster in denser materials. Also, the core of the earth must be liquid (or bounded by a liquid layer) because S-waves will not travel through a liquid. The core of the earth may well be liquid throughout, and still denser than the outer portions, because P-waves generally travel faster in a solid than in a liquid of equal density.

Most geophysicists favor an earth model in which the core is all liquid, perhaps of molten iron or nickel-iron, resembling in chemical composition the iron meteorites (Chapter 14). At the center the iron core may be compressed to a density twelve times that of water. No one knows how hot the core is, but most estimates are above 2000° C. There are a few dissenters to the iron-core hypothesis. W. H. Ramsey argues that the 2900-km discontinuity necessitates a change, not in composition but in the physical state of ordinary earth materials. He suggests that enormous pressures, greater than two million atmospheres, compress this rocky material into a new type of solid, which has not yet been duplicated in the laboratory. Extremely high pressures were first produced in a laboratory by P. Bridgman of Harvard. Pressures above 100,000 atmospheres are now possible, but they are still short of those at great depths below the earth's surface. At enormous pressures, Bridgman discovered several modifications of a number of familiar materials, such as ice, which assume new molecular arrangements. Even though we may provisionally accept the iron-core hypothesis for the earth, we must keep an open mind for new developments by geophysicists and physicists.

Near the surface of the earth seismic waves, both natural and artificial, provide us with information about rock formations of commercial interest in construction, mining, and oil prospecting. The outer crust consists of relatively low-density igneous rocks such as granites and, somewhat deeper, basalts. The sedimentary rocks are completely superficial, mostly in the upper mile or two. The top crust is thicker under continental areas, typically 30–60 km (20–40 miles), so that the continents "float," isostatically (see Sec. 10-9). In contrast, under oceanic areas the upper crust may be only 5–8 km (3–4 miles) in depth. The conspicuous seismic discontinuity identifying the outer crust is named for its discoverer in 1909, A. Mohorovičić. An important research project was proposed to drill a hole through

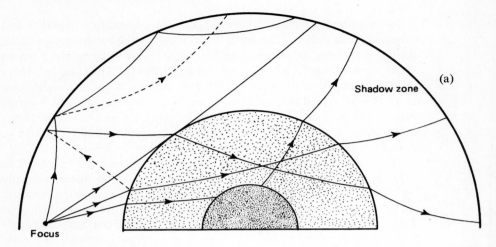

FIGURE
10-17
*Major discontinuities of the earth's deep interior. (a) Paths of seismic waves. (b)
Density variations with depth. (c) Depths of well-known discontinuities (in miles).*

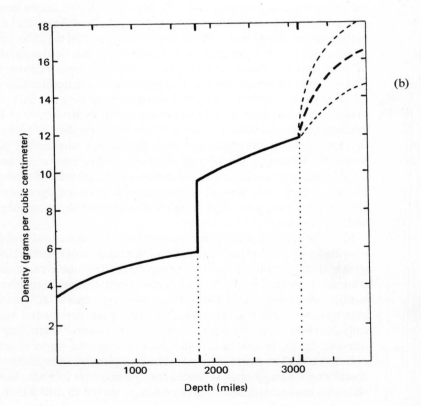

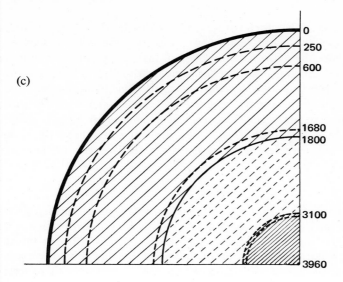

(c)

this layer where the ocean bottom is thinnest. The project carries the appropriate coined name, Mohole, to identify both the discoverer and the goal of the project in a form that is easily spelled and pronounced, at least by English-speaking people. The project, however, was discontinued because of the rising costs. Figure 10-17(c) indicates other discontinuities in the earth's radial structure indicated by careful seismic analysis. As mentioned above, some may result from phase changes in molecular structure rather than from discontinuities in composition.

Below the upper crust of the earth slow circulation patterns must occur in the mantle as well as in the core. The geological record shows clearly that buckling occurs to produce mountain ranges and ridges. The continents have been growing throughout geological history, while the evidence suggests that the total water in the oceans is not decreasing and is probably increasing with time. Thus the lighter rocky materials must be slowly concentrating in the continental areas. The existence of geological eras with huge variations in the rates of mountain building suggest that these internal circulation patterns are irregular with time.

Long ago geologists noticed that the continents, if moved around on the globe, could be shifted into a nearly perfect jigsaw pattern of one supercontinent. The east coasts of North and South America, for example, fit neatly into the pattern of the west coasts of Europe and Africa. This concept of *continental drift*, however, was long discounted because the continents seemed to be so well rooted into the upper mantle. Beginning in about 1960 geologists, oceanographers, paleontologists, and paleomagneticists began to pool their data and discovered that in each field their observations suggested continental drift. Ancient flora, fauna, and geological strata tend to match on the two sides of the Atlantic Ocean, just as the jigsaw suggests. The story is too long for this book, but there is strong evidence from the ocean ridges nearly bisecting the Atlantic Ocean that indeed a spreading of the Atlantic Ocean is taking place at a rate of perhaps a very few centimeters per year. The ancient story should be filled in by the completed records and interpretations of paleomagnetism—the direction and strength of the earth's magnetic field preserved in rocks through geological ages. The ancient magnetic-field changes are large in direction, including many reversals in sign, and are not consistent with a simple motion of the magnetic poles with respect to fixed relative continental positions.

10-9 **Polar Motions.** Motions of the earth's pole provide further information about the interior of the earth. We can watch the earth spin like a top in our study of the position of the equinoxes (see Sec. 5-3). The earth's shape and the gravitational attraction of the moon and sun provide a basis for calculating the rate of precession, which comes out larger than observed unless we assume that the density of the equatorial bulges is smaller than the average density of the earth as a whole. Thus the earth is not as dense near the surface as it is near the center, in agreement with the seismic data.

A spinning top always tends to "wobble" about its axis of rotation. The earth wobbles too. The position of the pole upon its surface shifts very slowly and by an extremely small amount. If the earth were spinning without any wobble, the astronomical latitude at any point on the earth would remain absolutely constant. Very precise measurements of latitude, made over many years on an international program at various stations around the earth, show that latitudes vary by a few tenths of a second of arc, or a few tens of feet on the surface. The earth's poles actually shift to produce these variations, since opposite changes occur on opposite sides of the earth. Figure 10-18 shows how the true pole moved over the earth's surface between 1958.0 and 1966.4.

We can divide this complicated motion into two parts: (a) a yearly motion, roughly circular, with a radius of about 10 feet, and (b) a 14-month motion, discovered by S. Chandler three-quarters of a century ago, with a radius varying between 10 and 20 feet. In addition, there may be small irregular displacements of the pole. It is noteworthy that the pole shows no strong, persistent motion away from its present position, although recent data suggest a slight tendency for it to shift some 4 inches (10 cm) per year. This rate is in perfect agreement with values suggested for continental drift. In addition, however, slow, progressive shifts of the pole might occur, if the outer skin of the earth is sliding around like the shell of a raw egg over its liquid interior.

The yearly motion in latitude undoubtedly arises from seasonal motions in the atmosphere and oceans, and particularly the growth and shrinkage of the polar ice caps. If the earth were absolutely rigid, the Chandler period

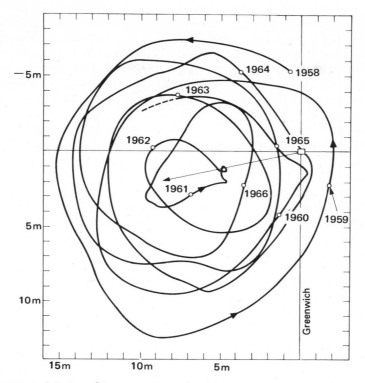

FIGURE 10-18 *Motion of instantaneous pole of rotation on earth's surface causes small variations of geographic latitudes with 12- and 14-month periods with an apparent systematic trend. Rectangle is mean pole of 1900–1905. (Smithsonian Astrophysical Observatory.)*

should be only 305 days. Hence the earth is not completely rigid, but as a whole has about the rigidity of steel, in agreement with the tidal data. H. Jeffreys finds that tidal friction within the earth should damp out the 14-month motion in about 15 years. However, observations show that the 14-month period has persisted over at least 200 years, and hence this motion is being maintained in some fashion. We may infer that the earth's rotation is being repeatedly disturbed by internal shifts in mass, perhaps associated with earthquakes or possibly with internal circulation.

10-10 How Old Is the Earth? Only through studies of radioactive elements such as uranium and thorium can we expect to answer this difficult but most intriguing question. Individual radioactive atoms explode at unpredictable times, but in a sufficiently large sample of atoms the fraction that will disintegrate within a given time interval follows a simple law. On the average, half of the atoms of a particular element will disintegrate within an ascertainable interval, called the *half-life*. Of a sample of radium atoms, for example, half will disintegrate in some 1600 years. The remaining atoms ignore the fate of their companions, but in another half-life period, one-half of them will disintegrate, and so forth. Hence, the fraction of the atoms left after x half-life periods is $(\frac{1}{2})^x$. If we calculate backwards in time from the present, we find, correspondingly, that the number of atoms doubles for every half-life, unless the atom, like radium, is itself the product of some radioactive atomic parent. Uranium 238, for example, is the parent for radium and has a half-life of 4.5 billion years.

When radioactive atoms disintegrate, they may disgorge alpha particles or helium nuclei, electrons (beta rays), neutrons, or gamma radiations, as well as other nuclear debris. At the end of a fairly complicated series of such disintegrations, uranium and thorium are transmuted into character-istic isotopes of lead. A few examples of naturally radioactive atoms and their isotopic molecular weights with their final decay products and half-lives for disintegration are presented in Table 10-3. Radiochemistry em-ploys literally hundreds of atomic isotopes today, but these few illustrate the basic methods for determining the ages of rocks.

TABLE 10-3   *Selected radioactive atoms*

Atom	Decay Products	Half-life[a]
U^{238}	$Pb^{206} + 8\ He^4$	4.51
U^{235}	$Pb^{207} + 7\ He^4$	0.71
Th^{232}	$Pb^{208} + 6\ He^4$	13.9
Rb^{87}	Sr^{87}	46
K^{40}	Ar^{40}	1.25

[a] In billions of years.

After a molten rock cools sufficiently, any helium from uranium and thorium decay will be trapped, as will the argon (40) from potassium (40). Hence measures of the uranium, thorium, and helium content of an igne-ous rock determine the age since it cooled; similarly for argon (40) and potassium (40). In the case of rubidium (87) and strontium (87) the age determination applies from the time that the mineral composition ratio of rubidium to strontium became fixed —that is when the rock first became solid, not necessarily cool. The oldest earth rocks by these methods are found to be somewhat older than three billion years. But how old is the earth since it formed? Obviously the rocks could have been molten any length of time before these age determinations apply. The methods used to find the real age of the earth are more complicated, involving tests of many rock samples for uranium, thorium, and lead content, including the other nonradioactive lead isotopes. C. Patterson, from an extensive study of many rock samples from many parts of the earth to make corrections for the original lead-isotope abundances, concluded that the earth itself was formed about 4.6 billion years ago. Other studies confirm this age to an accuracy of perhaps 0.1 billion years.

The same methods applied to meteorites (Chapter 15) indicate the same age of formation, indicating that the meteorites (and asteroids?) were formed concurrently with the earth.

The amount of radioactivity in earth rocks is much greater than in meteorites and too great to be consistent with the heat flow through the crust of the earth, if the entire earth is assumed to contain as much radioactivity as the crustal rocks. In other words, the radioactive elements of the earth must be highly concentrated in the crust. The conclusion is consistent with chemical properties of the radioactive elements and the evidence that the earth has been thoroughly melted. The radioactive elements happen to combine chemically with the lighter elements that floated to the top. Thus the average abundance of radioactive elements in the earth need be no greater than in the meteorities, consistant with a fairly uniform source of elements in the formation of both the earth and the meteorites.

CHAPTER 11

The Moon

A hypothetical visitor from space would very likely think of the earth as a double planet, for the moon, compared with its primary, is indeed a giant among satellites. The two objects interact in many ways. We are particularly interested in the effects that the moon has on the earth. Among these are the precession of the equinoxes and tidal phenomena already studied, and eclipses, which we shall discuss later in this chapter. First, however, we must account for the most simply observed feature of the moon—its phases.

11-1 Phases of the Moon. Figure 11-1 depicts the phases of the moon with respect to the earth as the moon passes through its regular sequence: new, crescent, first quarter, gibbous, full, gibbous, last quarter, crescent, and new again. The changing shape results from the varying angles at which we view the sunlit hemisphere from the earth. The interval between successive quarter phases is just over a week. The unusual term *syzygy* refers to the situation when the earth, moon, and sun are in line (apart from the inclination of the moon's orbit)—that is, at new or full moon.

As the earth-moon system revolves about the sun, the moon apparently revolves about the earth in $29^d\ 12^h\ 44^m\ 2^s.78$, on the average, through the

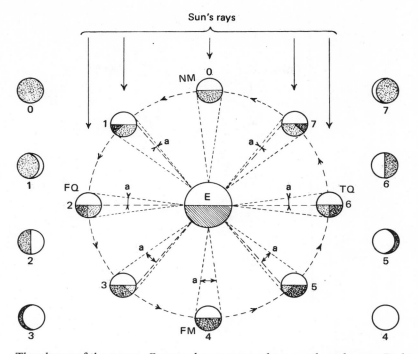

FIGURE 11-1   The phases of the moon. *Center: the geometry of sun, earth, and moon. Right and left: phases as seen from the earth; 0.* New; *1.* Crescent; *2.* First Quarter; *3.* Gibbous; *4.* Full; *5.* Gibbous; *6.* Third Quarter; *7.* Crescent. *The angle a indicates the maximum visible illuminated region.*

phases described above, the so-called *synodic month*. The *sidereal month*, the true period of revolution of the moon, measured with respect to the stars, is two days shorter than the synodic month, and takes 27^d 7^h 43^m $11^s.47$. The synodic month is longer because we measure it from new moon to the next new moon again. In the meantime, the earth has moved nearly 30 degrees forward in its orbit so that the moon completes its sidereal circuit before it returns again to the same phase. Figure 11-2 illustrates this phenomenon.

We can derive the relationship between the lengths of these two months and the year from Figure 11-3, where P is the length of the sidereal month, S the length of the synodic month, and Y the length of the year. As the sun, moon, and earth, starting from position 1, move into position 2, we find the relationship

$$\frac{1}{P} = \frac{1}{S} + \frac{1}{Y}.$$ (11-1)

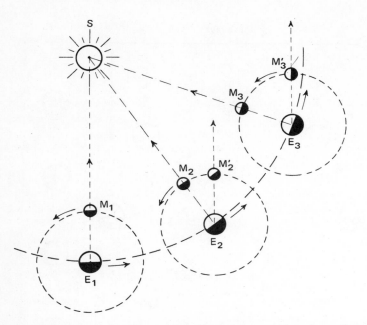

FIGURE 11-2 Two synodic months. *Note that one sidereal month would have elapsed at M_2', before a complete synodic month, at M_2.*

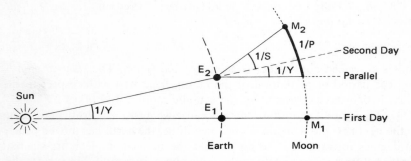

FIGURE 11-3 One day's motion of the moon. $Y = year$, $S = synodic\ month$, $P = sidereal\ month$. *Angles are in fractions of circumference.*

A similar relation applies to the apparent and sidereal periods of planets as seen from the earth and sun, respectively.

The average lunar day, from one sunrise to the next, at any point on the moon is *precisely* equal to the synodic month (29.5 earth days). That is to say, the moon rotates on its axis once a month, keeping the same face towards the earth at all times. We can imitate this rotation by holding a ball or globe in both hands at arm's length and turning around through 360°. The ball will also have turned through 360° but the same side will always remain in view.

11-2   The Moon's Orbit.   We are concerned here with the orbit of the moon about the earth. At all times the moon's path is concave towards the sun, and coincides with the earth's orbit to an accuracy of a quarter of one per cent (Fig. 11-4). Relative to the earth, however, the moon's orbit is a changing ellipse with an average eccentricity of about 0.05, at an average mean distance of 384,401 km (238,856 miles). Perturbations of the earth and sun cause the eccentricity to vary between about 0.04 and 0.06, while the actual distance of the earth to the moon ranges between the limits of 356,334 km (221, 463 miles) and 406,610 km (252,610 miles).

The perturbations also change the orbital inclination slightly about its average value of 5°, causing the nodes (the points where the moon crosses the ecliptic) to slide westward through a complete revolution in about 18.6 years, the period of nutation discussed in Chapter 5. At the same time, the point of nearest approach to the earth, the *perigee*, moves eastward through a complete revolution in only 8.85 years. The first of these two motions is called the advance in the line of apsides and the second is the regression of the nodes. The exact mathematical formulation of the many hundreds of periodic terms expressing the moon's motion constitutes one of the most fascinating and complicated problems of celestial mechanics, and one too involved for our consideration. The solution of the problem by the American astronomer E. W. Brown early this century occupied a lifetime and is one of the greatest triumphs of applied mathematics. Electronic computing machines today carry out the drudgery of such work at an enormously rapid rate, not only making the numerical calculations but actually deriving and printing out the expansions of complicated mathematical functions.

FIGURE 11-4   *Moon's orbit drawn to correct scale is always concave towards the sun.*

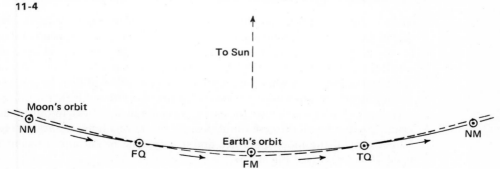

11-3 Librations.

The numerous perturbations in the moon's motion serve a useful purpose. They permit us to look over the moon's edge and, to a limited extent, see part of the "other side of the moon" without space vehicles. Those components of the lunar motion that allow us to see more than one-half of the moon's surface, called *librations*, may occur in several ways. Because the pole of the moon's rotation is inclined about $6\frac{1}{2}°$ to the pole of its orbital plane, we can see a bit beyond the poles of the moon.

FIGURE 11-5 *Libration effects. Two photographs at the same gibbous phase show differences in orientation caused by libration. (Lick Observatory.)*

Then, orbital revolution is nonuniform because of eccentricity, while the rate of axial rotation remains constant. As a result, rotation appears to lag (at perigee) and then to get ahead (at apogee), so that we see first around one limb and then around the other. Also, during the course of a day we can observe the moon when we ourselves are first west and then east of the line joining the moon and earth centers—a parallax effect. Furthermore, by traveling north or south over the earth, we can add still more to our lunar vista. Combining all of these librations, we can at one time or another see 59 per cent of the moon's surface and at any time be certain of seeing a fixed 41 per cent (see Fig. 11-5).

FIGURE 11-6 *The moon's back side televised by Orbiter IV, May 1967. (U.S. National Aeronautics and Space Administration.)*

A great achievement of the space age is the close-up photography of the moon, exposing the hidden side, first by the Russian space vehicle Lunik III in October, 1959. Immensely improved pictures of the unknown side of the moon are now available from the several U. S. NASA Orbiter spacecraft and Apollo missions (see Fig. 11-6).

11-4 Solar Eclipses. By a peculiar quirk of fortune the average length of the cone of shadow that the moon casts in the sunlight just about equals the distance of the moon. An observer within the shadow cone sees the sun in *total eclipse*. However, an observer within the partial shadow of the moon, the *penumbra* (Fig. 11-7), sees a *partial eclipse* in which the moon partly obscures the sun's disk. If the moon's dark shadow cone, the *umbra*, does not quite reach the surface of the earth, an observer within its extended cone sees an *annular eclipse*, or ring eclipse, in which a ring of the sun's disk appears to surround the moon completely (see Fig. 11-8).

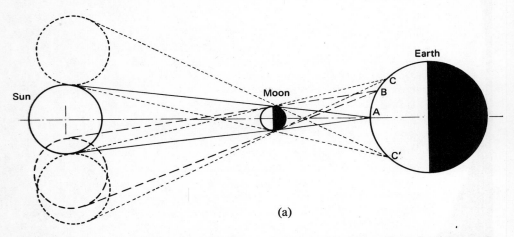

(a)

FIGURE 11-7 *Geometry of eclipses. (a) Observer A sees a total or an annular solar eclipse; observer B sees a partial solar eclipse; observers C and C' see no eclipse. (b) Zones of shadow in space: umbra (total eclipse), annular (solar eclipse only), and penumbra.*

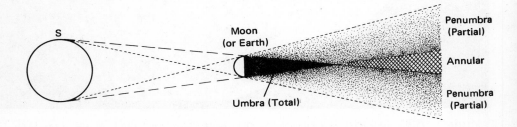

(b)

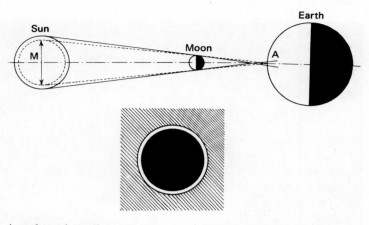

FIGURE   Annular solar eclipse. *Geometry for observer at A and appearance.*
11-8

A total eclipse of the sun is, without question, the most magnificent of all astronomical phenomena. For about one hour after *first contact*, when the moon's edge first encroaches on the sun's disk, the moon slowly eats away the west side of the sun until finally, at *second contact*, it completely covers the sun's disk. During this interval, the sun's brightness slowly diminishes, the air cools, and the sounds of animal life around slowly die away as the uncommon event unfolds.

Just before and after totality, ripplelike *shadow bands* move rapidly across smooth open surfaces on the ground, while crescent-like images of the sun appear under foliage. The shadow bands arise from irregular waves in the earth's atmospheric layers, much like similar shadow bands at the bottom of a swimming pool. The small spaces between the leaves of foliage act as so many pinhole cameras to project the crescent sun on the ground beneath.

The eerie quality suddenly intensifies a thousandfold as the solar corona flashes into view around the darkened sun. If the eye has been adapted to the diminishing light intensity by very dark glass, the corona registers as a brilliant spectacle even though it radiates no more light than the full moon. Totality can last for a maximum of $7\frac{1}{2}$ minutes, but most eclipses are of only several minutes' duration. The observer, who may have traveled many thousand of miles to see the eclipse. receives in those few minutes soul-satisfying reward for his efforts.

At *third contact*, the western edge of the sun emerges with extreme brilliance, momentarily breaking forth in a serrated crescent called Baily's beads. These brilliant points of light, visible at either second or third contacts, represent the thin rays of sunlight shining through the valleys of the moon's edge (Fig. 11-9). They were first recorded in 1780 by the American astronomer, Williams, 56 years before Baily's report. A single conspicuous bead and the inner corona show photographically as a magnificent "diamond ring."

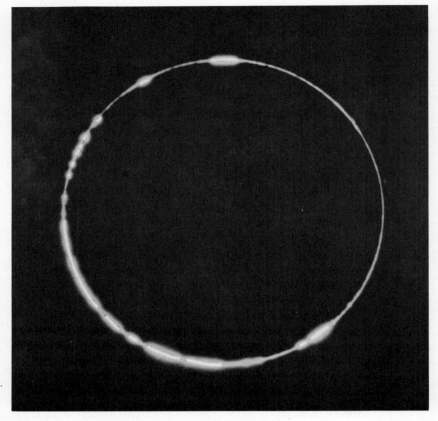

FIGURE
11-9 Baily's beads: *Solar eclipse of 1930. (Lick Observatory.)*

During totality we can see such extremely interesting features of the sun as its brilliant chromosphere, prominences, and the corona. Until recently, astronomers without special optical equipment (see Chapter 9) could observe these phenomena of the sun's atmosphere only at the time of solar eclipse.

11-5 **Lunar Eclipses.** Compared with total eclipses of the sun, lunar eclipses are relatively unspectacular affairs. As the moon moves into the penumbra of the earth's shadow, its light slowly dims. As it enters the umbra, the moon fades and assumes a coppery hue (Fig. 11-10). During the partial eclipse, when the earth casts its large circular shadow over part of the moon, the shadow boundary is not particularly sharp and the moon appears rather drab and distorted. During totality the moon is usually brighter at one edge than the other, but rarely does it become completely

(a)

**FIGURE
11-10**

*Partial phases of lunar eclipse of
October 1949. (G. de Vaucouleurs,
Péridier Observatory.)*

(b)

(c)

invisible to the eye. The earth's atmosphere refracts an appreciable amount of sunlight to the moon's surface. Since the earth's atmosphere absorbs more blue and violet light than it does yellow and red, its refracted light imparts to the eclipsed moon the characteristic coppery color, which changes in hue with the solar cycle, an unexplained phenomenon.

Lunar eclipses have made possible important observations concerning the cooling and nature of the lunar surface, since the sunlight is cut off and on again relatively quickly (see Sec. 11-12).

11-6   **Geometry of Eclipses.**   Eclipses can occur only at new or full moon. Because the moon's orbit is tilted 5°38′ to the ecliptic, eclipses occur only when the new or full moon happens near a node of its orbit. Otherwise the earth or moon shadow passes above or below its target (Fig. 11-11). The

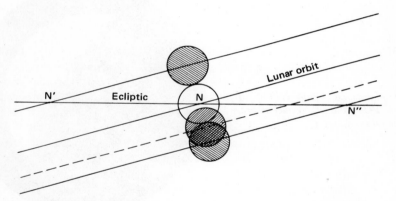

FIGURE
11-11
Ecliptic limits for eclipses. *If a new or full moon occurs within the ecliptic limits, N′N″, of the node N of the moon's orbit, an eclipse may occur.*

major ecliptic limit represents the maximum distance of the new or full moon from its node when an eclipse can possibly take place. The minor ecliptic limit represents the corresponding minimum distance, when an eclipse *must* take place. A partial lunar eclipse occurs when the edge of the moon just grazes the umbra of the earth's shadow, which on the average is some 9200 km (5700 miles) in diameter at the moon's distance. A partial solar eclipse occurs somewhere on the earth when the edge of the moon's penumbra just grazes the edge of the earth itself. Since the moon's penumbra is approximately 6400 km (4000 miles) in diameter and the earth's 12,700 km (8,000 miles), the sum is 19,100 km (12,000 miles), considerably greater than 9200 + 3500 = 12,700 km (5700 + 2000 = 7600 miles) for the earth's shadow at the edge of the moon. Hence, we see that the ecliptic limits for solar eclipses must be greater than the ecliptic limits for lunar

TABLE   *Eclipse data*
11-1

	Ecliptic Limits		Number Eclipses	
	Maximum	Minimum	Max/yr	Min/yr
Solar	11°50′	9°55′	5 or 4	2
Lunar	12°15′	9°30′	2 or 3	0
Total	—	—	7	2

eclipses. Table 11-1 contains values of these ecliptic limits as well as the numbers of solar and lunar eclipses per year, a subject that we shall now consider.

Since twice the minor ecliptic limit for solar eclipses, about 31°, is greater than the angular motion of the moon in a month, 29°.1, the moon cannot pass a node without producing some type of solar eclipse. Thus at least two solar eclipses must occur each year, one near each node. Furthermore, the moon moves less than twice the solar ecliptic limits per month, so it may possibly cause two solar eclipses for each nodal crossing, and even an extra one at the begining or end, because the nodes regress some 19° a year. A lunar eclipse must occur within each such pair because of the size of the lunar ecliptic limits. We see then that a maximum of seven eclipses may occur in one calendar year, either 2 lunar and 5 solar or 3 lunar and 4 solar. The moon, however, may slip by the minor lunar ecliptic limit, either just ahead of or just behind the earth's shadow, without producing a single eclipse during the year.

The longest possible total solar eclipse, up to more than 7 minutes, occurs when the new moon is simultaneously at a node and at perigee, while the earth is at aphelion. In this situation the moon attains its maximum, and the sun its minimum, apparent diameter. The observer is best placed as close to the moon as possible—that is, at the equator with the moon near his zenith. Furthermore, the earth's rotation gives the observer a maximum eastward velocity of about 1600 km/hr (1000 miles/hr) near the equator, helping him keep up with the moon's shadow, which travels at some 3200 km/hr (2000 miles/hr).

Why is it, then, that although solar eclipses outnumber lunar eclipses in the ratio 4 to 3, we rarely see a total solar eclipse without making a trip, whereas we see total lunar eclipses fairly frequently? The answer is that we can observe a lunar eclipse from more than half the earth, but a total solar eclipse only from within the narrow band of the eclipse shadow. The rarity of total solar eclipses at any one spot of the earth is apparent from Figure 11-12, which shows the path of totality for the eclipse of March 7, 1970. On the average, a total solar eclipse will be visible at a given point once in about 360 years.

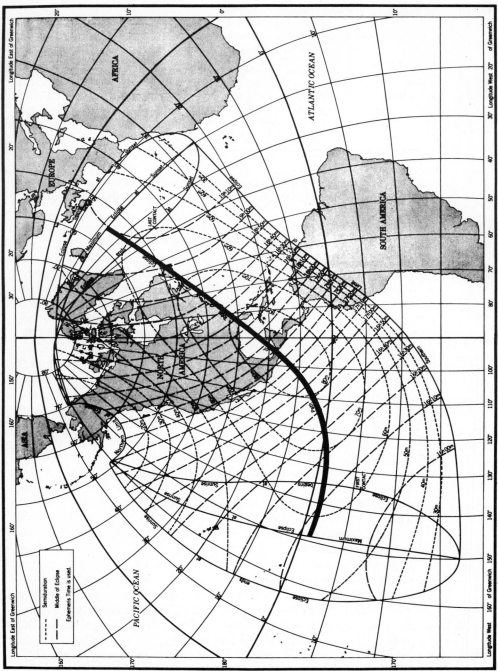

FIGURE
11-12

(*Left*) *Solar eclipse chart March 7, 1970.*
(*American Ephemeris and Nautical Almanac.*)

Eclipses repeat in a cycle known as the Saros, recognized by the Greeks and named by Edmund Halley three centuries ago. It happens that 19 eclipse years (earth revolutions measured from the lunar nodes), 6,585.78 days, are almost equal to 223 synodic months, 6,585.32 days. The latter period is the Saros. In the same interval the elliptical orbit of the moon has also swung around into nearly the same position it occupied at the original eclipse. Hence, similar eclipses recur at intervals of a Saros, except that the following eclipse lies about 8 hours further west in longitude on the earth, because of the 0.32-day difference from the integral number of days in 223 synodic months. After three Saros intervals, 54 years 33 days (or a day less, depending on how many leap days intervene), another eclipse occurs at much the same position of the earth (Fig. 11–13).

FIGURE
11-13

Paths of total solar eclipses, complete for North America from 1950 to 2000 A.D. The eclipses of 1918, 1936, 1954 and 1972 belong to the same Saros.

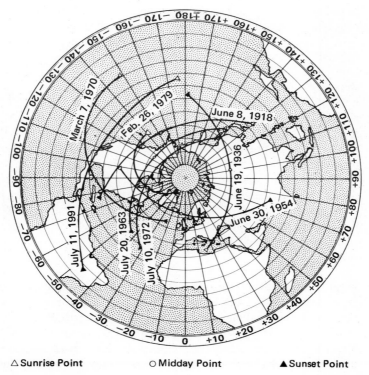

△ Sunrise Point   ○ Midday Point   ▲ Sunset Point

Solar Eclipses. Solar eclipses are valuable to science even in ʌ of orbiting observatories. They permit us to observe the sun's ʌd its surroundings without glare from the scattered light in our ʌere. They give us precise information as to the relative positions ʌarth, moon, and sun. Eclipse observations tell us about the sun's ʌg atmosphere of prominences, spicules, chromosphere, and corona, aʌ ʌey permit measurements of the heights of mountains at the edge of the moon and the positions of stars close to the sun. Einstein's theory of general relativity predicts that the light of stars appearing near the sun's edge should be deflected outward by 1″.76. Various astronomers, since 1919, have made these extremely difficult measurements from eclipse photographs and have generally confirmed the correctness of Einstein's prediction.

Even the radio astronomer finds solar eclipses extremely important. Since the sun's outer atmosphere strongly emits radio waves, the "radio sun" is larger than the visible sun. Hence, all radio eclipses are annular. From the rate at which solar radio emission fades during an eclipse, the radio astronomer can determine how big the sun appears to be at the radio frequency to which his receiver is tuned, and how it varies with general solar activity. Since the moon occults some of the ultraviolet and X-ray radiations responsible for the electrification of the ionosphere, an eclipse profoundly affects radio transmission of the daylight side of the earth.

Lunar eclipses are much less valuable, scientifically, than solar eclipses, but they do help us find out about the nature of the moon's surface (see Sec. 11-12) as well as the refraction, absorption, and scattering of light in the earth's atmosphere.

K. Fotheringham in England used the historical record of eclipses to prove that the speed of the earth's rotation is slowing down. Although the ancient observers of solar eclipses left no accurate records of the instants of the eclipses, a knowledge of the location where ancient eclipses had been observed on the earth was sufficient for Fotheringham's analysis. From Greek and Babylonian observations he calculated with surprising accuracy the times of the eclipses, and then compared the results with predictions from modern data. He found that the day is lengthening by 0.001 second per century, which adds up to an acceleration of 0.6 minute per century, a few hours of total change within historic times.

11-8 Why is the Earth Slowing Down? Emmanuel Kant presented the best answer in 1754, before the effect had yet been observed. Because the tides produce actual motion in the waters of the earth, they dissipate earth-moon orbital energy as heat of friction. In 1919 G. I. Taylor calculated that the tidal friction, mostly in the relatively shallow Bering Sea, could produce the observed lengthening of the day. Some forty years later, during which interval the tidal-friction theory was accepted generally, W. H. Munk and G. J. F. MacDonald showed that the known frictional effects account for only about one-third of the total power, some four billion horsepower, required to slow down the earth. Thus presently we have no entirely satisfactory answer to the problem. However, few if any geophysicists question that some form of tidal friction is responsible.

As the earth's rotation slows down, the *angular momentum* of the earth-moon system must remain constant, according to Newton's laws. We have already met this concept in Kepler's second law (Sec. 3-3). There we find that the product of the planet's mass, solar distance, and velocity perpendicular to the solar direction remains constant. If the products of these quantities, mass × distance × velocity, are added up for the earth and moon about their center of gravity (see Sec. 7-10) and for each particle of matter in both the earth and moon about their respective axes of rotation, the total sum is the angular momentum of the system. This sum remains constant unless some outside force changes it; for example, the sun might create tidal friction in the earth and slow its rotation rate. Thus lunar tidal friction that slows the earth's rotation reacts on the moon to move it into a larger orbit, transferring angular momentum from earth rotation into earth-moon orbital motion.

Sir George Darwin demonstrated that the moon will slowly recede from the earth and the day will slowly lengthen until the earth finally keeps the same face toward the moon as the moon already does for the earth. At that time the month and day will both equal 47 of our present days and the moon will be 550,000 km (340,000 miles) away. But this condition will not occur until some 50,000 million years in the future, according to Sir Harold Jeffreys. Carrying this process backward in time, Darwin concluded that the earth and moon may once have been almost in contact, with the day and the month both equal to about 4 hours. There is reason to doubt this latter conclusion, but certainly the moon must once have been much closer to the earth than it is now.

11-9   The Moon's Size and Shape.   We have seen that the moon is nearly a sphere, of radius of 3476 km (2160 miles), about one-fourth the radius of the earth. We should like to know its true shape, and particularly its rigidity, to understand its past history.

Ralph B. Baldwin reinvestigated the measured data on the shape of the moon and found that it bulges in the direction of the earth some 2200 meters (7200 feet). Darwin had earlier interpreted similar results as representing a "frozen tidal bulge," which developed as the moon cooled while closer to the earth. Still earlier, Newton had predicted such an effect. The bulge could measure the size of the earth tide on the moon when the moon "froze" and hence fix the distance between them at that time.

Jeffreys calculates this critical distance as about 90,000 km (60,000 miles), occurring some 60 million years after the Moon was formed. The reality of a frozen tidal bulge is questionable.

The motions of NASA's lunar orbiters led P. M. Muller and W. L. Sjogren of the Boeing Aircraft Company to a remarkable discovery. Gravitational concentrations of mass, or *mascons*, occur in perfect coincidence with the great Maria. Imbrium, Serenetatis and Humorum show the strongest effects, the Imbrium mascon being equivalent to 1/50,000 of the Moon's mass.

The lunar samples brought back by Apollo 11 from Mare Tranquillitatis contain high-density basaltic rock, strongly indicating that the maria are truly lava flows. The lava density of 3.3 times that of water equals the *average* lunar density! Surveyor VII, landing near the highland-crater Tycho indicates a lower density basaltic content for surface material there. Thus we have strong evidence for some isostasy (Sec. 10.7) on the Moon, the highlands floating because they are composed of somewhat less dense material than the average material at their bases. Further confirmation of this theory is provided also by Surveyor VII, whose position was measured to be about 3 km (2 miles) greater in altitude from the center of the moon then the other four Surveyor craft that landed on maria near the equator, confirming earth-based measurements of lunar topography. Thus the high-density mare surface lava appears to be depressed with respect to the lower density highlands.

With a mean density of 3.3 times that of water the moon may consist mostly of ordinary basic rocks, probably lacking a dense (iron?) core such as the earth appears to possess. Indeed the moon could well be composed from average materials of the earth's upper mantle, or materials drawn from the same source. Urey calculates that if the moon had the same average composition as the entire earth, it would be 4.4 times as dense as water. John Wood points out, however, that if the moon had never been heated greatly, the iron might remain in combination with other materials, to reduce the calculated density. Final resolution of the moon's composition awaits accurate evaluation of its internal temperature.

As the moon has only 1/81.3 the mass of the earth, its surface gravity is only $\frac{1}{6}$ that of the earth. The velocity of escape from its surface is some 2.4 km/sec (1.5 miles/sec) (see Sec. 7-13). Thus a high-powered rifle could almost shoot a projectile from the moon to the earth. A subsatellite in orbit near the moon's surface requires 109 minutes per revolution, slightly longer than the theoretical 83 minutes for a satellite near the surface of the earth.

11-10   The Moon's Lack of Atmosphere.   No one has yet found the slightest optical evidence for a lunar atmosphere. Stars occulted by the moon's limb disappear almost instantly [diffraction (see Sec. 6.4) scatters some light] and do not fade out as they would if the light passed through a scattering atmosphere. Furthermore, stellar occultations show no systematic lag, such as would occur if the starlight had been refracted appreciably around the limb of the moon. No one has observed an aurora borealis or night skylight near the surface of the moon. Meteors have not certainly been observed in the lunar atmosphere.

A more sensitive optical test of a lunar atmosphere is the search for twilight effects. A. Dollfus has made refined measurements of polarized (Sec. 8.2) light near the edge of the sunlit portion of the moon at first and third quarters and set the limit of a lunar atmosphere at less than 10^{-9} of the earth's atmosphere.

The time of obscuration of a radio star provides an even more sensitive measure of the moon's atmosphere. Observing a lunar occultation of the Crab Nebula at 3.7 meters wavelength, B. Elsmore set the limiting density of the moon's atmosphere at 2×10^{-13} times that of the earth's atmosphere at sea level, roughly equal to the terrestrial value at an altitude of 700 km (430 miles). At such a density the moon might show aurorae, if it had a magnetic field. But no aurorae have been observed. Magnetic measures near the moon's surface made by the U.S.S.R. spacecraft Lunik III indicated a magnetic field less than one-thousandth that of the earth, now reduced to less than a millionth by measures from U.S. spacecraft.

The moon must, of course, have some slight trace of atmosphere, but probably not a permanent one. The sun sends out a "wind" and also clouds of ionized gas that strike the moon, while cosmic rays contribute a trace of gas. Meteors knock out some gas in their explosive encounters, as do the solar particles and the cosmic rays. The moon itself exudes some gas from its interior, as exemplified by the famous observation of N. A. Kozyrev in the U.S.S.R. on November 3, 1958. While guiding the slit of his spectrograph on the central peak of Crater Alphonsus, he observed that the peak "became strongly washed out and of an unusual reddish hue," and two hours later he "was struck by its unusual brightness and whiteness." The spectra he took apparently showed bright bands of molecular carbon. There are a number of such observations of temporary cloudy effects on the moon.

Why does the moon have so little or practically no atmosphere? The answer lies in its low velocity of escape, only 2.4 km/sec or one-fifth the earth value. An atmosphere of light gases, such as oxygen, nitrogen, or carbon dioxide, would escape molecule by molecule because of their rapid molecular speeds even at moderate temperature. The average speed varies as the square root of the absolute temperature and the inverse square root of the molecular weight. Hydrogen would escape almost instantly from the moon, even at a temperature of 0° C, because the average speed of the molecules is 1.87 km/sec (1.2 miles/sec) and a large fraction of them moves faster than 2.4 km/sec at any moment. Even carbon dioxide would disappear slowly but surely from the moon, while oxygen and nitrogen, which are somewhat lighter, would escape more rapidly.

Among the inert gases—helium, neon, argon, krypton, and xenon—only xenon, of atomic weight 131.3, might remain for billions of years. In fact, however, the atoms in violent solar flares probably play an important role in "blowing away" and in determining the nature of the lunar atmosphere, which may actually be transient, replenished continuously by leakage of gas from within the moon and by capture of atoms from interplanetary space.

11-11   Surface Features of the Moon.   No celestial object surpasses the moon as a subject for contemplation and study with a small telescope. From hour to hour the *terminator*, or boundary between the dark and

**FIGURE
11-14**   *Lunar terrain near terminator. Craters Aristoteles (lower) and Eudoxus (upper) with tip of Alpine Valley (lower right). This region shows in opposite illumination near the bottom right of Fig. 11–15. (120-inch reflector, Lick Observatory.)*

FIGURE 11-15   *Moon at first quarter. (Lick Observatory.)*

sunlit hemispheres, sweeps at a slow but appreciable rate across the moon's surface. On the dark side of the sunrise terminator, between new and full moon, giant mountains first show as brilliant points against the blackness as the sun's rays strike a peak, while the valleys still lie in darkness. The mountains cast long shadows as the surrounding lower altitudes slowly rotate from darkness into daylight. The shadows rapidly shorten as the sun rises and, 24 hours later, a once conspicuous mountain may no longer stand out against the bright surface. Figure 11-14 shows the terminator near the large crater Eudoxus. This region is bracketed in Fig. 11-15, where the whole moon is shown at first quarter.

Thus the appearance of the lunar surface depends strikingly upon the solar illumination. Mountains, craters, and small surface irregularities stand out in bold relief near the terminator. At full moon, on the other hand, these objects change character or almost disappear (see Fig. 11-16). In a telescope, hazy naked-eye fantasies such as the "Man in the Moon" resolve into the great maria, regions once thought to be seas. Mare Imbrium (see Figs. 11-16 to 11-18) is one of the largest, an oval dark area stretching over more than 1100 km (700 miles) of the moon's surface. In all, the maria cover roughly a third of the moon's surface visible from

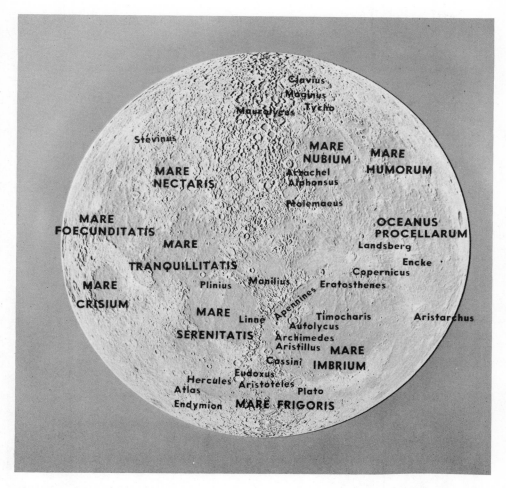

FIGURE 11-17   *Map of near side of moon.* (*Lick Observatory. From* Field Guide to the Stars and Planets, *D. H. Menzel.*)

earth. However, they are almost absent from the far side (Fig. 11-6). They appear relatively smooth as compared with the rest of the lunar surface. Although Mare Imbrium means "sea of showers," no rain has fallen on it for billions of years, if indeed it ever did. The upper layers of the maria must consist largely of lava flows (see Sec. 11-12).

At full moon, lunar photographs also show conspicuously the bright ray systems that emanate from some of the larger and more perfect craters. Rays fanning out from Tycho (see Fig. 11-16) can be traced almost completely around the moon's visible surface. These rays show up most prominently near full moon, whereas nearly all other types of details show much more clearly when they are near the terminator.

The lunar craters have probably created more controversy among students of the moon than all other lunar problems combined. The greatest clearly defined crater on the near side of the moon is Clavius, some 235 km (146 miles) in diameter. Small telescopes show numerous smaller craters. With large instruments we can count many thousands, down to small depressions less than 1 km (0.6 mile) in diameter; on TV photos taken by lunar spacecraft the number is fantastically great. The portion of

FIGURE 11-18   *Mare Imbrium. Encircled by the Apennines (left), the Carpathians (upper center), and the Alps (lower left). The large crater near the center is Archimedes. (Mount Wilson and Palomar Observatories.)*

the lunar surface visible from earth possesses eleven craters greater than 160 km (100 miles) in diameter, and about 1300 with diameters ten times smaller. Several hundred of the largest craters have been named after astronomers of the past. Others, such as Plato, Archimedes, and (on the far side) Lomonosov, honor philosophers and scientists.

The larger craters range in form from the most perfect, such as Copernicus (see Figs. 11-19 and 11-20), which probably appears much as it did when formed, to dilapidated structures with low, partly missing walls. The "most perfect" craters are nearly circular and cuplike and usually show a

FIGURE   *Copernicus. Diameter 57 miles. (100-in. telescope, Mount Wilson and Palomar*
11-19   *Observatories.)*

FIGURE 11-20 *Copernicus. View from south by U.S. Orbiter III showing central mountains in foreground. (National Aeronautics and Space Administration.)*

central peak. The outer walls rise gently above the surrounding plain and then fall steeply from the rim towards the interior of the crater, often with terraces and landslips. Many craters, however, may be described as *ring plains*, having nearly flat floors edged by relatively steep walls. Only a few of these ring craters contain central mountain peaks.

The central peak of lunar craters never rises above the outer walls, at least when we measure heights with respect to the curved surface of the moon. A number of craters, such as Clavius, are so large that the walls are not visible from the center because the curvature of the moon brings the walls below the horizon.

We see the "older" craters generally marred or partially obliterated by a confused array of newer and smaller craters. At times we can scarcely recognize the presence of the original crater walls. In many of the chaotic areas we can arrange the smaller craters from old to new, in terms of their relative perfection and superposition. Smaller craters occur at random on the larger ones.

Certain of the maria are outlined by mountain ranges. For example, the Alps, Caucasus, and Appenines surround Mare Imbrium. These latter mountains rise some 5.5 km (18,000 ft) above the level of the Imbrium plain, to compare favorably in height with their terrestrial counterparts. The Alpine range is distinguished by a great gash called the Alpine Valley, some 75 miles long and several miles wide (Fig. 11-21). It is radial to the apparent center of Mare Imbrium and probably represents a great crack formed with the mare and then filled with lava. An even greater rift has been discovered by the U. S. Orbiter spacecraft on the far side of the moon near the south pole (see Fig. 11-6).

Among the most striking of the lesser lunar features are the *rills* or *clefts*, which appear as long, shallow cracks. Note the long rill centered in the Alpine Valley (Fig. 11-21). We find that some rills extend for many

FIGURE 11-21   *U.S. Orbiter 5 photograph of Alpine Valley.* (*National Aeronautics and Space Administration.*)

kilometers, with occasional branchings sometimes much like forks in lightning flashes. Oftentimes the rills change into ridges as we follow their general course. These ridges look quite different in character from the *serpentine ridges* also visible on the maria. The serpentine ridges look like pressure markings on the surface of viscous materials such as tar.

We can determine the heights of mountains and the depths of craters and rills, and, in general, reconstruct the complete topography of the lunar surface, by measuring the length of shadows near the terminator and by stereoscopic pictures made in succession by lunar orbiters. In all cases the instant of the observation specifies the geometrical circumstances of the sun, moon, earth or spacecraft, and shadows to a high degree of precision. The reconstruction of the moon's surface from lunar photographs is a tedious but practical process. Comprehensive lunar photographs have been carefully collected by G. P. Kuiper into an invaluable lunar atlas. The reader will find a study of this atlas highly rewarding and may enjoy comparing a direct telescopic view of the moon, near quarter phase, with the best photographs. Immense projects for mapping the entire moon in three dimensions are now underway in the U. S. Space Program. Radar mapping from earth is an important adjunct to close-up photographs of the moon.

11-12   Physics of the Moon's Surface.   By direct inspection from the earth, we learn the general nature of the lunar features down to the limiting resolution of our telescopes and atmosphere, to details as small as 0ʺ.1 or 0.2 km (600 ft) in diameter. From lunar spacecraft the limit is almost microscopic. Study of the brightness variations of the moon with phase adds considerably to our knowledge. Figure 11-22 shows that the lunar brightness increases very rapidly near the full phase. We interpret this change of brightness in terms of reflection, not from a flat, reflecting surface but rather from an extremely irregular surface covered by craters of all sizes, rocks, and rubble down to microscopic dimensions. Such a surface is a poor reflector, and the moon's total average reflecting power, or *albedo*, is only 0.07—that is, only about 7 per cent. The surface is almost black, comparable to dark slate.

E. Pettit and S. B. Nicholson, with the 100-inch telescope at Mount Wilson, first made a thorough study of the temperature of the lunar surface. They used a vacuum thermocouple, which registered the total amount of radiant energy falling upon the thermocouple. By the radiation laws of Chapter 8 they calculated temperatures on the moon's surface.

At full moon, about half of the moon's visible surface is hotter than boiling water, with the maximum temperature, as we would expect, lying near the subsolar point. As seen at quarter phase, however, the temperature of the subsolar point is only 81° C when it lies at the limb. Therefore, the surface we see obliquely is some 30° cooler than the one we see perpendicularly.

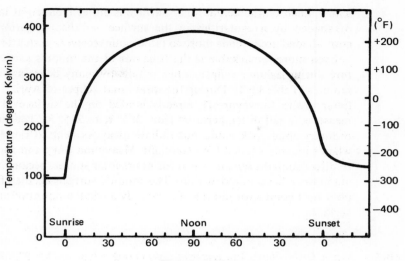

Elevation of Sun

FIGURE 11-22 *Temperature and light curve of moon. (a) Temperature on lunar equator during a lunar day. (b) The distance from the center of the diagram to the solid curve measures the relative brightness at the indicated phase. Near new moon the brightness is shown as magnified 10 and 100 times.*

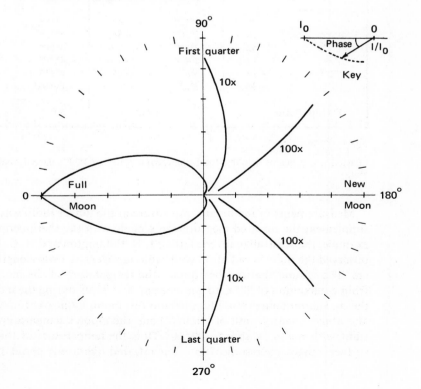

Thus, much of the actual lunar surface is either inclined at large angles, or shaded by irregularities in the surface, so that radiation from the cool, shaded areas tends more to fall on our receivers at quarter phase.

Even more remarkable is the behavior of the moon's surface temperature during a lunar eclipse, when it falls by many degrees in only a few minutes (Table 11-2). During the total lunar eclipse of April 24, 1967, the lunar probe, Surveyor III, already landed on the surface of the moon, measured a fall in temperature from $385°$ K to $215°$ K. The surface of an ordinary solid rock could not radiate away its heat so rapidly to space when suddenly shaded from sunlight. Heavy masonry and rocks exposed to direct sunlight remain warm for hours after sunset, because of the large amount of heat stored within. The moon's surface is, therefore, a very poor heat conductor indeed, storing only a small amount of heat near the surface.

TABLE 11-2   *Lunar temperature* (T *at center of disk*) *during eclipse of Oct. 27, 1939*[a]

Time	T (°C)	Time	T (°C)	Time	T (°C)
Before	96	1^h18^m	-76	3^h33^m	-89
Partial Begins[b]		1^h30^m	-78	3^h40^m	-60
0^h13^m	85	1^h42^m	-82	3^h55^m	-18
0^h25^m	75	1^h56^m	-86	4^h02^m	$+11$
0^h37^m	52	2^h14^m	-90	4^h11^m	$+36$
0^h49^m	22	2^h40^m	-93	4^h21^m	$+62$
1^h2^m	-14	3^h8^m	-95	4^h32^m	$+89$
1^h8^m	-30	3^h27^m	-95	Partial Ends	
Totality Begins		Totality Ends		After	$+100$

[a] By E. Pettit.

[b] Partial phase begins at 0^h0^m; total phase from 1^h13^m to 3^h33^m; partial phase ends at 3^h48^m.

Measurements of the lunar temperature at the longer radio wavelengths supplement the infrared measurements made with the thermocouple. For example, the Australian investigators, J. H. Piddington and H. C. Minnett, observed the moon's radiation with radio receivers at a wavelength of 1.25 cm. They found that on the equator the temperature of the moon varies from a maximum of $28°$ C to a minimum of $-75°$ C during the month, but the maximum temperature lags behind full moon by nearly four days. At and above a wavelength of about 10 cm, the moon's temperature is constant with phase at about $-50°$ C. Thus the temperature of the moon's surface ranges over some $290°$ C in infrared light, only about $100°$ C at wavelength 1.25 cm, and very little at wavelengths above 10 cm. A straightforward explanation accounts for these apparently contradictory results.

As the lunar Surveyors and Apollos have shown, the moon's top surface is covered by a thin layer of loosely packed dust or dirt, which need be only millimeters in thickness. Fine-grained material is an excellent insulator in air, and even more effective in a vacuum. Each space between the grains acts like a small thermos bottle. Because of its insulating qualities the topmost layer of porous material cools very rapidly as the earth's shadow crosses it during an eclipse, and afterwards heats quickly. But almost none of the heat flows into the probably more solid interior. In the range from heat radiation to the much longer radio waves, the radiation comes from increasingly deeper layers and hence shows smaller variations in temperature. Thus the temperature at a meter (yard) or more below the moon's surface, as observed by radio waves 10 cm or more in wavelength, is fairly constant, near $-50°$ C, about what we might expect if there is no appreciable heating from inside the moon. The persistent low temperatures observed at long radio wavelengths confirm this conclusion. On earth, we need go only a few meters (yards) below the surface to find a nearly constant temperature from summer to winter. The greenhouse effect (see Sec. 10-2), fortunately, keeps the earth's mean temperature considerably higher than the moon's, while internal heat adds a little to the temperature.

W. H. Wright and B. Lyot have each analyzed the polarization of the light reflected from the lunar surface to find out about its composition. Wright concluded that the lunar dust is high in silica content, like pumice or granite, while Lyot could satisfy his observations by a lunar surface of volcanic ash. The composition of the lunar surface is best determined by direct space exploration (see Chapter 16). The Soviet spacecraft Luna 10, April, 1966, measured gamma rays from the lunar surface in the wavelength range from 0.004 to 0.08 angstrom units (10^{-8} cm). The gamma rays produced in lunar material by radioactive atoms suggest that the lunar surface is more like basaltic volcanic magmas in composition than like stony meteorites or the deep interior of the earth. The chemical measures by the U. S. Surveyor V that soft-landed on Mare Tranquillitatis in 1967 confirm this suggestion as do the later Surveyors and the lunar samples returned by Apollo 11. The oxygen, aluminum, magnesium, silicon, and heavier-element composition of the Mare Tranquillitatis rocks is similar to that of the earth's outer mantle, typified by basaltic lava except that the lunar rocks contain more oxides of heavy elements and have the extraordinarily high density of 3.3 times water. Typical basalt has a density of about 3.0. Even more striking is the titanium abundance, more than ten times that characteristic of earth basalts, along with high abundances of zirconium and yttrium. Carbon compounds and water are conspicuously missing.

The spacecraft soft landings on the moon have clarified our understanding of its surface. The U. S. Surveyor III, in April, 1967, not only took

(a)

FIGURE
11-23
(a), (b) *U.S. Surveyor VII scoop on moon's surface. The trenches are about two inches wide. (National Aeronautics and Space Administration.)*

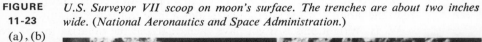

(b)

close-up pictures, as did Russia's Luna 9, in February 1966, but the U. S. Surveyors have also scooped up surface material (see Fig. 11-23). The Apollo 11 astronauts, Neil A. Armstrong and Edwin A. Aldrin, Jr., actually selected samples on July 20, 1969 and returned them to earth. These studies, from various points on the equatorial maria and one from the highlands, show a consistent character. The upper surface material is mostly porous except on rocks, but has some rigidity or coherence. The walls of the artificial trenches do not collapse, nor do the conspicuous footprints of the astronauts slump as they would in soft sand (Fig. 11-24). The direct exploration of the moon thus gives information also consistent with our earth-based deductions. The Moon's top layer is a rubble-heap of material redistributed by the explosive impacts of high-velocity meteorites. Because of the large energies per unit mass in explosions of ordinary meteoric bodies, the top layer consists almost entirely of pulverized lunar material. Large meteorites blast deep craters into lunar rock or lava and eject the rocks that are so frequent on the surface. The low velocity of escape from the moon permits

FIGURE 11-24   *Apollo 11 expedition on the moon. Edwin E. Aldrin, Jr. standing beside seismograph with laser retroreflector (center) and lunar excursion module in background. (By Neil A. Armstrong. U.S. National Aeronautics and Space Administration.)*

a fraction of the explosion products to escape from the moon forever. The moon appears to be losing rather than gaining mass by meteoritic bombardment. This process was suspected by the Russians from their Luna 9 pictures, which seemed to show some rocks on pedestals, like sand-eroded rocks in our desert regions. The Apollo 11 mission found more positive evidence. Partially buried rocks have shoulders at the lunar surface and some exposed rocks appear to be free of dust on their upper surfaces, where many tiny meteoritic impact craters can be seen. The meteoritic contribution is small, however, probably less than a meter in thickness per billion years. The matter overturned and "reworked," on the other hand, is a hundred to a thousand times greater, so that craters many meters in diameter have probably been obliterated during the moon's history. Thus small craters provide no measure of the rate of meteoritic bombardment.

The Apollo 11 samples presented an unexpected result, however. The lunar soil contains more than 50 per cent glass! The explanation is simple and should have been anticipated. High-velocity impacts produce a certain amount of glass, both by the quick cooling of impact melted mineral oxides and directly by shock without melting. Continuous reworking of the surface material by impacts thus increases the glass content to a high fraction.

The glazed or glassy surfaces seen on some of the lunar rocks in the truly remarkable close-up color stereo photographs made by the astronauts of Apollo 11 and on some of the rock samples seem readily explained by the heat produced in high-velocity meteoritic impacts.

The U. S. Ranger spacecraft photographs show that only 1.3 per cent of the lunar surface is tilted more than 12° over dimensions of 10 cm (4 in.) to 4 m (4 yd), and 30 per cent more than 4°. This result is consistent with radar measurements, which fail to indicate the extreme roughness observed at wavelengths of visible and infrared light. Radar echoes are highly concentrated to the central portion of the disk. The moon is still a poor reflector at radar wavelengths, with about the visual albedo of 0.07. Again this observation is consistent with an upper layer of broken and crushed lunar material that has partially reconsolidated over the ages.

A suggestion once made that fine dust on the moon hops about under electrostatic charge repulsion and actually flows to fill the lunar maria and older craters is disproved by the lunar landings. Also laboratory tests show that dust in vacuum generally tends to stick together, because no gas is present to prevent surface molecules from attaching to any material they touch. The extreme darkness of the lunar surface, darker than most blackboards, is not typical of rock dust or finely pulverized brittle material. Thorough analysis of the returned lunar samples should clarify this question.

The lunar rays, bright at full moon, appeared rough and rocky to Kuiper from earth-based observations. The U. S. Ranger and Surveyor probes confirm that the rays are pitted with secondary craters and with rocks.

Also they show that many rocks thrown about the moon are much whiter than average surface material. It follows that the rays are probably fairly thick debris excavated in crater formation from some depth in the moon where a whiter type of rock layer or layers exist. Whiteness alone, however, cannot account for the increased brightening of the rays as compared to other areas near full moon. Perhaps the increased roughness of "new" explosive debris as compared to "weathered" surfaces is adequate to account for the rays. Perhaps the rays contain a considerable fraction for fairly transparent glass beads that selectively reflect backwards, as do the tiny beads on some highway signs. In any case the whiteness persists for as yet unknown periods of time until the rays are obliterated by the slow destructive processes on the moon.

FIGURE
11-25

Hot spots on moon. Measured in infrared during a lunar eclipse. Compare to Fig. 11-16. (R. W. Shorthill and J. M. Saari, Boeing Aircraft Co.)

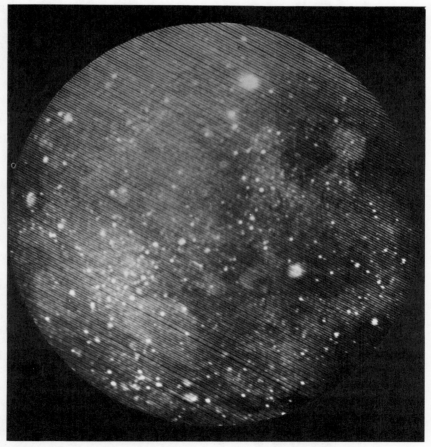

Some new craters with rays and many regions on the moon show as "hot" spots during lunar eclipses, when most of the surface has cooled to very low temperature (see Fig. 11-25). The higher temperature, however, does not signify the presence of any kind of volcanic activity. Heat conduction from the subsurface layers must be much better in these "hot" spots than on the average lunar surface. Large rocks, strewn about cold but relatively recent lava flows that have not been exposed long enough to develop a sufficiently thick layer of dusty rubble, provide a likely explanation for the "hot" spots. One such region occurs in and around the crater Tycho. That steep walls of new craters show the effect provides strong evidence to support this theory of the "hot" spots.

FIGURE 11-26(a) *The straight wall in Mare Nubium near first quarter. (Yerkes and Mount Wilson and Palomar Observatories.)*

(a)

11-13   The Nature and Origin of the Lunar Features. Much controversy has centered on the origin of the lunar craters. Prior to 1873, when
R. Proctor revived F. v. P. Gruithuisen's earlier suggestion of a meteoritic
origin for the craters, most astronomers favored geological processes as
the most likely explanation. Bubbles in cooling lava and volcanic action in
the later stages of a cooling moon were the favorite theories. In 1893, however, the great geologist G. K. Gilbert strongly supported the meteoritic
hypothesis and added enormously to the subject.

We find clear evidence for geologic processes, including volcanism, fos-
silized on the moon. Near the great crater of Copernicus and on other
parts of the moon *crater chains* occur, series of relatively small craters

FIGURE   *The straight wall in Mare Nubium near third quarter.* (*Yerkes and Mount Wilson*
11-26(b)   *and Palomar Observatories.*)

(b)

distributed like beads on a string as in Figs. 11-19 and 11-20. Almost everyone would agree that these must result from a vent type of volcano located along *faults* or cracks in the moon's surface. Some, of course, are secondary craters produced by debris from the great crater. The "Straight Wall" in Mare Nubium (Fig. 11-26) is universally accepted as a true fault, some 110 km (70 miles) long with a steep (41°) face rising some 350 m (1000 ft) above the almost level surface of the maria. A number of other faults exist on the moon. Orbiter photographs show that craters in large numbers occur along such faults. Similarly, the rills appear to be true cracks or crevasses, which have never been filled in by a terrestrial type of erosion on the moon, while some, such as the Alpine Valley (Fig. 11-21), show evidence of *dike*-filling by lava.

Numerous small domes, many with central craters, appear on certain areas in the maria (Fig. 11-27). These structures have clearly risen from their surrounding plains and show the characteristics of certain types of volcanoes, particularly by the presence of central craters, which would be ridiculously improbable statistically by meteoritic impact.

FIGURE 11-27 *Mare Procellarum near Crater Marius (upper right, 25-mile diameter). (Televised by U.S. Orbiter II, National Aeronautics and Space Administration.)*

The U. S. Ranger pictures show that in the maria and in small flat plains within some larger craters, many small dimple-shaped craters occur, in contrast to the "cup"-shaped craters characteristic of impact explosions. The dimpled formations must arise from subsidence phenomena in lava flows where "pockets" have permitted the surface to cave in. On a larger scale, evidence of subsidence phenomena appears almost anywhere in the maria and the flat areas of the moon, along with rills with craters threaded along them. On both highlands and the maria there exist several huge moonwide networks of parallel markings, both on small and large scales, indicating great strains and distortions over great areas of the lunar surface. The great serpentine ridges, some only tens of meters high and extending for many kilometers, show clearly that surface distortion occurred as the result of plastic flow, like that of a distorted tar surface. Examples of nearly all geological processes, except possibly those associated with water and wind erosion, will eventually be found on the moon. Immense "geological" activity has certainly taken place.

But is the moon geologically active today? No major changes such as new craters have been observed from earth, but there are many records of evanescent cloudlike phenomena seen in certain areas such as the crater Alphonsus. There, in 1958, N. A. Kozyrev of the U.S.S.R. obtained a spectrum showing bright bands. Probably gas is indeed seeping out from the depths of the moon, irregularly perhaps, because of earth-induced tidal strains, and from certain more active regions. Local exploration and the continuing seismic studies of moonquakes from implanted seismographs such as that left by Apollo 11 should decide how much activity is currently present.

Some high-resolution photographs taken by the various Lunar Orbiters show narrow valleys, meandering from higher levels on to the maria surfaces, where they finally disappear (Fig. 11-28). Various scientists have suggested that flowing water may have been the active agent in this type of erosion. If so, the moon must once have possessed a fairly extensive atmosphere. More likely these sinuous rills represent the flow of lava rather than the flow of water. Such *lava* gutters are frequent in terrestrial volcanic regions.

And how about meteorites and the formation of large craters? Again the evidence is clear. Even at only 2.4 km/sec (1.4 miles/sec), the velocity of escape from the moon, incoming masses have energies comparable to the chemical energies of similar masses of high explosives. Very fast meteoric bodies coming at from 11 to 72 km/sec (7 to 42 miles/sec) carry far more energy per unit mass than any chemical explosive. They release this energy immediately upon striking the lunar surface, much like a nuclear bomb. Hence they would produce craters nearly circular in cross section, even though they struck the surface at a low angle. Since the meteoric body would blow out a crater far larger than its own volume, the amount of matter immediately piled up above the surrounding surface level of the

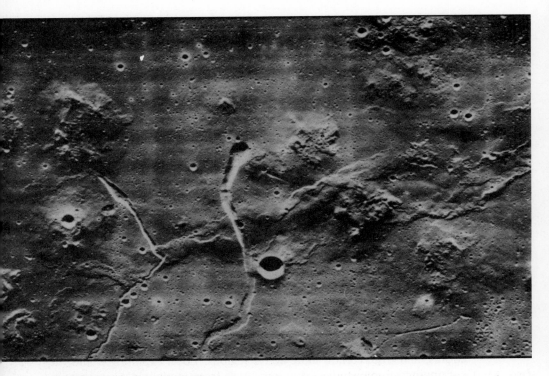

**FIGURE
11-28**
"Cobra Head" formations in the mare near Marius. North up. (National Aeronautics and Space Administration.)

moon should be comparable to that removed from below the surface. Schröter, about 1800, found that indeed such a relationship does hold approximately for many of the lunar craters.

Most volcanos on earth carry molten lava through a vent to the surface and pile up a mound of lava and ashes, with the crater nesting on top, much like the lava domes mentioned above. No such lava flows show definitely on the moon, even on close-up pictures of the walls of the larger lunar craters. Other types of volcanism or plutonic activity more nearly match the lunar craters, but do not produce very large craters on the earth. Furthermore, the number of large craters seen on the maria agrees quite well with the number calculated to be formed in four billion years by the present supply of "Apollo" asteroids (see Chapter 14). Comet nucleii may well add to the supply of large infalling bodies. On earth the now recognized meteorite craters, called *crypto-volcanic* features by geologists, have been forming at about the same rate. Some of the larger lunar craters, however, may have occurred by processes akin to those that produced terrestrial "calderas," of which some of the Hawaiian craters are well-known examples.

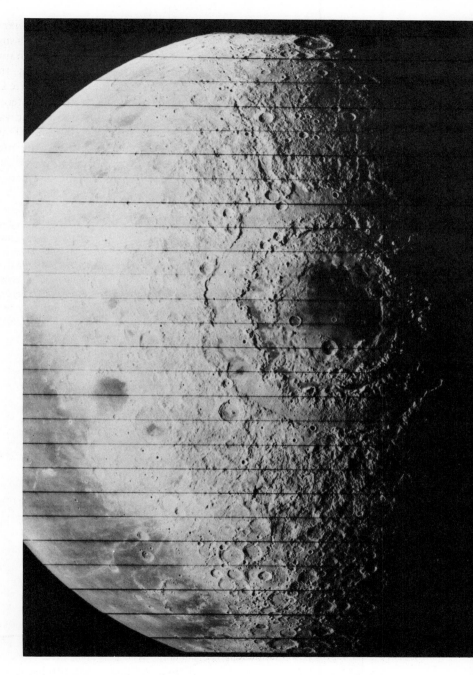

FIGURE *Mare Orientale, barely visible from earth, televised. U.S. Lunar Orbiter IV.*
11-29 *(National Aeronautics and Space Administration.)*

Accepting a meteoric origin for the majority of the larger lunar craters, we must face the question as to whether the maria, as Gilbert has suggested, originated in a similar manner. The existence of great mountain chains outlining Mare Imbrium, for example, and the splash marks and huge scars (such as the Alpine Valley) radiating from the center of Mare Imbrium over much of the moon's face, argue strongly for a titanic meteoric collision. The magnitude of the necessary explosion, however, dwarfs any extension of our experience, for the mare is some 1100 km (700 miles) in major diameter. Mare Orientale, a combination crater-mare mostly on the moon's far side, just beyond Oceanus Procellarum, shows an enormous crater 1000 km (600 miles) in diameter surrounded by three consecutive

FIGURE 11-30 *Observed and calculated changes of temperature during a total eclipse of the moon. Upper curve, rock covered with thin (0.6 mm) layer of insulating dust. Lower curve, rock covered with thick (7 mm or more) layer of dust. Actual measures of the temperature changes during the total eclipse of December 18–19, 1964 appear as dots (Tycho) or crosses (environs of Tycho). The latter appear to be covered with dust, whose insulating power reduces the total heat content and therefore permits more rapid cooling. Thus, during an eclipse or even after sunset, any area not so covered stays warm longer than its surroundings. (Harvard College Observatory.)*

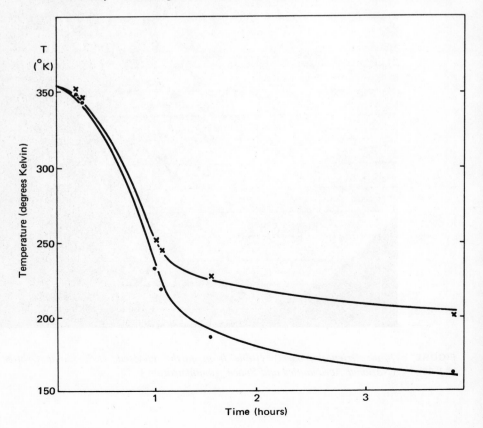

rings of mountains. The force of the explosion has devastated the neighboring area, splashing debris for hundreds of kilometers (Fig. 11-29). An explosion of similar scale on the earth, occurring let us say in Texas, would demolish cities as distant as Chicago, burying them under a pile of rubble. Early in its history, however, the earth apparently experienced similar encounters. Erosion and weathering have removed the conspicuous pock marks from the earth, while the moon is a museum of ancient scars and a library of ancient records.

But impact explosions alone cannot account for the many geological features observed on the maria and, indeed, in the flat-floored craters. Heated material from inside the moon clearly responded to the enormous holes catastrophically produced and welled up in them—but was this response slow or rapid? The great circular mountains thrown up were too heavy for parts of the lunar crust and subsidence occurred, sometimes irregularly or partially, sometimes symmetrically. The throes of gravitational readjustment, thermal expansion, and thermal contraction produced huge numbers of small and large tectonic effects, which will provide geologists with research data from the moon as complicated as those they have found on the earth.

A major puzzle concerns the lack of maria on the far side of the moon. We have seen that the moon was partially differentiated by limited melting, so that the highlands of the moon may truly be masses of lower-density rock that floated to the top as the continents did on the earth. If so, great meteorites may have been only small triggering devices that, falling at the right times and places, set off large tectonic processes. How thoroughly has the moon been heated? These and many related questions about ancient times make the exploration of the moon a modern scientific adventure.

CHAPTER *12*

The Terrestrial Planets

The nine principal planets fall into two sharply contrasting classes. The *terrestrial planets* resemble the earth in being relatively small and dense, with thin atmospheres, or in some cases practically none at all. This group includes Mercury, Venus, Earth, Mars, and Pluto. The Moon, though actually a satellite and not a planet, is physically rather similar. The other group comprises the *giant planets*, Jupiter, Saturn, Uranus, and Neptune, described in Chapter 13.

The *interior planets*, Mercury and Venus, show *crescent* phases between greatest elongations and *inferior conjunction*, and *gibbous* phases between greatest elongations and *superior conjunction*. At inferior conjunction the planet is nearest to the earth and between the earth and the sun (when exactly so, a transit in front of the sun's disk takes place); at superior conjunction it is beyond the sun and farthest from earth. The maximum elongations of the interior planets are defined by the tangents to the orbit drawn from the earth (Fig. 12-1). The *superior planets*, from Mars and beyond, are not so limited, and their elongations may reach up to 180°, at which time they come into opposition with the sun. When the elongation is ±90°, these planets are in *quadrature* with the sun. The phase angle between the earth and the sun at the planet varies from 0° to 180° for the interior planets and from 0° to some small angle much less than 90° for the exterior planets. Mars has the largest phase angle of the exterior planets, amounting to 48° at maximum. The main elements of the orbits and physical constants of the planets are listed in Tables 12-1 and 12-2.

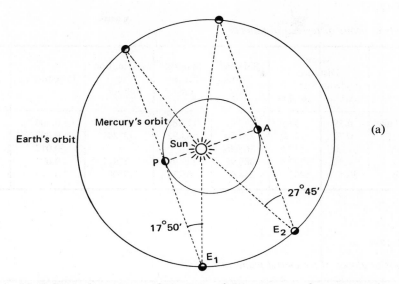

FIGURE (a) *Orbit and maximum elongations of Mercury, much greater near*
12-1   *Mercury's aphelion, A, than perihelion, P.* (b) *Transits of Mercury*
          occur only in May and November.

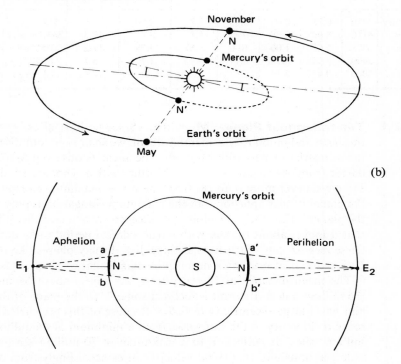

TABLE 12-1

Elements of orbits of terrestrial planets

	Mean Distance from Sun: A.U.	10^6 km	Sidereal Period, days	Synodic Period, days	Inclination to Ecliptic	Eccentricity	Mean Orbital Velocity, km sec^{-1}
Mercury	0.39	57.9	87.97	115.88	7°00′	0.206	47.9
Venus	0.72	108.2	224.70	583.92	3°24′	0.007	35.1
Earth	1.00	149.6	365.26	—	—	0.017	29.8
Mars	1.52	227.9	686.98	779.94	1°51′	0.093	24.1
Pluto	39.5	5902	248.4[a]	366.74	17°08′	0.249	4.7

[a] Years.

TABLE 12-2

Elements of globes of terrestrial planets

	Equat. Radius: km	$\oplus = 1$	Vol., $\oplus = 1$	Mass, $\oplus = 1$	Density, g cm^{-3}	Surface Gravity, m sec^{-2}	Escape Velocity, km sec^{-1}	Rotation Period	Incl. of Equat. to Orbit
Mercury	2420	0.38	0.05	0.05	5.2	3.80	4.2	59 ± 3 days	<28°
Venus	6110	0.96	0.88	0.82	5.1	8.7	10.3	243.16 days	88°
Earth	6378	1.00	1.00	1.00	5.52	9.78	11.2	23^{h}56^{m}4^{s}1	23°27′
Mars	3400	0.53	0.15	0.11	3.97	3.72	5.1	24^{h}37^{m}22^{s}7	25°12′
Pluto	3000?	0.47?	0.1?	0.8?	—	—	—	6.39 days	—

12-1 The Innermost Planet: Mercury. Since we have already considered the earth and moon in considerable detail, we shall begin our discussion of the terrestrial planets with Mercury. The ancients observed Mercury, as it darted from one side of the sun to the other with a synodic period of about 116 days, never more than 28° from the sun at maximum elongation when the planet is near aphelion, but the maximum elongation is only 18° when the planet is near the perihelion of its eccentric orbit (see below). We know that it moves about the sun with a true sidereal period of 88 days, and in telescopes we can watch its planetary disk display phases like the moon. Its brightness at maximum never quite equals that of Sirius.

The mean distance of Mercury from the sun is about 58 million km (36 million miles). Its orbit is inclined some 7° to the plane of the ecliptic and has a large eccentricity of 0.206. Because of this eccentricity the distance of Mercury to the sun varies from a minimum of 46 million km (29 million miles) at perihelion to a maximum of 70 million km (44 million miles) at aphelion. Its orbital velocity varies accordingly from 39 km/sec

(24 miles/sec) at aphelion to 57 km/sec (36 miles/sec). In keeping with its mythological attributes, Mercury is the fastest of all planets.

Its small eccentric orbit also furnishes a valuable check on Einstein's theory of relativity (Chapter 34). After all planetary perturbations have been allowed for in classical theory, the planet's perihelion moves eastward about 43″ per century. This observed effect has resisted all efforts to explain it by Newtonian gravitational theory. In the nineteenth century some astronomers tried to explain it by perturbations from a hypothetical planet, Vulcan, moving inside Mercury's orbit, but after some false alarms the postulated planet was never confirmed. Attempts to introduce ad hoc modifications to Newton's law were quickly rejected. Relativity theory, however, predicts such an eastward motion at a rate in almost perfect agreement with the observations. Thus Mercury provides one of the three astronomical proofs of relativity, a topic discussed in more detail in Chapter 34.

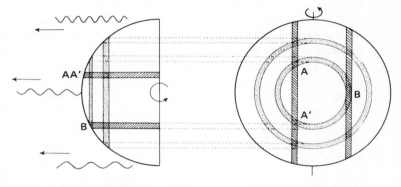

**FIGURE
12-2**   *Radar mapping of a planet uses differences in radial velocities and echo times of different zones. Hatched strips have constant radial velocity, circular zones constant echo time. There is ambiguity between A and A', except at equator in B.*

The intersection of the inclined orbit of Mercury with the plane of the ecliptic or line of nodes corresponds to positions of the earth's orbit that our planet crosses in May and November. Only during these two months can we see Mercury projected in front of the sun's disk (Fig. 12-2). *Transits* of Mercury occurred on May 5, 1957, and November 6, 1960; the next transits would be May 9, 1970, and November 9, 1973. During the May transits Mercury is near its aphelion, hence almost as near the earth as it can get, and its apparent diameter exceeds 12″; during the November transits it is near perihelion and its apparent diameter is under 10″. The disk of Mercury, observed in projection against the bright surface of the sun during transits, appears as a black spot, often shimmering and fuzzy because of telescopic and atmospheric effects. At such times the telescope is pointing straight at the sun, whose intense heat causes strong turbulence in the air outside and especially inside the tube of the instrument. Some strange luminous effects on and around the disk of Mercury caused by such disturbances were observed during transits in the nineteenth century and caused much needless speculation at the time.

When Mercury is observed outside transit near its maximum elongations from the sun, its apparent diameter is only about 6″. This angle is very small and difficult to measure precisely. For this reason the diameter of the globe of Mercury could not be measured precisely by optical means. About 4800 km (3000 miles) was regarded as probably within 100 km of the truth; this is 37 or 38 per cent of the earth's diameter. Radar techniques, however, can determine the diameter of a planet by the time difference between the earliest return of a pulse from the center of the disk and the latest return from the edge. The method is quite accurate for the moon. M. E. Ash, I. I. Shapiro, and W. B. Smith, with the 1000-ft fixed radio reflector at Arecibo, Puerto Rico, found the diameter of Mercury to be 4868 ± 4 km (3025 miles), within the range of optical uncertainty. Because Mercury rotates slowly (see Sec. 12-2), the centrifugal effect of rotation is negligible and the globe of the planet does not depart measurably from a sphere. The surface of Mercury is only 14 per cent and its volume only 5.5 per cent of that of the earth.

The mass of Mercury is difficult to measure because it has no known satellite; hence we cannot apply the straightforward method described in Chapter 7. Instead we must fall back on indirect methods, which derive a planet's mass from the minute perturbations it causes in the motions of other planets, particularly Venus, the earth, and the minor planet Eros, or more recently in the motions of space probes. By such methods the mass of Mercury was estimated at about 1/6,120,000 of the mass of the sun, or 5.4 per cent of the mass of the earth (or again 3.25×10^{26} g). With the radar radius, the average density of Mercury is about 5.4, close to that of the earth (5.517) and among the highest of all the planets (with the possible exception of Pluto, Sec. 12.3). This conclusion is surprising, because in many respects Mercury seems rather similar to the moon, which has a much lower density, similar to that of the surface rocks of the earth (Chapter 11). We can surmise only that during the formation of the planet, the proximity of the sun caused a drastic loss of light elements in the region where the planet formed, or else near the surface of the proto-planet, to leave only a thin mantle around a relatively large core of heavy elements (nickel-iron?).

From the mass and radius of Mercury we conclude that the acceleration of gravity at its surface is about 3.6 m sec^{-2} or 37 per cent of its value on earth. Also the escape velocity near the surface is 4.2 km sec^{-1} or 38 per cent of the earth's value. Hence all light molecules would have escaped easily from any atmosphere that Mercury might have possessed in the past, and only the heaviest elements such as the rare gases xenon and krypton could be retained in its atmosphere. Indeed a spectroscopic search has failed to detect the slightest trace even of relatively heavy carbon dioxide (< 0.5 meter atmosphere).

12-2 The Surface of Mercury. Through the telescope, Mercury resembles the moon as seen with the naked eye. Dark, permanent spots appear often in nearly the same relative positions with respect to the terminator at the same phase. From this fact the Italian astronomer G. V. Schiaparelli concluded in 1889 that Mercury presents always the same face to the sun, just as the moon does to the earth—that is, that the period of rotation was equal to the period of revolution, 87.97 days. This result remained unchallenged for 76 years, and in fact was repeatedly confirmed by more recent observations, until radar studies proved conclusively in 1965 that this result is erroneous and arises through a peculiar coincidence.

Radar echos bouncing off a planetary surface have their frequencies slightly changed by the Doppler effect caused by the motion of the surface relative to the observer. The wave reflected by the approaching half of the visible hemisphere has a slightly shorter wavelength than the wave reflected by the receding half. The amount of shift increases from the center of the disk to edge and in proportion to the projected distance to the rotation axis (a slight displacement results from the relative motion of the earth and planet). If the delay or echo travel time of the return signal is considered also, the surface of the planet is further resolved into concentric circular zones of constant range, so that the Doppler shift can be related to one or possibly two specific areas of the planet. In this way one measures the radial velocity of successive zones of the planet and from it the rotation period. In the simplest case, if $2V$ is the maximum difference velocity measured between opposite limbs of the planet (diameter D), the rotation period is $P = \pi D/V$.

In this fashion observations of Mercury with the radio reflector at Arecibo proved that the rotation of Mercury is direct and with a sidereal period of 59 ± 3 days around an axis inclined perhaps $28°$ to the normal to the orbital plane. This value approximately equals two-thirds of the sidereal revolution period. Further investigation of basic celestial mechanics showed that astronomers could have predicted this result theoretically. Because of the great eccentricity of the orbit of Mercury, most of the tidal braking of the original rotation period occurs when the planet is near perihelion, since the tidal forces are inversely proportional to the cube of the distance (Sec. 5-5). But at such times the angular velocity of the source of the tidal force, the sun, as seen from Mercury, exceeds the average value derived from Kepler's law of areas (Sec. 3-3) precisely in the ratio of 3 to 2. Hence the average rotation period locks on two-thirds of the revolution period. If this reasoning is correct, the exact sidereal rotation period of Mercury should be $87.97 \times 0.6666 \ldots = 58.65$ days.

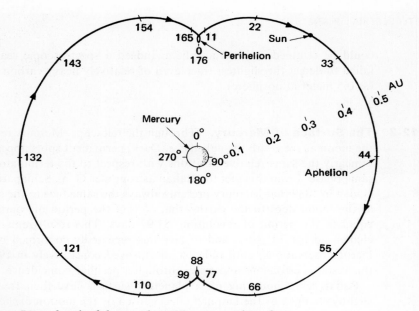

FIGURE
12-3

Diurnal path of the sun about Mercury resulting from combination of 88-day revolution and 59-day rotation. Relative positions of sun are marked at 11-day intervals. (After S. Soter and J. Ulrich.)

Furthermore, the day on Mercury is extraordinarily long. If the ratio of the rotation and revolution periods is indeed 2 : 3, the average interval from sunrise to sunrise is exactly twice the year, or 175 earth days. For certain longitudes and in the polar regions the sun's apparent motion can become quite complex, as a study of Fig. 12-3 will show.

A reexamination of the older data showed that most visual observations of spots on Mercury had been made at times that were multiples of both 87.98 and 58.65 days and could be satisfied by either period. And so we must discard the old belief that Mercury always turns the same face toward the sun. A map of Mercury based on visual and photographic observations reinterpreted in terms of the correct rotation period is shown in Fig. 12-4.

The amount of detail visible on Mercury even with large telescopes hardly exceeds what can be seen on the moon with the naked eye. The smallest details visible near quadrature are of the order of 0.″2 to 0.″3 or 150 to 250 km (100 to 150 miles). It is not surprising that mountains, if present, cannot be detected and that most of our information on the nature of the surface of Mercury must be deduced indirectly from its light-reflecting properties. Radar measurements do show that both the roughness and the reflecting power of Mercury's surface closely resemble those of the moon's surface.

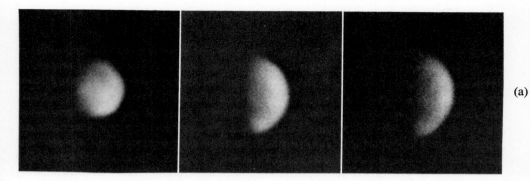

(a)

FIGURE 12-4 *Observations of Mercury by B. Lyot with 38-cm refractor in 1942, and A. Dollfus with 60-cm refractor in 1950 at Pic du Midi Observatory. (b) Map of Mercury from visual and photographic observations, 1942-1966. (A. Dollfus, Pic du Midi Observatory.)*

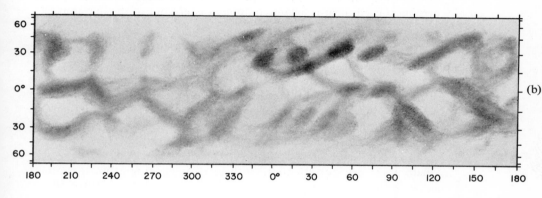

(b)

Photometry also shows that the variation of the intensity of light reflected by the planet sharply increases near full phase (that is, near superior conjunction), very much as moonlight does (Chapter 11). We conclude that Mercury, like the moon, must be covered with porous or granular rocks, or at least that it has a rough surface texture. The reflectivity of Mercury also resembles that of the moon. The visual albedo of the planet is about 0.063—that is, Mercury reflects 6.3 per cent of the incident sunlight, as compared with the moon's 7 per cent. Likewise the color of Mercury, measured by its color index (Chapter 20) of about +1.0 mag., is close to that of the moon.

The polarization of light reflected by Mercury and its variation with phase angle, measured by the French astronomer B. Lyot in the 1920's, is similar to that of the moon. Later polarization observations showing a slight excess in green light compared with red light have been attributed to molecular scattering in a very thin atmosphere, but this result is probably a gross overestimate because of a misinterpretation of marginal data. Reports of temporary haze veiling surface features have also been discounted.

The surface temperature of Mercury was measured in the 1920's and 30's by E. Pettit and S. B. Nicholson by means of highly sensitive vacuum thermocouples placed at the focus of the 100-inch telescope of Mount Wilson Observatory. The average temperature of the center of the illuminated disk (subsolar point) was 610° K, in good agreement with the theoretical estimate of 620° K computed from the laws of thermal radiation (Chapter 8). The eccentricity of the orbit should cause the temperature to vary from 560° K at aphelion to 650° K at perihelion. Lead, tin, and possibly even zinc would melt at the subsolar point on Mercury, the hottest point on any of the planets.

All in all, Mercury is a most unattractive prospect for human exploration, but inspection at close range by means of automatic space probes carrying rugged, heat-resistant equipment is not out of the question.

12-3 **The Red Planet: Mars.** Ever since G. V. Schiaparelli and P. Lowell suggested, toward the end of last century, the possible existence of higher life forms on Mars, this planet has captivated the attention of amateur astronomers and excited fantastic speculation among laymen. Mars is an unusually interesting planet because we can clearly observe its surface through the thin atmosphere, which in its basic meteorology sometimes resembles and more often differs from that of the earth. Mars is, of course, a high-priority target in the program of space exploration.

Before we turn our attention to these more exciting aspects, we must review the main elements of its orbit and globe. Mars is the fourth planet, the first beyond the earth. It moves around the sun in 1.88 years at a mean distance of 1.52 A.U. = 228×10^6 km (142 million miles). The orbit is noticeably eccentric (0.093). The distance from Mars to the sun varies from a minimum of 208 million km at perihelion to a maximum of 249 million km at aphelion. This orbital eccentricity causes an appreciable inequality in the length of the seasons, as follows:

	Spring	Summer	Autumn	Winter
Southern Hemisphere	146 d.	160 d.	199 d.	182 d.
Northern Hemisphere	199 d.	182 d.	146 d.	160 d.

As on the earth, but much more so, the southern hemisphere of Mars has a shorter and warmer summer and a longer and colder winter than the northern hemisphere (Fig. 12-5).

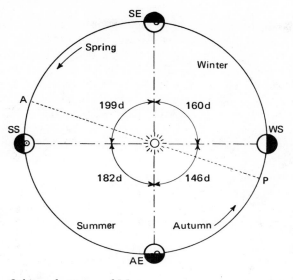

SE

Spring

Winter

A

199d   160d

SS   WS

182d   146d   P

Summer   Autumn

AE

FIGURE   *Orbit and seasons of Mars.*
12-5

The apparent diameter of Mars varies from a maximum of 25″ at peri-helic opposition, when it is nearest to us, to 14″ at aphelic opposition, and to a minimum of 4″ when in conjunction with the sun. Observations are more profitable when the planet is within a few months from opposition. The combinations of the motions of the earth and Mars cause the favor-able perihelic oppositions to return at intervals of approximately 15 years, during which seven oppositions take place. The synodic period is 780 days on the average. The opposition dates of the current cycle are as follows: September 11, 1956 (perihelic); November 16, 1958; December 30, 1961; February 4, 1963 (aphelic); March 9, 1965 (aphelic); April 15, 1967; May 31, 1969; August 10, 1971 (perihelic) (Fig. 12-6).

The mean diameter of Mars is known fairly precisely, being close to 6800 km (4200 miles) or 53 per cent of the earth's equatorial diameter. Its flattening, however, is somewhat in doubt. Visual and photographic ob-servations of the ellipticity of the disk indicate that the polar diameter is about 1 percent shorter than the equatorial diameter. This value is too large to be explained by the centrifugal force of rotation, which permits an equatorial bulge of only 0.5 per cent. The latter value is confirmed by an analysis of the perturbations of the satellite orbits, and the conflict with the optical estimates is as yet unresolved.

Mars has two small satellites, Phobos and Deimos, discovered in 1877 by A. Hall. Their main orbital elements are given in Table 12-3. Their ap-parent brightnesses indicate that these satellites are probably under 10

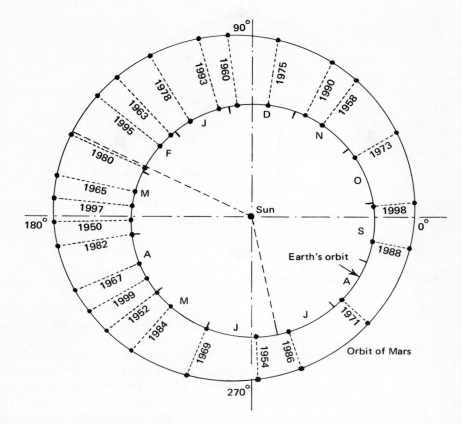

**FIGURE
12-6**
Orbit and oppositions of Mars 1958–1999.

miles in diameter. Phobos is so close to the surface of Mars (6000 km = 3700 miles) that its period of revolution is less than one Martian day. It is the only satellite in the solar system that, observed from the planet's surface, will appear to rise in the west and set in the east, traversing the sky in $4\frac{1}{4}$ hours at the equator. Deimos, moving very slowly from east to west, remains above the horizon for 60 hours at the equator and rises every 132 hours on the average. The light variations of the satellites suggest that they are irregular in shape, and they may conceivably be captured minor planets, although the probability of two such events is very small indeed.

The presence of satellites permits accurate calculation of the mass of Mars, which is 10.8 per cent of the earth mass or 6.4×10^{26} g. Since the volume of the planet is 15 per cent of the earth's, its mean density is 4.0, intermediate between the Moon (3.34) and the earth (5.5). Mars may possess a small nickel-iron core of some 1000 km (600 miles) radius. The surface gravity is about 38 per cent of the earth's, and the escape velocity

TABLE 12-3

Satellites of Mars

	Dist. from Planet, 10^3 km	Sidereal Period, days	Orbit Inclination[a]	Orbit Eccentricity	Radius, km	m_v[b] Discovery
Phobos	9.4	0.32	1°8	0.019	8	11.5 Hall 1877
Deimos	23.5	1.26	1°4	0.003	4	12.5 Hall 1877

[a] From planet's equator.

[b] Mean magnitude at opposition.

decreases from 5.0 km/sec at the surface to about 4.0 km/sec at a level some 1000 km above the surface, where atmospheric molecules can actually escape.

The surface markings of Mars can be clearly seen through its tenuous atmosphere. From observations of transits of the same markings at the central meridian of the disk during the past three centuries, the rotation period can be very precisely determined, most recently by J. Ashbrook, who fixed it at $24^h 37^m 22^s\!.669$ with an uncertainty of only a few thousandths of a second. We know the rotation period of Mars more accurately than that of any other planet except the earth. The axis of rotation is inclined 25° to the normal to the orbital plane, similar to the earth's 23°.5. Except for their longer duration the seasons on Mars are comparable to ours.

12-4   The Atmosphere of Mars.   The presence of an atmosphere around Mars has been known for a long time. The seasonal variations of the white polar caps require an atmosphere for the migration of the material (water or carbon dioxide) from one pole to the other. Occasional haze and clouds mask surface features, and an atmospheric veil obscures the surface most of the time in blue and violet light (Fig. 12-7a,b). The mass of atmospheric gases above a unit area of the surface was initially estimated from the scattering and polarization effects of this atmosphere. But the amount so derived was rendered suspect by the confusing effect of atmospheric haze. More reliable values, obtained since 1963 by spectroscopic studies of the CO_2 bands and by the atmospheric refraction of radio signals from the space probe Mariner IV at the time of its occultation by the limb of Mars on July 14, 1965, indicate a surface pressure of the order of 10 millibars, or about one per cent of the value at sea level on earth. Because of the weaker Martian gravity, however, the atmospheric pressure decreases more slowly with altitude. Whereas in the earth's atmosphere the air pressure decreases tenfold every 16 km (10 miles), one would have to rise some 21 km in the atmosphere of Mars to experience the same relative change. It follows that

(a) 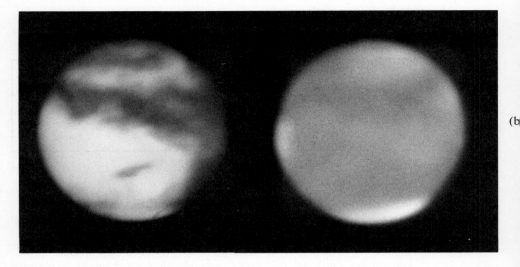 (b

**FIGURE
12-7** *Surface and atmosphere of Mars in red (a) and violet (b) light on November 22, 1958. (R. S. Richardson and A. G. Wilson, 60-inch reflector, Mount Wilson and Palomar Observatories.)*

atmospheric pressure on the surface of Mars approximates the pressure in our atmosphere some 35 km (22 miles) above ground (that is, in the upper stratosphere).

Another consequence of the small Martian gravity is that temperature must decrease more slowly with altitude than it does on earth.

Meteorological principles lead us to expect a temperature gradient of 4.5° C per km (13° F per mile) in the lower atmosphere of Mars as against an average of 6° C per km (18° F per mile) in the earth's atmosphere. These pressure and density factors indicate that weather phenomena and disturbances will be much more gentle on Mars than here.

The composition of the atmosphere is another basic consideration. The problem can be attacked by direct or indirect means. For instance, from general considerations of cosmic chemistry concerning the abundances of the elements in space (Chapter 17), coupled with the low escape velocity on Mars (Sec. 12-4), we can expect that neither hydrogen nor helium is present in significant quantities in the atmosphere of Mars. Of the common atmospheric gases, oxygen can be ruled out by the negative results of careful spectroscopic searches. The search for water vapor was also fruitless for a long time, but in recent years traces, perhaps variable with season, have been detected. Nitrogen, which cannot be detected in the spectral range transmitted through our atmosphere, was not observed by the U. S. spacecraft Mariner VI, and VIII in their 1969 close approaches to Mars. The only common gas definitely detected in the atmosphere of Mars is carbon dioxide, discovered spectroscopically in 1948 by G. P. Kuiper at McDonald Observatory. The amount present is just enough to account for the newer low values of the surface pressure. Other gases may, of course, be present, but probably only as traces, as in the earth atmosphere (Chapter 10).

The vertical structure of the atmosphere of Mars is still little understood. A dark haze layer obscures the surface details on photographs in blue, violet, and ultraviolet light (Fig. 12-7). We do not know how thick this layer is, or whether it floats high in the atmosphere or extends all the way to the surface. Various estimates of the altitude of its top vary greatly, ranging from less than 10 km to more than 100 km. The nature of the haze particles is something of a mystery. For a while many authors attributed it to merely a mist of minute crystals, possibly of water ice or perhaps of dry ice (carbon dioxide), at the deep-freeze temperatures of the upper atmosphere of Mars. However, the reflectivity of this assumed haze is surprisingly low, under 10 per cent in blue light and 5 per cent in the ultraviolet. This property is more suggestive of carbon smoke than of icy haze. In 1953 the Belgian physicist Rosen proposed that carbon particles might form in the Martian atmosphere, perhaps by decomposition of carbon dioxide, and in 1961 E. J. Öpik presented strong arguments in support of the idea that some unspecified material present in this atmosphere absorbs strongly and scatters light reflected by the surface. To make the riddle more puzzling, this blue haze seems to clear up almost completely at rare intervals, when the surface of the planet suddenly becomes visible on bluelight photographs. This phenomenon of the "blue clearing," discovered by E. C. Slipher at Lowell Observatory in 1937, has caused much speculation over the years, but no one has yet come up with a satisfactory explanation.

We recognize at least three types of clouds in the atmosphere of Mars, conventionally designated white, blue, and yellow (Fig. 12-8). The white and blue clouds appear usually near the edge of the planet, at the sunrise or sunset limb, or above the polar regions during autumn and winter. In the first position we are probably observing low-lying morning fogs (or possibly even hoarfrost on the surface) formed by nocturnal cooling, and on the evening side high-level condensations caused by convection during the day. In the polar regions we see a mist under which the polar caps are formed during the cold season. This mist must be very tenuous, because it does not appear on photographs in red and infrared light. The yellow clouds appear almost exclusively when the planet is near its perihelion and the atmosphere is often hazy. The accepted explanation is that near the time of maximum solar heating, exhanced convection whirlwinds carry surface dust into the atmosphere, much as the "dust devils" do in the deserts of the Southwest. On several occasions, especially in 1909, 1911, 1924, and 1956, spectacular and extensive dust storms blotted out surface markings over large areas for several weeks at a time. Explorers of Mars will have to be wary of such treacherous storms, which, from recorded cloud motions, may reach velocities in excess of 100 km per hour (60 mph),

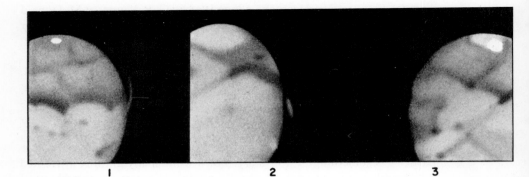

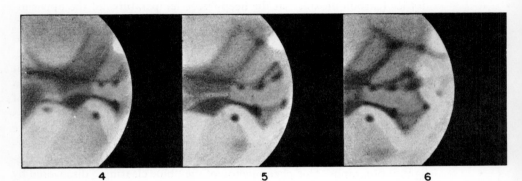

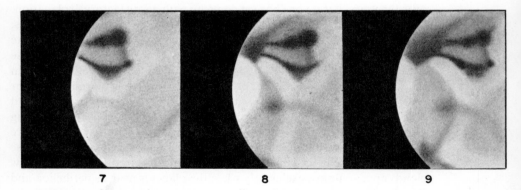

FIGURE 12-8

White and yellow clouds on Mars. 1,2. High-altitude clouds at sunrise appear partly or fully detached from the terminator. Elevation: 20 to 30 km. 3. Low-altitude clouds at sunset are detected by distortions of the terminator. Elevation = 5 km. 4,5,6. Hoar-frost or low-level icy-fog patch at the morning limb evaporates shortly after sunrise and does not appear to rotate with the topography over 2-hour time interval. 7,8,9. Motion of persistent white cloud system observed near the evening terminator in May 1937. (G. de Vaucouleurs, Physics of the Planet Mars, Faber, 1954.)

in their early phases. After a few days, however, they settle down to a gentle breeze.

The seasonal waxing and waning of the bright polar caps provide evidence of the possible presence of water, at least in the solid and vapor states, on the surface and in the atmosphere of Mars. This identification was tentatively confirmed about 1950 by G. P. Kuiper, who observed that the infrared reflectivity agrees with that of ordinary ice. A. Dollfus also matched the polarizing properties of the polar cap with those of a thin layer of water-ice crystals deposited on a surface at low temperature. Some physicists, nevertheless, argue that carbon dioxide (dry ice) is the major constituent of the polar caps, with perhaps a surface layer of water ice. At the end of winter the edge of the polar cap is near latitude 60° in the southern hemisphere and 70° in the northern hemisphere. The larger area of the south polar cap results from the longer and colder autumn and winter in the southern hemisphere, 381 days compared with 306 in the northern hemisphere. The thickness of the polar caps must be very small, probably a fraction of an inch, and even if their bulk is made up of ordinary ice the amount of water in them would hardly fill the Great Lakes. If so, Mars is drier than our most desolate deserts—an intriguing planet for exploration, but a most uninviting land for colonization.

12-5 The Surface of Mars. To earthlings Mars is not only a dry planet, it is also a very cold one. The average temperature of Mars can be computed through the laws of radiation, and because of its greater distance to the sun it is obviously much colder than the earth. The calculated temperature is about 220° K or −50° C (−60° F), a value confirmed by measurements in the radio spectrum, since radio physicists of the U. S. Naval Research Laboratory detected the thermal emission of Mars at centimeter wavelengths for the first time in 1956 and 1958.

Low resolution permits radio telescopes to give only the average temperature of the whole planet. Observations of infrared radiation with optical telescopes give a much more detailed picture of the Martian climate. The earth's atmosphere has a "window"—is semitransparent—for radiation of 8 to 14 microns wavelength, which happens to be the spectral region where bodies at ordinary temperatures have their maximum thermal emission (Chapter 8). Extensive measurements of the thermal radiation of Mars by means of sensitive thermocouples at Lowell and Mount Wilson Observatories in the 1920's and 1930's, confirmed by new observations of W. Sinton and J. Strong with the 200-inch telescope in the 1950's, provide detailed knowledge of the surface temperature of Mars, in good agreement with theoretical expectations (Fig. 12-9).

The diurnal variation of temperature near the equator resembles that observed in the desert regions of the earth, but some 50° C colder and with a larger range from day to night. This result accords with the fact that the

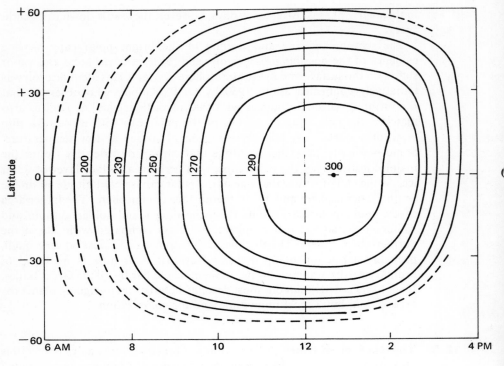

thin dry atmosphere affords little protection against nocturnal cooling. The temperature falls to a value as low as 200° K (− 70° C) by the end of the night, even in the tropical regions of Mars. During the day it is still freezing, even at noon when Mars is near aphelion. However, when Mars is near perihelion, midday temperatures of up to + 20° C and even + 30° C (70° to 90° F) have been recorded for the dark regions, which absorb sunlight more completely than the bright regions and are therefore 5 to 10 degrees hotter (Fig. 12-10).

The bright reddish areas that cover about three-quarters of the surface of Mars are, as all evidence indicates, vast desert expanses covered with fine dry dust. We do not know precisely the nature of this dust, but we may reasonably assume that it consists mainly of silicates colored red by iron oxide. Much discussion has centered in recent years on a similarity, noted by A. Dollfus, between the polarization curve of Mars and that of the mineral limonite, a common variety of hydrated iron oxide (2 $Fe_2O_3 \cdot$ 3 H_2O) which also has the same detailed color characteristics and visual albedo, about 0.15, as the bright regions of Mars.

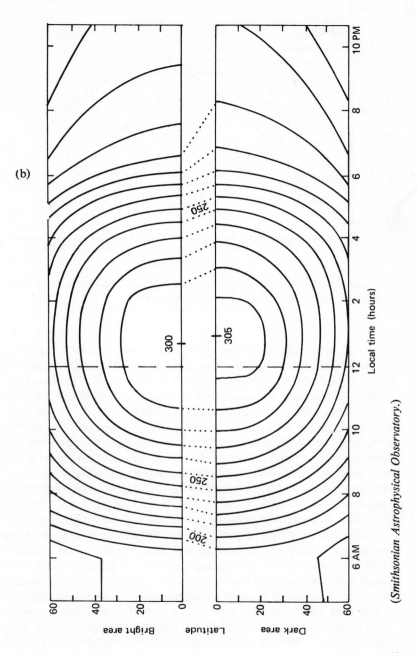

(b)

(Smithsonian Astrophysical Observatory.)

FIGURE
12-9(b)

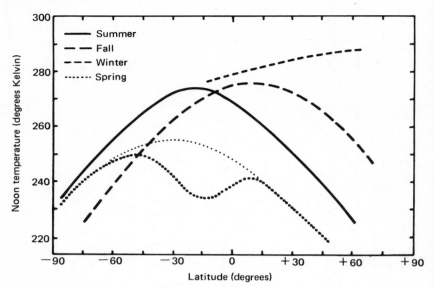

**FIGURE
12-10** *Latitude variations of noon temperature on Mars for different seasons of southern hemisphere. (After F. Gifford, Lowell Observatory.)*

In the polar regions during spring the snowy deposit hangs on a few weeks longer in some regions than elsewhere; whether this implies elevations or depressions is a moot point. D. B. McLaughlin has listed the geologic phenomena or processes that, from current knowledge, are definitely or probably absent on Mars: rain, hail, and snow; fluvial erosion and deposition; wave erosion and sedimentation; glacial erosion; folding and thrusting (mountain formation). He lists as possibly or probably active: chemical weathering (slow); exfoliation of exposed rock; gravity faulting, and, perhaps, volcanism. The only phenomena certainly present are frost, aeolian erosion, and wind transportation and deposition. We expect also the same slow deposition and meteoritic dust as on earth (Chapter 15), meteoritic impacts, and possibly occasional collisions with minor planets.

Little definite detail appears in the bright regions. Faint, irregular, streaky or patchy markings—the so-called "canals" and "oases"—are observed, whose exact structure and nature are unknown. The best modern visual and photographic observations do not confirm the existence of an extensive geometrical network of narrow lines in the bright regions.

The dark areas form a conspicuous and fairly stable pattern, which has been mapped in great detail (Fig. 12-11a). The "geography" of Mars [or "areography"—from Ares, the Greek name for Mars] forms a fascinating and tantalizing spectacle to watch through a good telescope as the rotation of the planet slowly brings different regions into view. Experienced observers have seen so much detail that names derived from Greco-Latin

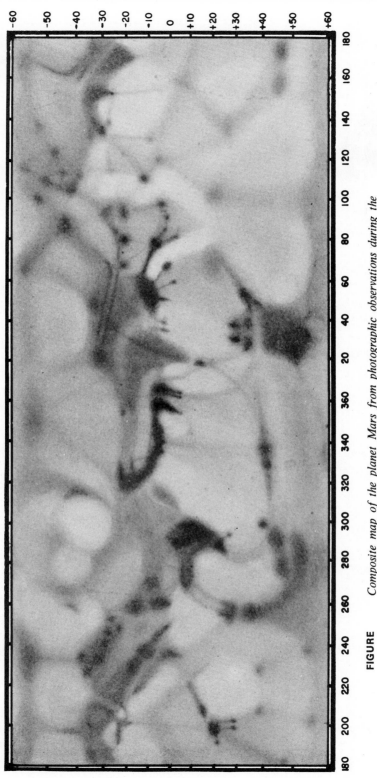

FIGURE 12-11(a) Composite map of the planet Mars from photographic observations during the perihelic oppositions of 1941 and 1958 (late summer in southern hemisphere, at top), for central longitude 0°. (Mars Map Project, G. de Vaucouleurs and J. Roth.)

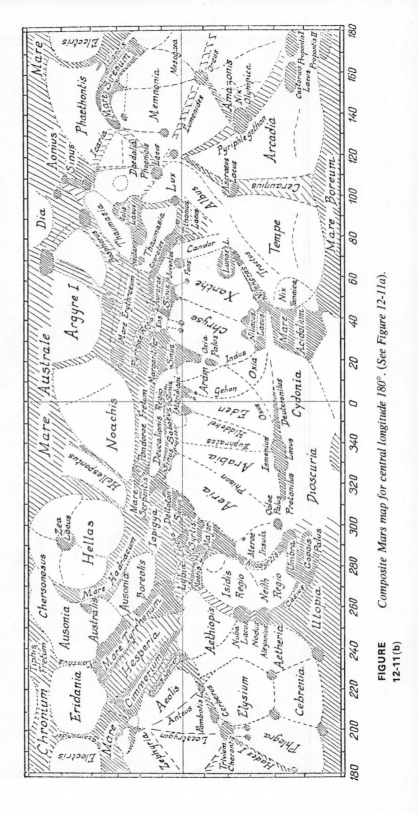

FIGURE 12-11(b)   Composite Mars map for central longitude 180°. (See Figure 12-11a).

geography and mythology, first introduced by G. V. Schiaparelli in 1877, are commonly used to designate the more stable features [Fig. 12-11b)]. For instance, recognizable records of the largest dark area, Syrtis Major, extend all the way back to the time of Hooke, Huyghens, and Cassini, some three centuries ago. The origin of the areographic longitudes is close to to another permanent spot, Meridiani Sinus, the "Greenwich of Mars." Nevertheless, conspicuous irregular changes sometimes take place in the Martian topography. On several occasions dark areas have encroached over previously bright territory for periods of from a few months up to several decades. As a rule these changes are temporary. Few, if any, are secular, and sooner or later the affected region returns to its "normal" appearance. Certain areas are more subject to change than others, and some of the major streaks or "canals" share in this activity (Fig. 12-12). Minor spots also vary conspicuously in intensity, occasionally to the vanishing point, often to reappear later. The cause of these changes is unknown, but must be related to the general activity of the dark areas.

An important characteristic of these regions is their seasonal variation of albedo, polarization, and possibly color,* all related to and apparently caused by the waxing and waning cycle of the polar caps (Fig. 12-13). During the Martian spring a "wave of darkening" spreads all over one hemisphere of the planet from the polar regions toward the equator and beyond, followed in half a Martian year by the corresponding wave moving in the opposite direction from the other pole. This activity has long been regarded as evidence for the existence of a vegetation, whose renascence after the long cold winter is, according to this theory, stimulated by the arrival of humidity through the atmosphere from the polar regions. It is difficult, however, to understand how the very minute amounts of water vapor involved can initiate and sustain such dramatic changes. One could speculate that Martian vegetation is highly adapted to ambient conditions and makes the best of the meager resources available.

Support for the vegetative hypothesis seemed at hand when in the 1950's W. Sinton, working first at Harvard Observatory and then at Mount Palomar, detected a new absorption band near 3.5 μ in the infrared spectrum of Mars. This band (or bands, since two or three seemed to be indicated) appeared to agree closely in wavelength with three bands observed in the reflection spectrum of terrestrial vegetation and attributed to vibrations of the C-H bond in complex organic molecules. Alas, a few

---

* The dark regions of Mars commonly look greenish when observed through a refracting telescope. This coloring, though often striking, results almost entirely from the contrast with the surrounding red areas combined with the "secondary spectrum" of the lenses spreading over the image the out-of-focus blue and violet light of the planet. The widely popularized conclusion that this green color indicates the presence of plants in the dark areas of Mars is misleading and unfounded. The true color or colors of the dark regions are as yet unknown. Close-up observation from space probes will be necessary for a definite solution.

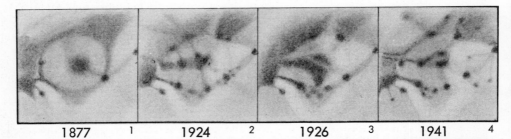

1877   1   1924   2   1926   3   1941   4

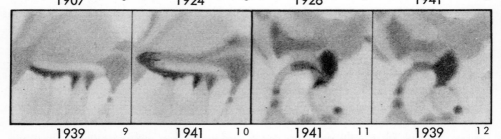

1907   5   1924   6   1926   7   1941   8

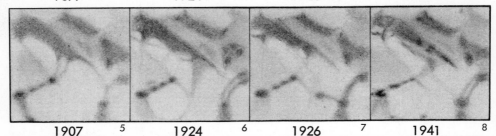

1939   9   1941   10   1941   11   1939   12

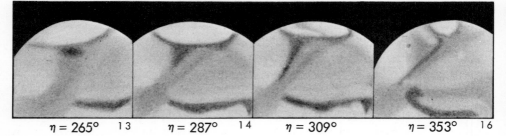

$\eta = 265°$ 13 $\eta = 287°$ 14 $\eta = 309°$ $\eta = 353°$ 16

FIGURE 12-12   *Irregular changes and seasonal variations on Mars. 1 to 4. Irregular changes of Solis Lacus. 5 to 8. Irregular changes in the Mare Cimmerium and Nepenthes—Thoth area. 9,10. Seasonal variations in Pandorae Fretum and Mare Serpentis area. 9,10. Seasonal variations in Pandorae Fretum and Mare Serpentis area. 11,12. Seasonal variations in Syrtis Major. 13 to 16. Seasonal variations in Hellespontus as a function of heliocentric longitude from the end of winter to mid-summer in the southern hemisphere.* (G. de Vaucouleurs, Physics of the Planet Mars, Faber, 1954.)

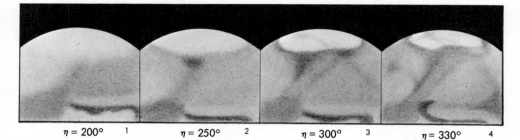

$\eta = 200°$ 1 $\eta = 250°$ 2 $\eta = 300°$ 3 $\eta = 330°$ 4

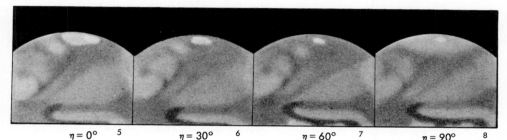

$\eta = 0°$ 5 $\eta = 30°$ 6 $\eta = 60°$ 7 $\eta = 90°$ 8

FIGURE *Seasonal variations in the south polar regions of Mars showing the decrease of the*
12-13 *polar cap from mid-winter to late summer. Note related topographic variations.*
(G. de Vaucouleurs, Physics of the Planet Mars, Faber, 1954.)

years later skeptical spectroscopists (including Sinton himself) pointed out that these bands could be more plausibly assigned to the semiheavy water molecules HDO in our own atmosphere! (D stands for deuterium, double-weight hydrogen.)

Even without this supporting evidence, many scientists still believe that life on Mars remains a distinct possibiility. Others prefer to rely only on basic physical, chemical, and meteorological effects to explain the complex Martian phenomena—a difficult task to say the least.

A curious controversy has developed in recent years around the question of the relative elevations of the bright and dark regions of Mars. Perhaps for no better reason than analogy with the moon, most astronomers believed for a long time that the dark regions are *lower* on the average than the bright regions. However, radar observations of Mars in 1965 by Goldstein, using the 84-foot reflector of the JPL Tracking Station near Goldstone, California, have been interpreted by C. Sagan and J. Pollack in terms of a sloping terrain arranged to be consistent with the observed high- and low-reflectivity regions of the planet. The disposition of the postulated slopes suggests that the dark regions may be extremely high mountains and the bright regions low flatlands filled with dust. Sagan and Pollack postulate that the seasonal variations of the dark areas or highlands arise from wind-carried fine dust that changes the optical characteristics of the high-alitude regions. These ingenious suggestions are not supported by the more direct radar-range measurements made in 1967 and 1969 by G. H. Pettengill with the powerful Haystack radar of MIT.

The remarkable TV photographs made at close range by the NASA space probe, Mariner IV, on July 14, 1965, provided the first definite information on the topography of Mars (Fig. 12-14a). Speculations as to artificial, linear "canals" should have ceased with the publication of these photographs supplemented by the even superior ones by Mariners VI and VII in 1969 (Fig. 12-14b). No such "canals" are present. Instead, the best photographs show crater formations that match qualitatively the crater topography of the lunar highlands. The crater frequency is less by a factor of two than on the lunar highlands but tenfold greater than on the lunar maria.

The largest recorded crater formation (Fig. 12-14a) is about 150 km (90 miles) in diameter and is rather similar to Clavius (Fig. 11-17) on the moon, though somewhat smaller. Smaller craters dot the greater one, its environs, and even its wall. All the craters appear to have lower walls on the side that is indistinct or missing from the large crater, as though some "geological" subsidence phenomenon had been at work.

The Martian craters appear to represent a "saturation" condition, as do those on the lunar highlands; that is, the destructive forces of new craters and "geological" processes kept the general nature of the topography in equilibrium, so that crater counts measure a combination of crater-formation rates and destructive forces. We expect the influx rates of meteorites (asteroidal debris) to be roughly 20 to 200 times greater on Mars, near the edge of the asteroid belt, than on the moon. Thus the "geological" destructive processes on Mars should be two to 20 times shorter than the average age of the lunar maria, since the crater frequency ratio is about ten times greater on Mars. If they should average, for example, 3×10^9 yr, the "geological" processes on Mars would have a crater destruction period of some 50 to 500 million years, somewhat greater than the geological periods on earth, roughly 100 million years.

The Mariner VII photographs show a striking absence of features on the large bright circular area, Hellas. Near its transition zone at the edge of Hellespontus small craters are indiscernible while an echelon of scarps and ridges appears. Preliminary analysis does not distinguish between a possible meteorological cause, such as a dust cover, or a tectonic crustal smoothing of the Hellas area. Atmospheric fog or haze both seem unlikely as an explanation. Another region in the Mariner VII photographs, west of Sinus Meridiani, exhibits a type of jumbled roughness not found either on the moon or the earth. Perhaps Mars is more active "geologically" than we have suspected.

The southern polar cap is littered with craters in the Mariner VII photographs with many large formations characteristic of the lunar highlands (Fig. 12-14c). The region between "snow" and bare ground is relatively

(a)

FIGURE 12-14 (a) (b) (a) *Mariner IV photographs of craters on Mars. Width of field 170 miles.* (b) *Mariner VII view of Mars. Dark Hellespontus region with corner (upper right) showing edge of the bright but featureless Hellas region. North up. Blue filter 700 × 1000 km. (U.S. National Aeronautics and Space Administration.)*

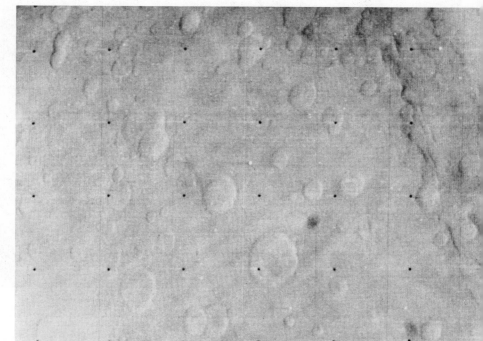

(b)

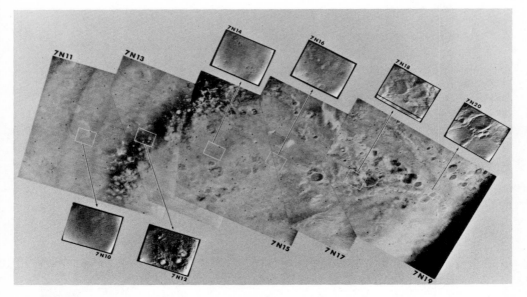

(c)

FIGURE
12-14
(c) (d) (c) *South polar cap region of Mars by Mariner VII. Edge of cap 60° S latitude: central left, over 90° span of longitude. South pole: lower right portion of 7N17. Evening terminator: right edge.* (d) *Mars by Mariner VII. Pictures spaced 12° for stereoscopic viewing. Nix Olympica: ring-shaped, upper center. Mare Sirenum: dark diffuse area lower left. North, 12° left of top. (U.S. National Aeronautics and Space Administration.)*

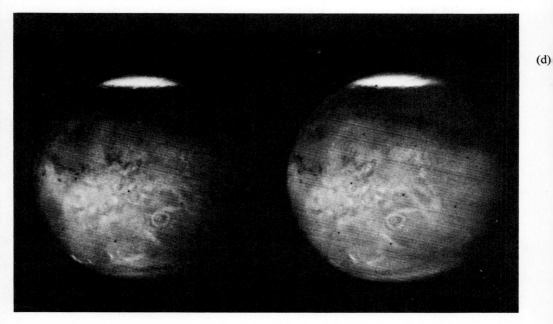

(d)

narrow, about 100 km (60 miles) across. Most craters in the transition region show white basins while many in the cap itself are relatively dark.

Many irregular and oddly shaped markings, both dark and light, can be seen on the Mariner photographs, which will require more time for interpretation.

Hundreds of lineaments—fairly straight ridges, edges of plateaus, furrows, or other linear markings (not "canals"!)—show in the Mariner photographs. They show two preferred, nearly perpendicular directions, again like similar lunar markings.

The Mariner photographs tell us nothing about the possibility of life on Mars. Even the works of man, if like those on the earth, could not be seen on such a scale of resolution.

But the prospect of meeting life on another planet is one of the most exciting scientific incentives for space exploration, one that may add a fascinating new field to biology: *exobiology* or biology of extraterrestrial life. It is clear that manned expeditions to Mars will have to approach the planet with more than usual precautions, in particular as regards possible contamination of Mars by terrestrial microorganisms and of back contamination of earth by Martian living things. For further discussion see Chapter 18.

12-6 **The Cloudy Planet: Venus.** Popularly called the "evening star" when it shines brightly in the western sky after sunset or the "morning star" when it appears above the eastern horizon before sunrise, Venus was known to the ancients as Vesperus or Hesperus in the evening and Phosphorus or Lucifer in the morning positions, before they were recognized as different apparitions of a single planet. At its mean distance from the sun of 0.723 astronomical unit or 108 million km (67 million miles), Venus describes its nearly circular orbit in a sidereal period of 225 days at a mean velocity of 35 km/sec (22 miles/sec). The eccentricity of the orbit, 0.0068, is the smallest of all the major planets; its inclination to the plane of the ecliptic, 3°.4, is one of the largest (after Pluto and Mercury). The synodic period is 584 days, of which 71 days is the mean interval between inferior conjunction and the maximum elongations of 48° (Fig. 12-15), and 221 days is the mean interval between elongations and superior conjunction. Venus appears at its brightest in crescent phase at elongation 38°.

At rare intervals Venus passes in front of the sun when it is very close to one of the nodes at the time of an inferior conjunction (as for Mercury, cf. Fig. 12-2). Through the combination of the motions of Venus and of the earth, favorable conditions happen at intervals of 8 years, $121\frac{1}{2}$ years, 8 years, and $105\frac{1}{2}$ years, after which the cycle is repeated. The last four transits of Venus took place in 1761, 1769, 1874, and 1882, and were extensively observed by astronomical expeditions sent from Europe and America to the far corners of the world. The parallactic displacement of

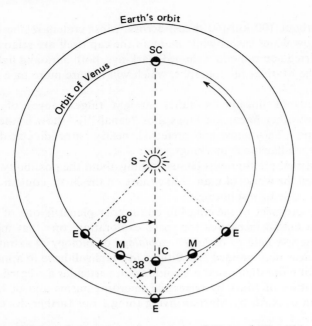

FIGURE *Orbit and maximum elongations of Venus (48°). The*
12-15 *planet reaches maximum brightness at elongation 38°.*

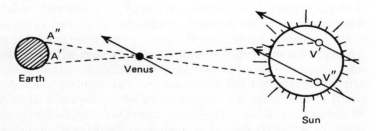

FIGURE *Transits of Venus across the sun occur at different*
12-16 *positions for observers A and B.*

Venus observed from different stations on earth (Fig. 12-16) was in the eighteenth and nineteenth centuries one of the best ways of measuring the solar parallax, and hence the astronomical unit of distance, by the method proposed by E. Halley in 1716.

It is most unlikely that the next two transits of Venus in 2004 and 2012 will be observed specifically for measurement of the solar parallax. Much more accurate methods of finding the sun's distance have been developed in the meantime. The most precise uses the reflection of radar signals by Venus. (For further details see Sec. 7-6.)

The apparent diameter of Venus varies from less than 10″ in superior conjunction to over 60″ in transit; it is about 25″ in quadrature (at maximum elongation). The diameter of Venus is difficult to measure precisely because of the dazzling luminosity and crescent shape of the visible disk when the planet is nearest to us (Fig. 12-17). In transit, when the disk of Venus appears as a large, round black spot on the sun's disk, the diameter of the visible cloud surface of Venus is estimated at 12,200 km (7600 miles), with an uncertainty less than 50 km. This is nearly 96 per cent of the earth's diameter. No flattening is perceptible and because of the slow rotation none is expected. The diameter determined by radar is 12, 112 ± 1 km (7526 miles) according to Ash, Shapiro, and Smith. It refers to the solid surface, while the optical value refers to the top of a planetwide cloud deck.

The mass of Venus was difficult to measure for a long time because, like Mercury, it has no observable satellite. Indirect determinations from the perturbations of the orbits of especially minor planets and of several space probes give a value that is 81.4 per cent of the mass of the earth (or 4.86 × 10^{27} g). The corresponding density is 5.3, close to the earth's, if the diameter of the globe is 12,110 km. From the radius and mass we find the acceleration of gravity at the surface, about 8.8 m sec^{-2} or 90 per cent of the earth's value. In the upper regions of the atmosphere the escape velocity for

FIGURE
12-17
Extension of horns of Venus in crescent phase by atmospheric scattering. (B. Smith, New Mexico State University.)

atmospheric molecules is about 10 km sec^{-1}; unless temperatures are very high, only the lightest molecules (H, He) will escape rapidly.

Visual observations of Venus by Schiaparelli had been interpreted by him as indicating a rotation period synchronous with the orbital revolution period. This conclusion, as reaffirmed by Lowell and then by several other astronomers, was more or less generally accepted until 1962, when radar observations at the Jet Propulsion Laboratory and the MIT Lincoln Laboratory led to positive results indicating a retrograde rotation in about 250 days. From several years of continuous studies, by these as well as U.S.S.R. and Arecibo investigators, I. I. Shapiro finds a more precise value of 243.1 $\pm$ 0.2 days for the sidereal rotation around an axis inclined by $-88°$ to the normal to the orbital plane (the minus sign indicates a retrograde rotation). As for Mercury, this rotation period might also result from a resonance effect, this time attributed to perturbations by the earth, which act most strongly when Venus is in inferior conjunction, hence closest to the earth. If this view should prove correct, the exact value of the rotation period should be 243.16 days. The synodic period, or "day," on Venus lasts for some 114^d.

Contrasting results, but not necessarily conflicting with the radar data, were obtained in recent years by French astronomers, who secured long series of ultraviolet photographs of Venus at the Pic du Midi Observatory. Consistent and repetitive cloud patterns (Fig. 12-18) were recorded that indicate a general rotation of the atmosphere in about four days. No one has yet found a way of explaining how a slowly rotating planet can be surrounded by a fast-rotating atmosphere.

FIGURE 12-18 *Ultraviolet photographs (21–22 May 1967) of Venus showing variable cloud pattern. (B. Smith, New Mexico State University.)*

12-7 **The Atmosphere of Venus.** The solid surface of Venus is hidden by
a heavy cloak of clouds (Fig. 12-18), which form the visible surface of the
planet and are responsible for its dazzling brightness. Photometric measurements indicate that the cloudy atmosphere of Venus reflects back into
space from 70 to 80 per cent of the light received from the sun. The reflectivity is especially high in visible and red light, but drops rapidly in the violet
and ultraviolet to about 50 per cent, either because of atmospheric absorption or because the reflecting particles of the clouds are intrinsically
colored. Kuiper finds evidence that the particles contain ferrous chloride
dihydrate ($FeCl_2 \cdot 2H_2O$). Possibly they are formed at the surface by aeolian
erosion and raised by atmospheric convection currents. Although Venus is
redder than sunlight at small phase angles, it turns bluer at phase angles
larger than about 130° (in narrow crescent phases) when atmospheric
scattering makes an increasingly important contribution to the light
reflected by the planet (see Fig. 12-17). Selective scattering of blue light by
small haze particles, perhaps 0.5 micron in diameter, floating in the upper
atmosphere of the planet above diffuse cloud layers, could account for this
blueing effect.

The spectrum of Venus shows strong absorption bands with heads at
λ 7820, 7883 and 8698 Å, which W. S. Adams and T. Dunham in 1932
identified with carbon dioxide (see Fig. 12-19). By comparing their intensities with laboratory spectra of CO_2 in long-path absorption cells, they
estimated that the amount of CO_2 in the Venus atmosphere was equivalent
to a layer about 1 km (0.6 mile) thick under normal atmospheric pressure
(there is only 2.2 m of CO_2 in the earth's atmosphere). More recent
interpretations by H. Spinrad and D. M. Hunten suggest that both the
equivalent thickness of carbon dioxide and its mean temperature increase
as one considers weaker rotation lines in the λ 7820 rotation-vibration band
(see Sec. 8-11). This effect occurs because in a strong line (that is, one of low
quantum number J) the absorption coefficient is large, and light of this
particular wavelength is almost completely absorbed before it can penetrate very deep in the atmosphere of the planet; thus the reflected light
comes largely from the low-density upper atmospheric levels. Conversely,
in a weak line of high quantum number J the absorption coefficient is
small, and light of this wavelength must penetrate to a great atmospheric
depth before a visible absorption line is formed. The mean temperature of
the absorbing gas in each case is derived from the intensities of the corresponding rotation lines when the pressure is known. The broadening of the
line profile results from the Doppler shifts of the random velocities of the
absorbing CO_2 molecules, which increase with temperature. High-disper-
sion spectra of Venus have been interpreted along such lines to show that
temperature and pressure increase as one penetrates deeper in the atmosphere of Venus (a pretty safe conclusion in any case). The upper cloud
layers are very diffuse, like thin haze, so that the levels applicable to the

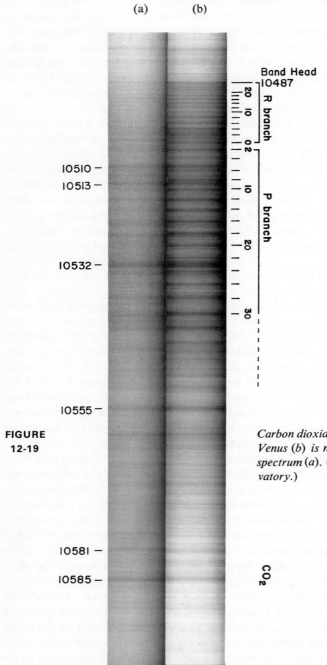

FIGURE
12-19

Carbon dioxide band at 1.05 μ in spectrum of Venus (b) is much stronger than in daylight spectrum (a). (E. Barker, McDonald Observatory.)

measured temperatures, even if they could be taken at face value, are poorly defined. The lowest temperatures are in the range 210° to 270° K and should apply to the top of the haze layers, in fair agreement with two independent sets of data: (1) the temperature of the very high atmosphere, $T = 210°$ to 300° K, derived by G. de Vaucouleurs from photoelectric observations of the occultation of Regulus by Venus in 1959, (2) the temperature of the top of the cloud layer, $T \simeq 235°$ K, derived from radiometric observations of the infrared thermal emission of Venus.

The thermal emission of Venus in the 8- to 14-μ semitransparent "window" of the earth atmosphere was first detected and measured in the 1920's at Lowell and Mount Wilson Observatories. More detailed measurement with improved apparatus made more recently by Strong and Sinton and also in the 3.75-μ "window" fully confirm the earlier data. The mean temperature of the emitting layer, $T = 235°$ K, is very stable and, surprisingly, it is almost exactly the same on the dark nighttime side and on the bright daytime side of the planet. This result indicates that a powerful thermostatic mechanism is at work in the emitting region, which is probably at or close to the top of the cloud layer. Here the atmospheric pressure is probably a few tenths of an atmosphere.

If the cloud particles were made of ice and the temperature appreciably above 200° K, lines of water vapor should appear in the spectrum of Venus. Attempts to find water vapor have led to conflicting conclusions, but place the H_2O/CO_2 ratio at less than 10^{-5}. Thus the cloud particles may be composed of dust or concentrated aqueous solutions rather than pure water.

Surprisingly, P. and J. Connes, W. S. Benedict, and L. D. Kaplan have detected lines of hydrochloric acid, HCl, in the spectrum of Venus, and probably a few of hydrofluoric acid, HF. The quantities of these gases are minuscule, less than one part in a million for HCl and one hundred times smaller for HF. Their physical significance in the atmosphere of Venus is not known, except that they are light molecules and very stable if not in the presence of basic materials. The search for other atmospheric constituents has been thorough and, except for traces of CO, so far fruitless.

The remarkable space probes, Venus 4 (1967), 5, and 6 (1969) by the U.S.S.R. and Mariner V (1967) by the U.S., proved that the atmosphere of Venus is almost entirely composed of carbon dioxide. The Venus 4, 5, and 6 command modules made ablation entries into the atmosphere and then ejected 380-kg (836-lb) instrumented canisters that parachuted safely into the lower atmosphere. The 240-kg (528-lb) Mariner observed the planet from a nearest approach of some 4000 km (2500 miles). The atmospheric sample obtained by the canister contained 90 to 95 per cent of carbon dioxide, less than 7 per cent nitrogen, 1.6 per cent combined water and oxygen, and 0.4 per cent water. No atomic oxygen could be detected at the 200-km (120 mile) level, indicating that any quantity present must be less than 10^{-8} of corresponding terrestrial altitudes. The Mariner V measures by a radio technique gave a carbon dioxide concentration of 72 to 87 per cent, in good agreement with the direct measures.

12-8 **The Lower Atmosphere and Surface of Venus.** The high surface temperature of Venus was discovered through its microwave emission in the 3- to 10-cm wavelength band, first detected in 1956 by Mayer and other radio astronomers at the Naval Research Laboratory near Washington. If the surface radiates like a black body, the radiation received requires a surface temperature of the order of 600° K. However, the surface of Venus is probably not completely "black," and the radar reflectivity of Venus, about 12 per cent, implies a true surface temperature in excess of 600° K. This is the temperature of the dark night side observed near inferior conjunction. At elongation the apparent temperature is still higher by some 50° K, suggesting that the apparent temperature of Venus in superior conjuction when the sunlit hemisphere is turned toward the earth may be as high as 700° K. Figure 12-20 shows how the radio temperature of Venus varies with wavelength. Since an ionosphere becomes increasingly opaque and therefore hotter with increasing wavelength, the high radio temperatures at longer wavelengths could have represented only an ionospheric temperature, not the true surface temperature of Venus. This possibility was excluded by the U. S. Mariner II spacecraft that passed Venus at 35,000 km (21,600 miles) in December, 1962. Its antennas were able to resolve the disk of the planet, showing that the radio temperature was *greater* at the center of the disk than at the limb. An ionosphere would give a higher limb temperature because of its greater depth when seen edge-on ("horizontally").

In 1967 the U.S.S.R. Venus 4 canister actually measured the atmospheric temperature as it slowly parachuted toward the surface. During its 94-minute descent from a reported radar altitude of 26 km (16 miles) the canister telemetered back a maximum temperature of 280° C (536° F), whereupon its transmission terminated. Whether the canister reached the actual surface (mountain top or valley?) or was incapacitated by the immense pressure while in the atmosphere could not be ascertained immediately. But later radar evidence favors the second alternative. This outstanding experiment, however, proved without question that the surface of Venus is very hot compared with that of the earth, whether or not the true surface temperature is as high as indicated by the radio measurements.

The Venus 4 canister also answered the most tantalizing question of all concerning Venus. How deep is the cloud-obscured atmosphere? The canister readings had reached a pressure of 15 to 22 atmospheres when it stopped transmitting. One may ask why a planet, otherwise so much like

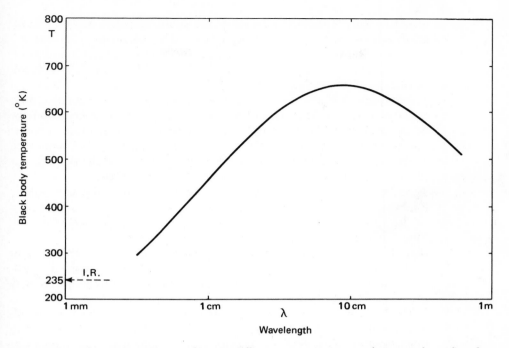

the earth, should retain so much more atmosphere. Actually the question should be rephrased, because the earth once supported a comparable atmosphere. Some 20 atmospheres of carbon dioxide, combined with silicate rocks to form carbonate rocks (limestone), were required to form the kilometer-thick layer of carbonate rocks now present on the earth. Here, as H. C. Urey has shown, water was the necessary catalyst. Apparently adequate water is and was missing on Venus to permit "fossilizing" the carbon dioxide now present in the atmosphere. The real question concerns the relative scarcity of water on Venus, not the large abundance of carbon dioxide.

The high surface temperature on Venus requires an atmosphere that strongly absorbs the infrared radiation of the planet's surface and prevents it from escaping to space. Since the atmosphere of Venus reflects some 70 to 80 per cent of incident sunlight back to space (Sec. 12-8) to begin with, it must retain close to 100 per cent of the thermal radiation in order to raise the surface to such high values. This action, somewhat similar to that of glass, is often referred to as the "greenhouse effect." If Venus had no atmosphere and radiated freely to space, its equilibrium temperature would be only 70°C in the center of the sunlit hemisphere. The known great abundance of carbon dioxide may be an adequate "greenhouse" window. In addition the atmosphere may contain enormous amounts of dust in suspension, as proposed by E. J. Öpik, which could help prevent the escape of surface radiation.

Radar observations have detected two main regions of high reflectivity to radio waves on the surface of Venus. These two features, provisionally designated "Alpha" and "Beta," stretch more than 1000 kilometers in the northern hemisphere of Venus. The Alpha feature appears to be a mountain chain somewhat comparable to our Rocky Mountains or the Andes. The radar observations also show that, on a small scale, the surface of Venus is smoother than the moon, and that it reflects like rocks, not like watery liquids.

The various Venus space probes detected no radiation belts around Venus, and the Mariner V found only a weak magnetic field, about $\frac{1}{300}$ that of the earth, at the boundary of the planet's ionosphere and the solar field. While we might expect Venus to possess a molten core much like that of the earth, its slow rotation presumably cannot set up internal currents to produce a strong intrinsic magnetic field.

The Mariner V also detected an ionospheric layer on the sunlit side of Venus with an electron density comparable to that of our ionosphere, but much thinner. The Venus 4 detected no ionosphere on the night side. Improved observations and more direct exploration by space probes are needed to lift some of the veil of mystery that still surrounds the cloudy planet.

12-9 The Outermost Planet: Pluto. Pluto in many ways resembles the small terrestrial planets and may be added to the group, even though its orbit lies beyond those of the giant planets Jupiter, Saturn, Uranus, and Neptune (Chapter 13). The successful discovery of Neptune, after gravitational theory (Chapter 4) had predicted its existence, led the American astronomer P. Lowell, at the turn of the century, to investigate anew the motions and orbits of Uranus and Neptune. Small residual irregularities in the motion of Uranus suggested to Lowell—and later to W. H. Pickering—that another unknown planet was still awaiting discovery at the borders of the solar system. Both Lowell and Pickering proceeded to calculate its expected position (or positions) from theory. Unsuccessful optical searches were made at several observatories during the 1910's and 1920's. Finally, in January, 1930, C. Tombaugh, at the Lowell Observatory, Flagstaff, Arizona, after a thorough photographic survey of the zodiacal belt discovered by means of the blink comparator (Sec. 6-10) a faint new planet. It lay within five degrees of the positions predicted by Lowell and Pickering but was much fainter than expected. The Lowell Observatory astronomers proposed the name Pluto and the symbol ♇, a monogram from the first two letters of the planet's name—and also, appropriately, from the initials of Percival Lowell.

The orbit of Pluto was soon determined with precision, thanks to photographs of the planet taken some years earlier at Lowell's request for the purpose of locating the planet. The images were so faint that they could not be identified until accurate orbital predictions were available. The semimajor axis of 39·5 astronomical units or 5.9×10^9 km (3.7 billion miles) corresponds to a period or revolution of 249 years. The orbit has an eccentricity of 0.25 and its plane makes an angle of some 17° with the ecliptic. All three of these orbital characteristics are larger than those of any of the known principal planets of the solar system. The orbit of Pluto presently marks the boundaries of the planetary system, although cometary orbits with their extreme eccentricities extend much beyond (Chapter 14). Because of its eccentric orbit, the distance from Pluto to the sun varies from a maximum of 49.4 A.U. = 7.4×10^9 km at aphelion to a minimum of 31.6 A.U. = 4.7×10^9 km at perihelion. The inclination of the orbit makes the perihelion fall slightly inside the orbit of Neptune ($a = 30.1$ A.U.) when projected on the plane of the ecliptic (Fig. 12-21). In space, however, the two orbits do not intersect, and of the two planets Pluto is always the more distant from the sun. Pluto is now approaching perihelion, which it will reach in 1989. The planet will then appear nearly half a magnitude brighter than at the time of its discovery, when it was close to its mean distance from the sun. At this distance Pluto's apparent stellar magnitude is 14.9 (at opposition), which is too faint to be seen through small telescopes; few people have actually *seen* Pluto. However, during the next fifty years or so it should be visible (provided its exact position among the stars is known in

FIGURE   *Orbit of Pluto in relation to those of the major planets. Note the*
12-21   *orbital tilt shown here in perspective.*

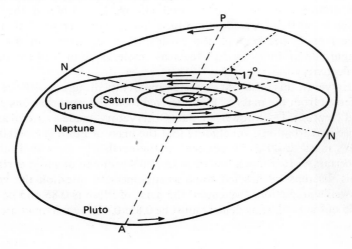

advance) through medium-sized telescopes of aperture 12 inches or larger.

Photoelectric observations of Pluto at McDonald Observatory by G. P. Kuiper in 1952–53 and by Walker and Hardie in 1954–55 showed that its brightness varies by about 10 per cent (0.11 mag.) in a period of 6.39 days. Obviously the planet must be spotted, and its brightness changes with the variable reflectivity of different regions of its surface turned toward the earth. Thus, even though the disk of Pluto is much to small to show any surface detail, we can still find its period of rotation (Fig. 12-22).

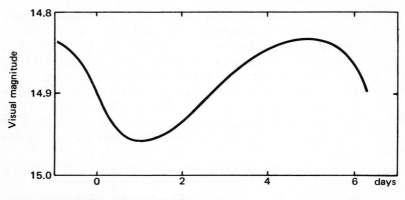

FIGURE 12-22   *Light curve of Pluto, the result of rotation.*

The diameter of Pluto was the subject of many inconclusive discussions during the 1950's after G. P. Kuiper succeeded in measuring it with the 200-inch telescope. By comparing the apparent disk of the planet with images of small luminous disks projected in the telescope field, and allowing for atmospheric and instrumental effects, he estimated the apparent diameter at $0''.2$, corresponding to a linear diameter of 5750 km (3600 miles) or 45 per cent of the earth's. This diameter and the apparent magnitude of Pluto lead to the reasonable value of 0.14 for the albedo (reflectivity).

However, this observed diameter apparently disagrees with the diameter inferred from the mass of Pluto derived from the perturbations of Uranus and Neptune. D. Brouwer, of Yale Observatory, made a careful analysis of these perturbations, concluding that the mass of Pluto is 1/400,000 of the sun's mass or 80 to 90 per cent of the earth's mass, with a rather large uncertainty. It is much less than Lowell's estimate of seven earth masses. This discrepancy has led some astronomers to question the validity of Lowell's prediction. Even so, if the mass of Pluto is 0.85 and its diameter 0.45 of the earth's, its density must be $0.85/(0.45)^3 = 9.3$ times greater than that of the earth, or about 50 times that of water. This density of course, is higher than that of any known terrestrial material, even if we allow for a density increase under high pressure in the planet's interior.

There are three ways of escaping the contradiction: (1) the measured diameter may not represent the true size of the planet; (2) the mass derived from perturbations may be too large; (3) the planet really has an unusually high density.

We may dismiss the third possibility, inspired by the hyperdense "white dwarf stars" (see Chapter 22). The possible central density falls far short of that required for degeneracy—that is, for the production of hyperdense matter.

The first possibility rests on the supposition that the surface of Pluto has peculiar optical properties. For example, it might reflect light as does a polished mirror, so that reflection occurs only from the central portion, providing a small apparent diameter. Other types of surfaces might be imagined, but they are highly artificial, unlike any planetary surfaces we know.

A probably definitive solution to the dilemma stems from the alertness of I. Halliday of Canada. He predicted a possible occultation of a faint star (magnitude 15.3) by Pluto on the night of April 28, 1965. He and other observers carefully watched for the event and made a series of photographs at observatories in Texas, Arizona, California, and Canada. No one observed an occulation. Very precise astrometric observation showed that, in order not to produce the occultation, Pluto must have had a diameter not greater than 0.25 second of arc or 6800 km (4200 miles), just greater than Kuiper's measured value.

We can conclude with some assurance that the second possibility listed above is correct. The mass of Pluto derived from perturbations is too large. This possibility is not surprising in view of the minuteness of the deviations in Neptune's motion ($< 1''$ of arc) from which the mass is determined. Unfortunately, we must further conclude that the prediction of Pluto was fortuitous. Let us call it a happy accident.

The calculated average surface temperature of Pluto, based upon its mean distance from the sun and its albedo, is about 42° K, and any atmospheric gases likely to be retained by the planet's attraction must have frozen out long ago. Pluto's spectrum does not differ from that of reflected sunlight, except for a slight yellowing caused by selective reflection of its surface. The color index, $+0.8$, is actually less (the light is whiter) than those of the moon and Mercury. The reflecting surface might consist of ice and other substances that are gases at room temperature, and it may contain deep pits formed by meteoritic impacts. But we must frankly admit that at the moment we know practically nothing about the nature of the surface of Pluto.

No clear-cut evidence supports the existence of trans-Plutonian planets.

However, the German astronomer K. Schuette called attention in 1950 to a grouping of the aphelia of cometary orbits at an average distance of 85 A.U., similar to the comet "families" of Jupiter and Saturn (Chapter 14), and suggested that a planet might exist in this region. Some astronomers have already picked a name (Proserpine) for the problematic planet, but Tombaugh's thorough survey of most of the sky has failed to disclose any such planet.

13

The Giant Planets

13-1 Telescopic Appearance. The telescopic appearance of the giant planets, Jupiter, Saturn, Uranus, and Neptune, sets them apart from the terrestrial planets. They all have thick, cloudy atmospheres and high albedo. Each has striking peculiarities of its own. Tables 13-1 and 13-2 give their orbital and physical characteristics.

TABLE 13-1

Elements of orbits of giant planets

	Mean Distance from Sun: A.U.	10^6 km	Sidereal Period, years	Synodic Period, days	Inclination to Ecliptic	Eccentricity	Orbital Velocity, km sec^{-1}
Jupiter	5.20	778	11.86	398.9	1°18′	0.048	13.1
Saturn	9.54	1427	29.46	378.1	2°29′	0.056	9.6
Uranus	19.18	2870	84.02	396.7	0°46′	0.047	6.8
Neptune	30.06	4496	164.79	367.5	1°46′	0.012	5.4

TABLE 13-2

Elements of globes of giant planets

	Equat. 10^3 km	Radius: $\oplus = 1$	Vol., $\oplus = 1$	Mass $\oplus = 1$	Density cm^{-3}	Surface Gravity, m sec^{-2}	Escape Velocity, km sec^{-1}	Rotation Period	Incl. of Equat. to Orbit	Oblate-ness
Jupiter	71.8	11.04	1347	317.8	1.30	23.01	57.5	9^h50^m	$3°4'$	0.062
Saturn	60.3	9.17	771	95.1	0.68	9.06	33.1	10^h14^m	$26°44'$	0.096
Uranus	23.5	3.70	51	14.5	1.58	9.72	21.6	10^h49^m	$98°$	0.06
Neptune	22.3	3.50	43	17.2	2.22	13.47	24.6	15^h+	$29°$	0.02

FIGURE 13-1   *Jupiter in yellow light showing rotation, satellite, and shadow. (a) November 22, 1964, 20^h 50^m U.T. Note satellite Io at left. (b) Same date at 22^h 15^m U.T. showing the Great Red Spot. (c) November 25, 22^h 23^m U.T., showing red spot at left and the shadow of the satellite Europa at right. The image of the satellite is lost against the planetary disk. (P. Guerin, 40-inch reflector, Pic du Midi Observatory.)*

(a

(b)

(c)

Jupiter, the giant among the planets, has an equatorial diameter of 143,600 km (89,000 miles), eleven times that of the earth (see Fig. 13-1). The yellow disk of Jupiter is crossed by a number of dark belts, parallel to the planet's equator. The most conspicuous belts lie near the equator, while fainter and more delicate ones extend all the way to the polar regions. They are rarely uniform, but change in width around the planet. Spots, both bright and dark, often interrupt the pattern. The belts, which must be cloud formations, exhibit a remarkable range of color, from deep chocolate brown to yellowish white. Occasional delicate blues and grays appear, and one pronounced feature, the famous "Red Spot," first seen in the seventeenth century and continuously observed since 1878, has usually been brick red.

This *great Red Spot* has varied in size, shape, and conspicuousness, but it usually appears as an elliptical marking about 48,000 km (30,000 miles) long and 11,000 km (7000 miles) wide. It lies in latitude 20° S, and its longer axis is roughly parallel to the planet's equator (see Fig. 13-2). The Red Spot drifts in longitude with a slow, irregular, oscillatory motion. The total

FIGURE 13-2   *Jupiter in blue light showing the Great Red Spot and satellite Callisto with its shadow. Because of its color, the Red Spot appears dark in blue light. (200-inch reflector, Mount Wilson and Palomar Observatories.)*

drift relative to other surface markings has been as much as 1180°, more than three revolutions. From 1830 to about 1890 the spot was rotating more slowly than its surroundings. When it was about $1\frac{1}{2}$ revolutions behind, in the early 1890's, the spot reversed direction and began to rotate faster than its surroundings. Since then, observations have indicated pauses, irregular movements, and partial reversals. One can only conclude that the great Red Spot, whatever it may be, has no direct connection to the solid surface. The visibility of the spot varies erratically; sometimes the object almost completely disappears.

The remarkable colored clouds of Jupiter must in some way depend on circulation in the atmosphere, and their parallel course strongly suggests zonal belts of circulation. Certain molecules probably form more readily in some regions and decompose in others as a result of differences in temperature and pressure from one belt to another. An alternative possibility is that dust, clouds, or vapors may arise from the solid surface of the planet and float into the uppermost atmospheric levels. For example, smoke from a terrestrial volcano, if not otherwise disturbed by horizontal air currents, would tend to form a dark band parallel to the equator.

Rapid changes sometimes occur along the ragged, tufted structure of Jupiter's cloud belts. These fluctuations suggest that the whitish cloudlike spots can expand, contract, or break up, in intervals of an hour or so. Presumably red light penetrates more deeply than blue into the atmosphere of Jupiter. The red photographs reveal a rougher and probably more turbulent cloud structure than do the blue photographs. Some of the more permanent features show irregularities of rotation not unlike those of the great Red Spot.

Saturn, the second largest planet, is 121,000 km (75,000 miles) in diameter (see Fig. 13-7). It is remarkable for the unique system of rings that encircles it. We shall discuss the rings in Sec. 13-7; here we consider only the globe of the planet.

The predominant color of Saturn, like that of Jupiter, is yellowish, but the shade is duller and grayer. The polar zones have a distinct bluish or greenish tinge, somewhat darker than the rest of the planet. The faint bands that run parallel to the planet's equator are by no means as well marked or as highly colored as those of Jupiter. They are, in fact, only slightly darker than the body of the planet, and rarely show any distinct structure. On a few occasions, small yellowish-white spots have been observed on Saturn's surface. They fade away in a few days or weeks. One such spot, which appeared in 1933, spread into a band three-fifths around the planet before it vanished.

The planet *Uranus* appears as a greenish disk, so much brighter at the center than at the limb that it is hard to tell where the planet's edge actually is. Few, if any, well-defined spots have ever been seen on the planet, but it shows vague belts parallel to its equator. The diameter of Uranus is 47,000 km (29,200 miles).

In the telescope, *Neptune* resembles Uranus except for its smaller apparent size, distinctly bluer color, and perhaps more sharply defined edge. Vague markings show up only rarely. Actually, Neptune is almost as large as Uranus, with a diameter of 44,600 km (27,700 miles).

13-2   Surface Temperatures.   The surface temperatures of the giant planets are extremely low, as thermocouple observations have shown. Most of the heat reaches us from the upper cloud layers and not from the solid planetary surface, whose temperature remains unknown.

The measured temperatures for Jupiter and Saturn are 130° K and 95° K, respectively. Uranus and Neptune send us so little radiation, we can say only that their surface temperatures are probably less than 100° K. These temperatures are extremely low. Liquid air at atmospheric pressure boils at a temperature of 80° K; air should therefore be liquid on Saturn, Uranus, and Neptune because of the higher expected pressures. Before these temperature determinations, first made in the 1920's, many astronomers described the giant planets as very hot, because of the steamy appearance of their clouds. Microwave radio temperatures tend to confirm these low temperatures, but the radio spectra of the giant planets are complicated, as we shall see in Sec. 13-5.

13-3   Internal Constitution.   The internal constitution of the giant planets distinguishes them markedly from the terrestrial planets. Each of these planets has several satellites, for which we can observe the orbits and measure the periods of revolution. Thus, by a simple application of Kepler's third law (Sec. 3-3 and Fig. 3-3), we can calculate the masses of the planets with high accuracy. Since we also know their diameters, we can compute their volumes. Dividing the masses by the volumes, we obtain the mean densities of the planets, as shown in Table 13-2.

These densities are extremely low compared with those of the terrestrial planets, which range from about 4 to 5.5. The low densities show that the giant planets cannot be composed of rock throughout. Saturn, the least dense and in many ways the most remarkable of all the planets, has a mean density less than that of water. Saturn poses a real problem because its mean density is so extremely low. We cannot picture it as composed even of ice, because ice is too dense. We must try to find some likely material of extremely low density as a major constituent of the planet.

In Sec. 9-2 we noted that the sun resembles the earth in chemical composition, so far as the heavier elements are concerned. However, it contains a much greater abundance of the lighter elements, particularly of hydrogen and helium. We have also noted that the earth, moon, and terrestrial planets cannot retain appreciable amounts of hydrogen. The low molecular weight of hydrogen, the weak gravitational fields, and the comparatively high surface temperatures of these planets have facilitated the loss of their hydrogen atmospheres into space. But the velocities of escape for the giant planets are much greater than for the terrestrial planets, being 61, 37, 22, and 25 km/sec for Jupiter, Saturn, Uranus, and Neptune, respectively. For the earth the escape velocity is only 11 km/sec (see Sec. 7-14). In addition, the outer layers of the giant planets are relatively cold, and accordingly they have been able to retain even the lightest gases, hydrogen and helium. Several investigators have devised theoretical models of the major planets, which they regard as mixtures of hydrogen, helium, and ices, overlying a small core of heavier material, such as rocks and metals.

P. W. Bridgman showed that the familiar form of ice, which exists at subfreezing temperatures under atmospheric pressure, is not the only variety of ice. Many other crystalline forms of ice can occur, and may well be formed under the high pressures that must occur at great depths in the giant planets. Some of these forms could exist even at the fairly high temperatures that might exist in the deep interior, and an icy mantle about the denser core is not an impossibility, although the temperature may well be too high for ice.

Both Jupiter and Saturn bulge appreciably at the equator because of their rapid rotation. The amount of this bulge depends on the distribution of material below the surface and on the thickness of the atmosphere. H. Jeffreys has concluded that the rotational characteristics of Jupiter and Saturn require a very thick atmosphere and a higher-density core. The perturbation rates of satellite orbits give additional information about the concentration of mass towards the center of a planet. Calculated model planets must fit such mass-distribution data as well as the physical laws governing the compression of solids, liquids, and gases under high pressures. The internal temperatures are *never* measured, adding another uncertainty to that of composition.

One gets some insight into the possible structure of the giant planets by studying, theoretically, the behavior of large masses of solid hydrogen, at the absolute zero, in gravitational equilibrium. The results of the analyses appear in Fig. 13-3. The diagram (after W. C. DeMarcus) shows how the computed radius of such a planet changes with increasing mass. The weight of the overlying layers tends to compress the hydrogen core until finally, for planets having a mass greater than 10^{31} grams, the collapsing interior atoms can no longer support the outer structure. The heavier bodies thus have smaller radii. Jupiter is about as large as any cold planet could possibly be. As H. N. Russell conjectured long ago, the heavier planets form a sequence including the highly compressed stars known as white dwarfs (see Chapter 20).

In this diagram, the actual planets indicated by the black dots all lie to the left of the theoretical curve. Their radii, therefore, are smaller and their densities somewhat higher than those of the cold-hydrogen model. The

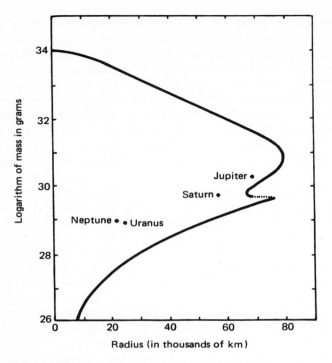

FIGURE
13-3 *Relation between mass and radius for model planets made of pure cold hydrogen. (W. C. De Marcus.)*

higher densities probably arise from an admixture of heavier atoms. Adding heavier substances moves the theoretical point to the left. The introduction of internal heat tends to inflate the object, increase its radius, and move the point to the right. Hydrogen must be the dominant constituent of these planets, but the proportion of heavier elements and the size of the earthly cores are still unknown. In Fig. 13-3 the shaded area surrounding the sharp point is imperfectly defined. This deviation from an otherwise smooth curve results from a supposed lower compressibility of the molecular phase of hydrogen, so that the radii tend to increase until gravity causes the heavier masses to collapse. In most models of Jupiter the earthly cores are compressed to some thirty times the density of water.

For Uranus and Neptune any large fraction of hydrogen reduces the mean density below the observed values. W. H. Ramsey obtains models composed mostly of a mixture he calls CHONNE, made up of methane, ammonia, water, and neon. He uses a dense core, essentially a duplicate of the earth, adding 10 to 15 per cent only of hydrogen. Models by R. T. Reynolds and A. L. Summers are somewhat similar; H, He, and Ne together comprise about 10 per cent of the mass of Uranus and 4 per cent of Neptune.

13-4   Chemical Composition. The chemical composition of Jupiter and Saturn, as we have seen, must be mainly hydrogen and helium. Observations of their spectra provide strong confirmation that the atmospheres of the giant planets are rich in hydrogen. The strongest absorption bands in the spectrum of Jupiter result from the molecules of methane (CH_4) and ammonia (NH_3). H. Spinrad and L. M. Trafton in 1963 discovered in the Jupiter spectrum a special type of absorption band caused by the hydrogen molecule. The absorption corresponds to some 100 kilometers of hydrogen at earth-atmospheric pressure above the cloud layer on Jupiter. The intensity of the methane bands increases progressively from Jupiter to Neptune. This increase probably results not so much from the higher abundance of methane as from the fact that light penetrates to greater depths in the atmospheres of the more distant planets. The observations accord with the idea that the apparent surface of a giant planet is a cloud layer that lies at lower and lower levels, the farther the planet is from the sun and, therefore, the colder its surface.

In 1952 W. A. Baum and A. D. Code observed an occultation of the star σ Arietis by Jupiter. The rate at which the star faded provided a measure of the density gradient of the planet's atmosphere. This observation could be interpreted if the atmospheric mean molecular weight is 3.3, in accord with the idea of an appreciable abundance of helium.

There has been much discussion of the probable chemical nature of the clouds of the giant planets. Terrestrial clouds consist of water droplets or (cirrus clouds) of ice crystals, which condense out of moist air at levels where the temperature is low. Similar condensations could occur in the atmospheres of the giant planets, but the substances will be those that "freeze out" at much lower temperatures. R. Wildt has suggested that in the hydrogen atmosphere of Jupiter, ammonia behaves as water does in the air. The clouds may consist of crystalline ammonia. However, just as the air is often most humid close to the surface of the earth, ammonia may be most abundant in the lower levels of Jupiter's atmosphere, where the temperature is higher.

Methane has a very much lower freezing point than ammonia, so it is not surprising that the ammonia bands decrease markedly in intensity from Jupiter to Saturn, whereas methane strengthens. Both ammonia and methane absorb ultraviolet energy. In spite of the low density of solar radiation at these great distances, a certain amount of both gases should be decomposed into other compounds by absorption of ultraviolet quanta. One of these is ethylene, $CH_2 = CH_2$, which in turn would react under the influence of radiation to produce the gas acetylene, $HC = CH$. If lightning occurs in these giant cloudy atmospheres, as seems quite likely, a number of organic molecules would be produced. C. Sagan and S. L. Miller have carried out such an experiment in the laboratory and found brightly colored molecules that would eventually decompose in the atmospheres of the giant planets. They have thus demonstrated the validity of an earlier expectation by H. Urey. Wildt, on the other hand, suggests that metallic sodium dissolved in liquid ammonia may be responsible for the colors of the giant planets.

13-5 Radio Noise from Jupiter. Three well-recognized types of radio noise reach us from Jupiter. The radio noise in the millimeter band of wavelengths appears to be of thermal origin, consistent with the infrared surface temperature, about 130° K, probably with some day-to-day variations. The radio temperature, however, increases with increasing wavelength and becomes clearly nonthermal in origin (Fig. 13-4a).

The decimetric radiation appears to consist of three distinct components. The first component, unpolarized, represents the contribution from the effective surface layer at a temperature of about 130° K. The intensity of the second component, also unpolarized, increases with wavelength. The origin of this second, non thermal component of decimetric radiation is uncertain. The high temperatures (10^4–10^5° K) required to explain its intensity render it unlikely that the radiation could be coming from layers of a hot atmosphere below the cloud surface. It may come from below the ionosphere or from a small inner "radiation belt" of energetic electrons around the planet.

FIGURE 13-4   *Jupiter's radio noise. (a) Energy distribution (solid curve) results from two components, thermal and nonthermal (dashed lines).*

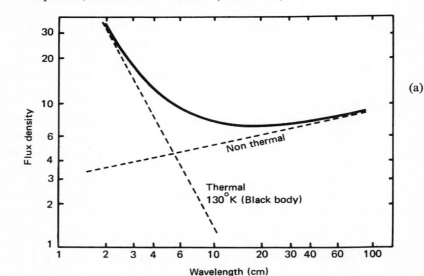

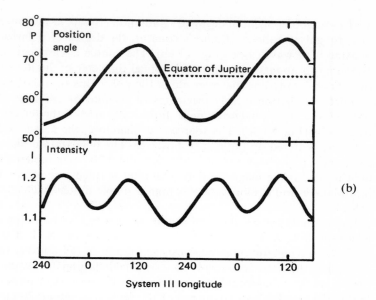

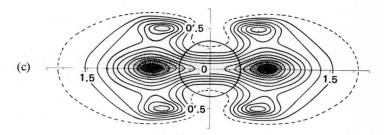

FIGURE
13-4
(b) (c)
(b) *Variations of polarization and intensity of decimetric radiation with longitude in System III.* (c) *Radiation belts controlled by Jupiter's magnetic field.*

The third component of decimetric radiation ($10 < \lambda < 170$ cm) is polarized, roughly parallel to Jupiter's equator, the degree of polarization increasing with wavelength (Fig. 13-4b). The source is elliptical in shape, extending to about three times the optical diameter of Jupiter along the equator, but only one diameter along the pole. This third component of decimetric radiation clearly originates in a cloud of energetic electrons trapped in a strong magnetic field, Jupiter's analog of the terrestrial Van Allen belts (Fig. 13-4c). The source and, therefore, the magnetic field are not quite centered on the planet, however. The polarization plane of this component rocks back and forth through about 20° near Jupiter's equator, indicating that the magnetic pole of Jupiter is tilted some 8° to Jupiter's rotation axis and is more complex than a simple dipole, such as the earth field.

Jupiter emits still another and very distinctive type of radiation at wavelengths near 15 meters, discovered by B. F. Burke and K. L. Franklin in 1955. The *decametric* emission consists of bursts of radio noise, somewhat resembling the static accompanying terrestrial lightning flashes, except that the Jovian lightning would have to be some 10^9 times more powerful than the terrestrial. The low frequency of the radiation—about 20 megacycles per second—complicates the observational problem. The largest of ordinary radio dishes, themselves only a few wavelengths in diameter, do not have sufficient aperture to resolve Jupiter and distinguish its radiations from other radio sources including man-made noise.

Interferometer observations, made with multiple antennas, do afford some spatial resolution, and these records enable the astronomer to separate the Jupiter emissions from other sources. Although the decametric radiation is intermittent, plots in terms of the rotation period of Jupiter show a marked tendency to recur when the same longitudes of Jupiter point earthward (Fig. 13-5a). The main radio sources, however, appear to rotate slightly faster than the equatorial features. Initially astronomers tried to identify the source with the great Red Spot or some other visible features. But this attempt was soon abandoned, when extensive observations in Australia and the United States demonstrated the existence of several fixed sources all sharing precisely the same rotation period of $9^h55^m30^s$, distinctly different from that of any visible feature. The decimetric variations also indicate this period, called the System III longitude system. Thus Jupiter must possess a fairly rigid internal structure supporting the magnetic field in constant rotation.

The decametric emission occurs only at frequencies below 50 megacycles per second (Mc/sec) with peaks in the range 18–24 Mc/sec. Estimates of the radiated power range up to 10^7 kilowatts, with bursts lasting from a few seconds to upward of an hour, and spikes that last only a few microseconds. The energy emitted appears to be highly concentrated in sharply focused or ribbonlike beams. We record this noise only when the earth crosses these beams—or, perhaps more accurately, when the beams cross

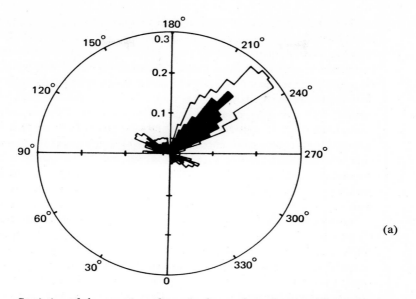

**FIGURE
13-5
(a) (b)**

Statistics of decametric radio noise bursts from Jupiter. (a) Longitude distribution in System III reveals three major sources. Radial distance measures number of bursts observed. (b) Occurrence of noise depends not only on longitude of Jupiter but also on orbital longitude of satellite Io.

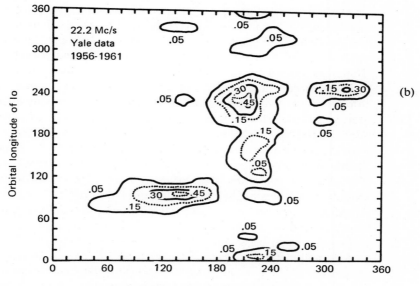

Longitude of central meridian (System III)

the earth. The frequency of this radio emission tends to vary rapidly in either direction.

The details of the phenomenon are still in doubt, but the emission that possesses right-hand elliptical polarization appears to be of a nonthermal type called *synchrotron radiation*, deriving its energy from fast electrons in a magnetic field (see Chapter 33).

In 1964 E. R. Bigg discovered an amazing correlation between the decametric bursts of Jupiter radiation and the position of Io, the innermost major satellite. The effect increases at wavelengths less than 10 m but does not show at all in centimetric wavelengths. Nor has the effect been certainly found for any of the other satellites of Jupiter. Bursts are almost all concentrated when Jupiter is turned to specific directions with respect to the earth and Io (Fig. 13-5b). The bursts occur while Io crosses the plane of Jupiter's magnetic field on only one side of the planet as seen from earth. We still do not understand how Io produces or modulates these radio bursts.

Saturn is a weak radio source, in contrast with Jupiter. It does appear to emit microwave radiation of two types: thermal radiation from the cold cloud surface, and some nonthermal emission.

13-6   The Rotation of the Giant Planets. In addition to radio observations for Jupiter, several optical methods are available for measuring the rotation periods of the giant planets. On the surface of Jupiter we see markings and spots, whose progress around the planet enables us to time the rotation with high accuracy. The method works less well with Saturn, where definite spots appear only rarely. It fails completely for Uranus and Neptune, which have never exhibited well-defined spots.

Visual observations show that at high latitudes the cloudy surface of Jupiter rotates with a period of 9 hours 55 minutes 41 seconds (System II), not far from the precise radio value of 9 hours 55 minutes 29.4 seconds (System III). The apparent rate of rotation, however, increases at lower latitudes, the equatorial belt moving fastest of all with a period about 9 hours 50 minutes 30 seconds (System I). Thus, like the sun, Jupiter has an "equatorial acceleration." Since the visible surface consists of clouds, this observation indicates that at low latitudes a strong east-moving wind is blowing around Jupiter at about 140 miles an hour. The earth has its trade winds, which blow towards the west. The observed easterly trend on Jupiter has never been explained.

A second method of determining rotation relies on spectrographic measurements of Doppler shifts (see Sec. 8-14) for the planet's spectrum lines. As a result of rotation, one side of the planet is approaching and the other receding. When the slit of the spectrograph is set along the equator of the planet, the spectral lines coming from the approaching limb are shifted towards the violet, those from the receding limb towards the red. The lines from the center of the planet are undisplaced, because the motion is perpendicular to the line of sight. The displacements in opposite directions produce a tilt of the spectrum lines, its angle proportional to the rotational speed (see Fig. 13-8).

The equators of Jupiter and Saturn are easy to locate, on account of the oblateness of the planets and the direction of the cloud belts. Uranus too has belts and is conspicuously flattened, but difficult to observe. Neptune shows no cloud bands.

Uranus, however, has five satellites, all of which move in the same plane, almost at right angles to the plane defined by the planet's orbit around the sun. Actually, the angle between the orbits of the satellites and the orbit of the planet exceeds 90°, so that the satellites revolve in the *retrograde* direction. The plane defined by the satellite orbits remains essentially constant in space—a very significant fact. Any rotating planet must be somewhat oblate, and the gravitational pull of the equatorial bulge would cause the plane of each satellite to wobble independently if it were inclined to the planet's equator. The fact that the orbits are all coplanar and do not wobble indicates that the planet's equator must lie in the plane of the satellite orbits. Spectrographic observations by Moore and Menzel at Lick Observatory have confirmed this orientation, establishing the rotation period of Uranus as 10 hours 49 minutes, with an uncertainty of about one-half hour.

A somewhat similar situation obtains for Neptune, which has two satellites. Only one of them, however, is well enough observed to establish the constancy of its orbital plane, and thus indicate the plane of the planet's equator. Spectrographic observations by Moore and Menzel fix the period of rotation at about 15 hours 40 minutes.

A rotating planet with spots on its surface should show variations of brightness. Changes in the visual brightness of Uranus were sometimes ascribed to the presence of spots on the rotating planet, but modern photoelectric observations do not confirm the reality of these periodic changes.

The results of various determinations of the rotation periods are shown in Table 13-2. The oblateness of the planets—the difference between the equatorial and polar radii, divided by the equatorial radius—is given in the last column.

13-7 Saturn's Rings. The rings of Saturn make this planet, when seen through a large telescope, one of the most beautiful objects in the heavens. We speak of "rings" because four well-defined concentric units exist in what might otherwise be termed a single ring. These are an outer ring (A), a bright ring (B), a semitransparent gauze or "crepe" ring (C), and a very faint inner ring (D).

Before the invention of the telescope, no one knew that such features existed. Indeed, after Galileo in 1610 had noted something peculiar about the appearance of Saturn, argument continued for many years as to exactly what that peculiarity was. The earliest telescopes were too imperfect to reveal the true form of the rings. Thus there were some who supposed that Saturn was a triple planet, others who thought that it had "ears," or was an ellipse with holes in it. The disappearance and reappearance of these peculiar appendages during the course of the early observations complicated the problem further. Christian Huyghens (Fig. 13-6), made the final identification in 1655: the rings of Saturn were a thin, flat disk, something like the brim of a hat.

The overall diameter of the ring system is 276,000 km (171,000 miles), and the bright, or middle, ring is 26,000 km (16,000 miles) wide (Fig. 13-7).

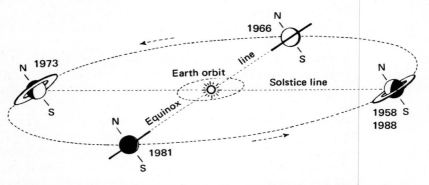

FIGURE 13-6 *Variable presentation of Saturn's rings to the sun and earth explains their changing appearance.*

FIGURE 13-7 (a) (b) (c) (d) *Saturn's rings. (a) Maximum opening, Mount Wilson Observatory. (b) Edge-on, on October 27, 1966. Disk overexposed to show rings faintly. (c) October 28, 1966, blue filter. (d) October 29, 1966, yellow filter. (Mount Wilson and Palomar Observatories and J. Texereau at McDonald Observatory.)*

(a)

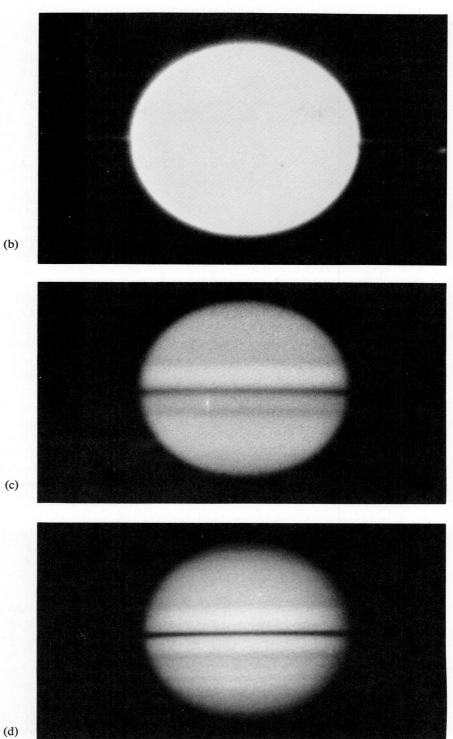

(b)

(c)

(d)

Its outer edge stops abruptly at what appears to be a true gap in the ring system, which separates the middle ring from the outer. This well-defined separation, called the Cassini division, looks like a dark line running around the ring. Encke's division is another less conspicuous line in the outer ring, perhaps not completely dark. The outer ring, some 18,000 km (11,000 miles) wide, is distinctly fainter than the second ring, and the crepe ring of about the same width is still fainter. The fourth ring, D, discovered in October 1969 by P. Guerin of the Institut d'Astrophysique, Paris, spans the distance between the planet and the inner edge of ring C, from which it is separated by a space about as wide as the Cassini division. The maximum surface brightness of D is about 1/18 that of B. It slowly fades inward to the edge of the globe.

The rings cast a definite shadow on the planet. The brightest ring is essentially opaque, a fact that becomes evident on the rare occasions when Saturn happens to occult a star. The crepe ring is considerably more transparent, and the planet shows dimly through its hazy structure.

The nature of these mysterious rings was a subject of speculation for more than two centuries after their discovery. The French mathematician E. Roche and the British physicist J. C. Maxwell solved the mystery about 1850. They showed that the rings could not possibly be solid disks, because the internal stresses that would occur within as they rotated in the gravitational field of the planet would disrupt them. Just as a planet near the sun revolves more quickly than one farther away, the inner portions of a solid ring would strive to rotate more rapidly than the outer parts, and the ring would quickly break into fragments. Maxwell thus demonstrated that the rings must be composed of small objects, though he could not determine their size. The spectrograph furnishes observational proof that these conclusions are correct, that the ring revolves in the same direction as the planet and that the inner edge revolves more rapidly than the outer (Fig. 13-8).

The rings are undoubtedly extremely thin (see below). They are tilted 27° to the plane of the ecliptic, so that twice during Saturn's 29-year circuit around the sun the rings present themselves edgewise toward the earth, and we see them, if at all, only as a thin line of light that extends on either side of the planet. Some observers have reported that the edge-on rings look like a dashed or broken line.

The chemical and physical composition of the rings is a major problem. The rings are whiter than the yellowish planet. The fact that Saturn casts sharp shadows on them shows that the rings shine by directly reflected sunlight. Kuiper measured their reflectivity in infrared light, and suggested that they consist of ice or hoar frost. L. Mertz and I. Coleman, however, find an infrared absorption feature at 1.66 μ which could not be attributed to ice, but possibly to paraformaldehyde $(HCHO)_n$.

Some investigators have suggested a meter or so for the average size of the "particles" and a kilometer or more for the thickness of the rings. Rings so constituted, however, would be unstable. Frequent and violent collisions would tend to break up the rocks into smaller fragments. The smaller the particles, the more efficient they are for blocking the light.

F. Franklin has observed the rings and analyzed the photometric data, especially the changing brightness in different colors when the planet is close to opposition. He has studied the dynamical stability of the rings to provide additional data. These studies are unable to provide an unambiguous description of the ring system except to say that its thickness to breadth ratio is uniquely small. The thickness is perhaps in the neighborhood of one kilometer.

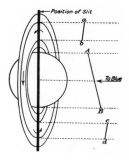

FIGURE 13-8 *Spectrum of Saturn and its rings, showing inclination of the lines due to rotation of the disk and the differential rotation of the rings. (Lick Observatory.)*

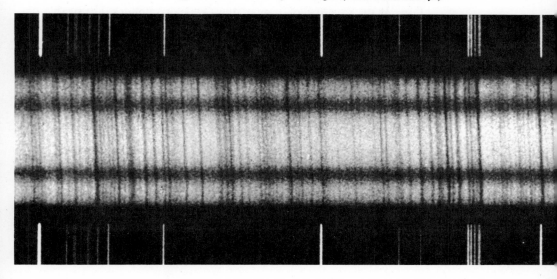

The apparent solidity of the planetary rings, therefore, is an optical illusion. They are actually a haze of small fragments. The semitransparency of the crepe ring indicates a thinning out of the material close to the planet. The Cassini and Encke divisions originate from gravitational perturbations of the ring particles by satellites of Saturn. The theory is complex and still not well understood. We note, for example, that a particle precisely in the Cassini division would make exactly two circuits of the planet while the innermost satellite, Mimas, made one. In consequence, every second revolution of Mimas would find the particle in the same position, and the satellite would exert a gravitational pull in the same direction on both occasions. Thus the particle would be subjected to repeated perturbations by Mimas, which would tend to change its period. Collisions may also play a role in clearing particles from the region of critical period. The other divisions of the ring system can be related to the periods of Saturn's inner satellites in a similar manner.

The problem of the origin of Saturn's rings is important from the standpoint of the evolution of the solar system. A *liquid* satellite, as close to the primary as the rings are, would actually disintegrate, and might form the type of ring observed. A solid satellite would have greater stability, by virtue of any tensile strength. Even so, it seems likely that the rings resulted from the disrupting force of gravitation that prevented satellites from forming so close to Saturn.

13-8 **The Satellite Systems of the Giant Planets.** Table 13-3 lists the various satellites of the solar system along with physical and orbital data. Those of the giant planets are decidedly in the majority. The table gives the number as well as the name of each satellite. Note that the radius can be measured directly only for the largest satellites; otherwise it is inferred from the brightness.

Four of the satellites of Jupiter, the Galilean satellites, are sizable objects. Together with Titan of Saturn and Triton of Neptune, they are really of planetary size, comparable to the planet Mercury or the earth's moon, which, as we have already noted, might be regarded as a planet rather than a true satellite. The two innermost Galilean satellites, Io and Europa, appear to be about as dense as rock, while Ganymede and Callisto are less dense, as though made of an ice-rock mixture. These Galilean satellites possess vague shadowy markings that can be mapped, but they are too distant for really detailed study.

Some of the outer satellites of Jupiter are interesting; they may originally have been minor planets, or asteroids, captured by Jupiter on the occasion of near approach. This view is strengthened by the fact that some of them are in retrograde motion, for Jupiter could capture another body more readily into a retrograde orbit than into a direct one.

TABLE 13-3

Satellites of giant planets

Name	Mean Distance, 10^3 km	Sidereal Period	m_v	Motion	Radius, km	Discovery	
Jupiter							
1 Io	421.6	$1^d18^h28^m$	4.8 var.	D	1,775	Galileo	1610
2 Europa	670.9	3 13 24	5.2 var.	D	1,550	Galileo	1610
3 Ganymede	1,070	7 03 43	4.5 var.	D	2,800	Galileo	1610
4 Callisto	1,880	16 16 32	5.5 var.	D	2,525	Galileo	1610
5	181.3	0 11 57	13	D	55?	Barnard	1892
6	11,470	250 14	13.7	D	40?	Perrine	1904
7	11,740	259 16	16	D	30?	Perrine	1905
8	23,300	738 22	18.8	R	<15	Melotte	1908
9	23,700	758	18.3	R	<15	Nicholson	1914
10	11,710	263 13	18.6	D	<15	Nicholson	1938
11	22,350	692 12	18.1	R	<15	Nicholson	1938
12	20,700	631 02	18.8	R	<15	Nicholson	1951
Saturn							
1 Mimas	186	0 22 37	12.1	D	240?	Herschel	1789
2 Enceladus	238	1 08 53	11.8	D	320?	Herschel	1789
3 Tethys	295	1 21 18	10.3	D	500	Cassini	1684
4 Dione	377	2 17 41	10.4	D	440?	Cassini	1684
5 Rhea	527	4 12 25	9.8 var.	D	750	Cassini	1672
6 Titan	1,222	15 22 41	8.4	D	2,400	Huyghens	1655
7 Hyperion	1,481	21 06 38	14.2	D	80?	W. Bond	1848
8 Iapetus	3,560	79 07 56	11.0 var.	D	400?	Cassini	1671
9 Phoebe	12,930	550 11	14.7	R	80?	Pickering	1898
10 Janus	160	0 17 59	14	D	?	Dollfus	1966
Uranus							
1 Ariel	192	2 12 29	14.4	R	400?	Lassell	1851
2 Umbriel	267	4 03 38	15.3	R	280?	Lassell	1851
3 Titania	438	8 16 56	14.0	R	480?	Herschel	1787
4 Oberon	586	13 11 07	14.2	R	400?	Herschel	1787
5 Miranda	130	1 09 56	16.5	R	150?	Kuiper	1948
Neptune							
1 Triton	354	5 21 03	13.6 var.	R	1,850	Lassell	1846
2 Nereid	5,570	359 10	18.7?	D	160?	Kuiper	1949

Titan, the brightest satellite of Saturn, is remarkable as the only satellite known certainly to have an atmosphere. Its spectrum shows the absorption bands of methane. Kuiper, who made this discovery, points out that Titan has a distinctly reddish hue, and suggests that the high reflectivity of all Saturn's satellites may be due to the presence of ice or hoar frost on their surfaces—a suggestion previously mentioned in connection with the particles that form Saturn's rings. Janus was discovered in 1966 when the rings were seen on edge; probably it will be observable only on such occasions.

The most distinctive feature of the five satellites of Uranus is the fact that they all move in the retrograde sense in coplanar orbits in a plane nearly perpendicular to that of the planet's orbit.

The satellites of Neptune are remarkable dissimilar. Triton is very large and its orbit sensibly circular but retrograde. Nereid is quite small and moves directly in a much larger and extremely eccentric ellipse. Is Pluto possibly a lost satellite of Neptune?

14

Asteroids and Comets

14-1   The Interplanetary Void.   The void between the planets contains a large number of small bodies that we can observe individually. They fall into two distinct classes: *asteroids* and *comets* (see Fig. 14-1a and 14-1b). Comets have long been the most mysterious and oftentimes the most feared visitors on the nightly scene. Their hazy or ominous appearance, their long, sweeping tails, their unexpected arrivals, and their apparently erratic comings and goings bewildered and frightened the superstitious minds of primitive men. More than eight hundred individual comets have now been recorded, and all present one distinguishing characteristic. When nearest the sun or when brightest, all comets appear slightly hazy, whether or not a starry nucleus is visible, and whether the comet is observed with the eye or photographed.

The asteroids, or *minor planets*, always appear pointlike in the largest telescopes, and, with one exception, their orbits do not appreciably exceed the distance of Jupiter. They are, in fact, large rocks or tiny planets, mostly moving in direct orbits of small inclination between the orbits of Jupiter and Mars.

Even though the combined mass of the observed comets and asteroids is probably much less than the mass of the earth, their very existence in such numbers and their distinctly contrasting characteristics suggest that they are in some fashion "leftovers" in the evolution of the solar system, containing vital clues about the details of how the planets and the sun came into being.

**FIGURE
14-1** (a) *Photograph of asteroid trail showing motion against stellar background.* (*Harvard College Observatory.*) (b) *Comet 1948 XI was discovered during an eclipse of the sun.* (*Boyden Station, Harvard College Observatory.*)

(b)

14-2 **Discovery and Orbits of Asteroids.** From about the time that the Copernican theory of the solar system was becoming generally accepted until the first night of the nineteenth century, astronomers suffered an aesthetic frustration because a conspicuous gap existed in the distribution of the planets. The so-called Bode's law (Sec. 17-2) predicted a planet at a mean distance of 2.8 A.U. from the sun, but none could be found. On the auspicious date of January 1, 1801, the Italian astronomer G. Piazzi was making routine meridian circle observations. The next night his second observation of a seventh-magnitude star in Taurus disagreed with the first; when he rechecked, he found that the "star" had actually moved. Many discoveries and observations have since proven that this "star," later named Ceres, is not just an isolated small planet filling the gap between Mars and Jupiter, but the largest of a host of smaller ones.

Olbers found the second one, now called Pallas, in March, 1802, and by 1807 the discoveries of Juno and Vesta had increased the list to four. The fifth did not turn up until 1845, but extensive search with constantly improved telescopes uncovered 322 by 1891. Max Wolf at Heidelberg found Number 323 by photography on December 20, 1891, and revolutionized the method of discovery. Figure 14-1a shows the typical trail left by an asteroid on a long-exposure photograph. Sometimes several asteroids produce such trails on a plate, making discovery a simple matter of inspection. Some 1600 asteroids with well-determined orbits are now followed systematically. Modern telescopes could follow many thousands more, if astronomers had enough time to spend on these microplanets.

The average asteroid moves in a nearly circular orbit with an eccentricity of only 0.15 and inclined $9°7$ to the earth's orbit. The mean solar distance of 2.80 A.U. is in perfect agreement with Bode's law. The orbital elements possess considerable range, but almost all have periods lying between 3.5 and 6 years. Some 2 per cent of the minor planets pass closer to the sun than Mars, while only 8 (0.4 percent) have their perihelion within the earth's orbit, and one, Icarus, discovered by W. Baade at Mount Palomar, comes closer to the sun than Mercury. The most unusual orbit, which is more like the orbit of a comet than of a minor planet, is that of Hidalgo, inclined 43° to the ecliptic with an eccentricity of 0.66 (see Fig. 14-2). No asteroid, however, moves in a retrograde sense from east to west.

The most distinctive asteroids are the fourteen Trojans, named after Homeric heroes. Their distinction lies in their orbits, practically identical on the average with Jupiter's orbit, except that the Trojans occupy average positions 60° ahead of or behind Jupiter, in equilateral triangles with respect to Jupiter and the sun. The French mathematician J. L. Lagrange proved two centuries ago that this surprising orbital position is stable gravitationally as a special solution of the "three-body problem." Actually the Trojans can wander more than 1 A.U. from the triangular point and average 15° in inclination to Jupiter's orbit without being perturbed into greatly different orbits. Troilus' orbit, for example, is inclined $33°7$.

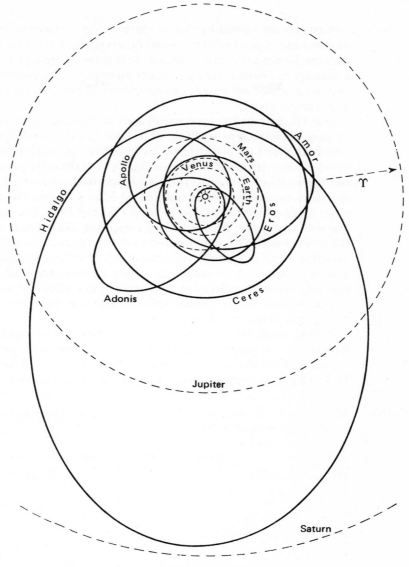

FIGURE
14-2   *Orbits of remarkable asteroids.*

No more than 50 to 100 Trojans probably exist, because very faint ones, less than 16 km (10 miles) in diameter, cannot be found. The largest, Hector, is perhaps 80 km (50 miles) in diameter. No one knows whether they are lost Jovian satellites, as G. P. Kuiper has suggested, original condensations in the making of the solar system, or possibly just bodies captured by Jupiter.

Asteroids are named by their discoverers, or nowadays, by the person who provides a good orbit for future observations. Except for the Trojans, feminine names are usually selected. At first they were mythological, until a shortage of deities forced a choice from more prosaic sources. Asteroids have been named not only for wives and sweethearts but for observatories and even pet dogs!

Eros has become the most useful asteroid because it can approach the earth to a distance as small as 17,000,000 km (13,900,000 miles)—only 0.15 A.U. (see Fig. 14-2). Hence, its rare close approaches make possible the most accurate observations for determining the solar parallax by geometrical methods (Chapter 7). The close approaches occur in late January, in pairs separated by 7 years and spaced 37 or 44 years. The most assiduously observed approaches occurred in 1901 and 1931. The next favorable one will be in 1975. Eros is shaped something like a brick, some 22 km (14 miles) long and 6 km (4 miles) in diameter, as evidenced by its light variations of 1.5 magnitude. It rotates in only $5^h 16^m$ and shows two maxima and two minima. As it moved by during its close approach in 1938, its changing aspect and changing light curve gave evidence of its direct rotation and shape. It is a whirling splinter, held together by tensile strength rather than by gravity alone.

In 1866, when only 88 asteroids were known, D. Kirkwood pointed out the existence of gaps in the zone of asteroids. None has a period close to one-half, two-fifths, or one-third the orbital period of Jupiter (see Fig. 14-3). Contrary to early theory, Y. Hagihara now concludes that Jupiter

FIGURE 14-3   *Kirkwood gaps. The frequencies of asteroids show gaps at periods of revolution that are simple fractions of Jupiter's period.*

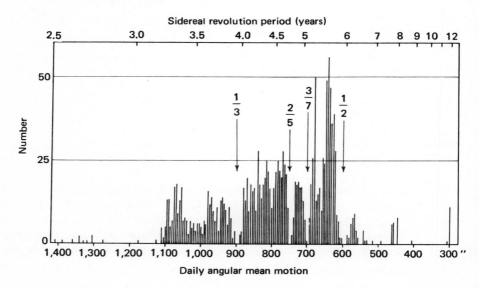

does not quickly change the orbital period of an asteroid that wanders too close to one of these critical values. The small perturbations caused by the attraction of Jupiter repeat at successive revolutions, but the resonance with Jupiter's period would produce no marked effect. Apparently the disturbing actions of many asteroids on each other, in resonance, force them out of period. The phenomenon is similar to the one causing the gaps in Saturn's rings, as discussed in Sec. 13-7.

K. Hirayama of Japan, early this century, discovered that asteroids group in "families" of similar and dynamically connected orbital elements. There can be no doubt that these families represent the fragments of larger asteroids that were broken or torn apart with low relative velocities, probably by collisions. The largest asteroid definitely associated with such a family is Themis, whose diameter is about 100 km (60 miles). According to D. Brouwer, at least twenty families are now recognized, comprising roughly 30 per cent of all the known asteroids.

The brightest asteroids whose diameters can be directly measured are listed in Table 14-1. Sample counts with large telescopes indicate that some 50,000 asteroids brighter than the magnitude 19.5 at opposition must exist.

TABLE 14-1   *Diameters and albedos of largest asteroids*

Name	Diameter		Albedo[a]	Absolute Magnitude[b]
	km	miles		
Ceres	770	480	0.06	4.0
Pallas	490	300	0.06	5.0
Juno	200	120	0.11	6.3
Vesta	380	240	0.20	4.2

SOURCE: Diameters after F. G. Watson; absolute magnitudes after G. P. Kuiper.

[a] Photographic; the moon's photographic albedo is 0.06 (visual, 0.07).

[b] Photographic; at full phase and hypothetical 1 A.U. distance from earth and sun.

The average asteroid of this brightness probably has a diameter of about 3 km (2.0 miles), possibly less if asteroids reflect light better than the moon. C. H. Schuette estimates that the total mass of all the asteroids down to fine particles cannot exceed $\frac{1}{10}$ of the earth's mass, and probably much less. A more realistic value may be $\frac{1}{500}$ earth mass or even less, since perturbations of planetary orbits by asteroids are too small to be measurable.

We have no precise information concerning the true mass or structure of any asteroid, but many astronomers believe that most are the broken fragments of two or (many) more small planets that were formed between

Mars and Jupiter and have undergone violent collisions. The degree of collisional breakup of the asteroids is quite poorly known, so possibly the asteroid belt has not changed much in character since it was formed (see Chapter 17).

E. Öpik has shown that both the earth and Venus would eliminate, either by collision or violent orbital change, essentially all of such bodies crossing their orbits within much less time than the lifetime of the solar system, some 5×10^9 years. Mars, however, could not have greatly depleted the number crossing its orbit. Hence we may suppose that the few asteroids whose orbits cross that of the earth have been thrown into these orbits by Mars relatively recently, probably within the last hundred million years. There is also the possibility that they are mostly the dead nuclei of very old and very large comets.

A number of asteroids besides Eros show appreciable variations in light. They are definitely rotating and must be small bodies of irregular shape or with mottled surface structure. G. P. Kuiper and his associates find that 20 out of 21 observed certainly vary, one with a period of only $4^h 09^m$. The brightest asteroids of known diameters exhibit low albedos (Table 14-1). The smaller ones must certainly have very rough surfaces. Their reflection of sunlight falls off very rapidly as they move from opposition in their orbits and we look at them at larger phase angles. Polarization and color studies reveal no conspicuous uniformity among the asteroids. In physical constitution they are probably pure stony silicates except for a few broken fragments of nickel-iron, as exemplified by meteorites on the surface of the earth (see Chapter 15).

14-3   Comets.   Comets, for some two thousand years, were believed to be phenomena in the earth's atmosphere. Even educated people, until the seventeenth century, believed them to be portents. Their appearance provoked alarms of pestilences, wars, and other undesirable worldwide occurrences.

The first step towards an understanding of comets was made by Tycho Brahe, who proved that the comet of 1577 was definitely farther away than the moon and could not be an atmospheric phenomenon. Newton believed that comets followed orbits according to his gravitational law, and indeed developed a method for calculating parabolic orbits for them. A comet in such a parabolic orbit must come from an infinite distance to perihelion in the solar neighborhood and then be lost again in space forever.

Edmund Halley, an ardent admirer of Newton, became convinced that the bright comets of 1531, 1607, and 1682, for which he calculated almost identical orbits, were in fact one and the same comet, returning to the solar neighborhood with a period of some 75 or 76 years. He then predicted that this comet would return about 1758, a forecast that caused considerable amusement among his critics who pointed out that Halley, born in 1656, could scarcely expect to survive and face a failure of his prediction. But the comet did return on schedule, and again in 1835; the third subsequent return in 1910 provided one of the best cometary spectacles of the present century (Fig. 14-4). We expect another grand display of Halley's comet in 1986.

FIGURE 14-4 *The head of Halley's comet on May 8, 1910. (Mount Wilson and Palomar Observatories.)*

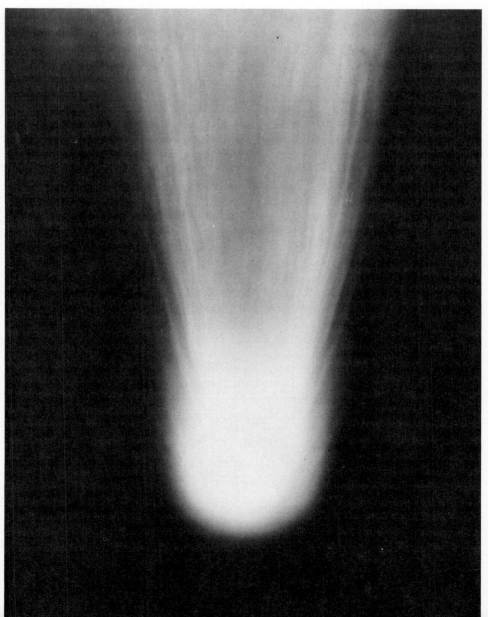

We have now observed more than 800 passages of more than 500 individual comets with sufficient accuracy to provide reliable orbital data. Some 300 move in nearly parabolic or in hyperbolic orbits, while about 200 move in elliptic orbits of measurable period. About 80 possess periods less than 200 years, of which half have been observed at more than one return. The record for returns rests with Encke's comet, with a period of only 3.3 years; it has made more than 50 round trips about the sun since its discovery in 1786. Halley's comet is second, with records stretching back certainly to 239 B.C. and possibly to 466 B.C., covering 31 periods or perihelion passages (Table 14-2) (see Figs. 14-4 and 14-5).

TABLE 14-2

Orbits of selected periodic comets

Name	P, yr	e	q, A.U.	Q, A.U.	i, deg	First Pass	Recent Pass	No. Obs. Passes	Notes
Encke	3.30	0.847	0.339	4.09	12.4	1786	1961	46	
Grigg-Skellerup	4.90	0.704	0.855	4.88	17.6	1902	1957	9	
Tempel (II)	5.27	0.548	1.369	4.68	12.5	1873	1957	12	
Pons-Winnecke	6.12	0.653	1.161	5.53	21.7	1819	1951	15	a
Giacobini-Zinner	6.42	0.729	0.936	5.97	30.9	1900	1959	7	b
Biela	6.62	0.756	0.861	6.19	12.6	1772	1852	6	c
Brooks (II)	7.10	0.469	1.963	5.43	6.1	1889	1911	4	d
Brooks (II)	6.72	0.505	1.763	5.36	5.6	1925	1960	6	
Holmes	6.86	0.412	2.121	5.10	20.8	1892	1906	3	
Whipple	7.42	0.356	2.450	5.16	10.3	1933	1955	4	
Oterma	7.88	0.144	3.388	4.53	4.0	1942	1958	3	e
Wolf (I)	8.43	0.395	2.507	5.78	27.3	1884	1959	10	
Schwassmann-Wachmann (I)	16.10	0.131	5.538	7.21	9.5	1925	1957	3	e
Temple-Tuttle	33.18	0.905	0.977	19.67	162.7	1366	1866	2	f
Halley	76.03	0.967	0.587	35.31	162.2	−239	1910	29	g
Herschel-Rigollet	156.0	0.974	0.748	57.22	64.2	1788	1939	2	
Grigg-Mellish	164.3	0.969	0.923	59.08	109.8	1742	1907	2	

SYMBOLS: P = period, e = eccentricity, q = distance from the sun at perihelion, Q = distance from the sun at aphelion, i = inclination angle.

[a] Not seen in 1957.

[b] Great Draconid meteor showers in 1933 and 1946.

[c] Split in two, 1846, and produced meteor shower in 1872.

[d] Period reduced from 29.2 years by close Jupiter approach in 1886.

[e] Observable at each opposition.

[f] Comet of Leonid meteors.

[g] Probably comet of Orionid and η-Aquarid meteor showers.

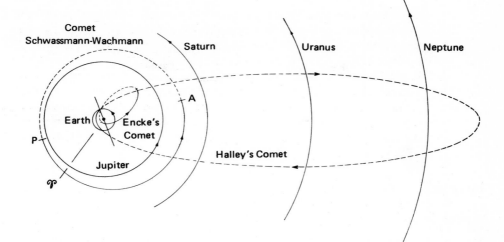

The orbits of comets Halley, Encke, and Schwassmann-Wachmann, as projected on the ecliptic. Note that Halley's comet moves in retrograde direction. Dashed parts of orbits are south of the plane of the ecliptic.

The orbits of nearly parabolic comets are tilted randomly in space, with no regard to the mean orbital plane of the planets. On the other hand, all but 6 of the 92 comets with periods less than 150 years are moving, like the planets, from west to east, and the orbital inclinations of only 13 of them exceed 33°.

Do any of the comets apparently moving in open parabolic or hyperbolic orbits truly come from outside the solar system? Or are the parabolic and hyperbolic orbits simply the result of inaccuracies and of planetary perturbations? In fact, a precisely parabolic orbit is only a mathematically convenient assumption used when the observations are not sufficiently numerous or accurate to establish the eccentricity with high accuracy. Studies by M. Fayet and E. Strömgren, however, show that the 22 comets with the most accurate parabolic or hyperbolic orbits (maximum $e = 1.002$) entered the solar system beyond the outer planets in definitely elliptical orbits or with hyperbolic orbits not differing significantly from parabolas. Perturbations by Jupiter and Saturn were responsible for the observed hyperbolic orbits. Hence we may conclude with assurance that *practically all comets belong to the solar system.* No clear-cut evidence for a visitor from external space has been found.

On the basis of these facts and a study of the distribution of cometary orbits, J. Oort of Leiden proposed, in 1950, a new theory for the replenishment of our comet supply. Oort showed, as had E. Öpik, that the sun can hold a family of comets, many of them going out to distances of more than 100,000 A.U., for as long as 5×10^9 years. During this period many passing stars would actually go through the swarm of comets much like a bullet

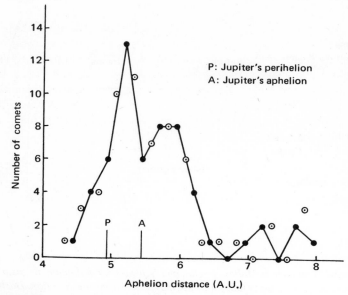

FIGURE 14-6 *Aphelia of short-period comets are clustered near Jupiter's orbit. Solid and open circles represent different ways of combining data.*

through a swarm of gnats. Some swarm members would be eliminated, but a moderate fraction of the swarm would still remain after 5×10^9 years.

According to Oort's theory, "new comets" come close to the sun for the first time when the gravitational action of passing stars perturbs their original orbits. As we shall see below, the lifetimes of comets appear to be quite short once their perihelion distances from the sun are reduced to 1 A.U. or so. At such small distances they start their external activity, so that they finally disintegrate and disappear. A very large fraction of the new comets are also lost again by perturbations of the planets, particularly Jupiter, while the perturbations shorten the periods of others.

Oort's theory for the distribution of comet orbits of longer periods received quick and widespread acceptance among astronomers. The theory has been extended by K. A. Shteins to account for the large number of short-period comets of low inclination. These orbits probably arise when passing stars bring the cometary perihelia close to Jupiter's orbit. Figure 14-6 shows how closely the aphelia of 63 comets of very short period cluster near Jupiter's orbital distance.

Before leaving the subject, we may well discuss a case of cometary idiosyncrasy. Comet Encke, the one with the shortest known period, 3.3 years, obtained its name from the great mathematical astronomer, J. F. Encke, who investigated its motion for several revolutions after its discovery in 1786. He was much surprised to find that the comet consistently came back to perihelion about $2\frac{1}{2}$ hours earlier than predicted. Since about 1860, the comet has been somewhat less spirited in its motion, arriving roughly an hour earlier on each successive revolution. This apparent defiance of Newton's law of gravitation appears to a less certain degree in the motions of a few other comets and has inspired a great deal of speculation as to mysterious forces, resisting media between the planets, and other possibilities. We have here an important clue as to the nature of comets, one that we shall explore further in a later section of this chapter.

14-4   Phenomena of Comets.   Extremely bright comets are rare. A few fairly bright ones have appeared in recent years, especially in 1957 (Comet Arend-Roland) and 1962 (Comet Seki-Lines), but the five extremely bright comets seen during the nineteenth century (1811, 1835, 1843, 1861, and 1882) have not been matched this century—although Halley's comet at its return in 1910, the unexpected comet 1910a, and the sun-grazing comet Ikeya-Seki in 1965 were spectacular enough (Fig. 14-7).

Comet designation first represents the order of their discovery (1910a, 1910b, 1910c, and so on) as temporary identification along with the name of the discoverer or discoverers (not more than three names). The permanent designation is decided later, the year followed by a Roman numeral in the order of perihelion passage. As an example, Comet 1932 V, discovered independently by Peltier and Whipple, was first called Comet Peltier-Whipple, 1932k. Periodic comets usually bear the names of their discoverers or occasionally of the orbit computer. Thus Halley's comet received its name because of Halley's extremely important prediction of its return in 1759. In its 1910 return his comet carried the designations 1909c and, permanently, 1910 II.

Extremely bright comets present a majestic appearance because of their huge tails, which occasionally can stretch more than halfway across the sky, and sometimes extend more than 100,000,000 km in space. The tail arches away from the head sometimes smoothly and sometimes with many irregularities. Occasionally the comet possesses multiple tails. Generally the tails point away from the sun in space (see Figs.14-7 and 14-8). The head of a very bright comet is nearly circular except where it merges into the tail. On fainter comets, the head may be nearly circular with only a slight suggestion of a tail, while very faint comets may be only a small hazy patch with or without a bright, starlike nucleus. Typical examples appear in Figs. 14-9 and 14-10.

The head of a comet often appears as a stellar nucleus surrounded by a fuzzy coma, which may extend for more than 100,000 km. In a bright comet, the coma may involve bright jets and streamers, the latter often sweeping backward in a graceful curve to form part of the tail. Dark lanes have been recorded from the nucleus into the tail. Rarely there may appear envelopes of trajectories of particles sent out from the nucleus towards the

**FIGURE
14-7**

Great sun-grazing comet Ikeya-Seki, past perihelion. Nov. 2, 1965. (Smithsonian Astrophysical Observatory.)

**FIGURE
14-8**

*Comet Ikeya, 1963 I, February 24, 1953.
(Courtesy Alan McClure.)*

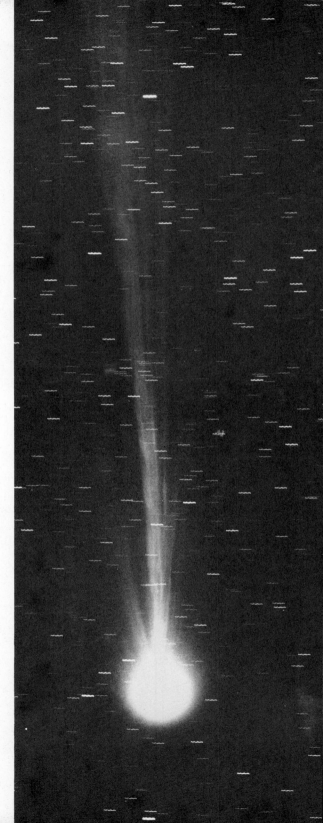

FIGURE 14-9   *Comet Cunningham, 1941 I, near discovery. (Harvard College Observatory.)*

FIGURE 14-10   *Typical faint periodic comet, Pons-Winnecke, March 21, 1964. Near perihelion. (E. Roemer, official U.S. Navy photograph.)*

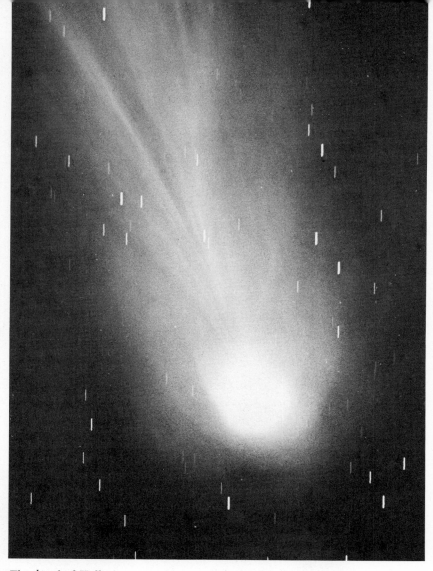

FIGURE
14-11 *The head of Halley's comet, June 5, 1910, showed envelopes about the head and streamers into the tail. (Mount Wilson and Palomar Observatories.)*

sun and forced backward into the tail (Fig. 14-11). Knots in the tails may change their appearance and brilliancy in days or hours (Fig. 14-12). Because comets move so rapidly across the sky, seldom more slowly than a degree a day when bright, and often several degrees a day, long-exposure photographs must be guided to follow the comet rather than the stars, hence giving an unblurred image of the comet while the star images are trailed.

Comets refuse to follow any simple law of brightening as they approach or recede from the sun. A solid of fixed dimensions would reflect sunlight

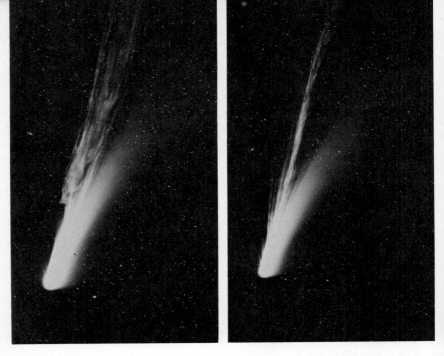

AUGUST 22   AUGUST 24

**FIGURE
14-12
(a)**
                Changes in comets (a) Comet Mrkos, 1957V, in August 1957.

AUGUST 26   AUGUST 27

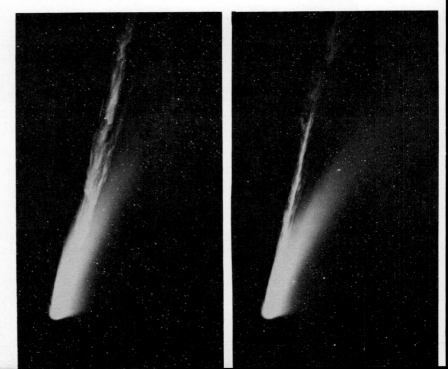

FIGURE 14-12 (b) (b) *Comet Humason, 1962 VIII at solar distances exceeding 2.4 A.U. Top: August 7, 1962. Bottom: August 28. (By K. Rudnicki and C. Kearns. Mount Wilson and Palomar Observatories.)*

inversely as the squares of the distances from the sun (r) and from the earth (Δ)—that is, would vary as $1/(\Delta^2 r^2)$. Failure of comets to follow this law indicates that the comets emit matter and radiate light, rather than merely reflecting sunlight, a fact that the spectrum confirms. N. T. Bobrovnikoff's intensive study of variations in comet brightness shows, however, that the power of r in the equation averages 3.3 instead of 2. In many cases the power is decidedly greater than 3.3, although it sometimes varies and in a few cases is actually negative: periodic comet Holmes, on its second return (as Comet 1906 III), actually faded to invisibility as it approached the sun. It has never been seen since.

Because of the rapid decrease in brightness as a comet recedes from the sun, relatively few are observed beyond 3 A.U. Among the notable exceptions, however, was the great comet of 1729, visible to the naked eye even though its perihelion distance was greater than 4.0 A.U. In 1927 Schwassmann and Wachmann, in Germany, discovered another remarkable comet whose orbit lies completely outside the orbit of Jupiter. The orbit is inclined only 9°.5 to the ecliptic and is nearly circular. Although the Schwassmann-Wachmann comet is extremely faint, about 18th magnitude most of the time, it occasionally flares up to become about a hundred times brighter. Such changes can take place in a day or so and constitute one of the many mysteries of comets.

Eleven comets have disappeared while under observation or have failed to return in their usual short-period orbits. The most famous is Biela's comet, which in 1846 was observed to split into two pieces, both of which returned in 1852 but faded out within a short time and have not been seen since.

The great comet of 1882 split into four components near perihelion, which was extraordinarily close to the sun (0.0078 A.U.). The comet actually passed through the solar corona. We shall have to wait about a millenium to find out whether these fragments will return. We do know, however, that six or seven other comets have appeared in almost the same orbit as this big comet of 1882, including the even greater comet of 1843 and the Ikeya-Seki comet of 1965. It seems probable that all of them were components of a still greater one that split on a previous perihelion passage, perhaps about a thousand years ago. J. G. Porter lists some nineteen examples of such comet groups, each consisting of two or more comets moving in similar orbits.

The spectrum of a comet is continuous at great solar distances, showing the Fraunhofer solar absorption lines under high dispersion when the comet is bright enough to permit observation. At a typical distance of 1 A.U. molecular emission bands are frequently conspicuous. Often comets show the bright yellow lines of sodium within a distance of about 0.8 A.U. of the sun and more rarely the lines of other atoms when very close to the sun. P. Swings and his associates made positive identifications of the molecules or atoms C_2, C_3, CH, CH_2(?), CN, NH, NH_2 OH, O, and Na

(a)

**FIGURE
14-13**

Slitless spectra of comets. (a) Brook's comet, 1911 V, showing molecular bands of ions in tail. (b) Comet Morehouse 1908 III. Note that some emission bands present in the head are not visible in the tail. (F. Baldet, Meudon Observatory.)

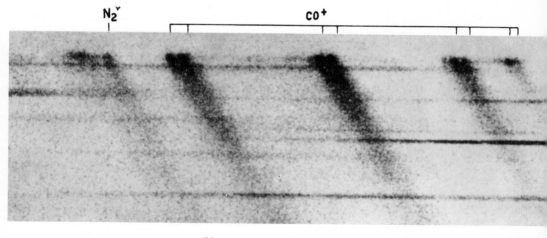

(b)

in the head of a comet; while the sun-grazing comet of 1965 near perihelion showed emission lines of Si, Ca, Na, K, Cr, Mn, Ni, probably Al, and many of Fe. Bands of CO^+, CO_2^+, and N_2^+ show exclusively in the tails of comets (Figs. 14-13 and 14-14). While CH^+ and probably OH^+ appear in the head, all the observed tail molecules are ionized. The strongest bands are the so-called Swan bands of C_2 and the CN bands in the head of the comet. Note that very simple molecules, made only from H, C, N and O, account for the entire identified molecular radiation.

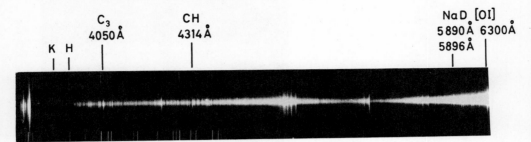

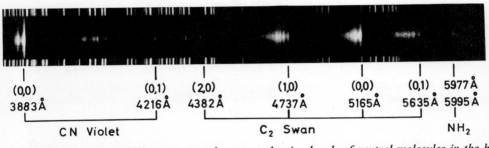

FIGURE 14-14 *Typical slit spectrum of a comet showing bands of neutral molecules in the head. Note in particular vibration sequence of C_2 Swan bands and below resolution of one band into rotational components. (P. Swings, Institute of Astrophysics, Liege.)*

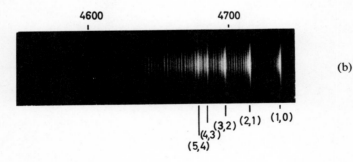

There are two major classes of comet tails: Type I, which stream out almost directly opposite to the sun, and Type II, which curve around backward in the orbit plane. Both are beautifully illustrated in Comet Mrkos, 1957 V (Fig. 14-12a). The Type-I tails, away from the nucleus, show emission-band spectra of ionized molecules and thus are called *ion tails*. The Type-II tails show scattered solar radiation—that is, the solar spectrum reflected by small grains—so that they are often called *dust tails*.

Studies of the polarization of the light in comets, begun by F. Arago in 1819, were continued only sporadically until recent times. The average amount of polarization is about 11 per cent for seven comets, and does not depend upon the angle between sunlight and the observed direction of the comet. A. Secchi found remarkable variations in the polarization of comets. In Comet 1861 III, particularly, the nucleus developed polarized light over an interval of two nights. Y. Öhman showed that molecular spectra as well as light scattered from small particles can exhibit polarization effects. Hence polarization alone does not prove the presence of dust in the heads of comets, although W. Liller has shown by the additional evidence of reddening in comets Arend-Roland (1957 III) and Mrkos (1957 V) that particles were present, with an average diameter of about 0.6 micron.

FIGURE 14-15 *The sunward spike of comet Arend-Roland 1957 III, April 1957. (Lick Observatory.)*

Comet Arend-Roland presented the startling phenomenon of a sunward tail or spike (Fig. 14-15). The maximum length of 14° or more occurred within minutes of the time that the earth crossed the plane of the comet's orbit. There is no doubt that the spike consisted of sunlight reflected from particles lost by the comet and spread out in its orbital plane.

We have, therefore, seen that comets show enormous and erratic variations in size, form, tail structure, brightness, polarization, spectra, and even in orbital motions.

14-5   The Physical Nature of Comets.   Their molecular spectra prove that comets expel gas, which could not be held gravitationally and which appears to stream into the tail. Faint reflected sunlight in the spectrum of the Type-II tails shows that fine dust is often present. Certainly the tail material must be lost forever from the comet. For many years most astronomers have been satisfied with the qualitative explanation that the pressure of sunlight on fine particles and gases expelled from the heads of comets forces them away from the sun into the tails. The theory is adequate for dust tails, which curve backward with respect to the orbital path, obeying Kepler's law of areas. Oftentimes, however, small knots in the ion tails move away from the sun with accelerations many times greater than that of the solar attraction. Bredichin, in the last century, made the first exhaustive study of these motions. Theorists have been hard pressed to explain the great values and wide variety of such accelerations by the very weak force of light pressure.

L. Biermann investigated the possibility that corpuscular streams of protons and electrons, shot out from the sun with velocities of hundreds of kilometers per second, may play the major role in determining the motions and shapes of ion tails. The spacecraft measurements of the solar wind (Chapters 9 and 10) led Alfvén and others to include a new physical process. The magnetic fields in the solar wind act as a net to drag ions from the head of a comet back to form the ion tail. The neutral molecules and atoms slip through the magnetic net undisturbed, except that fast electrons in the hot solar wind ionize some of the molecules around the comet head. Solar radiation itself is not enough either to ionize the molecules or to force them back to form a tail.

Comets divulge a clue to their nature in their clearly established ability to produce meteor streams. Even though the two components of Biela's comet disappeared in 1852, when the earth approached their orbits in 1872 at a point near the predicted positions, an enormous meteor shower occurred. Over 30,000 meteors were observed in 6.5 hours from one location in Italy. The numerous shooting stars in this great shower were small solid fragments of one or both of the components of Biela's comet, spread out along and near the original orbit. Today we know more than a dozen specific examples of comet-meteor associations. We conclude that the nuclei of comets consist not only of the gases shown by their spectra but also of solid particles ranging from fine dust to sizable pieces, perhaps larger than footballs.

We can also be certain that the original nucleus of a comet is concentrated within a volume not many kilometers in diameter. From the observed brightnesses and the known solar radiation we can calculate upper limits to the sizes of cometary nuclei by assuming a low albedo, like that of the moon. For Comet Pons-Winnecke, only 0.04 A.U. from the earth in 1927, F. Baldet observed no disk to the stellar nucleus under excellent seeing conditions with the large refractor at Meudon, France. He gave an upper limit of 5 km for the diameter of the nucleus. From the magnitude and assumed values of the albedo of 0.50 and 0.05, he calculated diameters of 0.2 and 0.6 km, respectively. At great solar distances comet nuclei appear stellar; diameters, calculated with an assumed lunar albedo, are in the range from less than a kilometer to a few tens of kilometers.

The early facts and deductions about comets led a century ago to what we shall term the "gravel-bank" model of a comet, in which the nucleus consists of fine gravel and dust as observed in meteors, with the gases *occluded* on (attached to) the surfaces of the solid particles. The quantity of gas that can be held this way is not large, although, in a vacuum tank, many hours or days may be required to eliminate the occluded gas that slowly evaporates from the outer molecular layers of a smooth solid.

Qualitatively, the gravel-bank theory seems to explain why comets brighten as they approach the sun. The gaseous coma surrounding the nucleus forms as the occluded gas evaporates from the warmed solids. The theory fails completely, however, to explain the great lifetimes of some comets, and particularly it does not account for the large quantities of gas observed to come from some cometary nuclei. As we shall show in Sec. 15-10, a comet could not pick up gas in space. In very few revolutions, the solar heating should largely eliminate the occluded gases. Thus a comet should fade in brightness very rapidly with time. But we have observed Comet Encke for more than 50 revolutions, while the meteoric evidence shows clearly that it has existed for at least 1500 revolutions.

These and other considerations led F. L. Whipple to develop a theory of comets wherein the nuclei are fairly compact solid bodies, made up of frozen gases or "ices," such as water (H_2O), ammonia (NH_3), and possibly methane (CH_4), less stable molecules and molecules ("frozen radicals") that are unstable at room temperatures, with earthy solid material—dust and possibly gravel—in the ices. We shall term this concept the "icy-nucleus" theory to distinguish it from the gravel-bank model. In space well beyond Jupiter the temperatures are so low that these ices and dust particles would remain little changed for practically infinite intervals of time. Furthermore, we have reason to believe that the lighter elements—hydrogen, carbon, nitrogen, and oxygen—are the most prevalent, chemically active elements in the sun, stars, and space, with a small trace of metallic and silicate or "earthy" compounds. Interstellar dust must be made of such materials. The noble gases, particularly abundant helium, will not be present except by adsorption, because they do not form compounds or freeze at the temperature expected.

We may visualize the lighter molecules as forming the ices, which vaporize at or below room temperature. The heavier metallic and silicate compounds will form meteoritic or earthy material, conglomerated with the ices, making dirty ice. Within a cometary nucleus, under gravitational pressures of a fraction of an atmosphere for thousands of millions of years, the mass of small particles will tend to form a weak structure that will mostly disrupt as the gases escape in sunlight. A few of the larger and stronger clumps will form meteoritic particles (see Chapter 15).

When the icy nucleus of such a comet approaches the sun, the solar heat vaporizes the ices, which escape against the negligible surface gravity of the comet. The gaseous molecules, which must occur in greater variety than the few indicated above, dissociate under the action of solar ultraviolet light to produce the molecular features observed in cometary spectra. We call such a process "photodecomposition."

The escaping gases also carry along the dust particles, which then spread out around the comet orbit to produce dust tails and meteor streams. During a single revolution the comet may lose only a few meter of radius, the amount that solar heat can vaporize as the comet swings by perihelion. Hence if the initial diameter of the nucleus is as great as a kilometer or more, it may last for hundreds or even thousands of revolutions.

The icy nucleus can also explain the deviant motions of comets. As the ices vaporize, a thin coating of rocky dust, meteoritic material, probably

FIGURE 14-16 *Whipple's "icy conglomerate" model for an old nucleus of a periodic comet.*

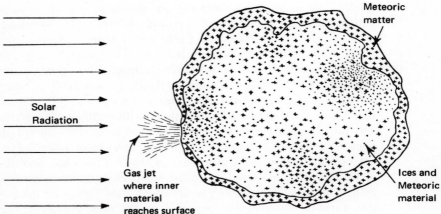

Meteoric matter

Solar Radiation

Gas jet where inner material reaches surface

Ices and Meteoric material

very fragile in character, may remain on the comet's surface. Thus the sun's heat must be conducted through the crust down to the ices beneath. If the nucleus is rotating, like all known planets, asteroids, and other varieties of celestial objects, the nucleus will have turned part way around before the solar heat can reach and vaporize the ices. Hence the outflowing stream of gas can act as a weak rocket jet, as in Fig. 14-16, either opposing or adding to the forward motion of the comet in its orbit. B. Marsden, using modern computing techniques, has analyzed carefully the motions of eighteen periodic comets that have been observed through three or more perihelion passages. Of these, fourteen show definite deviations from Newtonian motion, about half accelerating and half decelerating their motion. For Encke's comet, a loss of only 0.2 per cent of the mass per revolution could produce the maximum suspected change in period.

Without a fairly solid coherent nucleus the "sun-grazing" comets could not withstand the heat and tidal disruption at a fraction of the sun's radius above its surface. The fact that most of them *split* rather than completely disintegrate is evidence of a compact nucleus.

Probably, then, a comet is nothing more than a deep-freeze mixture of many kinds of ices and less volatile materials, which slowly evaporate like a piece of dry ice in warm air as the comet approaches the sun. The study of comets now provides clues concerning the origin of the solar system and the nature of the interstellar dust particles, which are so important in the structure of the universe.

CHAPTER *15*

Interplanetary Debris

15-1 **The Contents of Interplanetary Space.** Interplanetary space, even near the earth, is not quite empty. The earth's atmosphere constantly receives a barrage of meteors, meteoritic dust, and cosmic rays as well as a rain of ionized gas from the sun. Near the plane of the earth's orbit, dust particles and gas scatter sunlight to produce the *zodiacal light*, a faint diffuse cone of light visible after dark in the evening and before dawn in the morning. An even fainter glow, called the *gegenschein*, appears in the night sky opposite the sun.

Fortunately for us, few solid bodies large enough to survive their passage through the earth's atmosphere fall upon its surface. Such bodies are known as *meteorites*. The term *meteor* designates the luminous appearance as a small particle boils away by friction during its rapid flight down through the earth's atmosphere. The particle producing the meteor may be called a *meteoroid*. Meteors bright enough to cast shadows at night or be visible in the daytime are known as *fireballs* (Fig. 15-1), while those rare ones whose explosions in the atmosphere produce audible sounds are *bolides*.

The earth's attraction draws particles into the atmosphere with a minimum speed of 11.2 km/sec (7.0 miles/sec)—the velocity of escape (Chapter 7) acting in reverse. The relative motion of the particle and the earth adds

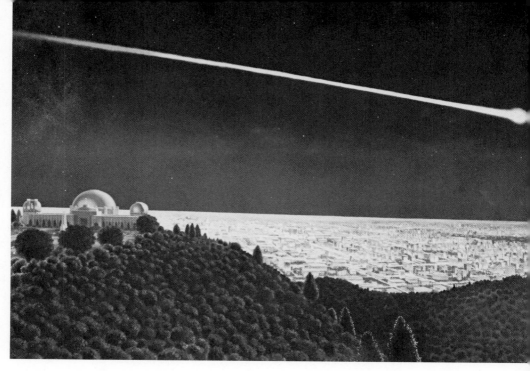

FIGURE
15-1

Painting of fireball by James A. Roth. (Courtesy Griffith Planetarium.)

to this speed. The earth moves about the sun at a speed of 29.7 km/sec (18.4 miles/sec) while the velocity of escape from the solar system at the earth's distance is some 42.4 km/sec (26.3 miles/sec), the parabolic velocity (see Sec. 7-14). If a particle in a nearly parabolic retrograde orbit at perihelion happens to strike the earth head-on, the combined velocities add up to as much as 72.1 km/sec (44.7 miles/sec). This speed is the maximum for a solar-system meteor (see Fig. 15-2). Particles coming in from interstellar space could encounter the earth with even higher speeds. Thus these small projectiles from space strike the earth's atmosphere with speeds many times that of the fastest bullets.

FIGURE
15-2

Meteorites.   A large meteorite enters the atmosphere with far greater kinetic energy than the explosive energy of an equal mass of TNT. Friction with the air melts and boils away half or more of the outer mass, while air resistance reduces the original velocity. Bodies of a few tons entering the atmosphere at these high speeds will finally land as slowly as if they had been dropped from a high-flying aircraft. The percentage of mass that a meteoroid loses during passage through the atmosphere depends upon its velocity, the loss increasing rapidly with increasing velocity. Slower meteoroids may lose from one-half to nine-tenths of their masses, while faster ones are almost completely consumed. Hence meteorites picked up on the surface of the earth generally represent the slower-moving bodies.

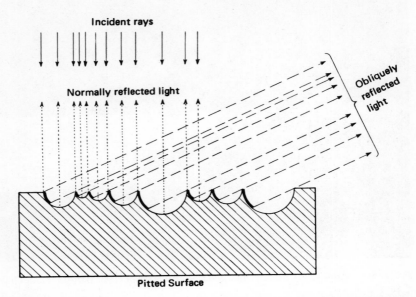

15-2 *Geometry of meteor velocities.* V *is velocity vector of* E *(earth),* M *(meteor in space) and in atmosphere 1 (evening), 2 (midnight) and 3 (morning).*

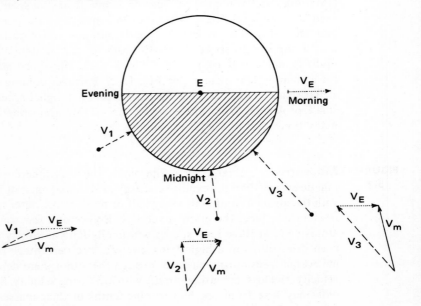

FIGURE
15-3 *Flight markings on a meteorite. Found in Bruno Saskatchewan, 1931, this iron meteorite retains the flow markings of its flight through the atmosphere. (Photograph courtesy of the Smithsonian Institution.)*

When found soon after their fall, all unbroken meteorities show a dark or blackish fused surface crust, which is usually very thin except where the atmospheric heat may have penetrated into cracks. Many meteorites show flow markings on their surfaces, mute witness to the violent heating, melting, boiling, and spraying that have eaten away inches or even yards of the solid material in the few seconds that the meteorite was in transit through the earth's atmosphere (see Fig. 15-3).

Some of the larger meteorites smash their way into the ground to form craters, while very small ones may land so gently that they leave no mark of impact. When analyzed in the laboratory, meteorites differ so widely in their characteristics that an expert can usually match broken and separated pieces from a single fall by their appearance and chemical composition. In spite of the apparent variety in structure and composition, classification is straightforward. Compositions vary from the *irons*, of amazingly pure nickel-iron (over 90 per cent iron), to *stones* of light silicate materials, often of very fragile structure. These silicates are compounded of silicon and oxygen, with some iron, magnesium, aluminum, and calcium. Intermediate between the irons and the stony meteorites are the rarer *pallasites*, which in cross section look like stone raisins set into an iron pudding (see Fig. 15-4).

A pallasite meteorite. This polished and etched cross section of the Brenham, Kansas pallasite shows the crystalline structure of the iron as well as the stony mixture. Natural size. (Courtesy of the Smithsonian Institution.)

In most stony meteorites a small amount of iron occurs as a fine honeycomb, which serves to distinguish cosmic stones from those of terrestrial origin. The polished and etched surface of an iron will usually reveal the *Widmanstätten figures*, large crystalline structures suggestive of modernistic art designs (Fig. 15-5). The crystalline layers may be quite large, often a few inches across. Such structures have been duplicated in the laboratory only in minute proportions. They appear rarely in geological formations, and then on a microscopic scale. The theory of the formation of Widmanstätten figures has become fairly well developed, showing that they were formed by slow cooling under moderate pressures, such as exist within the known asteroids. The irons show every evidence of having been subjected to large stresses under great pressures, as would occur in collisional breakup. Evidence of shearing is frequent in the exceedingly strong and rigid structure (Fig. 15-6).

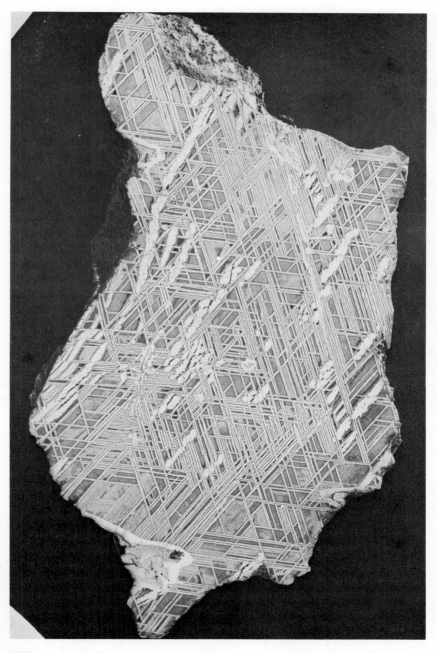

**FIGURE
15-5**   *Widmanstätten figures. After being polished and etched this section of the
Edmonton, Kentucky iron exhibits the low-nickel crystalline structure of kamacite
as white regular and irregular bands. (Courtesy of the Smithsonian Institution.)*

FIGURE
15-6

(*Left*) *A fault plane in a meteorite. This magnified* (50×), *polished, and etched section of the New Baltimore, Pennsylvania iron meteorite shows a fault or slippage plane.* (*Courtesy of the Smithsonian Institution.*)

FIGURE
15-7

(*Below*) *The Ehole Stony Meteorite. A fall in Southwest Africa, July 1961. Note the fusion crust formed by atmospheric heating, partly broken by natural fall and by hammer and chisel. Length 10 cm.* (*Courtesy of the Smithsonian Institution.*)

Of some 1600 known meteorites whose single or multiple pieces have been recognized, recorded, and studied, nearly half are so-called *falls*, in which the brilliant fireball was seen and the pieces recovered (Fig. 15-7). Among the falls, about 5 per cent are irons and 93 per cent stones, while among the *finds*, in which the meteorite was not seen to fall, some 66 per cent are irons and 26 per cent stones. Thus stones are far more prevalent in space, but on the earth's surface we find more of the irons because they are easily recognizable and also because some meteoritic stones deteriorate rapidly and are difficult to distinguish from terrestrial rocks.

TABLE 15-1   *Composition of meteorites and of earth's crust[a]*

Element	Meteorites			Earth's Crust
	Irons	Stones	9/1 ratio	
Oxygen	—	36.3	32.70	49.4
Iron	90.8	24.1	31.85	4.7
Silicon	—	18.0	16.28	25.8
Magnesium	—	13.9	12.88	1.9
Sulfur	0.04	1.8	1.87	0.05
Calcium	—	1.7	1.18	3.4
Aluminum	—	1.5	0.68	7.5
Nickel	8.5	1.5	2.08	0.02
Sodium	—	0.7	0.27	2.6
Chromium	0.01	0.30	0.12	0.03
Manganese	—	0.26	0.16	0.08
Potassium	—	0.18	0.06	2.4
Phosphorus	0.17	0.14	0.10	0.12
Cobalt	0.59	0.14	0.18	—
Carbon	0.03	—	0.15	0.09
Copper	0.02	—	0.01	0.01

[a] Per cent by weight; after F. G. Watson.

We have no clear idea as to which variety, stones or irons, loses more material during its flight through the atmosphere. Hence, we find it difficult to estimate the true average composition of meteoritic material in space. Table 15-1 represents F. W. Watson's estimate, based on the assumption that the ratio of stones to irons is 9 : 1 by mass. H. H. Nininger, the world's greatest collector of meteorites, believes this ratio is even higher. Values for the earth's crust appear in the last column of the table.

We note that oxygen and silicon are surprisingly abundant in both. Probably the relatively high abundance of iron with respect to magnesium and aluminum in the meteorites is real, but we do not know why. No valuable element is abundant in meteorites, although microscopic black diamonds occasionally appear, probably produced by impact shocks.

Most meteoriticists believe the meteorites are the fragments of planetary-type bodies, presumably broken by collisions that probably occurred in the asteroid belt. Marvelous new methods for studying meteorites have been developed since the 1940's. They depend upon mass spectrographs that can measure the isotopic composition of atomic species, upon electron guns that can measure the composition of a microscopic particle, and upon all the other advanced techniques of *radiochemistry*, the chemistry of radioactive atoms. The age of meteorites since they formed is almost the same as that of the earth (see Chapter 10), about 4.6 billion years—much older than earth rocks. Some have been solid and cool enough to hold argon for 4.3 billion years. Thus it appears that the bodies in which they were formed were heated and cooled to 500–600° C in *only* a few hundred million years. Calculations of cooling rates show that the parent bodies could not have been much larger than 200 km (130 miles) in diameter— that is, fair-sized asteroids. As these bodies heated, the nickel-iron collected in pockets or possibly flowed toward the centers to solidify as the iron meteorites. Then collisions broke up some of the asteroids or parts of them to produce the fragments that have fallen on the earth.

Some of the stony meteorites contain a considerable fraction of carbon, and many gases, trapped in the solid but weak structure. These *carbonaceous* meteorites are suspected of having formed in the outer regions of asteroids, where the temperatures and pressures were small, so that they were not chemically changed as much as the more typical irons and stones. Whether or not these carbonaceous meteorites contain rudimentary life-cell forms is a controversial question (see Sec. 18-12).

Since much of the early history of the solar system is written chemically in meteorites, we shall refer to them again in Chapter 17 with regard to the solar system's origin.

15-3   Meteorite Craters.   The largest meteorites so far discovered are irons, the Hoba West of South Africa and the Ahnighito brought back by Peary from Cape York, Greenland, and now in the Hayden Planetarium, New York. The Ahnighito weighs 34 tons and the Hoba West perhaps 60 tons (Fig. 15-8). The largest known stony meteorite, of which a one-ton piece was recovered, fell in Norton County, Nebraska, in 1948.

While these meteorites are the largest single pieces of their kinds found so far, larger irons could probably fall without being broken. Certainly much larger meteorites have entered the atmosphere. During this century, two huge meteorites struck Siberia. The first and larger, the Tunguska fall

FIGURE
15-8

Iron meteorites from the great Gibeon fall of Southwest Africa collected at Windhock. At least 65 masses, totaling over 20 tons, are known to have been found in an area of several hundred square miles.

in 1908, flattened the trees from the point of fall out to a distance of more than 50 km (30 miles). The explosion was "heard" all around the earth by means of delicate microbarographs that measured the pressure changes accompanying the very long soundwaves. Again in 1947, not far from Vladivostok, the Sikhote-Alin meteorite, weighing many tons, scattered iron fragments over a large area and formed more than 100 craters up to 28 m (30 yd) in diameter. Notable is the fact that the more violent Tunguska fall of 1908 left no large craters and no meteorites, except possibly tiny grains of nickel-iron. Fesenkov concludes that it was a comet nucleus, or part of one, and that its fragile structure was dissipated in the atmosphere by the violence of the encounter.

Still larger meteorites have struck the earth. As their dimensions become comparable to the thickness of the atmosphere compressed to their density (32 feet of water or 4 feet of iron), atmospheric resistance becomes less of a hinderance and they strike the earth's surface with nearly their original speed. They then form craters not unlike those of nuclear bombs. The largest certain example in the United States is the Arizona meteor crater (Fig. 15-9), whose meteoric origin was stoutly defended by D. M. Barringer against the opposing geological opinion at the turn of the century. Some 30 tons or more of meteoritic iron have been found in the area around the roughly circular crater, which averages 1200 m (4000 ft) in diameter and 200 m (600 ft) in depth from the upturned rim. J. H. Rinehart, extrapolating

(a)

FIGURE
15-9 *The Barringer meteorite crater. (a) From an altitude of 37,000 ft. Note the bright rim, not dissimilar to many lunar craters. (b) Closer view. (Photograph by the late Moreau Barringer.)*

(b)

from the many shallow borings around the crater, estimates that 13,000 tons of meteoritic dust are distributed in and about the crater. The largest meteorites found weigh about a ton. Some sizable pieces have caused difficulty in deep borings in the crater bowl, but there is yet no evidence for any huge buried mass of nickel-iron. From theoretical calculations, E. J. Öpik concludes that the original mass weighed 2.6 million tons. E. M. Shoemaker, extrapolating from craters formed in nuclear tests, places the mass at 63,000 tons for a velocity of 15 km/sec (10 miles/sec). One million tons of iron would form a cube 31 m (34 yd) on a side, the volume of a medium-sized apartment house.

One of the larger known craters of proved meteoritic origin is the New Quebec crater in Canada, discovered in 1950 by F. W. Chubb and studied by V. D. Meen (Fig. 15-10). This colossal bowl is about 3 km (2 miles) across, in a huge granite sheet, and is partly filled by a lake some 260 m (850 ft) deep. The highest part of the crater wall reaches vertically 410 m (1350 ft) above the bottom of the lake.

More than a dozen large meteor craters, or astroblemes, are now definitely recognized in various parts of the world. Others are being discovered by aerial photography, as field geologists become increasingly aware of the existence of such structures. Even craters that have been badly eroded can now be recognized through new techniques developed by C. S. Beals, in Canada, whereas their true nature escaped attention in the past.

Fossil meteorite craters on the earth, comparable to craters on the moon have only recently been recognized by geologists. The largest, identified by Hall and Molengraaff, is the Vredefort Dome in South Africa, a circular uplift some 120 km (75 miles) across, with a 40-km (25-mile) core of granite. The dome now tilts somewhat and considerable areas have eroded away. Huge volumes of crushed rock are in evidence, but not the slightest trace of lava flows. R. Daly placed the volume of the crushed and shattered materials possibly as high as 800 cubic kilometers (200 cubic miles).

The existence of such fossil meteorite craters provides an answer to the long-standing question concerning the possible meteoritic origin of lunar craters. Opponents of the meteoritic theory were prone to ask, "If such craters have resulted from meteoritic impacts on the moon, why is there no evidence of similar ancient activity on the earth?" There is! R. B. Baldwin counted twelve such fossil structures recognizable from geological descriptions. The list is growing and the true number must be much larger, in view of the difficulty in recognizing old ones that have been eroded, filled or covered with sediments, tilted and folded by earth forces, and then eroded again.

In spite of the violence of meteorite falls on the earth, we need not fear them as personal hazards. L. LaPaz estimates that once in 300 years a meteorite strikes a human being. In recorded history there are only one or two instances of deaths from meteorites.

FIGURE
15-10 *Meteorite crater in New Quebec, Canada. Diameter: nearly 3 miles. (Photograph Royal Canadian Air Force, courtesy the Dominion Observatory.)*

15-4   Observing Meteors.   While untrained casual observers report most of the fireballs, specialists—usually active amateur astronomers, led for many years in the United States by C. P. Olivier—have devoted themselves to observing the fainter, more frequent, meteors, These observers have laid the foundations of modern meteoric astronomy. They have found that most meteors, no matter how bright they may become, first begin to glow at an altitude of around 100 km (60 miles) and penetrate several miles lower before they are consumed. The brighter meteors penetrate deeper into the atmosphere. Meteors that appear somewhat brighter than Venus may persist to altitudes as low as 40 km (25 miles).

If two observers, located at stations several miles apart, track the same meteor against the background of stars, they can then calculate the precise path of the meteor with respect to the earth (Fig. 15-11). Under average observing conditions, a single observer may see 8 to 10 meteors an hour. He sees nearly twice as many per hour during the last six months of the year than during the first six months and in morning hours than in evening hours. The strongest meteor showers visible at night happen to occur mainly in the latter months of the year, while smaller but fast meteors show brighter in the morning (Fig. 15-2). In the morning hours the velocity of the earth's rotation adds on to the orbital velocity.

Photographic and radar techniques for observing meteors have tended to replace the visual observer. Harvard Observatory has carried out a very extensive program of simultaneous photography at two stations located some 30 km (20 miles) apart. Each camera is occulted 20 or 60 times per second by a rotating shutter driven by a synchronous motor. The photographed trail of a meteor is thus chopped into segments whose lengths can be measured to indicate the meteor velocity and how it varies.

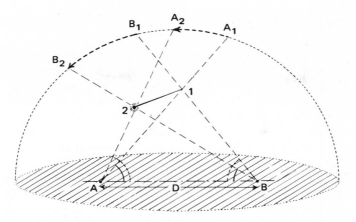

FIGURE 15-11   *Triangulation of meteor flight. Observers (A and B) see the ends of the meteor trail (1 and 2) projected onto the sky at A_1, A_2, B_1, and B_2.*

The Super-Schmidt camera, designed by J. G. Baker, represents the most effective camera yet produced for meteor photography. It has an aperture of 12 inches, a focal length of 8 inches, and a field of some 55°. Whereas the older cameras would record one meteor per 100 hours of exposure time, the Super-Schmidts consistently photograph more than three an hour (see Fig. 15-12). Two-station photography establishes the meteor's path with

FIGURE 15-12   *Super-Schmidt meteor camera. (Courtesy Harvard College Observatory.)*

respect to the earth with high precision, and the rotating shutter measures the velocity to an accuracy of within 1 per cent. These data can then be corrected for the earth's attraction, the motion of the observer caused by the earth's rotation, the resistance by the atmosphere, and the orbital motion of the earth, to provide accurate orbits of the meteors around the sun. Since the resistance of the atmosphere slows down the meteors measurably, the photographs also furnish data about the density in the upper atmosphere.

The Harvard program has included photographic studies of the persistent trains left behind the brighter meteors (Fig. 15-13). Motions of these trains, which usually persist for only a few seconds, furnish accurate measurement of wind velocities in the upper atmosphere.

FIGURE 15-13 (a) *Chopped meteor trail, splitting at end. 20 breaks/sec. (Courtesy, Smithsonian Astrophysical Observatory.)*

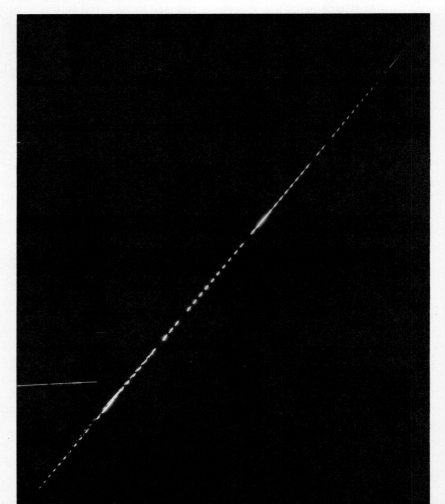

**FIGURE
15-13
(b)**

Meteor train, distorted by wind. Photographs at 2ˢ intervals by Super-Schmidt camera. Bright vertical trail is bright star. (Courtesy Harvard College Observatory.)

Since World War II, new radio techniques have been developed especially for meteor observations. A meteor produces, in its train, a cylindrical path of atmospheric ionization that can appreciably reflect radio waves of wavelengths from about 3 to 100 meters. Both continuous-wave and pulsed radar-type transmitters can be used for this research. Canadian investigators, headed by D. W. R. McKinley and P. M. Millman, the English group under Sir Bernard Lovell, and the Harvard group have made remarkable contributions by radar techniques (Figs. 15-14 and 15-15). Radar measures meteoric velocities and paths with less accuracy than the photographic method does, but it is more sensitive to faint meteors. It can be a thousand times more sensitive than the eye. Moreover, it has the tremendous advantage of detecting faint meteors in broad daylight!

FIGURE 15-14 *Radar meteor antenna. Radiating area: 21 × 61 m. Tall towers are microwave links to receivers at other five stations. (Havana, Illinois. Courtesy Harvard College Observatory.)*

FIGURE 15-15 *Pulse echoes from lengthening ionized meteor trail received at four of six receiving stations as in Fig. 15-14. Frequency: 41 MHz. Pulses: 750 per second. (Courtesy Harvard College Observatory.)*

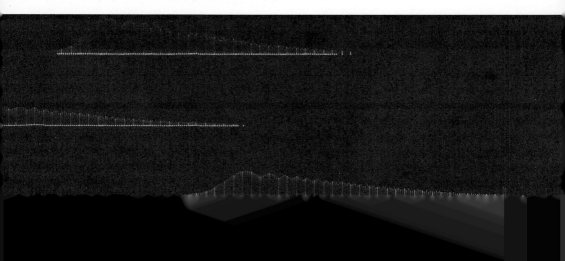

Spectra of very bright meteors can be recorded photographically if one places a prism or a transmission grating over the camera objective. One must also have enough patience to wait for a sufficiently bright meteor to appear in the region of the sky toward which the camera is pointing. Meteor spectra show mostly low-excitation lines, particularly of iron, with the H and K lines of CaII frequently strong. Lines of the atoms NaI, MgI, AlI, SiI, CaI, CrI, MnI, NiI, and SrI have been observed. The presence of Ba and FeO are in doubt, but P. M. Millman has found the lines Hα and Hβ of hydrogen. Thus we see that, apart from hydrogen, meteors show the same spectroscopically observable elements as the meteorites. This correspondence does not necessarily prove, however, that the small bodies producing meteors are identical in nature with the great meteorites. Later evidence will prove they are not.

A. F. Cook and P. M. Millman discovered N_2 molecular bands in a meteor spectrum, the first evidence for the presence of meteoric radiation that is probably of atmospheric origin. Some O lines also appear in the near infrared, while Z. Ceplecha of Czechoslovakia has observed the molecules C_2 and CN.

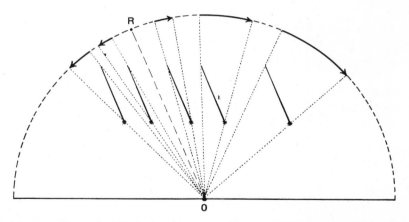

FIGURE   *Radiant of meteor shower. The projections of the parallel meteor paths seen from*
15-16   *0 appear to diverge from R on the celestial sphere.*

15-5   Meteor Showers.   Meteor showers occur when the earth encounters a stream or cloud of small meteoric particles in space. Because all of the meteors move through the atmosphere in parallel paths, the observer sees them apparently stream from a common point on the sky, the *radiant*. This effect of perspective, the apparent convergence of parallel lines, is familiar to us from the appearance of railroad tracks stretching far ahead, or parallel lines of telephone poles. The point of convergence represents the direction from which meteors approach the earth (Fig. 15-16).

In a great meteor shower a spectacular display of meteors occurs. As many as 20 per second stream from the radiant into all parts of the sky. A typical annual shower, however, shows a much weaker activity, with some 10 to 50 meteors per hour visible to a single observer. Meteor showers are named for the constellation containing the radiant. Thus in the strongest annual shower reliably observable, the meteors are called *Perseids*, because they radiate from the constellation Perseus. The shower lasts for nearly two

TABLE 15-2

Meteor showers and radiants

Stream	Date at Maximum	Extreme Limits	Radiant: R.A.	Radiant: Dec.	$V\infty$, km/sec	Hourly Rate[a]	Associated Comet
Quadrantids	Jan 3	Jan 1 ——4	230°	+48°	42.7	30	—
Lyrids	Apr 21	Apr 20 ——23	270	+33	48.4	5	1861 I
Aquarids	May 4	May 2 —— 6	336	0	64	5	Halley
Southern Aquarids	July 30	July 21 Aug 15	339	−17	43.0 ⎫	10	—
Northern Aquarids	July	July 14 Aug 19	339	− 5	42.3 ⎭		—
Perseids	Aug 12	July 29 Aug 17	46	+58	60.4	37	1862 III
Draconids	Oct 10	Oct 10	254	+54	23.1	periodic	Giacobini-Zinner
Orionids	Oct 22	Oct 18 ——25	94	+16	66.5	13	Halley
Southern Taurids	Nov 1	Sept 15 Dec 15	51	+14	30.2	5	Encke
Northern Taurids	Nov 1	Oct 17 Dec 2	52	+21	31.3	5	Encke
Leonids	Nov 17	Nov 14 ——20	152	+22	72.0	6	Temple
Geminids	Dec 14	Dec 7 ——15	113	+32	36.5	55	—
Ursids	Dec 22	Dec 17 ——24	206	+80	35.2	15	Tuttle

[a] Maximum visual rate.

weeks, with the heaviest concentration of meteors near the end, about August 12. In 1866, G. V. Schiaparelli first noticed that the orbit of the faint comet 1862 III nearly touches the earth's orbit in such a way that the calculated radiant coincides with the radiant of the Perseids.

The most famous meteor stream is that of the Leonids, from the constellation of Leo. Spectacular showers occurred on November 11, 1799, and November 12, 1833. Soon afterwards, H. A. Newton traced the Leonid shower back to A.D. 902 at 33-year intervals. According to expectations, an extremely brilliant shower occurred in 1866. Three astronomers simultaneously associated it with a new comet, 1866 I, which has a period of 33.18 years. Later, however, the orbit of the Leonids was perturbed so much by the planets that the earth no longer passed through the rich central part of the swarm. These perturbations caused great disappointment in 1899 and in 1932, when the expected great display failed to reappear, but the return of a great shower in 1966 was a pleasant surprise.

Most of the major meteor showers (see Table 15-2) have now been identified with comets, but a few important showers and hundreds of lesser ones remain unassociated. The Taurid shower, which was identified by F. L. Whipple with comet Encke, has the unusually great duration of six weeks each year (see Fig. 15-17). Calculations of the perturbing effects of Jupiter on the Taurid meteors show that Enche's comet has continued to move in an orbit somewhat similar to its present one for perhaps 5000 years or more.

FIGURE 15-17 *Orbits of Encke's comet and Taurid meteors, projected on plane of ecliptic. (a) Nighttime shower. (b) Daytime shower. (See Fig. 15-18.)*

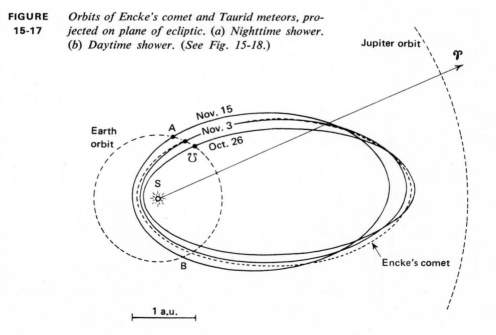

British radio-meteor observers, A. C. B. Lovell and his associates, found a remarkable amount of daylight meteoric activity during the months of May, June, and July (Fig. 15-18). They identified several new streams of short period, and observed the Taurid stream at its other possible intersection with the earth's orbit, marked in Figure 15-18. This latter observation confirmed a prediction from the photographic data that daylight showers should be observable from the Taurid stream at that time.

Meteoric particles can spread completely around a comet's orbit in relatively few revolutions, even without high-speed ejection from the cometary nucleus. Any slight differences between the motions of particles and comet change the period slightly, spreading out particles along the orbit. Planetary perturbations, particularly Jupiter's, affect various particles differently, broadening the stream. This breadth has become enormous for the Taurid meteors and the Perseids.

FIGURE 15-18 *Daytime and nighttime radio-meteor activity, May–July 1950.*

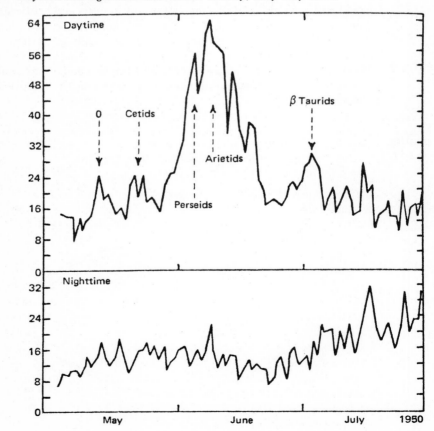

15-6 Sporadic Meteors. Sporadic meteors are those not belonging to any of the known showers. They include, as far as we know, all of the meteorite falls. R. E. McCroskey and G. S. Hawkins find that about half the photographic meteors are sporadic. Most of the radio meteors fainter than the eighth magnitude (on a corresponding visual scale) also are sporadic. Hence, the sporadic type predominates for extremely bright and for very faint meteors, whereas the shower meteors are mostly of intermediate brightness, corresponding roughly to pea-sized masses of less than a gram.

The total number of meteors bright enough to be seen by the naked eye, falling on the entire earth in 24 hours, is 200 million, according to P. M. Millman and M. S. Burland. We could now calculate the total mass of observable meteoritic material falling on the earth per day, if we knew the masses of the meteoric bodies. These are still uncertain. A zero-magnitude meteor moving at 30 km/sec probably weighs a little less than one gram. Adding up the masses from faint to extremely bright fireballs, we estimate that the total mass of meteorite falls on the earth per day lies in the range from 10 to 100 tons. The faint meteors and dust may add another 100 tons or more.

15-7 Meteor Orbits and Origins. One of the most controversial subjects in meteoritics for many years was the question of the existence or nonexistence of hyperbolic meteors from beyond the solar system. The Harvard photographic meteor program, with its very accurate orbits, has failed to demonstrate the existence (in more than 500 orbits) of any meteoric velocities above the hyperbolic limit.

McKinley has determined the velocities of more than 10,000 radio meteors somewhat fainter than the visual limit, without evidence for interstellar meteors. Almond, Davies, B. Lovell, and G. S. Hawkins have measured the velocities of fewer but somewhat fainter radio meteors, also without proof of a hyperbolic fringe. We may conclude safely that more than 99 per cent of meteors to the eighth magnitude belong to the solar system. Perhaps all meteors do.

Among the precise sporadic orbits determined at Harvard about 90 per cent are clearly cometary in nature, with about 20 per cent of long-period and random inclinations. The remainder of low inclination and short period are representative of the periodic comets. Less conclusive arguments place all but 1 per cent in the cometary class. Thus, bright meteors move like comets of both short and long periods, with a few orbits like those of the rare asteroids whose perihelia lie inside the earth's orbit.

The problem of determining the original orbits of meteorite falls is an exceedingly difficult one. Early observers concluded that many of the orbits were hyperbolic, but the careful work by C. C. Wylie on several well-observed falls indicates that their orbits are similar to those of the peculiar asteroids with perihelia within the earth's orbit. The fact that they are not associated with showers augments the evidence for an asteroidal origin, although, as E. Öpik suggests, many of them might be old and inactive comet nuclei.

Only one meteorite fall has been photographed. Recorded in Czechoslovakia by Z. Ceplecha on April 7, 1959, it had an orbit typical of the eight asteroids that cross the earth's orbit: velocity 20.9 km/sec (21.0 miles/sec), inclination $10°.4$, perihelion distance 0.79 A.U., and aphelion distance 4.1 A.U. A large program to photograph bright fireballs and meteorite falls is being conducted by R. E. McCrosky of the Smithsonian Astrophysical Observatory. Sixteen automatic photographic stations patrol more than a million square kilometers (400,000 square miles) in the western prairies of the United States. His preliminary orbital results are in line with expectations. Only fifteen out of the first one hundred orbits have aphelia beyond Jupiter's orbit, and the mean orbital inclination is 12°. He finds, however, a huge preponderance of fragile (cometary?) meteoroids.

Probably, then, about 99 per cent of ordinary meteors are of cometary origin, with a few asteroidal representatives. L. Jacchia and R. E. McCrosky have shown from the Harvard data that most photographic meteors arise from extremely fragile objects, dust balls, which tend to fragment under the relatively small pressures of atmospheric resistance. The evidence indicates further that they are of very low density, possibly one-half the density of water. Such particles are just what one would expect on the basis of the icy model for comets, in which the meteoric fragments would be the porous remnants after the ices had vaporized.

15-8   The Zodiacal Light and Gegenschein.   At the end of evening twilight in spring or just before morning twilight in fall, the faint glow of light near the ecliptic, the zodiacal light, stands out to its best advantage. At these times the ecliptic is nearly perpendicular to the horizon, so that the glow forms a roughly vertical triangle with a broad base at the horizon and a hazy tip of indefinite position at a considerable altitude (see Fig. 15-19). At low latitudes in extremely clear skies the zodiacal light extends as a faint belt completely around the ecliptic.

The spectrum of the zodiacal light is similar to that of the sun with Fraunhofer lines, indicating that its light is reflected or diffracted from small bodies concentrated to the plane of the ecliptic. In 1947 H. van de Hulst of Holland and C. W. Allen of Australia independently suggested an association of the zodiacal light with a part of the sun's corona. W. Grotrian and Y. Öhman have indicated that some of the coronal light is continuous like the sun's but does not show the Fraunhofer lines, while another portion does. This latter part is generally known as the Fraunhofer corona, in distinction to the bright lines of the corona and the continuous spectrum without lines. They further showed that the continuous corona must arise from electron scattering, which spreads out the lines because of the high

FIGURE 15-19 *The zodiacal light from an altitude of 5200 m in Bolivia. D. E. Blackwell and M. F. Ingham photographed sunlight scattered by dust in space. (The vertical line is instrumental.)*

kinetic velocities of the individual scattering electrons. The Fraunhofer corona, according to van de Hulst and Allen, consists of light diffracted and reflected from small particles near the plane of the ecliptic and identical with the particles that produce the zodiacal light.

By relating the diffraction scattering in the Fraunhofer corona with the reflection at large angles from the sun in the zodiacal light, both van de Hulst and Allen conclude that the scattering particles must be quite small, of the order of 0.001 cm (0.0004 in.), with relatively few particles greater than 0.03 cm (0.01 in). They then calculate that the space density of this fine dust must be of the order of some 10^{-22} to 10^{-24} g/cm^3 at the earth's distance from the sun. These deductions lead to the conclusion that the earth should intercept between 10 and 1000 tons of such dust per day. With considerable uncertainty this quantity is comparable to the amount calculated in the previous section for all fireballs, visual meteors, and fainter meteors.

At a point 180° from the sun, experienced observers can detect the so-called *gegenschein*, or counterglow, an extremely faint patch of light some 10° across (Fig. 15-20). The gegenschein has been variously attributed to an accumulation of zodiacal particles held opposite to the earth by perturbations, to an electromagnetic accumulation of gas, and, recently, to an earth tail of gas, blown by corpuscular radiation from the sun. The evidence for the low-density, porous structure of cometary particles brings again into favor an older suggestion by H. von Seeliger to explain the observed reflection by Saturn's rings. He showed how a porous, honeycomb surface of a dense cloud of particles would reflect light much more strongly in the direction exactly opposite to the incoming ray. Figure 15-21 shows the effect in the limiting ideal case. The phenomenon is the same in principle as the "bright shadow" seen on rough ground vegetation from high-flying

FIGURE 15-20 *The gegenschein. (Photograph by R. Roosen at the McDonald Observatory, October 28, 1967.)*

FIGURE
15-21 Reflection from a pitted surface *is strongest in the direction of incident light.*
Note that only a fraction of the incident light is available for scattering at oblique
angles.

aircraft (Fig. 15-21) or "heiligenschein." The latter can best be seen
early on a sunshiny morning when the shadow of one's own head on dewy
grass appears to be surrounded by an aureole of light. The light reflected
directly back along the incoming ray follows the same path to give a
maximum reflection. At other angles the outgoing ray weakens in strength
as it encounters blades of grass or foliage and branches. Observations from
space vehicles should soon establish definitively the origin of the gegenschein.

15-9 Interplanetary Dust. Dust and tiny meteoroids represent real hazards
to space exploration, as they etch or even penetrate space vehicles. Thus
new lines of evidence besides those from zodiacal and coronal data are
being actively pursued.

E. J. Öpik has shown that very small meteorites, or micrometeorites, at low speeds, can radiate away the energy of encounter with the atmosphere fast enough to escape vaporization. Hence zodiacal dust particles should fall into the earth's atmosphere and drift slowly down, eventually to form a layer of meteoritic dust on the surface. Evidence for this dust has been gathered by many observers. Magnetic methods can separate this iron-containing dust from ordinary terrestrial dust in many remote spots such as glaciers, arctic snow, roof drains in the desert areas, sheets of glass with sticky surfaces, cracks in the tops of exposed rocks, and so on. High-flying aircrafts and rockets are now collecting dust from the higher atmosphere less contaminated by terrestrial sources than that collected at the surface.

All these methods suffer from the facts that microscopic particles do not carry obvious labels telling where they originated and that by far the majority of such particles have a terrestrial origin. The subtle methods of radiochemistry, however, are now finding identification tags for these tiny particles—for example, helium atoms of atomic weight three, He^3, that must have come from radioactive tritium, H^3, and other by-products of cosmic-ray and solar-wind encounters with dust in space. Hence we shall soon have good measures on earth of this material from space. Deep-sea sediments can also be measured for their extraterrestrial content as the chemical space tags become easier to read.

Microscopic dust moving about the sun must slowly spiral in towards the sun unless it is more rapidly destroyed by collisions. The so-called Poynting-Robertson effect operates qualitatively as follows. Sunlight striking a particle moving in an orbit must either be scattered or reradiated more or less uniformly in all directions. The scattered or reradiated light carries momentum that the moving particle must provide. Hence the forward motion of the particle is slowly retarded. With loss of momentum in its orbit the particle slowly spirals in towards the sun. H. P. Robertson showed that a spherical rock 1 cm in radius would spiral into the sun from the earth's orbit in about 20 million years. The time required is proportional to the radius of a particle; hence smaller particles would spiral in much faster.

Applying the theory to small dust particles in the zodiacal light, we find that a few tons of dust per second, supplied to the zodiacal cloud, would replace the loss by the Poynting-Robertson effect and by collisions to maintain the zodiacal light and the Fraunhofer corona continuously. The rate of disintegration of comets on the basis of the icy-comet model amounts to hundreds of tons per second of such solid material. After many losses, the required few tons per second may finally reach the inner portions of the solar system to produce the zodiacal light. Hence, comets appear to supply a satisfactory quantity of material for this purpose, although asteroid dust may also contribute.

Since small meteoroids constitute a hazard to space vehicles and astronauts, the U. S. National Aeronautics and Space Adminstration has been urgently seeking information about small particles in space. The first direct method, that of telemetering back the "pings" of small particles striking high-altitude rockets and space vehicles, was first developed by L. Bohn in the late 1940's. This method led to high rates of impact inconsistent with the zodiacal-cloud data and the faint-radio-meteor studies. Very reliable experiments involving the punctures of thin surfaces on space vehicles have been incorporated into the U. S. Explorer and Pegasus satellites (see Fig. 16-4). R. J. Naumann has interpreted these measures into impact rates for particles of mass less than 10^{-5} g, as shown in Fig. 15-22. Other impact rates from radio and photographic meteors, meteorites,

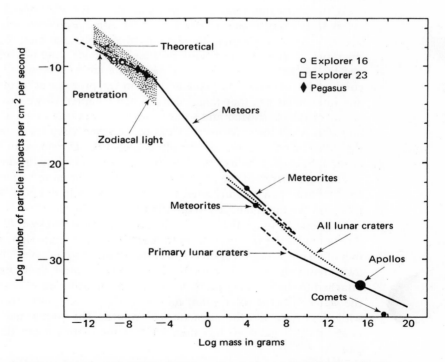

**FIGURE
15-22**   Impact rates of meteoritic material on the earth, *neglecting gravitational attraction. The rates are cumulative, i.e., including old masses greater than the indicated values. Data:* vd Hulst *early Zodiacal Light Calculations;* penetration *from Satellite puncture experiments;* meteors, *both radio and photographic;* GSH, *G. S. Hawkins' estimates;* HB *Harrison Brown's estimates;* lunar craters, *from impact calculations;* Apollos, *from Apollo asteroid calculation;* Comets, *from comet nucleii estimates.*

spacecraft, Apollo, asteroids, and comets are also shown in this diagram, compared with the calculated mass rates required to produce the lunar craters on the maria. All of this interplanetary matter adds up to an influx of about 100 to 1000 tons per day on the entire earth.

15-10   Interplanetary Gas.   Interplanetary gas is so rare that before the space program we had no good measure of its quantity. We knew only that electrons produced some scattering of light near the sun and that solar protons struck the high atmosphere to produce aurorae and magnetic storms, particularly when set off by solar flares and general solar activity (see Chapters 9 and 10). The artificial satellites and space probes show that the quiet sun blows off continuously about a million tons of ionized gas per second at a speed of some 500 km/sec (300 miles/sec) in all directions.

At the earth this solar wind blows at about the same speed as it does closer to the sun, filling nearby space with some 5 protons and electrons per cm^3 (80/$in.^3$). Helium, doubly ionized as alpha particles, is present in the ratio of about one to twenty protons. The solar wind not only ionizes and blows out molecules from the heads of comets; it blows all other gas completely out of the interplanetary system and prevents us from observing any interstellar gas that might drift by. In Chapter 10 we discussed the effects of the solar wind and solar storms on the earth's high atmosphere and Van Allen belts. Measures from rockets and satellites show too that the solar wind excites the far-ultraviolet line Lyman α in hydrogen around the earth, which extends out into a long tail in the anti-sun direction.

The high energy of the gas causes it to knock out more material from a solid surface than it contributes. Hence a comet could not possibly acquire gas from the interplanetary solar wind (see Sec. 14-5).

Magnetic fields in the solar wind form a spiral backwards from the direction of solar rotation. Even though the field changes sign, sometimes in half an hour, the direction is still closely aligned to the spiral, towards or away from the sun. Amazingly, the magnetic spiral turns like a solid attached to the sun in space. It is similar to a rotating water hose in which the water, like the solar wind, moves nearly radially away from the sun while the stream, like the magnetic field, rotates. The average magnetic field is about 10^{-5} gauss or less than 10^{-4} of the earth's magnetic field (0.5 gauss).

CHAPTER *16*

Space
Exploration

16-1   Introduction.   We live in a remarkable age, an era when man is fulfilling an ancient dream of conquering space. Long before the planets, gravity, and space were at all understood, man developed a desire to fly above the earth to the unknown and imagined realms beyond. Our instruments have now roamed vast distances through the solar system and sent back information unattainable from the surface of the earth (an example: Fig. 16-1). Man, himself, has made forays into these hostile regions. His accomplishments in space exploration depend upon a myriad of social, psychological, military, commercial, and political factors lying outside the realm of astronomy and physical science. Setting aside such matters, we shall consider some of the basic problems and scientific advances that accrue from space exploration.

New ventures are frequent and are reported in the public press. Basic principles and first successes, however, remain unchanged and deserve a record here. Successive technological achievements and scientific results remain a continuous source of admiration and astonishment not only to the lay public but to scientists and engineers, even including those who are involved in these exploits.

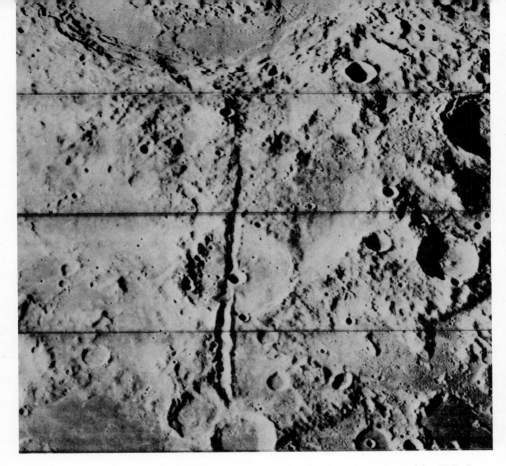

FIGURE 16-1   *A great gash, 240 km long on the far side of the moon. Televised by U.S. Lunar Orbiter IV. Compare Fig. 11-6. (National Aeronautics and Space Administration.)*

16-2   The Rocket Principle.   Above an altitude of only a few miles the earth's atmosphere becomes so thin that it cannot support a winged vehicle. No matter how fast it moves, an airplane cannot fly in a vacuum. To move in a circular orbit about the earth a vehicle must attain a tangential speed of 7.9 km/sec, and, to escape completely from the earth's attraction, a speed of 11.2 km/sec. No rapid type of acceleration, such as by a cannon or catapult at the surface of the earth, can be used to attain these velocities. The heat resulting from friction with the lower atmosphere would melt away the forward surfaces of the projectile while the resistance would demand still higher velocities. The high accelerations would be prohibitive for human occupancy. Only jet propulsion—that is, the rocket principle— can carry man beyond the gravitational barrier of the earth. Newton's third law of motion applies directly to rockets. Matter ejected at high speeds from the rear of the rocket imparts a forward acceleration to the main mass of the vehicle.

To simulate the operation of a rocket, imagine an iceboat on a frictionless, level expanse of ice. We have loaded the iceboat with suitable projectiles, say snowballs, and propose to propel it forward by throwing the projectiles backward with a small catapult. Suppose that the iceboat without "fuel" weighs 100 kg and has an improbable "fuel" load of 900 kg, 1000 kg in all; let the catapult eject balls weighing 1 kg at a speed of 100 km/hr. Release of the first ball will set the vehicle moving forward at a velocity of $\frac{100}{999}$ km/hr, since the total momentum (mass × velocity) must remain constant. The second ball adds a velocity of $\frac{100}{998}$ km/hr, and so on, until the last one adds $\frac{100}{100}$ or 1 km/hr. The final velocity is then

$$V = 100(\tfrac{1}{999} + \tfrac{1}{998} + \tfrac{1}{997} + \cdots + \tfrac{1}{101} + \tfrac{1}{100}) \text{ km/hr.} \qquad (16\text{-}1)$$

The sum of the series in parentheses comes out close to 2.3 ($= \log_e 10$), so that our iceboat attains the remarkable theoretical velocity of 2.3 times the velocity of ejection or 230 km/hr (143 mph). Note that the last ball added nearly ten times as much velocity to the empty iceboat as did the first.

Equation (16-1), governing the rocket principle, can be written more generally. Let M_0 be the initial mass and M the final mass of the rocket. We suppose that the rocket starts from rest. Then, if the velocity of fuel ejection is v and the final velocity of the rocket is V, the rocket equation becomes

$$M = M_0 \times 10^{-V/(2.303v)}. \qquad (16\text{-}2)$$

This equation frequently appears in the inverted form in terms of the *mass ratio*, $R = M_0/M$, so that the final velocity V becomes

$$V = 2.303\, v \log_{10} R. \qquad (16\text{-}3)$$

Applying these equations, we find that by the time the rocket has attained a velocity 2.303 times the ejection velocity of the fuel, only one-tenth of the original mass will remain, confirming our iceboat calculation. The graph in Fig. 16.2 relates M/M_0 and V/v for values relevant to rockets. These results are independent of the nature of the fuel, be it snowballs, gasoline, TNT, or ions. Friction, of course, reduces the efficiency of any actual rocket used in the earth's atmosphere, while gravity reduces the upward thrust proportionately to the mass of the rocket and fuel at each instant. Rocket motors, however, give greater thrust in vacuum. From Fig. 16.2, neglecting these factors, we find that a rocket can reach the ejection velocity of its fuel if the ratio of initial to final weight equals 2.72, twice the ejection velocity if the ratio is 7.39, and three times the ejection velocity if the ratio is 20.1.

To carry a rocket vertically against gravity, the initial impact acceleration must exceed the acceleration of gravity, g—that is, the initial thrust must exceed the weight. Furthermore, air resistance, greatest in the lower atmosphere, cuts down the final velocity. Let us indicate the total effect of

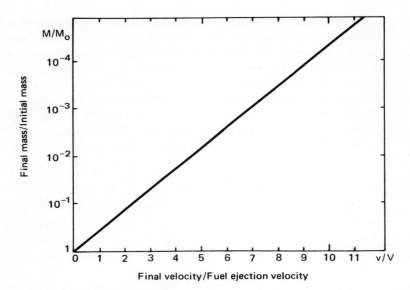

FIGURE 16-2   *The rocket formula. The graph shows the relation between final velocity and mass ratio.* M_0 = *initial total mass;* M = *final mass;* V = *final velocity,* v = *fuel ejection velocity.*

air resistance on velocity as V_R. Then if the total time of vertically powered flight is t, we can represent the final space velocity, V, approximately as

$$V = 2.303v \log_{10} R - gt - V_R. \tag{16-4}$$

In fact, g will decrease with height, but much more slowly than air resistance. Note that once a satellite is in orbit about the earth, neither air resistance nor the g loss of equation (16-4) reduces the efficiency of the rocket. An extremely weak thrust from an ion rocket could spiral the spacecraft into a larger orbit or into deep space. Hence, it is imperative from the practical point of view to place a spacecraft into a low orbit as expeditiously as possible, and then more leisurely propel it into the desired orbit about the earth or into an intermediate orbit about the sun, according to the ultimate objective of the mission.

A practical exhaust velocity for present-day chemical fuels is a little over 2 km/sec. Well-designed rockets usually work quite efficiently. Thus a rocket having a 6.4 ratio of fuel to final weight might be expected to attain a final velocity of nearly 4.0 km/sec, taking into account the inefficiencies indicated in Eq. (16-4). We need, however, about twice this velocity, a minimum of 7.9 km/sec, to attain a satellite orbit about the earth. Figure 16.2 shows that for a final velocity four times the ejection velocity and 100 per cent efficiency, the weight ratio is 54 times, so that the initial rocket must contain more than 98 per cent fuel. This is impractical, but we get around the difficulty by staging two rockets—the first, say with a 6.4 weight ratio, but carrying a much smaller, similar rocket that can add another 4.0 km/sec to the velocity attained by the first. The second rocket, fueled, might weigh a third of the final weight of the first stage. Suppose that 1000 kg of initial weight is 840 kg fuel and 160 kg of load including the second rocket, which might weigh 60 kg. If its weight ratio were also 6.4, the final weight in orbit would be 10 kg, of which, again, one-third could be useful *payload*, 3 kg. Of our total intial weight, including fuel, 0.3 per cent of useful payload goes into orbit.

FIGURE 16-3

The great Saturn V rocket powering the Apollo 4 test mission. Height: 111 m; thrust: 7.5 million lb. (National Aeronautics and Space Administration.)

For higher velocities the number of rocket stages can be increased as necessary. The multiple-stage rocket (Fig. 16-3) possesses the advantage of discarding the heavy motors and empty fuel tanks of the first stages while the velocity is still relatively small.

Wernher von Braun, who was in charge of the first successful U. S. satellite launching, many years earlier planned a three-stage rocket to carry 36 tons of payload into a satellite orbit about the earth at a height of 1730 km above the earth's surface. The velocity required for this effort is 8.3 km/sec. Using a liquid fuel, such as hydrazine and nitric acid, he chose a three-stage system with a total initial mass of 7000 tons for a 36-ton payload, a mass ratio of about 200 times. In practice, his Explorer I, January 31, 1958, required 34 tons to put 14 kg into an earth orbit. Of this amount, only 8.2 kg was really payload, as the fourth-stage casing was left attached. Thus the effective mass ratio was 3780. The Mercury Atlas VIII, October 3, 1962, carrying astronaut Walter M. Schirra for six revolutions about the earth, weighed some 130 tons on the launching pad, while the orbiting load

FIGURE 16-4 *The Pegasus meteoroid detector satellite. In space the detector unit expanded to a 30-m span. (National Aeronautics and Space Administration.)*

was 1.5 tons, a mass ratio of 86. Explorer XVI, December 16, 1962, the four-stage solid-fuel micrometeoroid satellite (Fig. 16-4), required a lift-off weight of 18.3 tons to put 101 kg of payload into an orbit of perigee 755 km and apogee 1172 km, a mass ratio of 165.

With improved fuels such as hydrogen and liquid oxygen this ratio can be greatly reduced—but even so, enormous initial masses are required to place appreciable loads in a satellite orbit. If, however, we make the rocket boosters of the heavy first stages return to earth (or water) safely by wings or parachutes and recover them undamaged, the cost of space experiments can be enormously reduced, since most of the expense is involved in the huge motors and control devices of the first stages. Such vehicles are called *space shuttles*.

If fuel and power were of no consequence in space rocketry, we could return to the earth by reverse-rocket techniques, as must always be done on the airless moon. Our astronauts have used the reverse-rocket principle only enough to bring them within the range of the atmosphere. At an altitude somewhat below 100 km the atmospheric resistance rapidly slows down the vehicle, but heats the forward surface enormously. The vehicle actually becomes a slow meteor. Careful design of the nose cone has overcome the major problems of great heating and ablation (actual loss of material by friction with the atmosphere). However, a stubby-winged, recoverable satellite represents a desirable solution to the problem.

The space shuttle, coupled with the recoverable satellite, then makes *rendezvous* in a *parking orbit* about the earth a practical routine operation. Standard shuttle rockets can carry materials, fuel, supplies, and men into the parking orbit and bring them together. There new vehicles can be assembled, equipped, fueled, and manned for expeditions to the moon and to greater distances in the solar system, or a space laboratory and telescope can be operated. Use of rendezvous in a parking orbit drastically reduces the weight of the initial rocket stages required to send large payloads on distant missions and provides much more flexibility in planning and execution. Many trips of standardized shuttle rockets, made reliable by experience, can replace the gigantic specialized rockets needed for direct trips, and greatly reduce the hazards.

For landings on the moon and planets, parking orbits about these bodies can reduce the complexity, weight, and hazards of direct landings and takeoffs from and to hyperbolic orbits. Until recently we had hoped that the atmosphere of Mars would be dense enough to permit a parachute or ablation soft landing. The highly successful U. S. Mariner 4 space probe that took closeup pictures of Mars in 1965 (see Chapter 13) unfortunately indicated that the surface pressure on Mars is less than 1 per cent of that on earth. Thus a reverse-rocket landing will probably be necessary, as for the moon. This procedure will considerably reduce the payload possible for a soft landing on Mars. Hence experiments are still in progress, to find some type of parachute that may act in the rarefied Martian atmosphere.

16-3 **Exploration in Space.** The planning of rocket flights to the moon, to Mars, and to Venus is both exciting and exacting. Even to calculate the energy or power required for the effective total velocity to make the trip poses a number of detailed questions. What kind of a trip is to be made? Is it to be manned? If so, the accelerations must not exceed 10 g or 10 times the earth's gravity, even for trained astronauts, and must be considerably reduced for less sturdy individuals. Low accelerations cost fuel and money because gravity reduces the efficiency. Will there be rendezvous in a parking orbit about the earth? Will our space vehicle simply pass by the moon or planet, or will it go into a parking orbit? Will a landing be made? If the expedition is unmanned, will we wish to return some equipment to the earth, or will we be satisfied to telemeter information by radio? How much acceleration can our payload withstand upon landing? Must we use reverse rockets, or can we land by glider or parachute?

Then, too, we must consider the length of time we can afford for the trip. For a trip to the moon this interval does not make much difference, but to Mars or Venus we may be able to shorten the trip greatly by the expenditure of more fuel. For an instrumented payload, time is of little importance. For a manned expedition we must also consider the length of the return trip. Finally, we must cope with a host of questions concerning communications, remote control versus man-control within the vehicle, computing for landings and trajectories, problems of determining precise positions and motions in space, power supplies, and other engineering problems too numerous to mention.

Consider the specific example of sending a payload directly to the moon, landing it softly by reverse rockets, returning part of it to the earth's high atmosphere, guiding it into the high atmosphere for a well-controlled reentry using ablation, and, finally, letting the precious remaining payload float gently to the surface of the earth by parachute.

To send a vehicle to the moon requires only about 0.1 km/sec less than the velocity of escape to infinity from the earth, 11.2 km/sec. The saving, however, is not significant, because we must not accelerate a human passenger to more than about 10 g and we must compensate for losses by atmospheric resistance. We require a takeoff velocity capability of approximately 12.5 km/sec instead of only 11.1 km/sec.

The velocity of escape from the moon is only 2.34 km/sec. Even so, we must execute a completely controlled reverse landing, so that the rocket payload will make a low-g landing on the moon. This minimum landing velocity, however, can be attained only if our space ship makes an exceedingly slow approach from the earth to the moon, which requires about 116 hours. If we wish to shorten this time appreciably, we must plan to approach the moon faster than the minimum velocity of escape. About 3.0 km/sec represents the absolute minimum for a practical slow approach, with some allowance for inefficiency in the descent.

Once on the moon, our space explorers will eventually remain for some days, carrying out scientific measurements and general exploration. Hence they can remove much surplus weight from the landing vehicle, allowing a minimum return load. Another 3.0 km/sec will be required to carry the remainder of the equipment from the moon's surface back to the earth's atmosphere. Some fuel for guidance corrections and proper approach to the earth must be left in the return vehicle, which strikes the earth's atmosphere at a speed of somewhat more than 8 km/sec, to be slowed down by air resistance (without too much ablation) to a final parachute landing.

Neglecting the maneuvers near the earth, we have already used a minimum of 18.5 km/sec. Thus, fuel must be carried for an equivalent capsule velocity totaling about 22 km/sec, a practical minimum.

For a mass ratio of 7.4 (see Table 16.1) a minimum of three stages would be required for the round trip if oxygen-hydrogen fuel is used with an exhaust velocity of 3.6 km/sec. The minimum mass at takeoff from the earth would then be $(7.4)^3$ or about 400 times the weight of the final capsule entering the earth's atmosphere. In practice, the actual ratio will be even greater because the expedition would jettison any dead weight, such as fuel tanks, motors, and so on, when they had fulfilled their original purpose. Thus the overall mass ratio might be of the order of 1000:1. For a final minimum capsule weight of 1.5 tons including one human passenger, the inital rocket weight with fuel might be of the order of 2000 tons or more.

TABLE 16-1   *Total velocities for payloads on various missions*

	Velocity	
	Theoretical, km/sec	Realistic, km/sec
A. One-way Journeys		
Orbit around Earth	8	10
Earth to Moon (L)[a]	13.5	16
Earth to Mars (L)	17.3	20
Earth to Venus (L)	22.2	26
B. Round Trips		
Earth-Moon-Earth (NL)[a]	22.4	25
Earth-Moon-Earth (L)	27	32
Earth-Mars-Earth (NL)	23.2	26
Earth-Mars-Earth (L)	34.6	40
Earth-Venus-Earth (NL)	23	26
Earth to orbit about Jupiter	20	23

[a] (L) = landing, (NL) = no landing.

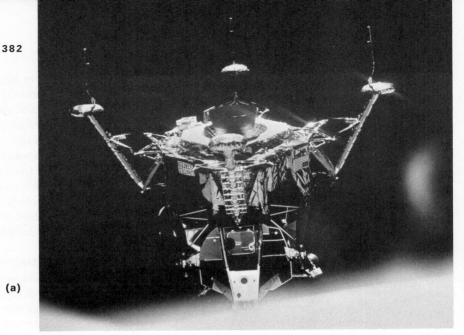

(a)

FIGURE 16-5 (a) (*Above*) *Lunar excursion module of Apollo 11 over moon.* (b) (*Below*) *Edwin E. Aldrin, Jr. of Apollo 11 on moon with lunar excursion module (by Neil A. Armstrong). (National Aeronautics and Space Administration.)*

(b)

The highly successful Apollo program of the U. S. Aeronautics and Space Administration was calculated to send three men to the moon, one remaining in lunar orbit while two descended to the lunar surface for exploration. The return involves a delicate rendezvous while in a lunar orbit. The total spacecraft assembly on the ground is 111 m (363 ft) in height (Fig. 16-3), 10 m (33 ft) diameter at the base, 2700 tons (6×10^6 lb) of fueled weight, and powered by a Saturn V rocket with 3400 tons (7.5×10^6 lb) of thrust. After $2\frac{1}{2}$ minutes the second stage separates from the first and fires $6\frac{1}{2}$ minutes at 450 tons (10^6 lb) thrust. The third stage at 90 tons (2×10^5 lb) thrust attains earth orbit, is checked out, and approaches the moon. The *service module*, weighing 25 tons (55,000 lb) with the *lunar module* weighing 16 tons (35,000 lb) attached, then separates from the third stage and fires into a lunar orbit. Later the lunar excursion module, leaving one astronaut in the service module, carries two astronauts to a soft landing on the lunar surface by reverse rockets (Fig. 16-5) controllable at thrust from 0.5 to 5 tons (1050 to 10,500 lb); note that moon gravity is only one-sixth of the earth's. The lunar module is supported on the moon by four carefully designed legs. Finally the cone-shaped *command module* (Fig. 16-6)

FIGURE 16-6 *Apollo 10 Command module. 100 km above moon. By Thomas P. Stafford and Eugene A. Cernan. (National Aeronautics and Space Administration.)*

weighing 5.4 tons (12,000 lb), separates from the service module and executes reentry maneuvers to carry the three astronauts back to a parachute landing in the Pacific. The entire flight lasts just over eight days.

In the epoch-making Apollo 11 mission of July 1969, the first men to land on the moon, Neil Armstrong and Edwin E. Aldrin, Jr., explored the moon in space suits for more than two hours. After nearly 22 hours on the moon, they left behind the lower part of the lunar excursion module, two scientific experiments, and all equipment not necessary for their return to earth. Their 1.6-ton (3500 lb) thrust rocket carried them again into lunar orbit where they rendezvoused with the orbiting service module carrying Michael Collins. They then abandoned the lunar module and returned to a near-earth orbit.

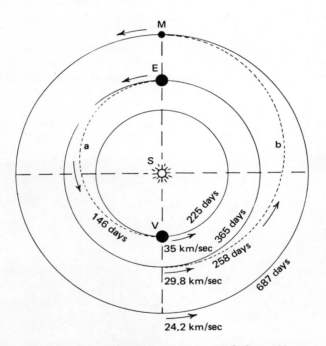

FIGURE 16-7   *Hohmann's ellipses, for minimum energy flights to Venus and Mars.*

An interplanetary flight to Mars or Venus involves much possible variation in flight course. The minimum-energy flight (Hohmann's ellipse, Fig. 16-7) involves an orbit that is tangent to the earth's orbit at perihelion and to Mars' orbit at aphelion. Let us neglect the eccentricities and relative inclinations of the two orbits, adopt mean values of the velocities involved,

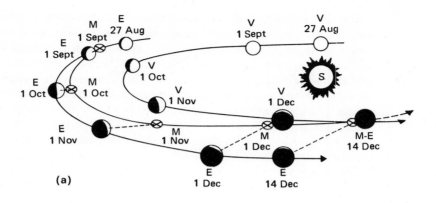

(a)

FIGURE 16-8 *Mariner II. (a) Orbit to Venus. Period of revolution: 348 days. (b) Payload to Venus. Weight: 447 lb. Total lift-off weight: 275,000 lb. (National Aeronautics and Space Administration.)*

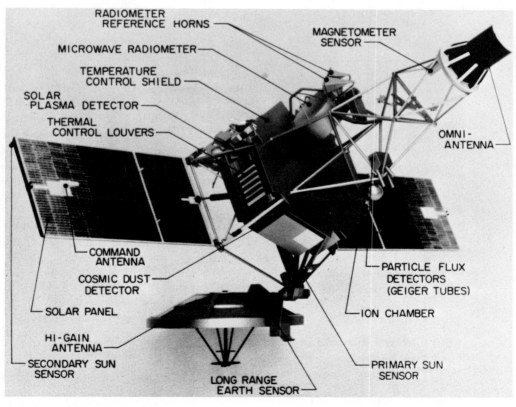

(b)

and calculate for an average circumstance. Our basic celestial mechanics (see Chapter 4) tells us that in such an ellipse the velocity at perihelion would be 32.7 km/sec versus 29.8 km/sec for the earth and at aphelion 21.5 km/sec versus 24.2 km/sec for Mars. Thus the velocity differential at each end amounts to approximately 3 km/sec. From the earth this velocity could be obtained with an initial surface velocity of 11.6 km/sec, only 0.4 km/sec greater than the minimum escape velocity. Remember, in making the direct step, we need consider only the total energy involved. Hence, the end velocity equals the square root of the sum of the squares of the two velocities to be so added. Thus, the velocity required at the earth's surface to send space probes close to planets, not matching their velocities or making reverse-rocket landings, requires very little more energy than the velocity of escape from the earth.

The minimum-energy flight to Mars is rather slow, however, as the orbital semiaxis is 1.25 A.U. and the period 1.41 years. Thus the one-way trip requires half of this, or 258 days. Shortening this travel time would require higher velocities at each end relative to the planets. The corresponding minimum-energy flight to Venus requires 146 days one way, with approximately the same velocity differentials at the beginning and end as for Mars.

The first successful Venus probe, Mariner II, launched by the National Aeronautics and Space Administration on August 27, 1962, made much better than maximum time. It passed some 34,800 km from Venus on December 14, after only 109 days. Figure 16.8 shows the orbital path. Perihelion, at 0.705 A.U., is just within Venus' orbit and aphelion well beyond that of the earth's, at 1.230 A.U. To attain the orbit the Agena B rocket was fired after the Atlas D had placed it in a parking orbit about the earth. A midcourse maneuver was executed 8 days after launching; without it, the probe would have missed Venus by 375,000 km on the dark side. A velocity correction of less than 0.1 per cent, some 30 m/sec, made this enormous improvement in the trajectory. The payload weight was 203 kg and the mass ratio 615.

We note that the minimum velocity at the earth's surface to reach Jupiter is only 14.3 km/sec, but the journey would take 33 months. Table 16-1, based upon calculations by Arthur C. Clarke, indicates the effective minimum total velocity that must be available for a payload to execute the space maneuvers indicated.

16-4 Man's Survival in Space. Solving the problems of man's survival in space requires an enormous effort. We can deal with only a few broad aspects of these problems. Four major subdivisions of these bioastronautical problems deserve attention here.

1. Effects of high acceleration and zero gravity on man and his activities.

2. Problems of man's survival in a sealed container.
3. Effects of the external environment on man and equipment.
   (a) Temperature control.
   (b) High-energy radiation.
   (c) Meteors.
4. Psychological factors.

We can readily study the problems of high acceleration on the surface of the earth by means of centrifugal forces in rotating machines and by other direct means. As noted earlier, trained men, in specially designed suits, can stand gravities up to 10g or more for short intervals of time. They can tolerate much higher accelerations for fractions of a second without serious effects. On the other hand, space travel for the layman will require radically reduced accelerations at correspondingly increased costs in fuel and power supplies.

Proper engineering can solve the physical problem of zero gravity. Space structures require special design for human occupancy. The experience of weightlessness will certainly be exciting for everyone, although many suffer from space-sickness akin to its terrestrial analogue, seasickness. Perhaps drugs, and certainly centrifugal motion of space vehicles to produce an artificial gravity, can remedy this ill. At best, weightlessness is a nuisance requiring special training and equipment for its solution. Exercising devices will be essential, especially in long voyages, to avoid muscular atrophy and to counter other effects of physical inactivity.

The problems of survival in a closed system where no external food, air, supplies, or assistance can be available for long periods of time represent an extension of the problems encountered in submarines. In space these problems are aggravated by minimum-weight requirements that are almost unbelievably rigid, restricting both the volume of space for occupancy and the materials and equipment to be carried. Reuse and purification of air and water are required, as the food and oxygen supplies alone become enormous weight loads for extended flights.

Fascinating progress is being made in the development of closed *ecological* systems, where plants and animals live a mutually balanced existence indefinitely with no material exchange to the outer world. Only an energy source, internal or external, is required for the transformation of waste matter into useful matter. Hydroponic gardens, utilizing the plant's ability to make chlorophyll with sunlight or an energy source, or the development of tasty yeastlike foods from bacteria, may play major roles in man's conquest of space. Such developments, too, may lead to greatly increased efficiency in food production on earth.

The environmental problem of temperature control in space vehicles presents no basic physical difficulties, but will always demand engineering attention to detail. One side of a space vehicle normally is exposed to the sun's heat while the other side radiates to space, practically at a temperature of absolute zero. Surface materials that reflect visual light (where solar radiation is strongest) and radiate well in the infrared (at low temperatures) can combat overheating the sunlit side, while materials that radiate poorly in the infrared can reduce heat losses on the dark side. Proper selection of surface materials, coupled with proper insulation and air circulation, lead to satisfactory heat control.

Three types of damaging high-energy radiations will always constitute a hazard in space travel. The engineer of space vehicles must allow for radiation in the Van Allen belts near the earth, cosmic rays everywhere, and the radiations in occasional solar flares, generally at great expense in terms of vehicular weight or power.

The Van Allen belts, first discovered by space probes near the earth (see Chapters 9 and 10) consist of high-energy protons (and some other charged particles) trapped within the earth's magnetic field. Exposure to these high levels of radiation cannot be tolerated for long by human beings. Fortunately the belts are limited in dimensions and position. They can be avoided if one uses the relatively expensive polar route, traveling along the extensions of the earth's magnetic poles. Alternatively, very rapid transit through the belts can reduce the danger.

Cosmic rays present a hazard in all parts of space. Inadequate shielding may even increase the hazard because of secondary rays produced in the shield. How serious cosmic rays may be to the individual during prolonged exposure can be evaluated only by extended tests in space.

The high-energy particles in short-lived solar flares can be extremely dangerous. Fortunately, a careful and continued study of the sun's surface can give warning of the times when flares may be expected. Protecting devices of thick shields or magnetic fields induced in the space vehicle may be adequate for prolonged flights where the flares cannot be avoided. Probably the fuel and fuel tanks will be used for protection. Experience will probably give us the only clear-cut answer to whether adequate protection can indeed be found in space against the solar flares. Fortunately the solar wind presents no real hazard.

Meteoroids in space cannot be avoided by overt evasive actions, but continued exploration of space will provide increasing knowledge of regions to be avoided to reduce the probability of impact. Large meteoroids that could seriously damage equipment or produce large holes in a space vehicle appear to be so rare that they represent no greater statistical hazard than lightning on the surface of the earth. On the other hand, the small dust particles weighing about a milligram or less will certainly puncture the skins of vehicles from time to time. Figure 15-22 in the preceding chapter gives some indication of this hazard. A *meteor bumper*, consisting of a second thinner skin or skins outside the major skin of the vehicle (see Fig. 16-9), can greatly reduce the number of punctures. Around fuel tanks and in vehicles where air pressure must be maintained or a fluid coolant

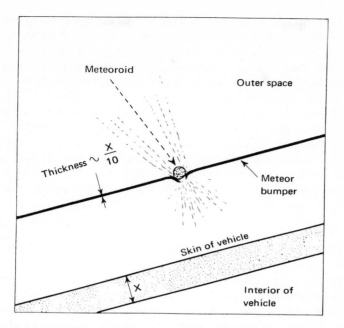

FIGURE 16-9   *Principle of meteor bumper. A high-velocity meteoroid explodes on collision with bumper so that the concentration of momentum and energy is reduced on the skin of the space vehicle.*

circulated, some method of preventing excessive losses from punctures will almost certainly be required for all extended flights into space.

Energy sources for man's survival in space appear to present no more than a normal engineering problem. Batteries are heavy and limited in power, but solar radiation is always available to provide ideally about one horsepower per square meter, at the earth's orbit, if efficiently used. *Solar cells* can transform a fraction of this energy directly into electrical energy for immediate use. Also many types of solar engines have been devised. Concentrated radioactive isotopes provide an additional compact low-weight source of fuel, so that long flights into space need not be vitally curtailed because of energy sources, at least within the inner portions of the solar system. Power in space, however, is always expensive and may represent the actual limiting factor.

To man in space, the psychological problems may loom greatest of all for prolonged flights. Depressing factors of isolation and continuous potential danger will be lessened as technology improves. Nevertheless, it may be many years before deep space flight can be considered routine or becomes generally available to the layman.

No hope is yet in sight for human travel to planets belonging to other stars than our sun. Even though the velocities required, a minimum of only 12.6 km/sec from the earth's orbit or 20.2 km/sec from the earth's surface, are within our reach, the times involved are excessive. The nearest star is 4.3 light years away, involving a round-trip time of 8.6 years if we had the energy and the technology to travel with the velocity of light. At a velocity of 1000 km/sec the round trip would be 300 times as long, or 2000 years. Propulsion systems to obtain such a speed are still far beyond our present-day technology, although not at all inconceivable. Even two-way radio communication is marginal with the nearest stars, and carries the 8.6-year minimum time lapse between message and reply. Thus, we cannot predict when, if ever, we can expect to know much about highly developed cultures (if any exist!) outside our solar system.

16-5 Science from Space. Already our ventures into space have led to surprising new results. No one can set limits on the increased knowledge and understanding that space exploration can bring to us. In Chapter 10 we discussed geophysical results that could not have been obtained without conquering space. The Van Allen belts are physical laboratories measuring interactions of ionized gases from the sun with the earth's magnetic field, while the solar wind now explains the previously mysterious ion tails of comets. The exploration of the ultraviolet and X-radiation of the sun provides clues to intricate physical processes of the solar atmosphere, particulary the interplay of radiation, gas dynamics, and magnetic fields that produces the corona and the spray of high-energy ions at the earth. This prelude to stellar exploration in the ultraviolet with large telescopes in orbit assures us of spectacular progress in understanding stellar processes, origins, and fundamentals of the universe. The direct measures of cosmic rays, gamma rays, and X-rays in space give us further insight into highly energetic processes that occur in a larger section of the universe. New types of X-ray sources have already been found (see Chapter 33). New discoveries will certainly be made.

In the solar system the Mariner series gave us vital close-up data about Mars and Venus. The amplification of planetary research with future probes will answer basic questions about the planets and asteroids, questions both of today's conditions and of the basic origins of our system, eventually answering the vital question of other life in the solar system. Knowledge of the moon is pouring in at an accelerated rate. Already many

FIGURE 16-10 (*Right*) *The U.S. orbiting solar observatory, OSO. The lower section rotates to provide stability so that fine guiding can be obtained in the fixed section.* (*National Aeronautics and Space Administration.*)

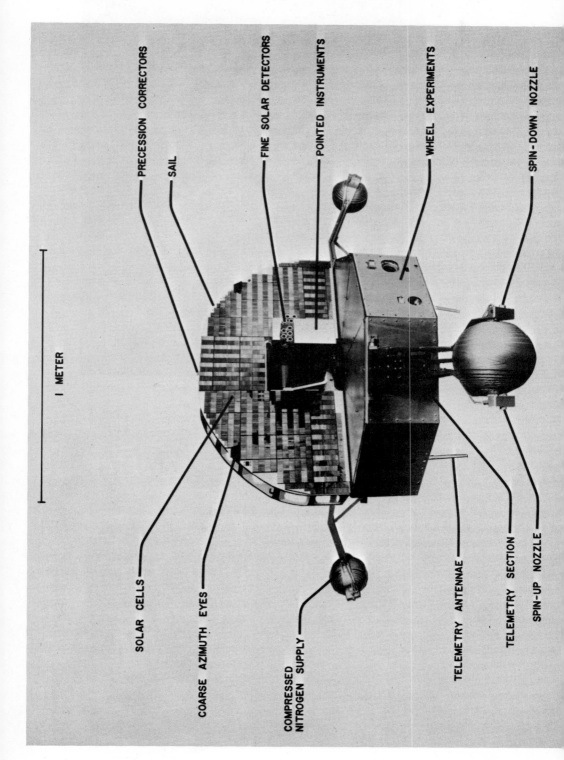

I METER

PRECESSION CORRECTORS

SAIL

FINE SOLAR DETECTORS

POINTED INSTRUMENTS

WHEEL EXPERIMENTS

SPIN-DOWN NOZZLE

SOLAR CELLS

COARSE AZIMUTH EYES

COMPRESSED
NITROGEN SUPPLY

TELEMETRY ANTENNAE

TELEMETRY SECTION

SPIN-UP NOZZLE

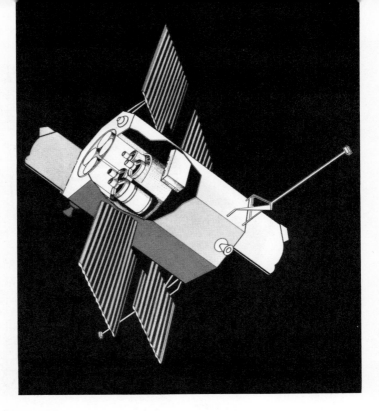

**FIGURE
16-11**

The orbiting astronomical observatory, OAO—
launched December 7, 1968. (a) (Above) Cutaway
diagram showing solar panels and four telescopes
mounted in the 1-m diameter cylinder. Seven other
telescopes point in the opposite direction. (b)
(Right) Details of one telescope and television
camera sensitive only to far ultraviolet radiation.
(National Aeronautics and Space Administration.)

old questions have been answered and more new ones have been raised. The "geology" of the moon promises to be as complex and exciting as that of the earth, and all the scientific tools developed here can eventually be used there.

On the practical side, our meteorological satellites are revolutionizing not only the forecasting of weather, but the basic science of meteorology. Navigational satellites have added a new dimension of speed and precision in the measurement of positions of ships on the earth, while communication satellites make possible all of the radio communications that our increasingly technological and complex civilization requires for its continued growth and development. Already this area of space science is commercially profitable.

Astronomical telescopes in space are not limited by the earth's atmosphere. No longer does the hazy, unsteady, opaque, and illuminated atmosphere prevent us from measuring the radiation from distant objects in space. The satellite observatory (Fig. 16-10) represents a step forward in the progress of astronomy comparable possibly to the invention of the telescope itself. The fundamental radiation from most atoms, and most of the radiation from hotter stars and gas clouds, lie in the far ultraviolet region of the spectrum. The light elements, hydrogen, helium, carbon, nitrogen, and oxygen, constitute more than 99 per cent of the universe and radiate most of their light in the far ultraviolet, forever unobservable from the earth's surface. Orbiting or moon-based telescopes can not only measure these fundamental radiations from space, but they can also observe with greatly increased resolving power by the elimination of atmospheric turbulence and scattering.

Many years ago Henry Norris Russell, then the "dean" of astronomers, said, "When old astronomers die, they should be allowed to go to the moon, because it is the ideal site for an astronomical observatory." His wish is being fulfilled in principle, without requiring a lunar afterlife for the scientist.

The advantages of space are not confined to short-wavelength optics. Our atmosphere is opaque to radio waves longer than about 300 meters; the extreme infrared spectrum to the millimeter radio region is also entirely closed to us at the earth's surface. As commercial radio communication increasingly clutters up the spectrum, radio telescopes in space, and particularly on the far side of the moon, can open up a new regime of information concerning the interplanetary medium, stars, the interstellar medium, galaxies, and the basic structure of the universe. Old theories can be tested while new ones are developed, to make the future of astronomy brighter tomorrow than ever before.

17

Origin and Evolution of the Solar System

17-1 **Statement of the Problem.** We cannot hope to discover the *ultimate* origin of the solar system, if indeed the term "ultimate origin" has any meaning. As a reasonable goal, we should like to trace the development of the solar system back to a time when it was radically different from its present condition. If we go back far enough, we may even find a time when the earth had not yet come into existence. Thus, the question of origin and evolution of the earth and planets is a legitimate one for scientific inquiry.

We assume that the physical laws of nature have remained as they are today. Certain shreds of evidence support this assumption. In mica of ancient rocks, small concentrations of radioactive materials have been breaking up ever since the crystals formed, ejecting alpha particles (helium nuclei) that are stopped in short distances by collision with crystal atoms. These alpha particles have altered the original appearance of the mica so that we see small rings, so-called *pleochroic halos* (Fig. 17-1), which are in effect stains marking the extreme ranges of the alpha particles. The penetrating power of the alpha particles has not changed during the past 3×10^9 years since the rocks were formed.

The spectral lines of atoms in radiation sent out by distant galaxies billions of years ago are like the lines of modern atoms, except for the shift to the red (see Chapter 34), which may result from a velocity of recession.

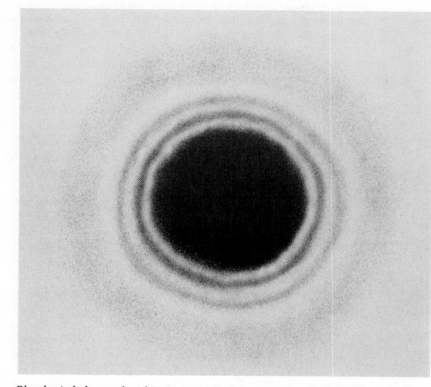

FIGURE
17-1 *Pleochroic halos produced in ancient mica by α-particles of different energies from uranium and other radioactive atoms at center. Maximum diameter ~0.05 mm. (After J. Joly.)*

Also the "constant" of gravitation, certain properties of gases, and the laws governing the release of nuclear energy deep in stellar interiors could scarcely have changed a great deal. Otherwise we should encounter inconsistencies between theory and observation for radiating stellar bodies.

The assumption of unchanging physical laws over some 5×10^9 years may well be wrong. However, no valid basis exists for predicting how or why the laws of nature could change with time.

17-2   Evidence of Order in the Solar System.   Whenever we meet exceptional order, whether in a pack of cards or in the universe, the probability is high that the arrangement did not occur accidentally. Nine planets, thousands of asteroids, and most of the satellites in the solar system revolve about the sun in the same direction in orbits not greatly inclined to one another. That such a configuration could come about by chance is inconceivable. These bodies must either have been originally set in orbits much like the present ones, or else some directive force has systematically produced this order from an originally chaotic arrangement.

We note further that, among the sun and the eight planets whose rotation has been observed, only the axes of rotation of Uranus (and its satellite system) and of Venus are tilted more than ninety degrees with respect to the common plane of motion. We must eliminate Mercury from this discussion, however, since solar attraction has evidently controlled its rate of rotation (see Sec. 12-2). Jupiter is able to capture comets from random orbits and deflect them into orbits of low inclination and direct motion. But this type of perturbation could not change the axes of rotation. Nor could the attraction of Jupiter have brought fully developed planets into their present orbits from orbits initially distributed at random.

Original retrograde motion of a few bodies would substantially increase their chances of collision, so that such objects would undoubtedly disintegrate rapidly. Furthermore, erratic perturbations would seriously alter the paths of any bodies moving in retrograde orbits within the orbit of Uranus. Eventually, the forces would either eliminate such planets or change the sense of their motion.

Thus our thinking requires that the motions of the original planets must preponderantly have been direct, although we can by no means be certain that some planets that originally had retrograde motions may not have been eliminated. The very small inclinations and eccentricities of the planetary orbits constitute a further strong argument that few retrograde planets of considerable size could have existed originally and that their elimination did not greatly disturb the motions of the remaining ones.

Pluto's orbit appears to be basically unstable, since it crosses in projection the orbit of Neptune (see Chapter 12). But E. Öpik has shown that an unusual type of self-maintaining synchronism exists between their periods and perturbations to prevent the occurrence of a collision.

Bode's law represents another evidence of regularity in the solar system. In essence, this law provides a simple device for recalling the distance of the planets. To reconstruct the relation, Table 17-1, proceed as follows: (a) Write down the names of the planets in order of increasing distance from the sun, leaving one space for the asteroids between Mars and Jupiter. (b) Write down a 4 under each. (c) In the second line place a 0 for Mercury, 3 for Venus, 6 for the earth, 12 for Mars, and so on, doubling the number each time. (d) Add the sums in each column and move the decimal point one space to the left. The result represents the distances of the planets from the sun in astronomical units. For comparison the measured distances in A.U. are also given.

The agreement is fair up to and including Uranus. However, Bode's law fails completely for Neptune and Pluto unless we disregard Neptune and let Pluto follow Uranus. Similar relations exist for certain of the satellite systems.

TABLE   *Bode's law and the mean distances of the planets[a]*
17-1

Planet	Bode's Law	Actual Distance
Mercury	$4 + 0 = 0.4$	0.39
Venus	$4 + 3 = 0.7$	0.72
Earth	$4 + 6 = 1.0$	1.00
Mars	$4 + 12 = 1.6$	1.52
Asteroids	$4 + 24 = 2.8$	2.8
Jupiter	$4 + 48 = 5.2$	5.20
Saturn	$4 + 96 = 10.0$	9.54
Uranus	$4 + 192 = 19.6$	19.19
Neptune	—	30.07
Pluto	$4 + 384 = 38.8$	39.46

[a] In astronomical units.

The laws of evolution of planetary orbits do not require that the planets must have developed near their present orbits. Mutual gravitational perturbations or changes in the mass of the sun could have greatly altered their original orbits. However, we still cannot demonstrate that random perturbation of the planets would produce the type of order given by Bode's law. Hence, most astronomers prefer to believe that the planets did develop somewhere near their present orbits, or, if the sun has blown away considerable mass, at about the same relative distances from the sun.

17-3   Disorder in the Solar System.   Some disorder exists in the solar system. Uranus' equator and its satellite system lie in a common plane tilted somewhat more than 90° to the plane of the ecliptic while Venus' slow rotation is distinctly retrograde. All planetary orbits in the system show small eccentricity and low inclination to the mean plane, but Jupiter, Saturn, and Neptune possess retrograde satellites. Also, the "new" comets move in extremely long orbits, oriented almost completely at random. Since comets differ physically in so many ways from the planets, we must suppose that they originated in an entirely different manner. But any acceptable theory of the origin of the solar system must include the comets, because they constitute an extensive cloud of materials controlled by solar attraction.

Both the icy model and Oort's theory of the cometary cloud (Chapter 14) indicate that the comets formed largely at very low temperatures in the deep freeze of space, either at the periphery of the known planetary system or even farther out, unless the inner regions of the solar system were once much colder than at present. If comets formed in the region of the planets, perturbations by the latter may have expanded the cometary cloud to its present distance.

Thus, four general possibilities for the origin of comets are: (a) the comets developed in interstellar space, or at solar distances much greater than Pluto, from frozen interstellar gas and dust of an original solar cloud; (b) comets collected from frozen gases at the periphery of the planetary system essentially at the same time that the planets formed; (c) many comets coalesced in the inner regions of the solar system (Oort), when temperatures were very low, and a small fraction of them survived in very large orbits after they were perturbed by the planets (Öpik); and (d) comets formed more recently than the planets from captured interstellar dust, by a condensing process suggested long ago but recently advocated by R. A. Lyttleton. In this process the sun's passage through a dense cloud of interstellar matter concentrates the matter gravitationally along the line of motion to the rear of the sun. Since the interstellar matter converges toward this line, it loses its motion perpendicular to the line and the material coming close enough to the sun is trapped.

The authors favor hypothesis (b), in which the comets froze out, perhaps beyond Saturn, while the outer planets were being formed. Hypothesis (d) meets with various difficulties and appears not likely to produce icy comets. Processes (a) and (c) appear highly speculative in our present state of knowledge.

17-4   Angular Momentum.   Kepler's law of areas is the two-body statement of a profound dynamical principle, the conservation of angular momentum. Approximately, for many small bodies controlled gravitationally by a much more massive body such as the sun, the law becomes: the sum of all the areal velocities* about the center of gravity of the system cannot be altered by any internal forces within the system. The angular momentum, or total areal velocity, can, however, be exchanged among the various smaller members, with the very real possibility that some of them may leave the system completely, collide and break into small pieces, or collide and coalesce.

This law of the conservation of angular momentum plays a critical role in all discussions of the dynamical evolution of the sun and planets. It is indeed a remarkable fact that the planets carry more than 97 per cent of the angular momentum of the solar system but only 0.14 per cent of the total mass. The sun should be spinning much more rapidly than it is to equalize rotation in the solar system.

As we shall see in discussions to follow, most theories of the origin of the sun or of the planets lead to a rapidly rotating sun. Should such theories be eliminated on the grounds that they contradict the observed fact of a slowly turning sun? Furthermore, we know of *no star* of the same type as

---

* This area is that swept out per unit of time by the radius from object to center as projected on any plane.

the sun that rotates rapidly. The question concerning the significance of conservation of angular momentum deserves more consideration before we attempt any critical discussion of theories of evolution.

We have seen in the earth-moon system how the angular momentum of the earth gradually transfers to the moon, slowly increasing the size of the lunar orbit. This constancy of angular momentum applies only to an isolated system. If a physical process can be proven capable of transferring the angular momentum of the rotating sun to dust, gas, or stars, within or

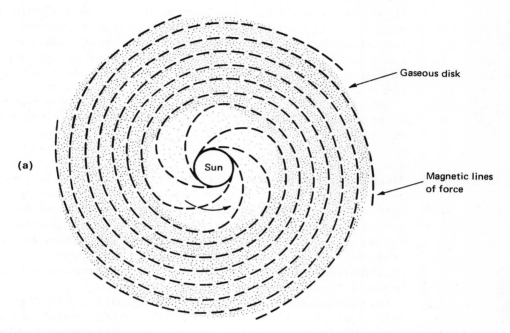

FIGURE
17-2
Magnetic lines of force in solar nebula. (a) Seen in plane of gaseous disk. (b) In meridian cross section of disk. (According to F. Hoyle.)

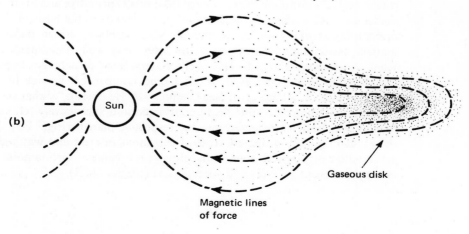

beyond the solar system, then we need not reject as untenable theories of the evolution of the solar system that start with a rapidly rotating sun.

H. Alfvén, D. ter Haar, and F. Hoyle have, indeed, discussed physical processes whereby the sun may lose angular momentum (see Fig. 17-2). They show how very strong magnetic fields on the young and rapidly rotating sun can be wound up in surrounding clouds of ionized gas and transfer the sun's rotation to this gas. As a result the gas spirals away from the sun, and the sun spins more slowly. These theories are difficult and have not been thoroughly criticized by experts in the field, but they profoundly change our attitude toward evolution theories. Most students of solar evolution go so far as to assume that some such process *must* have occurred whether or not these theories are correct in detail, on the grounds that no solar-type star rotates rapidly. Alternative types of magnetic brakes exist, including some that operate at the time matter is ejected from the sun, as in solar flares or in the solar wind. Let us class all these processes as the *magnetic-transfer* process or theory. The nearly rigid rotation of the magnetic-field lines with the sun in the solar wind (Chapter 15) strengthens one's confidence in this process.

17-5   The Nebular Hypothesis.   The more likely or important of the many theories that have been proposed for the origin of the solar system separate conveniently into three broad classes: the *nebular*, the *encounter*, and the *intrinsic* hypotheses. In the first, the planets develop from a large, extended, disklike mass of gas rotating about the sun; in the second, the planets develop as the consequence of the sun's encounter with another star; in the third, the rapidly rotating sun transfers its angular momentum and mass (magnetically) to the gas that finally forms the planets.

The philosopher, I. Kant, on the basis of suggestions by T. Wright, in 1775 made the first serious attempt to visualize the beginnings of our solar system. He pictured the system as starting from an extended volume of gas, which condensed under gravitational and chemical forces to the sun and the planets. Even though the physics of his day was primitive and his theory inadequate, Kant's concept is so broad that it stands as the basis of much present-day thinking on the subject. P. S. de Laplace, the great mathematical astronomer, some forty years later, was well acquainted with Kant's hypothesis and with Herschel's observations of nebulae. Laplace visualized these nebulae as in rotation and comparable in size to our planetary system, a huge error (see Chapter 32). He postulated that such a nebular disk about the sun slowly contracted and successively left rings behind, which then condensed into the respective planets (Fig. 17-3). The theory as originally proposed aroused enormous interest and was widely accepted throughout the nineteenth century; it became a starting point for many subsequent theories and speculations (see Sec. 17-8).

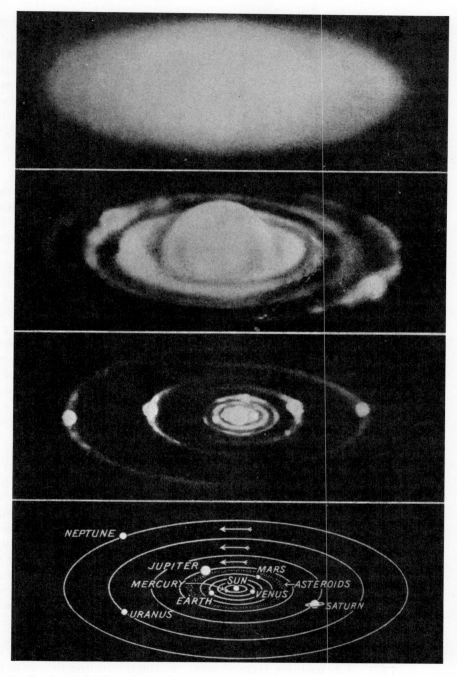

**FIGURE
17-3** *Laplace's nebular hypothesis. (Drawings by Scriven Bolton.)*

In its original form, however, Laplace's theory violates so many basic physical principles that detailed critique of the hypothesis is no longer necessary. He gives, for example, no demonstration that rings will theoretically be left behind by such a rotating condensing nebular disk or that such rings, if left, will condense into planets. Nor does he suggest how the sun could slow its rotation rate. Today we demand much more refined and quantitative study in any physical theory of the origin of the solar system—and, frankly, the physical complications in collapsing interstellar clouds still make the theory too formidable for satisfactory development.

17-6 **Encounter Hypothesis.** The encounter hypotheses derive the matter of the planets as the consequence of a close encounter (or collision) of the sun with another star.

T. C. Chamberlain and F. R. Moulton, early this century, presented the first detailed stellar-encounter theory, in which the planetary material springs from the sun by extremely violent tidal ejection. Their theory has generally been known as the *planetesimal hypothesis*, because they assumed that the gaseous material from the sun cooled quickly into small aggregates called planetesimals, which in turn collected into the planets. Thus the earth formed from cold matter, not hot gases—an important concept. This theory greatly spurred thinking in the field of planetary evolution.

Although the planetesimal and tidal hypotheses were presented as separate units, the first presumably being a possible origin of the planets independent of the method whereby the planetesimals were formed, the two are usually considered together as one overall hypothesis. We distinguish the first as an *accretion type* of hypothesis, applicable in any of the three general types of theories discussed here.

The encounter aspects of the Chamberlain-Moulton theory were completely discredited by: (a) H. N. Russell's proof that the tidal action of a passing star could not produce sufficient angular momentum in the planetary material, and (b) L. Spitzer's demonstration that the hot gaseous material of the sun, when violently removed by a tidal action, would expand so rapidly that it could not condense into either planetesimals or planets. A tidal scoop into the sun would remove gas under high pressure and at a temperature of over a million degrees. The difficulty pointed out by Russell applies to variations of the single-star encounter hypothesis, including a tidal hypothesis of J. Jeans and H. Jeffreys, as well as R. A. Lyttleton's suggestion that the sun was originally a double star and that the intruder encountered this companion.

17-7 **An Intrinsic Hypothesis.** F. Hoyle visualizes the sun as having been formed from the collapse of a large gas cloud. After it attains a diameter comparable to Mercury's present orbit, its rotation is very rapid. Hoyle proposes that in losing its angular momentum by ejection of matter via magnetohydrodynamic processes, the sun also transfers angular momentum to the planetesimals that are condensing and collecting in the nebular ring. Thus they move outward and eventually collect into the planets (see Fig. 17-2).

The process fails because the outward pressure of the coiled magnetic lines of force acts on the gas only. As the gas spirals outward, it moves under a smaller effective gravitational field than the planetesimals. Hence the gas, in nearly circular motion, moves more slowly than the planetesimals. The latter, therefore, meet effectively a resisting medium and must slowly spiral *inward* towards the sun, not *outward* as Hoyle concluded.

17-8   The Cosmochemical Approach.   H. N. Russell of Princeton initiated a valuable, modern method to assist in reconstructing the processes of origin of the solar system and contributed markedly to it during the first half of this century. More recently H. E. Suess, H. S. Brown, and especially H. C. Urey have made profound advances by this approach.

Spectroscopic techniques in the study of stellar atmospheres show that most of the material in the universe is hydrogen (over 60 per cent) and helium (over 30 per cent), while carbon, nitrogen, and oxygen (about 1 per cent together overbalance the remaining heavier elements (about 0.5 per cent). Thus almost all matter separates naturally into three distinct classes, *gaseous*, *icy*, and *earthy*, on the basis of boiling temperatures at fairly low pressures. The *gaseous* class of matter, mostly hydrogen and helium, boils at extremely low temperatures, only a few degrees above absolute zero. Except under extremely high pressures, as in large planets, possibly no solid hydrogen or helium occurs naturally in the universe. At higher but still rather cold temperatures the *icy* class, carbon, nitrogen, and oxygen, and its stable compounds with hydrogen, such as methane (CH_4), ammonia (NH_3), and water (H_2O), can all be frozen. At still higher temperatures, such as room temperature, however, the *icy* matter vaporizes at moderate pressures. Much higher temperatures are required to boil the *earthy* matter such as compounds of silicon with oxygen, the metals, and so on that form the rocks and minerals of the earth.

Thus the terrestrial planets represent a phenomenally high concentration of the rare earthy elements, while, as D. H. Menzel first showed, in the giant planets the proportion of hydrogen increases in the order of increasing mass (see Chapter 13). Table 17-2 indicates this remarkable distribution of these three classes of elements among members of the solar system. The abundances are poorly known except for the sun, earth and meteorites, but the trends shown by the table are significant.

Whereas the sun consists almost entirely of the gaseous elements, hydrogen and helium, only Jupiter and Saturn contain a large percentage. The terrestrial planets and the asteroids (measured by the meteorites) contain

TABLE 17-2 *Relative quantities of various atoms in the Solar System[a]*

Groups of Elements	Light	Medium	Heavy
Elements and atomic weights	Hydrogen (1) Helium (4)	Carbon (12) Nitrogen (14) Oxygen (16)	Magnesium (24) Silicon (28) Iron (56) etc.
Sun	1.0	0.015	0.0025
Terrestrial planets and meteorites	Traces	0.3	0.7
Jupiter	0.9?	0.1?	Trace?
Uranus, Neptune, and comets	Trace?	0.9?	0.1?

[a] By mass.

practically none of the gaseous elements and some oxygen of the icy class, while Uranus, Neptune, and the comets may be made almost entirely of the icy and earthy classes, including some oxygen and hydrogen in compounds.

The present solar heating allows only the earthy class of elements to freeze in the zone of the asteroids and terrestrial planets. The icy class and their stable compounds of water, ammonia, and methane all freeze out under moderate pressures at the distance of Uranus and beyond.

These facts tempt one to postulate (1) that the original gas from which the planets formed was much like the sun in composition, (2) that the temperatures were much as they are today in the regions where the planets formed, and (3) that the bodies of the system, except the sun, Jupiter, and Saturn, froze out of a rotating disk of gas about the sun as planetesimals and later collected into planets, asteroids, comets, and some satellites. This modern version of the planetesimal theory is consistent with most of the evidence—but, of course, it leaves enormous detail and many fundamental questions to the imagination. Presumably, Jupiter and Saturn started to form early and received most of their mass from the original gas before freezing produced the planetesimals in quantity.

Beyond Saturn, comets could well have been the planetesimals, because their mean composition appears similar to that of Uranus and Neptune. We know nothing about the composition of Pluto. If a huge number of comets did freeze out beyond Saturn, we can visualize their fate as follows: (a) A considerable fraction collected to form Uranus and Neptune (and Pluto?). (b) After Uranus and Neptune became massive, they perturbed many of the comets into the inner part of the system, where some were captured by Saturn, the other planets, and the sun. (c) Some disturbed comets decayed as comets do today, by solar heating. (d) Some comets were thrown completely out of the gravitational control of the sun to be lost forever. (e) A few, perhaps 1 per cent, were thrown into the great Oort comet cloud, now occasionally perturbed back in by meandering stars to become modern comets.

If this picture is true, there may still exist a "comet belt" beyond Neptune (Fig. 17.4). In this region the comets were not numerous enough to collect into a planet. Solar heat could not have vaporized them, while planetary perturbations could not have disturbed their orbits. They cannot be observed directly from the earth. The most likely method of proving the existence of such a comet belt, short of direct space exploration, is by observing a gravitational effect on the orbits of the outer planets or on longer-period comets. Thus far the evidence from comet orbits is negative. A possible comet belt to 50 A.U. cannot contain more than one earth mass.

Great progress in understanding the detailed formation of the terrestrial planets is now being made, as a more detailed picture of asteroid formation is pieced together from meteoritic data. Moderately large samples of stony meteorite materials show nearly identical gross chemical composition,

FIGURE 17-4 *Possible comet belt near the plane of the ecliptic and beyond Neptune. (After F. L. Whipple.)*

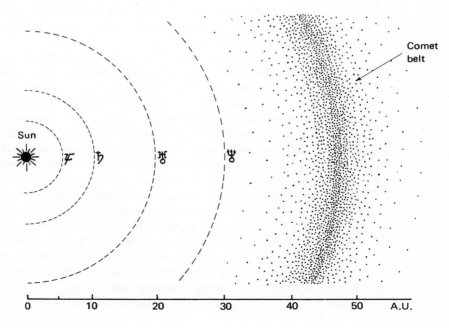

whatever the source. The individuals, however, show a mixture of many diverse minerals, whose character would be altered by melting. Large variations occur in the abundances of some trace elements and particularly, as Urey showed, of more volatile substances.

Micrometallurgy now shows that iron meteorites were not cooled under extremely high pressures. In fact they could have formed in asteroids of diameter only the order of 200 km (125 mi), according to studies by J. I. Goldstein and J. Wood. Furthermore, the fragmentary diamonds found in some iron meteorites were caused not by extremely high static pressures in large bodies such as the moon but by violent shocks such as those produced by collisions of small asteroids in space. Hence there is no need to postulate large moon-sized asteroids as incubators for iron meteorites. Furthermore, the ages since cooling for meteorites indicate that some were heated and cooled within some 0.4×10^9 yr after the earth and meteorites were formed. Thus many investigators now lean towards the concept that the asteroid belt may not have violently altered by collisions since its formation; there probably never were any asteroids much larger than those we see today.

17-9 Problems of the Earth and Moon. Drawing upon the well-known composition of the earth's crust and considerable knowledge about its interior, Urey finds no single temperature at which the earth could have coalesced to produce the chemical compounds in the relative abundances now present in the earth. The evidence leads him to believe that most of the earth's formation occurred at temperatures not much above 1200° C (2220° F), although a long cooler period must have passed before this temperature was reached.

Furthermore, some yet unknown violent process affected the earth before it settled down to about its present temperature. F. Aston, Russell, and Menzel developed an important argument that shows conclusively the secondary origin of the earth's atmosphere. The light noble gas, neon, is practically absent from the earth's atmosphere, although cosmically neon is perhaps more abundant than C, N, and O together. Furthermore, all of the noble gases are much less abundant than one should expect, even the heavier krypton and xenon, although these are better represented than the lighter gases. Since these heavy gases could have escaped from the earth's atmosphere only at temperatures high enough to remove almost all the gas, most of the present atmosphere and oceans must once have been stored within the earth and released later. The rocks of the earth appear to have contained an appreciable percentage of water and gases, which, through volcanic action, were sufficient to make up the present oceans and atmosphere. This evidence points, then, to some process or event that removed the earth's primitive atmosphere.

We are certain that the earth once possessed a very extensive atmosphere, because much of its formation period must have occurred while smaller bodies were being aggregated within Jupiter's orbit. Without a buffering nebular-type atmosphere, collisions among dust particles and planetesimals would be dissipative, preventing the accumulation both of asteroids and of terrestrial planets. Furthermore, the earth could not have escaped the acquisition of a considerable primitive atmosphere, if only from the gases absorbed or combined in the planetesimals. To purge the earth of this atmosphere by its own heat or by solar radiation requires conditions extreme beyond any reasonable physical expectation. Even though Hayashi has shown that new stars of the solar type become very luminous for a short period after they form, direct solar radiation alone appears inadequate to dissipate the earth's primitive atmosphere. However, the evidence is strong that the young T-Tauri stars of solar type (see Chapter 35) quickly blow out 0.1 to 0.4 of their masses in what appears to be a highly intensified solar wind. Perhaps, as A. G. W. Cameron has suggested, the early solar "gale" with its entwined magnetic fields blew away the earth's original atmosphere in the same fashion that the present weak solar "breeze" blows away comet atmospheres to form comet ion tails.

Another explanation, appealing but highly speculative, follows from D. U. Wise's modern version of George Darwin's early theory that the moon evolved by fission of earth. If the earth formed as a *very* rapidly rotating planet near the limit of rotational stability, it may quickly have heated to the melting point by internal radioactivity. The flow of molten iron to form the high-density core would have reduced its moment of inertia with no change in angular momentum. Hence it would have spun so rapidly as to become gravitationally unstable at the equator, throwing off considerable matter to form the moon and losing its atmosphere simultaneously. After the moon coalesced at a safe distance beyond the Roche limit (Sec. 13-7), tidal friction would then have further separated the earth and moon and reduced their rotation rates eventually to the present values.

This fission theory of lunar origin is tempting, because, as Darwin noted, the moon would have been made of the lower-density earth-mantle material rather than of high-density material typical of earth and Venus. The Apollo 11 moon samples, however, practically eliminate the fission theory. On Mare Tranquillitatis the basaltic rocks contain an extraordinarily high fraction of titanium, accompanied by an excess of zirconium and barium as compared with earth rocks. The moon rocks, being distinctly unearthlike in composition, could not have been scooped out of any part of the earth's mantle as we know it without later chemical differentiation. The fission theory for the moon's origin and for the elimination of the earth's primitive atmosphere appears untenable.

A second possibility of lunar origin envisages its accumulation from lighter silicate materials in an early ring about the earth. Suppose that in a cooling solar nebula, metallic compounds condensed out or aggregated before the lower-density silicate compounds. Thus Mercury, near the sun, may have formed primarily of metallic compounds to acquire its high uncompressed density, in excess of 5 g/cm^3. If, then, silicate compounds accumulated at the earth more slowly than metallic compounds, a greater concentration of the lighter silicates might have been collected by a slowly growing ring around the earth than by the earth itself. In such a ring a larger moonlet near the earth would, by tidal friction, work its way slowly outward through the ring, accumulating the entire outer contents of the ring. Such an origin for the moon, however, does not remove the earth's primitive atmosphere unless we postulate rather special circumstances.

A third method of possible lunar origin involves its capture by the earth in an unusual encounter. Besides being highly improbable, if not quite impossible, the suggestion fails to explain where or how the moon accumulated such low-density materials. On the other hand, the unusual composition of lunar rocks may have arisen elsewhere in the solar system, even though no meteorities show such characteristics. Furthermore, the complicated gyrations of moon capture would probably have been adequate to remove the earth's primitive atmosphere.

Cosmochemical arguments play a more and more vital role in the final selection of an acceptable theory of lunar and planetary origin. At the moment, they are not complete enough to define specifically the detailed processes whereby materials collected into the various planets and asteroids.

17-10 Nuclear Physics. Another new tool, nuclear physics, finds striking application in deciphering evolutionary records in the universe. Transmutations of elements into other elements and isotopes occur under bombardment by very high-energy particles such as cosmic rays, under conditions of extremely high temperatures, greater than $10^{6\,\circ}$C, and by natural radioactive decay. The relative abundances of the elements and their isotopes in any sample of the universe are clues to the past history of the matter (see Chapter 35).

To illustrate the application of nuclear physics to evolutionary problems in the solar system, we note that the temperature in the sun only a small distance beneath its surface is sufficient to destroy lithium. In earth rocks and in meteorites, lithium is rare, but relatively at least ten times more abundant than in the sun, whereas in cosmic rays its abundance is greater still.

These facts suggest that the material of the earth and meteorities could not have been taken directly from the sun after it had become a normal star without some subsequent transmutation process. The planets and sun might, however, have developed from the same cosmic mixture of elements. The solar "gale" could also have been a contributor to the formation of these elements.

J. Reynolds has introduced a revolutionary possibility in his studies of rare-isotope chemistry. He finds an excess of the isotope xenon of mass 129 (Xe^{129}) in some meteorities. So far, no one has found a reasonable source for Xe^{129} except as the daughter product of radioactive iodine of mass 129 (I^{129}). Note that I^{129} has a half-life of only 17 million years! Where did it come from? Reynolds suggests that I^{129} was made in a star that exploded less than 100 million years before the meteorites were formed!

17-11 Today's Preference in Theories. Present-day thinking almost invariably points to the formation of the sun and stars by the rapid condensation of large volumes of interstellar gas and dust. Most striking are present advances in infrared detection of dust clouds and cool nebulae in space, particularly in regions where new stars appear to be in the process of incubation. E. E. Becklin and G. Neugebauer at the Mount Wilson and Palomar Observatories observed an infrared star in the Orion Nebula, a dusty gaseous region where new stars are clearly being made. The star has a temperature of only 700° K. Then D. E. Kleinmann and F. J. Low found (at wavelength 22 μ) a 30″-diameter nebula, which appears to have a temperature of only 70° K! At last, perhaps, the Laplacian type of nebula is coming into the range of direct observation (see Fig. 17-5).

C. F. von Weizsacker and D. ter Haar, in their planetary evolution theories, both start with the Laplace nebular disk rotating about a fairly complete sun. They apply turbulence theory to the rotational pattern of the gaseous disk and attempt to derive Bode's law in the formation of the planets, through a hierarchical arrangement of larger and larger eddies around the sun. Satellite development follows in the same fashion as planetary development, each partially formed planet having its Laplace disk in which the satellites grow. The application of turbulence theory, however, remains too difficult for this explanation of the origin of the solar system to be presently acceptable.

G. P. Kuiper's theories follow along similar lines, but condense each planet from a ring so wide that each planetary mass could just hold its material gravitationally against the sun's tidal distortion. As a consequence the original spacing of the planets might be brought into agreement with Bode's law. These original condensations, or *protoplanets*, contained far more mass than the present-day planets, representing a typical cosmic mixture of elements. The earth, with its present high density, subsequently lost some 99.9 per cent of its original mass, the noble gases, and practically all of its H, C, N, and O. Jupiter and Saturn kept larger percentages of their original masses.

The physical processes whereby the protoearth and the terrestrial protoplanets generally lost such huge quantities of hydrogen and helium present overwhelming difficulties for all theories that do not gather the inner planets by accretion of solids. The sun, too, is left rotating very rapidly unless some special mechanism is postulated to transfer the angular momentum away. The solar "gale" of the young sun now appears to solve both problems—both dissipating the solar nebula and retarding the solar rotation.

Thus many pieces of the giant jigsaw-puzzle picture of the evolution of the solar system have been found and some of them seem to fit together. The picture is by no means well outlined yet, but progress is rapid. In brief summary, the authors (speaking only for themselves) tend to favor the following sequence:

The sun and planets formed as one of several or many stars and systems from the condensation of a large (parsec-sized?) interstellar gas and dust cloud. The time was very nearly 4.6×10^9 years ago.

The subcloud collapsed rapidly (1000 years?) after it reached dimensions of several hundred A.U., and residual rotation of the original cloud produced a Laplacian type of diskoid about the rapidly rotating protosun, in total considerably more massive than the present-day sun.

Jupiter and Saturn coalesced quickly (a few hundred years?) to most of their present masses, with mean compositions much like that of the cloud and the sun.

Beyond Saturn the icy and earthy material, on cooling from moderate temperatures ($\sim 1000°$ K or less) induced by the collapse, froze out and aggregated into cometesimals, comets, Uranus, Neptune, and possibly Pluto. Comets, perturbed by the planets, were mostly captured by collisions, thrown to infinity, and destroyed by heat near the sun, except for a very small fraction that were thrown into very large orbits to form the present comet cloud.

Within Jupiter's orbit the temperatures never dropped sufficiently for icy material to freeze, so that the asteroids and terrestrial planets collected from planetesimals of earthy material.

Satellite formation occurred largely in disks of gas and dust about the planets, much as planets collected mostly from small solids about the sun. Planet and satellite formation occurred almost entirely in the gaseous disk before the sun had blown away the gas and drastically reduced its period of rotation by coupling magnetically with the gaseous disk that it eliminated from the system. The entire process of final collapse of the subcloud, planet formation, and ejection of gaseous material and slowing of solar rotation occurred in less than a few tens of millions of years, perhaps much less. The sun became much brighter than today shortly after it formed, a T-Tauri type of star, and the solar wind was many orders of magnitude more intense than today, making possible the rapid elimination of the gaseous cloud.

Let us now consider one last question. "Is our planetary system rare or commonplace in the universe at large?" If the encounter-type theory of solar-system evolution were accepted, then our planets would be almost unique in the universe. Close encounters are extremely rare events even among the 10^{11} stars in our galaxy, because relative distances are so great. On the other hand, a nebular or magnetic-transfer process strongly favors the formation of planets. Some young stars actually show excess infrared radiation, as though surrounded by a Laplace-type nebula. Kuiper regards planet formation as resulting from the failure of a stellar companion to form about a primary star. He estimates, from statistics of the separation and mass ratios in double and multiple stars, that planets might occur as frequently as once in a thousand stars. If we accept his conclusion, a plausible one, then the Galaxy may contain from ten to a hundred million planetary systems, quite a satisfactory number to keep us from feeling unique, lonely, or superior.

CHAPTER 18

Extraterrestrial Life

18-1 **Belief in Life beyond the Earth.** Imagination and science fiction have contributed much to the widespread public expectation that at least some of the planets must be inhabited and that life must exist somewhere in the universe. Many people, aware of man's imperfections, hope that somewhere exists a type of being so enlightened, so intelligent, so much closer to perfection than we that the mere contemplation of this faultless existence reveals a goal for which we should all be striving.

This type of thinking, and the unbridled imagination of science-fiction writers, have contributed in large measure to the widespread belief in a phenomenon called "flying saucers," representing the visits of these super-beings to earth in mysterious saucer-shaped craft. Flying saucers—Unidentified Flying Objects, or UFO's—are, however, often the rags and tags and sometimes, literally, the old paper bags of meteorology.

When a material object is involved, weather balloons, kites, pieces of paper, and so on carried to great heights by violent updrafts of air have caught the eye and confused the populace. Most of the remaining "sightings" are reflection and refraction phenomena—reflections from distant planes, birds, or from fog, mist, or haze, searchlights playing on the clouds

—from spider webs, floating feathers, milkweed seeds, or swarms of insects; refractions by alternating cold and warm layers of air, which produce mirages of distant lights, bright planets, or stars just on the horizon. Radar also experiences mirages, producing false images or "angels." Refractions by crystals of ice in the upper atmosphere give rise to solar halos, "sundogs" or *parhelia*. In many cases, glowing meteors are also responsible. Physiology of the eye also contributes. The afterimage produced by some bright flash can appear real. There are hallucinations and, of course, many hoaxes.

Large numbers of reliable persons have reported seeing something. And the apparent ability of the saucers to move at fantastic speeds and outmaneuver any pilot attempting to cut them off has strengthened the belief that the UFO's are actual machines from outer space. However, many of the phenomena listed above also possess the same ability. A careful study of the best cases in the Air Force files does not provide evidence that beings from outer space are buzzing the earth with saucer-shaped craft.

The philosophical approach to science has been ineffective on the whole, especially when it tries to be independent of experiment. The concepts that the early philosophers proposed concerning the world and its intricate workings were, with rare exceptions, far from correct. The nature of physical laws and atomic structure, for example, emerged mainly from experiment, interpreted by theory. Many people seem to feel intuitively that life processes probably are not uniquely confined to the planet earth. Look at all the planets in our solar system! Note the 100,000,000,000 stars in our Milky Way, many of which may possess planetary systems! Look at all the Milky-Way systems or galaxies scattered through the universe! Why should life not be a universal phenomenon?

In a sense, this argument appears to carry great weight. Its force, however, depends upon whether planets are abundant or rare, a question we cannot positively answer until we know definitely how solar systems develop. But if planets are abundant, we may well expect to find elsewhere the peculiar chemical conditions that produced the first organic molecules and one-celled plants and animals in the warm, primitive seas of the earth.

18-2 Exobiology.   There was a time when the various fields of science were completely separate from one another. Nowadays we encounter many borderline fields, so important that they have received compound names to describe their nature. Thus we have physical chemistry, biophysics and biochemistry, mathematical physics, astrophysics, geophysics, radio astronomy, and so on. A logical extension of this nomenclature suggests that the science dealing with life in the universe should be called exobiology, a designation that has gained popularity in recent years. Whether this topic is important or not depends upon how widely disseminated life is throughout the universe.

It would help us to know how life began on the earth. The primitive atmosphere probably consisted largely of marsh gas (methane) CH_4, ammonia NH_3, nitrogen N_2, and cyanogen CN, in addition to water vapor H_2O. In that stormy atmosphere ultraviolet light and high-energy processes in lightning flashes began to produce various complex chemical substances —the very substances universally recognized as necessary for the existence of life.

Only six elements go to form these substances: hydrogen, oxygen, nitrogen, carbon, phosphorus, and sulphur. And of these atoms, carbon is the most important, even if it is by no means the most abundant element. Carbon possesses the uncanny ability to string atoms together by the hundreds and thousands, to form molecules of the highest complexity. Of all the hundred atoms recognized by the chemist, carbon is the only one that possesses this particular property. The element silicon somewhat resembles carbon chemically. It can form chains of molecules long enough to serve in the manufacture of special kinds of rubber or lubricants. The well-known bouncing putty (Sili-putty or "silly putty," Sec. 10-7) is perhaps the most familiar example. But such compounds are still a far cry from the tremendously complex molecules strung together with the help of carbon.

Over millions of years, lightning flashes aided by ultraviolet light from the sun gradually built up these molecules until the seas, warmed by volcanic heat, came to have the consistency of weak bouillon. This sort of evolution sounds reasonable, because scientists have carried out similar experiments in the laboratory. S. L. Miller and H. C. Urey found that electric sparks, flashed through a mixture of water, methane, ammonia, and hydrogen, could indeed produce amino acids and other complex organic molecules basic for living organisms.

Biochemists have identified twenty-nine such substances. Of these, four bases are especially important: Adenine, Thymine, Guanine, and Cytosine. All four, when combined with a special sugar (pentose) and a phosphate group, are called nucleotides, because they are vital components of the nucleus of a living cell. Hereafter, we shall refer to them by their initial letters. Of these substances, A and G are bigger than T and C.

Twenty of the remaining substances are amino acids, the elementary blocks for building proteins, the essential components of living matter. The other five are fats or sugars, which serve as sources of energy for life processes and as components of the membranes and walls that surround cells.

Time can accomplish much that the scientist cannot do in his laboratory. Nature is a patient experimenter. She is willing to wait millions of years, if necessary, for an experiment to be successful. Some time after a few million years—or perhaps after a few hundred million years (it does not matter)—the warm bouillon, like Goldilocks' porridge, became "just right." These elementary organic molecules began to cling to one another, forming longer and more complex molecules.

18-3   The Origin of Life. The four nucleotides, A, T, G, and C, began
to develop into a very complex substance indeed, now commonly referred
to as deoxyribonucleic acid (DNA). The molecular structure of DNA em-
bodies a ladderlike configuration. It has two uprights, each composed of
long strings of nucleotides, with one rung of the ladder connecting each
pair on opposite sides. The order of the nucleotides along one upright
determines completely the order along the other upright, in the sense that
A always lies opposite to T and G opposite to C. Thus, if the order along
one upright were GTGATCGA, the order along the other would be

FIGURE   *Section of a bacterium, electron micrograph. The nucleus contains visible fibers of*
18-1   *DNA. The surrounding cytoplasm contains the machinery for synthesizing vital*
                components of the cell. A thin but rigid wall encases the cell. (A. Ryter, Pasteur
                Institute, Paris.)

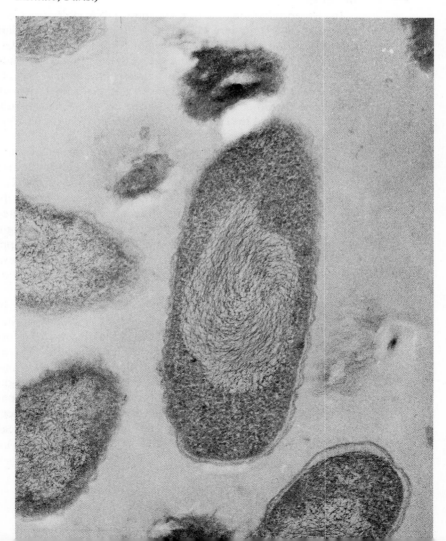

CACTAGCT. Actually, the DNA ladder contains millions of rungs. Furthermore, X-ray crystallography has proved that the ladder is highly twisted to form a sort of spiral staircase, a "double helix."

A complex of molecules of DNA lies within every living cell. The length of the ladder and the ordering of the nucleotides may differ from one animal to another or even from one individual to another. But the significant fact is that in every cell of your body the DNA is precisely the same. Your DNA will differ even from your brother's unless you happen to be an identical twin. Indeed, DNA contains the blueprint that distinguishes man from monkey, or monkey from amoeba. This is why identical twins look alike: they come from the same original egg and thus have identical forms of DNA. It carries the code that determines the pattern of one's development from egg to maturity. Note, however, that the individual nucleotides are the same in all living creatures: virus or bacterium, plant or animal.

In cell reproduction, one of these spiral DNA molecules splits neatly in two, as if each rung were sawed down the middle. The half-ladders mysteriously separate and then each begins to pick up nucleotides, carefully following the plan of matching pairs to reproduce two ladders, each identical with the original. In the simplest unicellular forms the cell then divides, each part with a DNA nucleus, and we have two single-celled organisms instead of one. During the process, the organism builds various other molecules, at the expense of foods derived from the environment, and thus grows.

The splitting process occurs again and again, and we get four cells, then eight, then sixteen, and so on. This may not seem to be a very rapid means of reproduction, especially if we assume that in very primitive organisms splitting may have occurred only rarely, say once a year on the average. However, such is the power of compound interest that in a mere 160 years, an original cell would have multiplied to a mass far exceeding that of the earth, if sufficient food and space were available. The food supply, including the amino acids, would finally limit the production of new members of a population to the rate at which food became available, through conversion of the atmospheric constituents.

DNA contains the blueprint, not only for a specific cell, but for the entire creature. But if DNA is the blueprint, a similar substance called RNA (ribonucleic acid) is the contractor, carrying the information to places in the cell where the proteins are manufactured. There are many kinds of RNA, each with a specialized skill. You might say that each belongs to its own trade. Since this book is not a treatise on biological science, we need not try to visualize the details of cell growth, except to realize that DNA sets the pattern in the form of a code. Somehow or other, the RNA reads this code and builds the cell appropriate for its own location in the body: the eye, the brain, or a muscle.

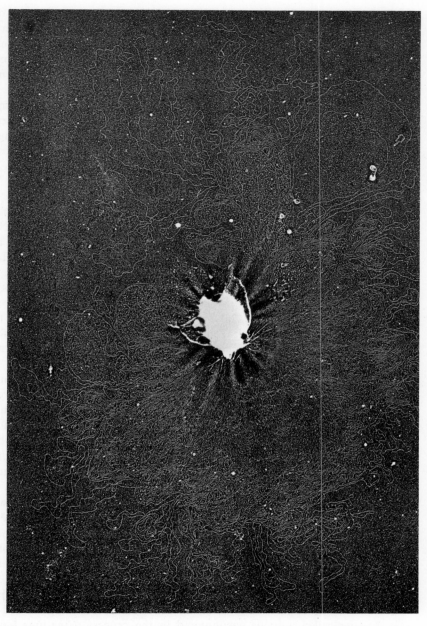

FIGURE   *Electron micrograph of a bacterium, exploded gently so that the entire chromosome*
18-2   *spreads out in a flat layer. Each visible fiber is a single chain of double-stranded*
              DNA some 20 Å in diameter and about 1000 times as long as the bacterium.
              (L. MacHattie, Harvard Medical School.)

We have to account for the basic facts of the reproduction of species. All species breed true. In short, bacteria generate bacteria; spiders produce spiders; and human beings give birth to human beings. Years ago, scientists used to imagine that a spider egg contained a miniature spider and a human egg a miniature human being. During gestation these miniatures were supposed to swell up like balloons.

We now know that DNA contains the key. Some specific arrangement of the four nucleotides along the rungs of the ladder spells out not only the kind of organism but also the physical details of the mature being: color of eyes and hair, shape of nose and ears, and so on.

18-4   Mutations. An amoeba, such as lived in the warm seas of primitive earth, continues to reproduce its kind, duplicating its own DNA, which provides for more amoebas. But nature can occasionally make a mistake. Once in a while the duplication process fails. For example, a T molecule will slip into the DNA where a C molecule should have gone. This accident introduces an error into the blueprint, which causes a change in the product.

Such a change we call a mutation. Most of these mutations tend to be deleterious and the weakened offspring tend to die out. Some of the mutations, however, are beneficial, and the new strain flourishes. In this way complex forms of life gradually evolved from the simpler.

One of the most important mutations, which must have occurred in the primitive seas, changed the instructions that ordered the primitive daughter cell to separate from its parent. When the organism failed to subdivide, a new creature began to develop, consisting first of two cells, then four, then eight, and so on.

By the time this new animal grew to contain 1000 cells or so, it might well have encountered some difficulties. When separate, these cells had derived their nourishment directly through their membranes. But now the organism, if spherical in form, although larger by a thousand times, would have a surface only one hundred times greater. The cells on the inside might die for lack of food, unless the organism could alter its shape, perhaps stretching itself to form a thin pancake instead of the more normal sphere. Or it might take on the shape of a long tube. Thus might have been born an elementary worm.

The primitive worm could further increase its food intake by actively sucking water through its digestive system. Presumably it could assimilate one-celled animals encountered in the process. Creatures having some sensing organs for heat, light, or sound near the mouth, feelers, or claws to aid in the capture of larger prey, and a means of propulsion, would have had a much better chance for survival over those less well equipped.

18-5   Biologic Evolution.   Darwin was the first to describe the principles of biologic evolution, the development of more complex forms of life from the simpler. His doctrine, "survival of the fittest," still is a valid description of the controlling factor in evolution, the role of natural selection.

Change in the primitive organisms was probably slow, in part because of the absence of environmental diversity. But this uniform primeval condition did not continue, for tectonic forces altered the face of the earth. Mountains rose and fell. Continents formed. Some of the warm seas became cold. The earth consisted of dry land, shallow seas, deep seas, and vast oceans. These provided a host of environments to test the stamina of living things and allow the "fittest" to survive, or at least to eliminate the least fit.

In time, creeping and crawling things emerged from the watery slime to try their energies under the more difficult environment of dry land. In further time some grew wings to become progenitors of the modern insects. Other forms came to be fish, reptiles, and birds. And lastly came the mammals. There are now millions of species comprising a great number of genera. The evidence is compelling that all of these sprang from common ancestry in the primitive life forms of a warm ocean. Mutations led to the differentiation of species, and natural selection kept the number under control, weeding out the weaker types.

The mutations themselves, changes in the order of nucleotides in the DNA ladder, arise in a number of ways. Radiation is a primary cause. X rays, cosmic rays, or natural radioactivity, such as potassium 40 in the rocks, can produce such changes. So also can artificial radioactive atoms, particularly those such as strontium 90, a product of atom-bomb explosions. This is why radioactive fallout is dangerous to the human race. Chemicals such as mustard gas, or even temperature extremes, can alter the DNA structure.

Six elements—hydrogen, oxygen, nitrogen, carbon, phosphorus, and sulphur—dominate the life processes. Other elements, such as iron, a component of hemoglobin, and various other inorganic ions, especially potassium and magnesium, are also essential for cell function. They often impart to the molecule entirely different and significant properties. Just as a small addition of chromium can change corundum into ruby or beryl into emerald, these special molecules are important to the development of the more complex forms of life. And the forces that bind the components together are closely related to those that result in crystal formation.

The development of living forms is a sort of cosmic game of cards: the twenty amino acids, the five molecules that represent sources of energy, and the four "royal" cards, already identified as A, T, G, and C. But if our cosmic deck of cards contains fewer distinctive units than does an ordinary bridge deck, the primitive oceans of the earth and other planets contained duplicates of these basic twenty-nine, billions and billions of times over.

The card game more nearly resembles Rummy or Old Maid than it does Bridge. The element of matching is there, a rapid and almost automatic shuffling that finds duplicates or, rather, affinities. There are two stages. The first results in the building of DNA from the four royal cards. The second stage begins as DNA controls the addition of other cards to build proteins and other components of the living cells.

We do not imply that the twenty-nine basic cards in the cosmic deck are themselves living forms. Nature itself shuffles these decks through energy acquired from the sun. The cards come in contact with one another perhaps as do autumn leaves swirling in a current of warm air. At what stage, then, does the complex supermolecule come alive? Mere growth and mere change are not of themselves sufficient indicators. Crystals can grow and their form may change, but crystals are not alive. The answer has already been indicated. Life begins when the game starts to play itself—when a given highly complex pattern of cards splits in two and then starts to reproduce by rebuilding the lost halves in terms of cards drawn from the deck.

The line of descent from the first living form to man himself is incredibly long, covering several thousand million years. During this span the earth itself has undergone profound changes to which life has had to adapt.

18-6 Life and the Earth's Atmosphere. Living requires energy. Merely having energy stored in a molecule is not enough in itself, however. The organism must, by some process or other, be able to get this energy out and use it. In other words, the energy-containing molecules must enter into a chemical reaction that releases energy. Two basic processes exist: *aerobic* and *anaerobic*.

The aerobic process is the more efficient. It is a type of combustion, wherein oxygen of the air combines with organic matter to form carbon dioxide (CO_2), water (H_2O), and other waste products. Thus glucose, a form of sugar, $C_6H_{12}O_6$, combines with molecular oxygen (O_2) to give

$$C_6H_{12}O_6 + 6\ O_2 = 6\ CO_2 + 6\ H_2O + \text{heat}.$$

Note that the number of carbon, oxygen, and hydrogen atoms—respectively 6, 12, and 18—is the same on both sides of the equation expressing this reaction. The substances on the right, however, are more stable than those on the left. The difference in energy becomes available as heat, which amounts to 1800 calories per gram of the reacting substances, glucose plus oxygen. Ordinary respiration is a type of aerobic process.

The anaerobic process is one that occurs in the absence of oxygen. Fermentation is the most familiar example. Various anaerobic organisms, such as yeast cells, derive energy from reactions such as the following, in which glucose breaks down into two molecules each of ordinary ethyl alcohol (C_2H_5OH) and carbon dioxide:

$$C_6H_{12}O_6 = 2\,C_2H_5OH + 2\,CO_2 + \text{heat.}$$

Cellular enzymes of the yeast or other organism mediate the reaction. The inefficiency of this reaction appears when we note that the evolved heat is only 100 calories per gram. We could, however, recover this energy and use it.

Most scientists believe that carbon dioxide was not an original, abundant constituent of the earth's atmosphere. For a long period of the earth's existence, covering many millions of years, primitive life forms continued fermentation, which introduced a lot of carbon dioxide into the atmosphere. Ultimately, another type of chemical reaction came into play, which enabled a growing plant to extract the carbon dioxide from the atmosphere, at the same time capturing some energy from the sunlight. In some circumstances energy from light causes carbon dioxide and water to combine into molecules of glucose, which may in turn serve as a building material for the more complex molecules required to form a living cell. The active agent in capturing solar energy in this process of photosynthesis is chlorophyll. The significance of the reaction lies in part in the fact it releases free oxygen into the earth's atmosphere, which had no such free oxygen originally. (Most of the original hydrogen would have escaped into space.)

Thus began the third stage in the evolution of the earth's atmosphere, which now contains about 77 per cent nitrogen, 21 per cent oxygen, and 2 per cent of everything else. It was this third stage that made possible most of the organisms we now know. Fish, reptiles, birds, and mammals all depend upon oxygen.

Oxygen is an extremely active substance; it is continually combining with other materials. It would disappear completely in a few thousand years, if the earth's vegetation did not continue to replenish the supply through photosynthesis. Similarly, carbon dioxide would vanish in only 20 years if decomposition of organic matter by microbes were to cease.

18-7   Homo sapiens.   We have only to look to the earth to see how varied creatures from single-celled animals to man have existed over time. Fossils of creatures no longer living, such as dinosaurs, giant butterflies, and many varieties of marine forms, indicate the blind alleys often followed by natural selection. We find fossil remains of many organisms not closely related to those now living.

One of the surprising features of the evolutionary chain, in view of the slowness of the development of species, was the rapidity with which primitive man became *Homo sapiens.* All this happened within less than 300,000 years, a twinkling of an eye in the cosmic time scheme. One almost has to conclude that man himself, perhaps unconsciously, greatly accelerated the natural processes of his own evolution.

We have to postulate a series of genetic changes that led to the highly organized central nervous system of man. His enhanced intelligence must have given him an enormous advantage over his fellows. Increased selection of females, for example, probably favored the production of intelligent offspring. In a relatively short time, breeding within the newly arisen intelligent type produced a race very different from its primitive ancestors.

But man is probably not the acme of intelligent life. Indeed, man's ignorance is a positive deterrent to his further advance. There may well exist, somewhere in the universe, other beings more intelligent—perhaps far more intelligent—than we. As for the possible structure of such beings, we speculate that they will probably resemble some combination of terrestrial forms, which indeed provides an enormous range of possibilities. The creatures may live on land, or perhaps they will be amphibious. They will have limbs. They will have a mouth and digestive tracts resembling those of some terrestrial creature. Their eyes and ears will occur in pairs near the mouth, where they served primitive ancestors in the detection and capture of food. They probably will stand upright, though they may not closely resemble terrestrial primates.

It seems unlikely that on planets other than our earth, whether in our solar system or revolving around some distant star, the individual playing cards in a cosmic deck of life will differ much from those twenty-nine briefly discussed above. Trace elements may vary from one place to another, for the abundance or accessibility of heavier elements may depend markedly on local conditions. However, it is difficult indeed to picture a life that would begin except in extensive warm seas.

There is one point we must particularly remember. In this primitive universe, some form of energy is necessary. On earth the primary source of this energy is clearly the sun itself. The elementary molecule of glucose, consisting of six atoms each of carbon and oxygen and twelve of hydrogen, has stored up the energy from sunlight. But we would certainly expect comparable energy storage on Mars or any other planet that derived its energy from its central star.

18-8 **Organic Chemistry.** Those persons whose philosophical outlook leads them to reject or question the spontaneous development of life wherever the proper conditions exist, should consider a simple fact of history. There was a time—more than a century ago—when the chemist despaired of ever producing artificially those substances that plants or animals made so efficiently. The belief was widespread that such substances, the organic molecules, possessed a "living spirit," or "vital force," which was impossible for man (or nature) to produce without the aid of a living organism. And then, one day in 1828, an enterprising chemist, F. Woehler of Germany, produced urea $CO(NH_2)_2$, an organic compound, by heating ammonium cyanate NH_4OCN, an inorganic compound. Let us not make the similar mistake of believing that life forms derive only from living progenitors. The elementary forces appear to be the same for molecules in the living creature as in nonliving matter. These are the "vital forces" necessary to the development of life. Recent experiments indicate that certain "viruses," artificially constructed in the laboratory, appear to be "living."

18-9   The Role of Water. We reasonably conclude, therefore, that life may come into existence on any planet having a supply of water on its surface. The temperature must not be too high or too low. The water must be liquid, not steam or ice. The atmosphere must have a suitable chemical composition.

Water has one very remarkable and almost unique property, among all of the substances necessary for the existence of life. When water freezes, the molecules push slightly against one another, so that the solid tends to expand. As a result, ice is less dense than water; this is why ice floats. Almost all other substances contract when they turn from the liquid to the solid state. If ice did the same, it would sink in the winter to the bottom of the ocean, to depths that sunlight could not reach during the summer. As a result, the oceans would be frozen to the bottom. Summer warmth would melt small lakes on the surface of a vast, icy glacier. The earth would be cold and inhospitable to the maintenance of life, let alone its development.

18-10   Life In Our Solar System. In our solar system, the greatest planet, Jupiter, possesses an extensive atmosphere containing large quantities of methane and ammonia. The surface temperature of the outer layers is extremely low, although hotter regions may exist beneath. Some scientists have speculated that ammonia, which remains liquid at temperatures very much lower than does water, might actually replace water as a medium for culture and growth of life forms. But, as G. Wald emphasized, solid ammonia is denser than the liquid substance. It would tend to sink to the bottom of planetary oceans. There it would form a layer of solid ammonia that would be difficult to melt, except by the weak internal heating of the planet resulting from its own radioactivity. The low temperatures of the planet would further have slowed down the development of life processes. It is questionable that Jupiter, Saturn, or the still colder planets, Uranus and Neptune, all of which possess similar atmospheres, could have developed forms of life. However, a few astrobiologists disagree, pointing out that the unobserved lower levels of Jupiter's atmosphere may be warmer than we think and might simulate the primitive conditions of earth, when life came into being.

The planet Venus bears a superficial resemblance to our earth. The two planets are about the same size, and each has an extensive atmosphere and about the same force of gravity. But there the similarity ends. The atmosphere of Venus consists mainly of carbon dioxide. Dense clouds cover the planet, and the temperature at the top of the cloud layer is about $-40°$ C. The surface seems to be broiling hot, far too hot for life to exist. The atmosphere seems to act like the glass of a greenhouse, allowing the sunlight to enter but resisting the escape of the planet's infrared radiation.

There is one other problem. Remember that the earth's primitive atmosphere probably consisted mainly of water vapor, ammonia, nitrogen, and methane. Fermentation, the result of life processes, eventually introduced quantities of carbon dioxide. Finally, vegetation used this substance, replacing it mainly by oxygen. The presence of carbon dioxide suggests that Venus must once have possessed living forms, because it reached at least the second stage. However, this gas tended to entrap the sun's heat, boiling away the oceans and leaving a hot, dry surface. Whether some form of life still exists in the temperate levels of the Venusian atmosphere remains to be seen. However, one would scarcely expect such life to be "intelligent."

18-11   Mars. Most speculations about extraterrestrial life have dealt with Mars. The red color of Mars we ascribe to rocks or deserts, somewhat resembling those of the southwestern United States. The dark markings, gray with perhaps a tinge of green, are permanent features of the Martian scene, although they do change somewhat with the seasons. Shading into the dark areas and apparently related to them is a finer, evasive network, the famous and controversial "canals." White areas, probably some form of ice, mostly carbon dioxide, mark the polar areas. The polar caps recede almost to the vanishing point during the Martian summer and advance during the winter (see Chapter 12).

The astronomer Percival Lowell, writing during the early years of this century, argued that the "canals" represented artificial waterways, constructed by an advanced civilization to carry water from the polar caps to the arid equatorial regions of the planet. In this view he stood essentially alone among experts. The canals are not as straight, narrow, or regular as he drew them. The Mariner photographs disposed of any remaining doubts.

Mars certainly possesses some atmosphere, in which storms occasionally occur. The winds whip up great clouds of reddish desert dust, whose progress we can follow from day to day. Sometimes the winds die out and the dust settles to form a layer that obscures the underlying gray areas. But within a few days or weeks the dark regions gradually revert to their original color and shape. The gray areas clearly have the power of regeneration. One proposed explanation is that dust, deposited on plants, would gradually shake off in the winds, allowing the original color to show through. Another, more prosiac view is that dust deposited in exposed rock faces is blown away into cracks or down the slopes.

The pictures of Mars recently transmitted to earth from the Mariner series revealed a surface strongly resembling that of the moon, heavily pockmarked with craters, large and small. The craters are by no means as rugged or as deep as those on the moon. Perhaps they have been partially filled in and obscured by windblown dust.

Let us try to visualize what scene would greet the first astronaut to land on Mars. He would have to wear his space helmet, because the Martian atmosphere is very thin, consisting mainly of carbon dioxide with only traces of other gases. The pressure at the surface would be less than that in the earth's stratosphere. The scene would be colorful indeed. Before him would stretch a desert of fine, red dust, smooth as wind-driven snow except for occasional dunes and crater rims. There would be few high mountains or elevations. The sky above would be dark blue, almost blue-black. Here and there a wispy semitransparent cloud would appear, although through yellow or red goggles it would be invisible. When a windstorm occurred, extensive clouds of fine red powder would rise aloft on the violent breezes; visibility might be limited to less than a mile. The daytime temperature might rise above freezing in early afternoon, but after sunset the thermometer would drop rapidly to as low as $-100°$ F before dawn.

The stars shine brightly in that clear, nighttime sky. The astronaut sees the same constellations familiar to earthlings: The Great Bear, Cassiopeia, Orion. But what is that bright blue star and its fainter yellow companion? With a touch of nostalgia, the astronaut suddenly realizes that the bright star is planet earth and its companion is the moon.

Will he find any form of life? The probabilities are excellent that some forms of life may exist, in spite of the adverse conditions, even if the gray-green areas prove to be geologic formations rather than vegetation. An unmanned space probe will first send us the answer.

It seems not unlikely that Martian "bugs" will resemble terrestrial bugs. At any rate, terrestrial "bugs" set in an artificial Martian-type environment—the so-called "Mars Jars"—not only survive but manage to propagate. But it is doubtful that any form of highly intelligent life exists on Mars. The profound silence that greets our ears when we point our giant radio receivers toward the planet indicates that no one is broadcasting to us from there.

18-12   The Moon and Meteors.   The moon and meteorites remain to be considered as possible abodes of life in our solar system. Meterorites—chunks of rock and metal orbiting interplanetary space—occasionally fall to ground after swift fiery passage through the earth's atmosphere. Some of these contain hollows filled with black carbon compounds. Study of two such specimens has revealed the presence there of long molecular chains, similar to those associated with living organisms. From such observations some scientists have concluded that perhaps meteorites are the fragments of an earthlike planet, on which life developed. Evidently some terrible catastrophe befell this world, causing it to explode. Some of the surviving micro-organisms went into orbit inside the various pieces of the planet. Or perhaps life formed spontaneously near the surfaces of small asteroids.

The Swedish chemist, S. Arrhenius, suggested, about the turn of the century, that life propagated from one spot to another in the universe in the form of tiny spores, perhaps adhering to the surfaces of meteors. This theory, sometimes called the *pan-sperm hypothesis*, raises a question concerning the surface of the moon, into which meteors have crashed over the ages. Is there indeed a layer of fossil or dormant microbes on the moon, arising either from the meteorites or from the moon itself? The possibility is interesting and illustrates one reason why man should carefully avoid biological contamination of (or by) the moon and planets. The successful Apollo missions placed the astronauts in quarantine, to ensure that they would not accidentally contract some form of lunar disease. Thus far, the lunar surface appears to be completely antiseptic.

18-13 **Life in the Universe.** If the chances of finding intelligent life within our own solar system appear to be vanishingly small, what are the chances elsewhere? In the entire universe some 10,000,000,000,000,000,000,000 stars exist, give or take a couple of zeros. This is indeed a tremendous number. If these stars, each a sun in its own right, were reduced to the size of a basketball, the resulting pile of balls would fill a volume the size of the earth. If, on the average, one star in a thousand has planets (Sec. 17-8), the number of planets in the universe must be enormous.

For a planet to develop life, many special conditions must exist. The planet must be not too far away or too near its primary source of energy. It must be massive enough to retain an atmosphere. The planet must have oceans. Finally, if advanced life is to develop, continents or dry-land areas must arise above the level of the seas.

Even if only one star in a million has planets to meet these conditions, there are still enormous numbers of places in the universe favorable to the development of life. Our own Milky Way might contain up to 10^5 such planets, some possibly inhabited by intelligent life at the present time ($\pm 10^5$ years) but the chances are that even the *nearest* such inhabited planet is so far away that it takes light about 1000 years to reach us. In other words, if we sent out a radio signal to this planet (supposing we knew where it is), we should have to wait 2000 years for the reply (assuming "they" were listening). For all practical purposes we are alone in space.

Other astronomers may have given quite different estimates. But the point is, all such figures are only estimates or, rather, wild guesses, tempered by our ignorance of many basic factors. The more sanguine astronomers have discussed the possibility of tuning in on radio broadcasts from a planet revolving around a nearby star. And thus Project Ozma was born, as a wild gamble, in the early 1960's at the U. S. National Radio Observatory in the mountains of West Virginia. Special recording equipment transcribed the radio noise, static and all, collected by a large reflector pointed at frequent intervals in the direction of two nearby solar-type stars, tau Ceti and epsilon Eridani, some eleven light years distant. After a long unsuccessful search, the program was abandoned. No one can say what the probable chances of success were for Project Ozma, perhaps one in a thousand, or one in a million. But how exciting it would have been if, over the cosmic static, we could have heard a regular pattern, indicative of intelligent origin. Despite the great distances and lack of direct contact, most scientists believe that we should be able to decipher the message and eventually communicate with these beings, if they are at all mentally related to our type of civilization and technology.

Some peculiar objects do send out sharp radio pulses at regular intervals, of the order of a second or two. Some scientists suggested that these *pulsars*, as the objects are known, represent some sort of radio beacon used for interstellar navigation. The British radio astronomer, M. Ryle, facetiously called these objects, when first discovered, the "LGM's," an abbreviation for "Little Green Men." However, the unbroken pattern of dots seems unlikely to carry any intelligence. Hence, most astronomers believe that some special natural phenomenon is involved, probably the rotation of highly dense neutron stars (see Secs. 27–7 and 33–8).

Meanwhile the question of life—especially intelligent life—elsewhere in the universe must remain open to speculation. The most hopeful argument rests on the high abundance of solar-type stars and the probable existence of planets in a favorable situation. The quest for evidence of life will continue and, who knows, someday it may even succeed.

CHAPTER 19

Stellar Distances and Luminosities

19-1 **The Study of the Universe.** Even in primitive times the stars were thought to be more remote than the sun, moon, and planets, but the ancients pictured them as fiery spots on a very distant heavenly sphere. Huygens, in the seventeenth century, surmised that they might be bodies like the sun, and used this assumption to roughly estimate their distances. But the modern idea of a star-filled universe is only two hundred years old.

When William Herschel began his study of the universe in the later eighteenth century, he already believed that all stars were truly enormous bodies like the sun, appearing faint only because of their tremendous distances. But at that time nobody had measured the distance to any of the stars, and consequently the true brightness of the stars was unknown. Herschel then made the same simple assumption that Huygens had made, that the stars are all roughly similar to the sun in absolute luminosity.

Herschel's large reflecting telescopes, which he built himself, confirmed Galileo's discovery that the Milky Way (Fig. 19-1) consisted of myriads of faint stars, and as he extended his search to fainter stars, he found that their numbers increased rapidly. Not only were faint stars most numerous in the plane defined by the Milky Way, but also the rate of increase of fainter stars was much more rapid in the plane than perpendicular to it. Assuming, then, that all stars resemble the sun, he concluded that the system of stars in the Milky Way was shaped something like a grindstone or millstone, as shown in Fig. 19-2.

FIGURE 19-1 *Star field and cluster in the Milky Way. (E. Barnard, Yerkes Observatory.)*

FIGURE 19-2 *Cross-section of the Milky Way perpendicular to its plane, according to W. Herschel, who supposed that the sun (S) was near the center.*

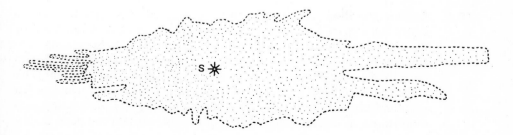

The conclusions Herschel reached as to the structure of the system were basically correct. They have been verified in more recent times by precise measurement of the distances, brightness, spectra, motions, and so on of many of the component stars of the Milky Way. But in Herschel's day such information was almost entirely lacking, and the picture that he drew was merely qualitative.

If we are to grasp the structure of our great stellar universe, and the nature of the stars that make it up, we must first turn our telescopes on individual objects such as stars, double stars, clusters of stars, and the bright hazy patches called nebulae, and measure as many of their properties as we can.

If, further, we wish to understand the behavior of stars and stellar systems, we must determine the structure of stellar atmospheres, learn how some stars vary in brightness, and investigate the laws of stellar energy generation. In this fashion we accumulate knowledge about the intrinsic nature of the individual objects and the physical relationships between them.

We cannot hope to measure every single star or nebula. We must choose certain samples for detailed study, hoping that they are representative of the universe as a whole. The samples will include not only stars, but aggregates of stars, and the interstellar material that lies between the stars. From this point of view we are interested in the samples mainly as tools for the study of stellar behavior, building-blocks for the construction of a model of the universe. Thus we make our model step by step. We begin with individual objects and build their properties into the structure of the Milky Way. Then we consider the other great stellar systems beyond our own, and finally we can speculate about the universe that comprises them all.

We have already discussed the sun and the solar system in detail. The scale of the solar system provides a starting point for measuring the distances and brightnesses of stars, and the sun provides a basis of comparison for the intrinsic properties of stars. From the baseline furnished by our small planetary system we can expand our viewpoint to embrace our stellar system, the Milky Way, and later the similar vast systems that make up the astronomical universe.

19-2 Stellar Parallax. As we saw in Chapter 5, Bradley was not able to detect stellar parallax, the apparent annual motion of stars furnished by the earth's revolution about the sun. As the earth moves about its orbit (Fig. 19-3), the direction of a nearby star relative to much more distant stars should change through the angle $2p$, indicated in the diagram. This angle proves to be so small that we can measure it directly for only a few thousand of the nearest stars. By statistical procedures we can then estimate the much smaller mean parallax of the more distant comparison stars

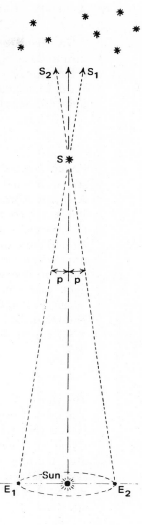

**FIGURE
19-3**

The parallax, p, of a star represents the angle, at the star, subtended by the radius of the earth's orbit. As the earth moves from E_1 to E_2, the star shifts its apparent position from S_1 to S_2 by twice the parallax.

and so correct the observed *relative* parallax of the nearby star to obtain the *absolute* parallax, that which would be observed if the background stars were infinitely distant. The small angle p, called parallax and expressed in seconds of arc (actually p is always less than $1''$), is equal to the angle subtended by the radius of the earth's orbit, as viewed from the star.

Not until 1837 was a stellar parallax actually measured. In that year the German astronomer Bessel succeeded in determining the parallax of 61 Cygni as $0''.3$. The best present-day determination is $0''.293$, close to Bessel's value. Actually, 61 Cygni is not the nearest star; the first-magnitude double star, α Centauri, has a larger parallax of $0''.754$, still less than a second of arc. A fainter star, close to it and probably associated with it, Proxima Centauri, has a parallax of $0''.765$, the largest on record.

To calculate the distance of a star from its known parallax, we must turn to the geometry of Fig. 19-3. Our baseline is the radius of the earth's orbit, the earth's mean distance from the sun (or astronomical unit, A.U.), about 93,000,000 miles (Chapter 7). Hence the distance, d, in astronomical units of a star having a parallax p seconds of arc is

$$d = \frac{206,265}{p} . \tag{19-1}$$

For convenience we define a unit of distance, the *parsec*, as the distance of a star with a parallax of $1''$. Hence, if $p = 1''$, a parsec is equal to 206,265 A.U., or 3.086×10^{13} kilometers (1.92×10^{13} miles). The distance of a star, expressed in parsecs, is therefore simply $1/p$.

A parsec is so great a distance that one cannot readily visualize it. A more picturesque unit, often used, is the *light year*, the distance that light travels in a year. Since light travels 299,793 km (approximately 186,000 miles) in one second, and since a year contains 3.156×10^7 seconds, a light year equals $299,793 \times 3.155 \times 10^7 = 9.46 \times 10^{12}$ km (5.9×10^{12} miles), or 6.324×10^4 A.U. One parsec is equivalent to 3.262 light years. Note that the light year is a unit of distance, not of time.

We can now write the equation of stellar distances:

$$d = \frac{1}{p} \text{ parsecs} = \frac{3.262}{p} \text{ light years}. \tag{19-2}$$

The star 61 Cygni is then $3.262/0.293 = 11.1$ light years distant from us. The nearest bright star, α Centauri, is $3.262/0.754 = 4.3$ light years away. Hence when we look at α Centauri we see it as it was 4.3 years ago, and 61 Cygni 11.1 years ago. For comparison, we see the sun only 8.3 minutes after the light left it, and light traverses the total diameter of the solar system defined by the orbit of Pluto in less than 12 hours. These figures more eloquently than words illustrate the vastness of the void between stars and the isolation of our solar system is space.

19-3   Proper Motions of Stars.   If the "fixed" stars were really fixed, we could measure a stellar parallax by making only two observations—for example, two photographs taken six months apart (Fig. 19-3). But while the earth moves halfway around its orbit, the star itself moves. Hence we must take a third photograph, say a year after the first, to measure the star's motion across the sky. The angular motion of a star in the course of one year is called the *proper motion*; it is usually designated by μ and is expressed in seconds of arc per year.

Proper motion has one great advantage over parallax from the measurer's point of view. The apparent shift of the star increases progressively with time, becoming larger year by year. Hence, photographs of the same

TABLE 19-1   *The ten nearest stars*

Stars	Coordinates (1900): α	δ	m_v	μ^a	p	M_v
1 Proxima Cen	$14^h22^m8 - 62°15'$		11.3	3″85	0″765	15.7
2 α Cen {A B	14 32.8 − 60 25		0.02 / 1.39	3.68	0.754	4.5 / 5.9
3 Barnard's	17 52.9 + 4 25		9.56	10.25	0.545	13.2
4 L726-8 {A B	1 34.0 − 18 28		12.45 / 12.95	3.36	0.500	16.2 / 16.7
5 Wolf 359	10 51.6 + 7 36		13.5	4.72	0.420	16.5
6 Lal 21185	10 57.9 + 36 38		7.5	4.78	0.398	10.5
7 Sirius (α CMa) {A B	6 40.7 + 16 35		−1.43 / 8.6	1.32	0.375	1.5 / 11.4
8 Ross 154	18 43.6 − 23 56		10.7	0.67	0.347	13.4
9 Ross 248	23 37.0 + 43 39		12.3	1.82	0.316	14.8
10 ε Eri	3 28.2 − 9 48		4.2	0.99	0.303	6.2

^{a}Seconds of arc per year.

TABLE 19-2   *The twenty brightest stars*

Stars		Coordinates (1900): α	δ	m_v	μ^a	p	M_v	Notes
1 Sirius	α CMa	$6^h40^m8 - 16°35'$		−1.43	1″324	0″375	+1.5	
2 Canopus	α Car	6 21.7 − 52 38		−0.73	0.022	0.006?	−5.0	
3 Rigil Kent	α Cen	14 32.8 − 60 25		−0.27	3.682	0.754	+4.1	b
4 Arcturus	α Boo	14 11.1 + 19 42		+0.06	2.287	0.102	+0.2	
5 Vega	α Lyr	18 33.6 + 38 41		+0.04	0.348	0.125	+0.6	
6 Capella	α Aur	5 09.6 + 45 54		0.09	0.437	0.073	−0.5	
7 Rigel	β Ori	5 09.7 − 8 19		0.15	0.003	0.005?	−6.5	
8 Procyon	α CMi	7 34.1 + 5 29		0.37	1.242	0.287	+2.7	
9 Achernar	α Eri	1 34.0 − 57 45		0.53	0.093	0.023	−2.6	
10 Agena	β Cen	13 56.8 − 59 43		0.66	0.039	0.024	−2.5	b
11 Betelgeuse	α Ori	5 49.8 + 7 23		0.7	0.032	0.012	−4.1	c
12 Altair	α Aql	19 45.9 + 8 36		0.80	0.659	0.210	+2.4	
13 Aldebaran	α Tau	4 30.2 + 16 19		0.85	0.205	0.048	−0.6	d
14 Acruz	α Cru	12 21.0 − 62 33		0.87	0.048	0.020	−0.4	b
15 Antares	α Sco	16 23.3 − 26 13		0.98	0.032	0.008	−4.0	e
16 Spica	α Vir	13 19.9 − 10 38		1.00	0.051	0.014	−3.3	
17 Fomalhaut	α PsA	22 52.1 − 30 09		1.16	0.367	0.144	+2.0	
18 Pollux	β Gem	7 39.2 + 28 16		1.16	0.623	0.095	+1.0	
19 Deneb	α Cyg	20 38.0 + 44 55		1.26	0.004	0.002?	−7.0	
20 Becrux	β Cru	12 41.9 − 59 09		1.31	0.054	0.007	−4.5	

^{a} Seconds of arc per year.

^{b} Magnitudes for both components.

^{c} Variable m_v: 0.4 to 1.0.

^{d} Variable m_v: 0.75 to 0.95.

^{e} Variable m_v: 0.90 to 1.06.

stellar region taken fifty years apart, for example, show a shift of the star's position fifty times as great as photographs taken a year apart. Relatively few stars move more than 1″ per year; Barnard's "runaway" star at present holds the proper-motion record with a rate of 10″.25 per year. The proper motions of the ten nearest stars together with their parallax, apparent magnitude, and absolute magnitude are given in Table 19-1. Table 19-2 gives the same data for the twenty brightest stars. See Fig. 19-4 for proper motions in the Big Dipper.

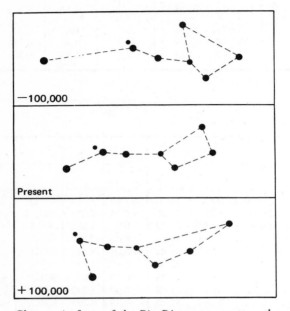

FIGURE 19-4   *Changes in form of the Big Dipper: present; as the result of proper motion during 200,000 years. Note that the six central dipper stars maintain their relative positions, while the two outermost stars show radical motion. The six central stars are part of a moving cluster.*

Proper motion measures only the apparent motion of the star across the line of sight. Figure 19.5 shows the components of the space motion of a star moving from S to S'; μ is the proper motion, p the parallax, and $d = 1/p$ the distance in parsecs. RS' is the *tangential motion* and SR the *radial motion* along the line of sight; in the triangle ORS' we have $RS' = \mu \times d = \mu/p$ in astronomical units if μ and p are in seconds of arc. If a star should move with a speed of 1 A.U. per year across the line of sight, then its proper motion would equal its parallax. But a velocity of 1 A.U. per year

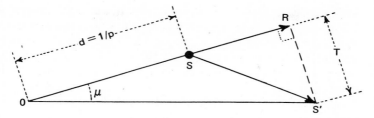

**FIGURE
19-5**

Components of space motion of a star, moving from S to S' during some given interval. The observer at O breaks the space motion SS' into two parts, the radial velocity SR toward or away from the observer, and RS', the tangential velocity, T, at right angles to the line of sight. The proper motion, μ, is the angular drift subtended by the component RS', in the course of one year.

corresponds to 4.74 km/sec. Hence the tangential velocity T, in km/sec, is given by the equation

$$T = 4.74 \frac{\mu}{p}. \tag{19-3}$$

Barnard's star, with $\mu = 10''.25$ and $p = 0''.545$, has a tangential velocity of 89 km/sec (55 miles/sec). The proper motions of stars tell nothing about their velocities in the line of sight. We measure the radial velocity by means of a spectrograph, as described in Chapter 8.

19-4   Measurements of Stellar Distances.   Direct measurements of stellar parallaxes, commonly called *trigonometric parallaxes*, are today made photographically. As we have seen, a minimum of three photographs would suffice for this purpose, but, in practice, to attain a higher degree of accuracy and derive the proper motion, we use a dozen or more photographs made over an interval of several years. Few branches of astronomy require more care, both in observational techniques and in measurement, than the determination of stellar parallaxes. A small number of active astronomers in observatories specially equipped for such work devote their time to this painstaking and slow task. At present the parallaxes of some 6000 stars have been measured within an accuracy of $0''.003$ to $0''.010$.

Parallax observers often select stars of high proper motion to measure, since [from equation (19-3)] for a star of given tangential velocity, T, the parallax will be $4.74\mu/T$, and thus will be proportional to the proper motion. If, therefore, a star has a large proper motion, it will probably also have a large parallax.

Stars of large proper motion are most easily discovered by a comparison of two photographs taken a number of years apart. A *blink microscope*

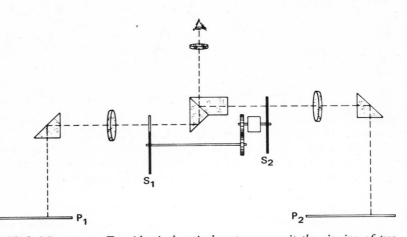

**FIGURE
19-6** *The Blink Microscope. Two identical optical systems permit the viewing of two
plates P_1 and P_2. The complex prism, nearest the eye, superposes the images from
the two plates. The shutters S_1 and S_2 alternately show first one plate and then the
other. A star that has moved between the times the photographs were taken, will
appear to jump back and forth.*

(Fig. 19-6) makes this task easy, as it permits one to look alternately in
rapid succession at the two photographs by the turn of a knob. When the
photographs are adjusted so that the images of most stars coincide, stars
that have moved appreciably betweeen the two exposures appear to jump
back and forth as one turns the knob.

The parallaxes, proper motions, and other data for the ten nearest stars
and for the twenty brightest stars are given in Tables 19-1 and 19-2. For
double stars that consist of two observable components, we have listed
both members of the pair. Note that only two of the ten nearest stars
(Sirius and α Centauri) are particularly bright to the naked eye. Many
bright stars are more distant. Vega, for example, is (counting the sun) the
fifth brightest star in the sky, yet its parallax is only $0''.125$. The list of the
nearest stars will undoubtedly be changed by the discovery of extremely
faint nearby stars. The very faint red double star that stands fourth on the
list was discovered only recently.

We see from Table 19-1 that the ten nearest stars extend to a distance
of 10.75 light years, corresponding to a parallax of $0''.303$. They thus occupy
a volume of 5160 cubic light years. Each star possesses on the average,
as its own private volume of space, one-tenth of this figure, or 516 cubic
light years. This figure is not representative, of course, of the average
free space alloted to each star. If we consider larger volumes of space, the
star density is somewhat greater. Nevertheless, we may say that the stars
are not closely crowded.

19-5   Stellar Magnitudes.   The scale of "magnitudes" used by astronomers to express stellar luminosities has a history going back to antiquity. Hipparchus (second century B.C.) and after him Ptolemy (second century) divided the 1022 visible stars of his catalogue into six groups numbered from 1 to 6 according to the order of decreasing apparent brightness, with first-magnitude stars the brightest.

In 1859, the physiologists Weber and Fechner showed that equal differences in brightness as seen by the eye correspond, in fact, to approximately equal *ratios* in luminous intensity. Expressed mathematically, this law states that human perception of light is proportional to the logarithm of the brightness, rather than to the brightness itself. A similar formula applies for the perception of sound. John Herschel had discovered the logarithmic relation earlier in his study of the brightness of stars, observing that a star of first magnitude was about a hundred times brighter than a star of sixth magnitude.

Pogson, in 1856, assuming that the law is exact (which it was not), defined the present system of stellar magnitudes on the basis of this mathematical relationship. He defined an interval of one magnitude as a ratio in brightness of exactly $\sqrt[5]{100} = 2.512$. Thus a star of any given magnitude would be 2.512 times brighter than a star one magnitude fainter. Pogson adjusted his scale so that it agreed with Ptolemy's at the sixth magnitude, near the limit of naked-eye visibility. The steps turned out to be roughly the same as those of Ptolemy's scale. Pogson maintained the early convention established by Ptolemy, that a *decrease* in numerical value of the magnitude corresponds to an *increase* in brightness. The scale was extended to negative magnitudes for extremely bright objects. Sirius has a magnitude of -1.4, the full moon of -12.5, and the sun of -26.8 (see Chapter 7).

If two stars have magnitudes m_1 and m_2, and if their apparent brightnesses or luminosities are l_1 and l_2, the ratio of luminosities follows from the equation:

$$\frac{l_2}{l_1} = 10^{-0.4(m_2 - m_1)} = -2.512^{(m_1 - m_2)}. \qquad (19\text{-}4)$$

To obtain a simpler equation, merely take the logarithm (to the base 10):

$$\log\left(\frac{l_2}{l_1}\right) = -0.4(m_2 - m_1). \qquad (19\text{-}5)$$

The 200-inch telescope at the Palomar Observatory can photograph stars fainter than the twenty-third magnitude. About fifty magnitudes separate the faintest observable stars from the brightest (our sun). The corresponding light-ratio, calculated from equation (19-4), is 10^{20} or one hundred billion billions. This is truly an enormous range in apparent brightness, but we shall see later (Sec. 19-8) that the range in intrinsic or absolute luminosities of stars is in fact far smaller, though still great.

We can roughly estimate the energy flux density or power—that is, the number of ergs per second carried by the radiation crossing an aperture of unit area, say 1 cm²—exposed to starlight. For the sun, whose apparent magnitude is −26.8, we know (Sec. 9-1) that the energy flux is 2.0 calories per square centimeter per minute, equivalent to 0.14 watt cm^{-2} or again 1.40×10^6 erg cm^{-2} sec^{-1}. From this we find through equation (19-4) or (19-5) that a star (of solar type) of magnitude 0.0 produces a flux density

$$\frac{l_2}{1.4 \times 10^6} = 10^{-0.4 \times 26.8}$$

or

$$\log l_2 = 6.15 - 10.72 = -4.57, \qquad l_2 = 2.7 \times 10^{-5} \text{ erg cm}^{-2} \text{ sec}^{-1}.$$

Similarly we receive from a solar type star of magnitude +23.2 a power

$$\frac{l_2}{1.4 \times 10^6} = 10^{-0.4 \times 50.0}$$

or

$$\log l_2 = 6.15 - 20.0 = -13.85, \qquad l_2 = 1.4 \times 10^{-14} \text{ erg cm}^{-2} \text{ sec}^{-1}.$$

If this power could be accumulated and used exclusively to raise the temperature of one gram of water, it would take a time T greater than the age of the Earth (5×10^9 years, see Chapters 17, 35) for it takes 1 calorie of energy to raise the temperature of 1 gram of water by 1 degree Celsius, and we know that 1 calorie $= 4.185 \times 10^7$ ergs, so that $T = 4.185 \times 10^7 / 1.4 \times 10^{-14} = 3 \cdot 10^{21}$ seconds $\simeq 10^{14}$ years. This hypothetical experiment gives some idea of the extraordinary minuteness of the radiant energy that can be detected by large telescopes.

Our calculation assumes that the stars have the same color or spectral energy distribution as the sun. Actually they vary a great deal (Chapter 20), and to make valid comparisons we should consider the energy carried by a specified narrow range of wavelengths, say 1 angstrom, or of frequency, say 1 cycle per second, if we remember that frequency = velocity of light : wavelength ($v = c/\lambda$) (Chapter 8). For example, we receive from the sun a *monochromatic* energy flux of about 200 ergs cm^{-2} sec^{-1} per angstrom near the maximum of its spectral energy curve, around 5000 Å (cf. Chapter 8). To make valid comparison we need, therefore, stellar-magnitude data based on specified wavelength intervals, or colors.

19-6 Visual and Photographic Magnitudes; Color Index. Thus far we have talked of stellar magnitudes without specifying how to measure them accurately. Before the advent of photography, the eye was the only practical instrument for measuring the brightness of stars. But we know from Chapter 8 that the brightness of an object depends on the color, or color range, we use to observe it. A red-hot poker, for example, may appear bright to the eye and extremely bright to an infrared detector. However, it is a very weak emitter in the ultraviolet. The human eye is most sensitive to yellow-green light, and thus is best suited for observation in this range of radiation. We employ the term *visual magnitude* to denote the brightness of a star, as estimated by the human eye in the yellow-green, or the region of $\lambda = 5500$ Å.

The development of photographic methods for measuring stellar brightness required another system of magnitudes. The early photographic emulsions, though sensitive to violet and ultraviolet light, did not respond to radiation of wavelength greater than about 5000 Å in the blue-green region. Some of the sensitivity in the ultraviolet was wasted, because most glass lenses absorbed light of wavelengths shorter than about 3700 Å, and silvered mirrors lose reflectivity at about 3400 Å. Hence, the early photographic plates recorded the brightness of the stars in the blue-violet region, and the adopted term, *photographic magnitude*, defined a system of magnitudes for wavelengths near $\lambda = 4500$ Å. Photographic magnitudes continue to play an important part in astronomy, and much work is still done with emulsions having this type of spectral sensitivity.

Astronomers had to adjust the zero point of the photographic scale to correspond to the visual scale, much as the Pogson scale was adjusted to correspond to the Ptolemaic scale. Since the difference between the photographic and visual magnitudes depends on the color of the star, the two were arbitrarily set to coincide for stars of some particular color. The magnitudes were arbitrarily made equal for the white stars of spectral class A, such as Sirius. (Chapter 20 deals with the classification of stellar spectra.) Since the eye is relatively more sensitive to red light than the classical photographic emulsion, a red star, which radiates relatively more energy in the red than in the blue-violet, appears brighter visually than it does photographically. Stars bluer than Sirius, on the other hand, are brighter photographically than visually.

These differences between the two systems of magnitudes permit us to measure the colors of the stars numerically. We define the *color index* of a star as the photographic magnitude minus the visual magnitude. Because the magnitude scale is such that smaller numbers denote brighter stars, color index is *positive* for all stars redder than Sirius and *negative* for the bluer stars. The blue star Rigel (β Orionis) has a visual magnitude 0.14, a photographic magnitude -0.03, and therefore a color index of -0.17; the red star Betelgeuse (α Orionis) has visual magnitude 0.70, photographic magnitude 2.14, and color index $+1.44$. A star's color index depends on the temperature of its surface. The index provides a useful measure of temperature as long as passage through interstellar dust has not reddened the starlight (see Chapter 29). Thus color index furnishes a simple tool for statistical studies of stars and of the structure of the Milky Way, since it enables the astronomer to determine the temperatures of individual stars.

We can measure the photographic magnitude of large numbers of stars on one photograph. In practice, the stellar images are compared with those of a few stars on the same photograph whose magnitudes have been carefully measured beforehand. The comparison can be made most simply through visual estimates of the relative sizes of the star images (the brighter the star, the larger the blackened area on the plate); with a little training a good observer can easily intercompare stars to within 0.1 mag., or about 10 per cent in relative luminosity.

More precise measurements require that we employ a more objective device than the eye, such as a photoelectric photometer, to measure the blackness or density of the photographic images. Here again the magnitudes must be referred to a series of stars of known magnitudes on the photographs, which have been carefully determined beforehand.

The determination of these known or *standard* magnitudes is laborious and difficult. They must be calibrated by photometric techniques such as would be used in the laboratory for the measurement of brightness. The first such photographic standard is a group of stars near the north pole of the sky, known as the *North Polar Sequence*. Other standard photographic sequences are ultimately referred to this one. Successive photographs of the north pole and the region to be standardized are made with the same telescope, at the same altitude (to eliminate the absorption of the earth's atmosphere), and under as nearly identical observing conditions as possible.

Photoelectric photometers (Chapter 6) today provide the most precise measurements of standard sequences and of individual stellar magnitudes, but the photographic plate is still unsurpassed for the measurement of the magnitude of large numbers of stars.

19-7 Other Photometric Systems. Ordinary photographic plates are always sensitive in the blue and ultraviolet regions. They can also be made sensitive to other wavelengths by treatment with dyes that absorb certain colors. Thus it is now possible to take photographs in yellow and red light, and even in the infrared beyond the range of the eye. If an "orthochromatic" emulsion, sensitive to green and yellow light (as well as blue violet, and ultraviolet), is combined with a yellow filter (that blocks the violet and ultraviolet), the combination closely simulates the color sensitivity of the eye. Magnitudes measured with this arrangement are known as *photovisual* magnitudes. The high sensitivity of photographic emulsions, in conjunction with long telescopic exposures, enables us to extend the visual system far beyond the limits that could be attained with the eye itself.

Similarly, photographic red and infrared magnitude systems have been used for special purposes. Any such magnitude system is determined by the combination of plate and filter that it uses, and the possible variety is almost limitless. For some problems, one chooses a filter that admits a wide range of wavelengths. In others, where a small range is desired, one employs a narrowband filter, or perhaps combines a plate having a sharp cutoff in sensitivity toward the red with a filter whose sharp violetward cutoff overlaps slightly. Even smaller is the wavelength range of the narrow-passband filters described in Chapter 9, used in the study of selected solar spectrum lines.

Photographic magnitudes are accurate within a few hundredths of magnitude at best, but photoelectric photometry is capable, in principle, of a precision of about 0.001 mag., or 0.1 per cent in luminosity. Systems of photoelectric standard magnitudes have been set up at different ranges of wavelength in the spectrum, from the violet to the near infrared. J. Stebbins and A. E. Whitford, for instance, have made measures at six different wavelengths, and their studies furnish not only an accurate survey of the spectral energy distribution of stars, but also a means of determining how much the light of a star is reddened and obscured by interstellar dust. The most commonly used photometric standard magnitudes have been established in three colors, near ultraviolet, blue, and yellow, by H. L. Johnson and his collaborators. This U,B,V system of magnitudes, as it is called, and the relations between the color indices B-V and U-B, have provided precise apparent magnitudes of stars and valuable data on space reddening and stellar evolution.

The systems just described provide magnitudes in definite ranges of wavelengths. Some energy receivers, such as the thermocouple, bolometer, and radiometer, respond to light of any color. They furnish the nearest approach that we have to measuring the total radiation of a star, or *bolometric magnitude*. However, this term implies that the measures have been corrected for the part of the energy curve cut out by absorption in the earth's atmosphere. For red stars the correction, though appreciable, is not too large, but most of the radiation of hot stars is in the ultraviolet part of the spectrum, which the earth's atmosphere does not transmit. For these stars, we cannot measure the bolometric magnitude directly, except from space vehicles.

The earth's atmosphere introduces many inaccuracies into the measurement of stellar magnitudes by any method. Haze, dust, clouds, molecular absorption, and scattering in the atmosphere all contribute their share of the uncertainty. An observatory, to be useful for photometric work, must possess both a uniform atmosphere and high atmospheric transparency. An apparently clear sky can sometimes be almost useless photometrically because of variable haze too faint to be detected visually. The best photometric sites are generally on top of mountains in arid, dry regions, as in the southwest of the United States or in the Andes.

19-8   Absolute Magnitudes. Since the illumination or flux density from a point source of light falls off inversely as the square of the distance from its source, we can calculate the intrinsic or absolute luminosities of stars, once we have measured their apparent magnitudes and their parallaxes.

By definition the absolute magnitude of a star is the magnitude it would have if seen at an adopted standard distance of 10 parsecs. If, then, we know the parallax p, and the apparent magnitude m, we can calculate the absolute magnitude M. For example, if a star has an apparent magnitude of $+10$ at a distance of 100 parsecs, it would appear $(100/10)^2 = 10^2 = 100$ times as bright at the standard distance of 10 parsecs, and since a factor of 100 corresponds to five magnitudes by Pogson's formula (Sec. 19-5), its absolute magnitude is $M = +10 - 5 = +5$.

More generally, if L is the absolute luminosity of a star at distance 10 parsecs and l its apparent luminosity at a distance $1/p$ parsecs, then by the inverse-square law and equation (19-4) we may write

$$\frac{L}{l} = \frac{(1/p)^2}{10^2} = 10^{0.4(m-M)}.$$

Upon taking the logarithms of the second and third members of the equation, we find

$$-2 \log p - 2 = 0.4(m - M).$$

If we multiply by 2.5 and rearrange, we obtain the following equation:

$$M = m + 5 + 5 \log p. \tag{19-6}$$

The calculations can be made directly, or one can make use of the graphical method of Fig. 19-7. As before, the ratio of the absolute luminosities L_1 and L_2 for two stars takes the form

$$\log \frac{L_2}{L_1} = 0.4(M_1 - M_2). \tag{19-7}$$

The quantity $m - M$, called the distance modulus of a star, is used very often to express distances when the absolute magnitudes M can be determined indirectly and the parallaxes are unknown or cannot be measured.

To establish the absolute magnitude of the sun, we note again that the parallax in seconds of arc is equal to the inverse distance in parsecs. The distance from the earth to the sun, or 1 A.U., is $1/206,265$ parsecs, so we find the sun's parallax $p = 206,265''$. The sun's apparent visual magnitude is -26.8; hence, by equation (19-6), we find $M = -26.8 + 5 + 5 \times 5.314 = 4.8$ for the absolute visual magnitude of the sun. That is, the sun would appear as a fifth-magnitude star, if seen at a distance of 10 parsecs.

The absolute visual magnitudes of the ten nearest stars and the twenty brightest stars are given in the last columns of Tables 19-1 and 19-2. The sun, with $M = +4.8$, is more luminous than its near neighbors, except for

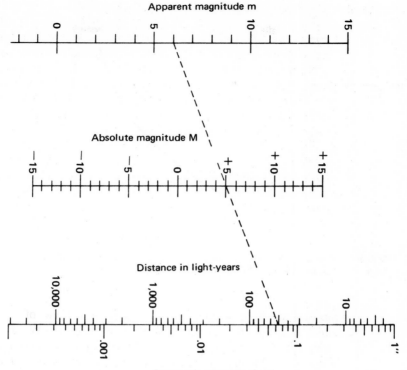

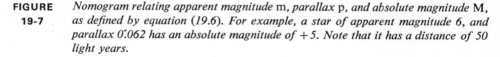

FIGURE 19-7 *Nomogram relating apparent magnitude* m, *parallax* p, *and absolute magnitude* M, *as defined by equation (19.6). For example, a star of apparent magnitude 6, and parallax 0".062 has an absolute magnitude of +5. Note that it has a distance of 50 light years.*

Sirius and the brighter component of α Centauri. On the other hand, all of the twenty brightest stars in the sky are intrinsically brighter than the sun. Rigel (β Ori) and Deneb (α Cyg), the most luminous of them all, are more than ten thousand times as bright as the sun in absolute luminosity, and more than a billion times as bright as the faintest star known, whose absolute magnitude is about +18. Thus the total range in the true luminosities of stars exceeds 10^9. The early attempts to estimate stellar distances on the assumption that all stars have the same intrinsic brightness were gross oversimplifications.

The fact that all the apparently bright stars are also intrinsically bright, whereas the nearest stars generally are faint, is very significant. If we drew our conclusions from the absolute luminosities of the naked-eye stars, we might infer that stars intrinsically very bright are much more prevalent in the universe than in fact they are. Figure 19-8 shows the numbers of stars of

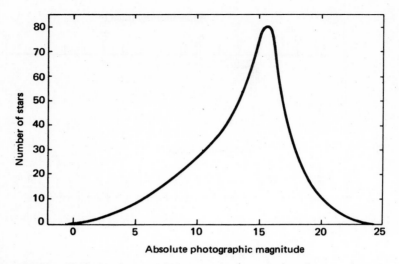

FIGURE 19-8 *Average numbers of stars of various absolute magnitudes in a large volume of space in the neighborhood of the sun.*

various absolute magnitudes in a sizable volume of space about the sun, defining the so-called *luminosity function.* W. J. Luyten, who compiled the table, applied corrections to the numbers of faint stars to allow for difficulty of discovering the faintest stars. Stars of negative absolute magnitude—a hundred times brighter than the sun or more—prove to be extremely scarce in any given volume of space, and the sun is considerably brighter than the average. The most common stars in space around the sun are red dwarf stars, having but one ten-thousandth of the luminosity of the sun ($M_{pg} \simeq +15.5$ vs. $M_{pg} \simeq +5.4$). Bright stars are, of course, more conspicuous and can be seen at great distances, and thus appear to be more frequent than they actually are.

CHAPTER *20*

The Spectra and Temperatures of Stars

20-1 **The Spectrum of the Sun.** The sun's spectrum which we have already discussed in considerable detail (Chapter 9), contains many thousands of dark lines, some weak and some strong, caused by the absorbing atoms and molecules in the reversing layer. These dark lines, we have seen, give us information about temperature and pressure in that layer, and about its chemical composition.

If the sun were so far away that we could not see a perceptible disk, even with a high-power telescope, we should not be able to isolate the light from a small area of its surface, such as the center of the disk or the limb. We should receive a mixture of radiation from the center, the limb, and the intermediate portions of the stellar disk. If the star were rotating rapidly, we should expect a certain amount of blurring of the spectrum by the Doppler effect, as the star's rotation caused one limb to recede from us, the other to approach us. For the sun the Doppler effect is small, because the velocity of rotation is small. But when we examine the spectra of stars, we must not forget the effect that rapid rotation might have on the appearance of the spectrum lines. Turbulence and convection currents in the solar atmosphere can also cause some regions to rise toward the observer and others to recede: such motions may conceivably exist in the atmosphere of stars, and they, too, can cause blurring of the lines in the spectrum.

Our ability to isolate different regions of the sun has permitted us to observe sunspots and flares, the chromosphere and the corona, all having spectra that differ from that of the reversing layer. But their contribution to the spectrum of integrated sunlight is small, and if the sun were not so close to us, they might not have been discovered. The sun is a star, and other stars may have similar properties. Some stars, perhaps, have spots and flares, chromospheres and coronas that contribute more to their integrated light than the corresponding features do for the sun. In our study of stellar spectra we must be on the lookout for evidences of such surface phenomena.

20-2 Stellar Spectra. Stellar spectra were first investigated visually as early as 1814 by the German optician J. Fraunhofer, who found that they displayed dark lines similar to those he had previously observed in the solar spectrum. W. Huggins in England and A. Secchi in Rome carried on extensive visual examination of steller spectra during the 1860's. A very important conclusion emerged from Secchi's work. Although he found that the spectra of many stars differed markedly from that of the sun, he was able to classify them in several groups that had features in common.

The fact that stellar spectra differed from one another encouraged many speculations. Did stars actually differ in chemical composition? Were there different chemical elements in different parts of the universe? Or, as Norman Lockyer suggested near the end of the nineteenth century, could the differences perhaps be due to the splitting up of atoms at very high temperatures? Lockyer believed that actual chemical transmutations were taking place in the atmospheres of the stars. We know now that stellar atmospheres are not hot enough to cause actual transmutations, but, up to a point, his reasoning was sound. Indeed, he devised and executed ingenious laboratory experiments to study the spectra of atoms under conditions that we should today call low, intermediate, and high excitation —in the flame, the electric arc, and the electric spark. He noted that these three modes of excitation produced variations in the spectrum of a given element very similar to those that he observed in stellar spectra.

Today we know that Lockyer was not achieving the true transmutations of elements, as he supposed. He was merely changing the excitation condition of the atoms, by tearing off electrons and thus producing ionized atoms, whose spectra, as we have seen in Chapter 8, are very different from those of the corresponding neutral atoms. When the excitation ceased, the electrons recombined and the atoms reverted to their original state.

Although Lockyer was mistaken in his interpretation of the laboratory experiments, his work furnished a foundation for the science of astrophysics. Some of his methods are still fruitful. Similar studies of a number of the chemical elements, by A. E. King of Mount Wilson Observatory, have in fact yielded valuable data, which have not only helped the physicist to interpret atomic spectra and understand the nature of the atom but also have provided the stellar spectroscopist with the means to interpret stellar spectra.

Secchi found that he could match the great majority of spectral lines, even in nonsolar stars, with the lines of known chemical elements. There were exceptions, of course, and the reasons for these exceptions have contributed some of the most interesting chapters of modern astrophysical research.

The application of photography to the study of stellar spectra revolutionized the subject completely. William Huggins in England and Henry Draper in America were the pioneers in this field. E. C. Pickering in

FIGURE 20-1   *Objective prism spectra, Nova Aquilae 1918, near center. (Harvard College Observatory.)*

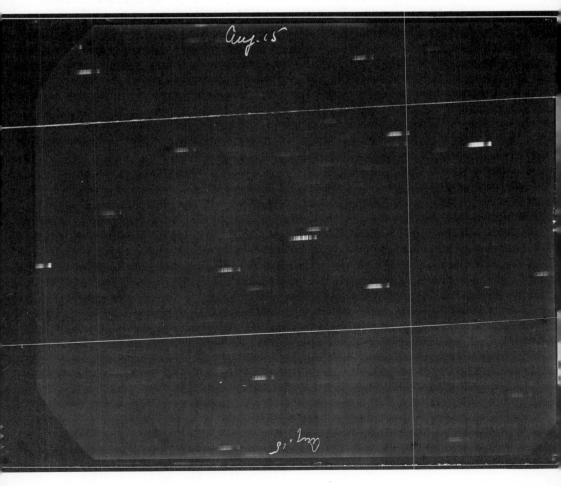

the 1880's instituted at Harvard Observatory a special survey of stellar spectra by photography, with Miss A. Maury in charge of the interpretation. Starting with Secchi's basic types, Miss Maury found that she could arrange all spectra in a continuous series, such that any one spectrum differed only imperceptibly from the one preceding or following. Secchi's types, therefore, were actually related and formed a sequence.

Miss Annie J. Cannon, who continued this work at Harvard, revised the classification, and developed it essentially in the form used today. The spectra used in these great surveys were made by means of an *objective prism*, not a spectrograph. The telescope, which ordinarily would produce photographs in which the stars appeared as small points, had a thin glass prism placed before the objective. As a result the light of each star was spread out into a small spectrum. The telescope was slightly trailed at right angles to the dispersion to widen the spectra. Some of the plates thus obtained (Fig. 20-1) may show more than a thousand individual spectra. Working with such photographs, Miss Cannon devoted a lifetime to the production of the most extensive catalogue of stellar spectra in existence, containing 225,000 stars. Filling nine large volumes of the Annals of Harvard Observatory, it is known as the *Henry Draper Catalogue*, in recognition of the pioneer observations made by Draper in the photography of stellar spectra. The catalogue contains nearly all the stars, down to about the ninth magnitude, in the northern and southern skies.

Since then, other students of stellar spectra, notably W. S. Adams and A. H. Joy of Mount Wilson Observatory, and W. W. Morgan of the Yerkes Observatory, have found it possible to subdivide and refine the classification still further.

20-3   The Henry Draper Classification.   As developed by Miss Cannon, the Henry Draper classification designates the different types of stellar spectra by the series of letters O, B, A, F, G, K, and M. These classes contain most of the stars; a few others will be mentioned later on. One notes immediately that these letters do not fall exactly in alphabetical sequence. Some of the missing letters were originally assigned to classes that later had to be dropped. Also, the order of the letters originally assigned did not arrange the spectra in a physically continuous sequence. Rather than rename the classes, which would have caused confusion, astronomers retained the original letters, but adopted the order given above.

The seven letters indicated only the broad characteristics of the spectra. Miss Cannon showed that the system was capable of much greater refinement, and achieved finer classification by the use of numerical subclasses from 0 to 9. This notation allows in principle for a total of seventy classes, ranging from O0 to M9. Thus, B8 denotes a spectrum about eight-tenths of the way from B0 to A0.

Astronomers sometimes refer to spectral classes O, B, and A as "early," and classes K and M as "late," but this practice implies nothing about the ages of the stars concerned.

Although high dispersion helps in the detailed study of a stellar spectrum, the outstanding features can be recognized on spectra of very low dispersion. This consideration is important, for it enables us to extend our catalogues to very faint stars. One can naturally secure a low-dispersion spectrum of a faint star much more readily than one of high dispersion, where the light is more spread out and requires a longer exposure.

20-4 Spectral Sequence. The spectral sequence is primarily a temperature sequence, with the highest temperatures associated with the earliest stars, lowest with the latest types.

Class O: Lines of ionized helium and of multiply ionized oxygen, nitrogen, carbon, silicon, and so on characterize the spectra.

Class B: Between B0 and B5, ionized helium disappears and neutral helium takes its place, along with lines of singly ionized oxygen, nitrogen, and other atoms.

The lines of hydrogen occur throughout the entire spectral sequence, though their intensity is low at both extremes. The Balmer lines of hydrogen increase in intensity as we pass from the early O classes toward class A.

Class A: The hydrogen lines reach maximum strength and completely dominate the spectrum in class A0 to A3. Neutral helium is absent. The early A stars show ionized lines of the more abundant metals, such as calcium, magnesium, and iron. These lines grow stronger with advancing class.

Class F: Ionized lines of metals are strong. The lines of ionized calcium become extremely intense, but reach maximum intensity later in the sequence. Around class F3, the stronger lines of neutral metals begin to appear.

The strength and persistence of the lines of ionized calcium, as compared with ionized iron, have three causes. First, these H and K lines (in terms of the Fraunhofer notation) are the two strongest lines in the spectrum of ionized calcium. The strongest lines of ionized iron lie in the far ultraviolet, where they appear in spectra taken from rockets. Second, the spectrum-making ability of iron is spread over hundreds of lines rather than just two. Third, the energy required to ionize calcium, the *ionization potential* (see Sec. 8-13), is considerably less than that needed to ionize iron. Thus as temperature and excitation decrease along the spectral sequence, the lines of ionized iron fade out and those of neutral iron appear, while those of ionized calcium still remain intense.

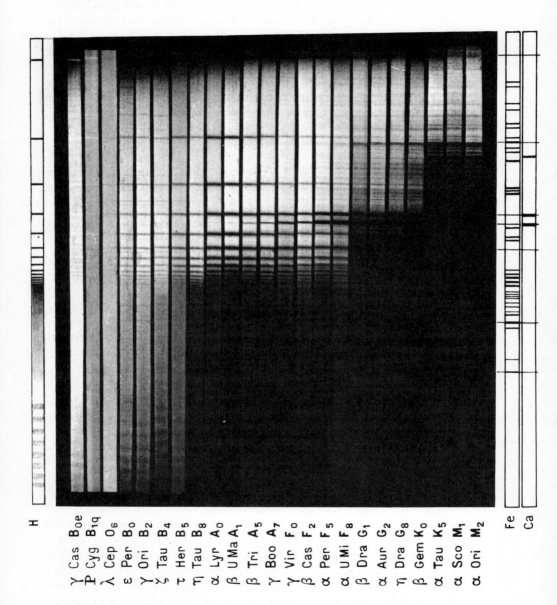

H

γ Cas B0e
β Cyg B1q
λ Cep O6
ε Per B0
γ Ori B2
ζ Tau B4
τ Her B5
η Tau B8
α Lyr A0
β UMa A1
β Tri A5
γ Boo A7
γ Vir F0
β Cas F2
α Per F5
α UMi F8
β Dra G1
α Aur G2
η Dra G8
β Gem K0
α Tau K5
α Sco M1
α Ori M2

Fe
Ca

450

FIGURE 20-2 *(Above) Spectral sequence, from λ 3100 Å to 4900 Å. (D. Barbier and D. Chalonge, Institute of Astrophysics, Paris.)*

FIGURE 20-3 *(Below) Spectrum of MIRA CETI, January 11, 1907, from 4100 Å to 7000 Å. The lower spectrum is a vertical enlargement of the upper, original spectrum. The bright lines bordering the stellar spectrum are from the comparison spectrum of iron, vanadium, and sodium. Note the emission lines of hydrogen Hα to Hδ and the dark bands of TiO with heads at 5450 Å, 5167 Å, and 4956 Å.*

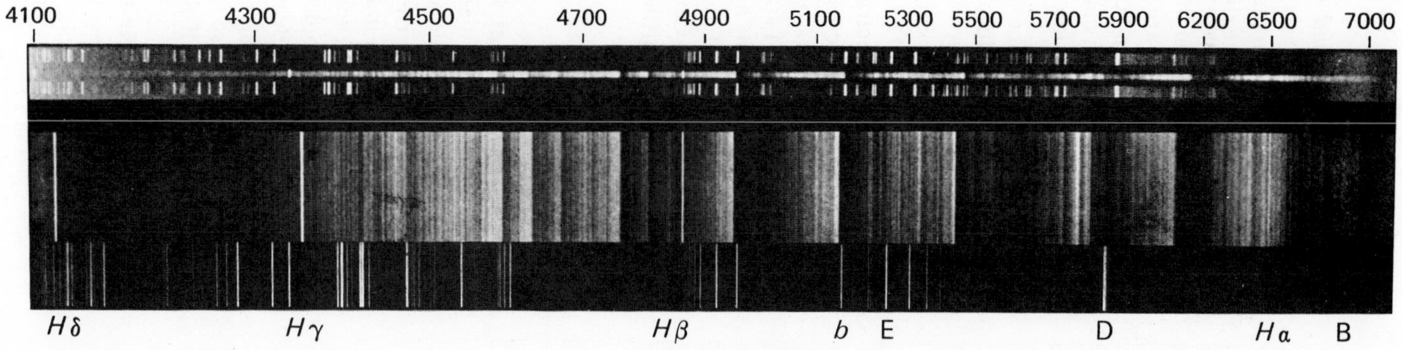

Classes F and G: Through these classes the hydrogen lines weaken rapidly, fading to negligible intensities for the coolest stars. The neutral metal lines dominate the spectrum, except for the lines of ionized calcium, which reach maximum intensity near F5.

Class K: The lines of neutral metals intensify, especially the low excitation lines that are strong in flame spectra. A few molecular bands, such as those of the radicals cyanogen (CN) and carbon hydride (CH), already fairly strong in class G, increase in intensity. About class K5, weak bands of titanium oxide (TiO) appear.

Class M: The TiO bands become very intense and furnish the major clue for classification. Metallic lines requiring low excitation are also characteristic.

Figure 20-2 illustrates the spectral sequence for the principal types from B0 to M2.

That the spectral sequence is primarily one of temperature is shown not only by the decrease in excitation in spectra from O to M, but also by the observed colors of the stars of different spectral types, the O stars being blue and the M stars red. The blue stars, whose early-type spectra point to high excitation, are hot. The red stars, with late-type spectra that show lines of neutral atoms and bands from molecules, are cool. Some of them, such as Mira Ceti, are variable (Fig. 20-3) and show emission lines of hydrogen. (Further discussion of stellar temperatures will appear in Sec. 20-6.)

20-5   Some Special Classes and Refinements of Classification. A few peculiar objects were designated by the letters Oa, Ob, and so on in the original Henry Draper classification. Such stars did not fit readily into the sequence. They included, among others, some stars with broad, intense emission lines, usually superimposed on a continuous spectrum, but occasionally accompanied by absorption lines. Secchi designated them as Type V. Astrophysical analysis of these objects by J. S. Plaskett, C. H. Payne, C. S. Beals, and others, demonstrated that these stars represent very high excitations. The French astronomers Wolf and Rayet were among the first to study these stars in the 1870's, hence their common name, the Wolf-Rayet stars, to which we now assign the special spectral class W.

A remarkable feature of the W stars is the apparent existence of two separate groups. One of these (WN) displays intense lines of nitrogen in various stages of ionization, whereas the other (WC) has similar strong lines of carbon. The ionization ranges from singly to quadruply ionized nitrogen, and from singly to triply ionized carbon. The fact that both groups show strong emission lines of neutral and ionized helium indicates that these stars must be extremely hot. A few remarkable W stars, themselves at the center of a vast cloud of gas known as a planetary nebula (see Chapter 28), contain both carbon and nitrogen in abundance, as the Belgian astrophysicist P. Swings has shown. Hence, O. Struve suggested that the apparent abundance effect may result from a peculiar stratification in the stars' atmospheres, rather than from a fundamental difference in composition and evolutionary history.

At the low-temperature end of the spectral sequence we find another abundance anomaly in stars of Henry Draper classes R and N. These stars show extensive band spectra, the carbon molecule contributing the most outstanding features. The normal molecular bands of the C_2 spectrum appear to have one or sometimes two weaker edges, of somewhat longer wavelength than that of the ordinary C_2 molecule. R. F. Sanford showed that one of these satellite bands comes from the molecule $C^{12}C^{13}$—a molecule composed of one atom of the C^{12} isotope and one of the rarer C^{13} isotope. D. H. Menzel showed that the still fainter component arises from the molecule $C^{13}C^{13}$. The presence of these isotope bands raises interesting questions about the relative abundance of the two carbon isotopes, a subject that we postpone to a later chapter (see Sec. 23-8). Probably the C^{13} isotope is relatively much more abundant in some of these stars than it is on the earth.

Another variety of star, apparently requiring separate classification, has been studied most extensively by P. W. Merrill at Mt. Wilson Observatory. These stars resemble the M stars, to which they are probably related. We classify them under the letter S. Like the M stars, the S stars are red, and their spectra contain low-temperature metallic lines. However, bands of zirconium oxide (ZrO), rather than those of TiO, dominate their spectra. We note that zirconium is the next heavier atom in the same column of the periodic table as titanium. Whether this anomaly is one of abundance is not yet certain. However, as Merrill discovered, at least one S star displays the lines of technetium, a short-lived radioactive atom with no known stable isotope. Hence, the possibility exists that nuclear reactions may be responsible for some of the spectral peculiarities of S stars. One well-known variable star, χ Cygni, displays M and S characteristics in varying proportions at different parts of its light cycle, suggesting that something more than mere abundance, perhaps atmospheric stratification, is involved.

Additional symbols are sometimes used to indicate special features observed in certain spectra. Such features are not usually visible in spectra of low dispersion, but some were recognized and described in the Henry Draper Catalogue. For a star that possesses emission lines of appreciable intensity, the letter e is placed after the letter and number denoting the spectral class (except for W stars, where emission is a normal feature). A star whose peculiarities make classification difficult carries the designation p (signifying peculiar) following the best classification that can be assigned. Such stars deserve special study.

20-6   Stellar Temperatures.   In our studies of the sun (Chapter 9) we employed the Stefan-Boltzmann and Planck radiation laws for determining the temperature of the solar surface. Two data that we know accurately for the sun are its distance and its diameter. Thus, when we measure the total radiation from the sun, either in all accessible wavelengths or over some limited range of wavelength, we find it a simple matter to calculate the actual amount of energy that comes from each square centimeter of the solar surface. The application of the Stefan-Boltzmann law or the Planck law fixes the surface temperature for us.

When we try to do the same thing for a star, however, we are handicapped. We can measure the energy that reaches us from the star, as we do for the sun. But even though we may know the star's distance, we generally do not know its diameter, and we cannot use the Stefan-Boltzmann law as we did for the sun. However, we have another relation at our disposal: Wien's law, a specialized application of Planck's law that expresses the quality, rather than the quantity, of the radiation. Even a casual inspection of the brighter stars with the naked eye reveals differences of color, and in Sec. 20-10 we shall note the close correspondence between color index and spectral class: the bluest stars are the hottest ones. According to Wien's law (Sec. 8-7) the wavelength of maximum energy $\lambda_{\max}$ depends on the temperature as follows:

$$T = \frac{2900}{\lambda_{\max}}, \tag{20-1}$$

where $\lambda_{\max}$ is expressed in microns. For the sun, as we have already noted, $\lambda_{\max} = 0.5\ \mu$, so that T is about 5800° K.

FIGURE   *Color index, C, vs. surface temperature of a star.*
20-4

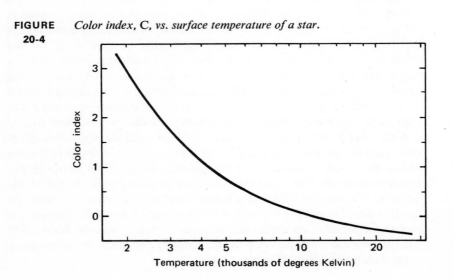

The color index C (Sec. 19-6) provides a more precise measure of temperature, as shown in the graph of Fig. 20-4. One should note that the color indices of the very hot stars (the very blue ones) show a tendency to approach the value -0.64. Thus a theoretical limit exists for the blueness of a star, as expressed by its color index. The red stars, on the other hand, have no such limit. A star whose surface temperature is 2000° K should, by Fig. 20-4, have a color index $+2.96$. The larger the value of C, the lower the temperature and the redder the star.

The use of all the preceding formulas—the Planck, Stefan-Boltzmann, and Wien laws, and the equation that relates color index to temperature—depends on the assumption that the stars behave like perfect radiators, or *black bodies*. This assumption is not quite true for any star, because the presence of absorption lines in the spectrum distorts the energy distribution somewhat, but for all except the coolest it is a good approximation. However, the spectra of M and N stars are so much distorted by the absorption of molecular bands that the approximation is no longer a good one. If we employ, instead, the corresponding formula that relates visual magnitude and magnitudes measured in the red or infrared region of the spectrum, the approximation is much better. Determination of temperature from color index is subject to other uncertainties. As we have already noted, the starlight may have suffered an unknown amount of reddening in its passage through interstellar space.

TABLE 20-1 *Stellar temperatures*

Main-Sequence and Dwarf Stars						Giant Stars		
Sp	C	T	Sp	C	T	Sp	C	T
B0	-0.33	23,000°	G0	0.57	6,000°	G0	0.67	5,500°
B5	-0.18	15,000	G5	0.65	5,600	G5	0.92	4,700
A0	0.00	11,200	K0	0.78	5,100	K0	1.12	4,100
A5	0.20	8,600	K5	0.98	4,400	K5	1.57	3,300
F0	0.33	7,400	M0	1.45	3,400	M0	1.73	3,050
F5	0.47	6,500				N	2.6	2,200

Ionization temperature may often be a better gauge of the physical state of a stellar atmosphere than temperatures calculated by other means. In Table 20-1 appears a temperature scale for the various spectral classes. One should note that the temperatures are listed in three columns, two for main-sequence and dwarf stars and one for giants. The giants are somewhat cooler than main-sequence stars of the same spectral class, a property whose physical significance we shall discuss later. The tabulation does not include the W stars. Estimates of their temperatures by C. S. Beals, mainly from excitation conditions and spectroscopic characteristics, indicate values as high as 200,000° K.

20-7 **The Luminosity Classification.** In her pioneer work on classification Miss Maury noticed a number of stars whose spectral lines were exceptionally sharp and well defined, and some others having lines unusually strong. She used the prefix c for such objects. The stars in this group are now known to be of high intrinsic luminosity—that is, supergiant stars.

Later, W. S. Adams and A. H. Joy added the prefix g (for giant) to stars that were bright but not of exceptional luminosity, and the prefix d (for dwarf) to designate stars of relatively low luminosity. We shall see later that usually these names are also appropriate to the sizes of the stars.

These developments made it clear that certain characteristics of spectra are closely related to temperature and others to absolute luminosity. W. W. Morgan and P. C. Keenan, at Yerkes Observatory, established in the 1940's the following five luminosity classes, which now replace the letters c, g, and d:

Ia: most luminous supergiants,   Ib: less luminous supergiants,
II: bright giants,   III: normal giants,
IV: subgiants,   V: dwarfs or "main sequence."

The criteria for luminosity classes differ from those for spectral classes depending upon the appearance of the feature lines. Figure 20-5 gives an illustration of the luminosity effect at class A0. Thus the complete notation for the classification of a stellar spectrum is, for example, B0 III, which indicates that the star is giant of type B0 and of a certain range of absolute magnitude.

α Cyg

ζ U Ma

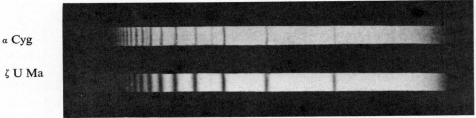

FIGURE 20-5 *The spectra of two stars of class A2. Note the sharpness of the lines belonging to the supergiant, α Cyg (upper), as compared with the broad, fuzzy lines of the dwarf, ζ U Ma (lower).*

An excellent atlas of stellar spectra in this "MK system," published by the University of Chicago, provides a series of spectra of standard stars for all basic spectral types. Even though the study does not establish quantitative measures of the intensities of pairs of lines, a trained observer can easily recognize and assign accurate classes by visual comparison with the standards. In some examples, a half subclass proved to be necessary at some stages—for example, at B0.5 between B0 and B1. For best results the spectra must be obtained with dispersion similar to that used for the standard stars.

Absolute luminosities can also be assigned, but not to a high degree of accuracy. Even so, such a two-dimensional classification of spectra, in terms of types (temperature) and luminosity, provides information useful for studies in stellar statistics.

20-8   Spectroscopic Parallaxes.   The initial estimates of absolute magnitudes of stars and the calibration of Morgan's luminosity classes were made, for the most part, from distances obtained from trigonometric parallaxes (Sec. 19-4). We know, however, that for a great majority of stars the parallaxes are much too small to be measured. One of the most important applications of the spectral and luminosity classification was therefore toward deriving distances of stars indirectly.

The absolute magnitude of a star may be estimated from the two-dimensional classification of the spectroscopic features. The difference between the apparent magnitude and the absolute magnitude, $m - M$, or the apparent distance modulus (Sec. 19-9) gives the star distance. From relation (19-6) we have

$$5 \log p = M - m - 5. \qquad (20\text{-}2)$$

The parallax so computed is called "spectroscopic parallax," a concept first introduced by W. S. Adams in 1914. For some objects only the distance modulus $m - M$ in magnitudes is used without conversion to parallaxes in seconds of arc, especially if the resulting parallax is very small.

This method furnishes a powerful means of determining stellar distances of faint stars as long as they are bright enough for their spectrum to be observed. Some uncertainties result from the methods of calibration of the scale of absolute magnitudes and spectral classification. If the error of the absolute magnitude of a star determined from its spectrum is ± 0.5 mag., the corresponding error of the computed parallax is ± 20 percent. For a star with small parallax, the accuracy may greatly exceed that of a measured parallax.

20-9   The Hertzsprung-Russell Diagram.   Named after its co-inventors, the Hertzsprung-Russell diagram has proved to be one of the most significant graphs ever plotted by astronomers. The H-R diagram is simply a plot of the absolute magnitudes of stars of known parallaxes against their spectral types. Figure 20-6(a) presents the H-R diagrams of the nearest and brightest stars. It shows the relationship between spectral class, plotted horizontally, and the luminosity (measured by the absolute visual magnitude), plotted vertically.

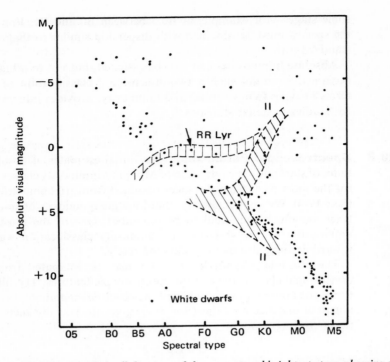

FIGURE 20-6

The Hertzsprung-Russell diagram of the nearest and brightest stars, showing visual absolute magnitude vs. spectral type. The shaded area represents that generally occupied by stars of so-called population II. Note the white dwarfs at the bottom of the diagram.

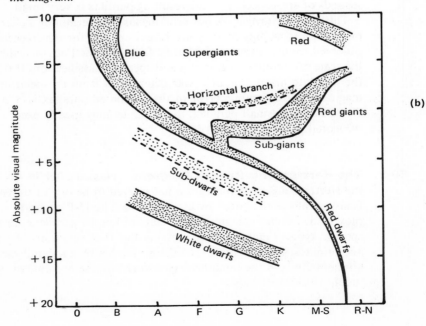

The remarkable feature of the diagram is that the stars are by no means distributed at random. Each spectral class appears to have a preference for certain ranges of absolute magnitude. Since the spectral class of a star is closely related to its temperature, we must conclude that a star of given temperature tends to be confined to some special range, or ranges, of luminosity. In other words, temperature and luminosity are not independent but actually closely related.

We note the remarkable tendency of stars to congregate along a line sloping downward to the right. These stars define the so-called *main sequence*, or branch of dwarf stars of luminosity class V, which, as we shall see, comprises the majority of all stars, including the sun. The members of the main sequence appear to be physically related, as though they belonged to one family. The second important concentration of stars is the branch of the giant stars of higher luminosity than the main-sequence stars and of luminosity class III in Morgan's classification. They are especially numerous from class G and later. Scattered above the giants are the supergiants of even higher luminosity, classes Ia and Ib. Supergiants are found in all classes from O to M but are very rare and not represented in the sampling of the diagram. Between the giants and the main-sequence stars are also scattered a few subgiants of luminosity class IV. Figure 20-6(b) depicts, somewhat schematically, the location of various types of objects on the H-R diagram.

Some stars are extremely faint. Their absolute magnitudes place them far below the vast number of stars that make up the main sequence. Many of these excessively faint stars have moderately high temperatures and are not extremely red, as the main-sequence dwarfs are. Hence Russell called them *white dwarfs*.

Astronomers thought that the H-R diagram, when first drawn and discussed over half a century ago, represented an evolutionary sequence. They theorized that the youngest stars were the luminous red giants and that the right-hand side of the giant branch must represent the beginning of a star's history. The stars were imagined to contract gravitationally, growing progressively denser and hotter to a maximum temperature, then cooler down the main sequence from left to right. This concept now possesses only historical interest. We still think that the H-R diagram defines an evolutionary sequence (or rather, long-lived stages in a series of evolutionary sequences)—but, as we shall see in Chapter 35, an entirely different one. The evidence for our conclusions will appear in the following chapters, especially Chapter 30, where H-R diagrams for clusters represent stars that originated together at the same time.

20-10   Color Index and Spectral Class.   Color index and spectral class are closely related, as shown in Fig. 20.7, which gives the mean color index of stars of various spectral and luminosity classes. It is not surprising that such a relationship exists, since we have already noted that the spectral

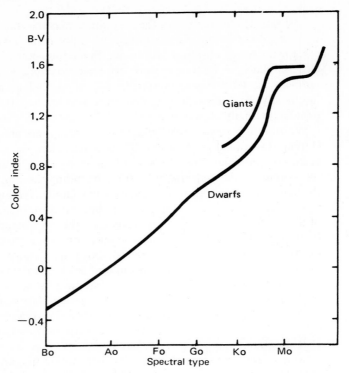

**FIGURE
20-7** *Color index $(B-V)$ vs. spectral type. Note that giants, for a given spectral class, have a higher color index than the corresponding dwarfs.*

type of a star depends on its temperature, and the temperature, in turn, determines the star's color.

We should remember, however, that while the spectral type expresses some unique intrinsic property of a star, conditions other than surface temperature can affect the color. If, for example, the starlight has passed through a region of space that contains considerable interstellar dust, *selective scattering* of the light by the dust may redden it appreciably, although the spectrum remains essentially unchanged (see Chapter 29).

We can measure the colors of faint stars much more readily than we can determine their spectral types. But many faint stars are distant, hence the most likely to have their colors reddened by interstellar absorption. As a result, simple color indices, such as the one defined in Sec. 19-6, convey less information about stars than a knowledge of their spectra, and a diagram based on the ordinary color index is less significant than an H-R diagram. However, observations of the colors of stars have been greatly refined by the precise techniques of photoelectric photometry (see Sec. 19-7). If, instead of determining a simple color index by comparing the

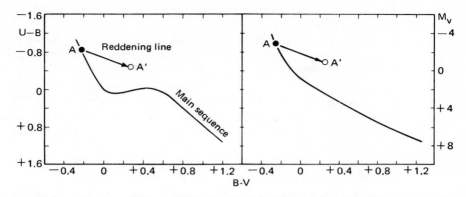

FIGURE 20-8 *Color-color array (left), with $U - B$ color index (ultraviolet minus blue magnitudes) plotted against the standard color index, $B - V$ (blue minus visual). A normal star at A, reddened by cosmic dust, slides downward to A', along the "reddening line." There is a corresponding shift in the color-magnitude array (right) similar to the H-R diagram.*

star's light at two wavelengths only, we use methods similar to the six-color photometry of Stebbins and Whitford or the three-color U,B,V system of Johnson and Morgan (Sec. 19-7), we can often detect the effect of interstellar reddening and place a star of known distance correctly in a color-magnitude array, such as the one shown in Fig. 20-8. Such techniques yield information whose significance is comparable to that of the H-R diagram, and can be extended to stars whose spectra are too faint to record (see Chapter 30).

20-11   Limitations of the H-R Diagram.   Both the H-R diagram and the color-magnitude arrays for stars of known distances are affected by unavoidable errors. The tendency of the points plotted in the H-R diagram to lie in vertical columns results from the fact that we classify a star as either F5 or F6; our usual measures of spectral class are not yet refined enough to permit such classification as, for example, F5.23. The stars that tend to fall in a vertical column should actually be spread out to fill in part of the gap between that column and its neighbors.

Measurements of apparent magnitude, especially those made photoelectrically, can be very precise; hence they do not contribute appreciably to the scatter. However, we plot not apparent, but absolute magnitudes, and can calculate the latter from the former only if we know the star's distance, or we can estimate the absolute magnitude from the luminosity class (see Sec. 20-7). Here there is scope for considerable error because of the uncertainties in stellar parallaxes, especially for the more distant stars, and in the calibration of luminosity classes. In consequence, the position of a star on the vertical scale is subject to an uncertainty that may be considerable.

These sources of error are obvious. But a third effect would convey a very false impression, if we were to make the mistake of thinking that the H-R diagram represents the actual population of interstellar space. To find a place on the chart, a star must appear bright enough for us to determine its spectral class. Second, we must know its distance. Hence, if a star should not be near enough for direct measurement of parallax, we can place it on the diagram only if it is a member of one of special groups of stars whose distances we can determine by indirect means (see Chapter 30). Consequently, the H-R diagram is affected by a marked *selection factor*.

The H-R diagram contains a fairly complete census of the nearby stars. But it overemphasizes the number of the most luminous objects, simply because they can be seen at greater distances. The diagram, therefore, does not represent a true census of the different kinds of stars in space.

It is possible to evaluate the effects of the selection effect on the H-R diagram and to produce a corrected diagram, called the *Hess diagram*, which represents the density of population of the various kinds of stars, defined by their spectra and luminosities. Such a diagram, resembling a contour map of a mountain range, better illustrates that the majority of stars are faint members of the main sequence (the G, K, and M dwarfs), which are by far the most common, if inconspicuous, inhabitants of space around us.

Double Stars and Stellar Masses

21-1 **Discovery of Double Stars.** The discovery of double stars provides one of the more interesting and important chapters of astronomical history. Riccioli, in 1650, was the first person to notice that a certain star, which appeared single to the naked eye, split into a double star under telescopic magnification. Even with the primitive telescopic equipment of his day, he noted that Mizar (ζ Ursae Majoris), the bright star at the bend of the handle of the Big Dipper, appeared double. Mizar has also a faint companion, Alcor, barely visible to the naked eye, which both the Arabs and the North American Indians have traditionally used as a test of keen eyesight. Mizar, when viewed through a telescope (Fig. 21-1), shows this companion widely separated from the bright star, which itself appears as an extremely close pair of stars, nearly equal in brightness. Mizar was also the first double star to be photographed (by G. P. Bond at Harvard Observatory in 1857), and its brighter component was the first star recognized as a spectroscopic binary system, as discussed in a later section.

Halley's discovery in 1718 that the stars are in motion with respect to one another, so that the constellations are gradually changing their shapes, focussed attention indirectly on the relative motions of pairs of stars. By that time a number of double stars had been discovered, but most astronomers assumed that these apparent close associations of two stars were the results of chance. They assumed that the brighter star was near to us, and

Historical (1886) photographs of double stars. Multiple exposures from top down, γ Vir, 58 Crv, ζ U Ma, and Boo (P. and P. Henry, Paris Observatory).

that the fainter was farther away, and just happened to lie in the same direction. Such close pairs of stars, actually far apart in space and lined up by chance, are known as *optical pairs* or *optical doubles*.

As early as 1767, the Englishman J. Michell recognized that most close pairs are probably not all optical. He pointed out that the number of double stars was far too great to be a result of chance. Hence, he reasonably concluded that many pairs are true or *physical binaries*.

In 1779, the German astronomer C. Mayer published the first catalogue of double stars, which listed eighty pairs. But astronomers did not yet appreciate the force of Michell's arguments, for Bode suggested that the relative motions of such pairs would be valuable for the measurement of *proper motion* (see Chapter 19). He, too, apparently believed that the brighter components were near, and should hence show more rapid motion than the fainter ones.

William Herschel also believed that the study of motions of double stars would be important, not only for the determination of proper motions but also for the possible detection of the small annual oscillation of position caused by the motion of the earth about the sun. Such movements, if observed, would lead to a determination of stellar parallax. With this end in view, Herschel began a systematic search for double stars, of which he discovered nearly a thousand. In 1802 he finally decided that many of the double stars were true binaries, since they showed neither large relative motions nor relative oscillations at yearly intervals, but instead in several cases had changed their relative positions as if in orbital revolution.

Thirty years later his son, John Herschel, established a temporary observatory near the Cape of Good Hope in South Africa, and produced in a few years the first extensive list of double stars in the southern sky. Meanwhile the study of binaries continued in the northern hemisphere, especially through the work of Wilhelm Struve and his son Otto, who, at Pulkovo Observatory, catalogued more than 3500 double stars. The modern study of double stars began with the precise measurements of position angle and separation, made possible by the filar micrometer, described in the next section.

21-2   Study of Visual Binaries.   The study of visual binaries has occupied the devoted attention of some of the keenest observers in astronomical history. In modern times, S. W. Burnham, R. G. Aitken, and G. van Biesbroeck have been the leading American contributors.

Approximately 23,000 double stars have been discovered and the list is still growing. The most complete search survey, made by Aitken at the Lick Observatory, yielded 5400 binaries among the 100,000 stars brighter than visual magnitude 9.0. On the basis of visual observations, Aitken estimated that one star in nineteen is a visual double. W. H. van den Bos, working in the southern hemisphere, came to the slightly higher frequency of one in seventeen.

For over a century the human eye remained unexcelled for the measurement of double stars. The observer must be keen-sighted and must train himself for many years. The work requires a large and perfect telescope, operating under atmospheric conditions as nearly ideal as possible to produce steady, well-formed images. The measuring instrument attached to the telescope is the *filar micrometer*. On looking into the eyepiece of the micrometer, the observer sees, in addition to the pair of stars he wishes to measure, a long straight "spider thread," and two shorter threads perpendicular to the first. One of these is stationary, the other movable by means of an accurate screw with a graduated head. To measure a double star, the observer first sets the longer wire, a-a in Fig. 21-2, along the line that connects the two stars. The angle of the line a-a, measured from the north around to the east, defines the *position angle* of the pair. The observer next sets the stationary vertical wire, b-b, on the star that he judges to be the brighter of the pair. Finally, he adjusts the movable thread, c-c, until it crosses the image of the fainter star. Noting the number of revolutions through which he has turned the screw, he can thus compute the spacing, d, between the two wires. Then, knowing the focal length F of the telescope, he can calculate the angular distance $\rho = d/F$ between the two components. The two quantities, position angle θ and distance ρ, together with any information about the colors and relative brightnesses of the two stars, the seeing conditions, and so on, constitute a measure. Position angles vary all the way from 0° to 360°; distances often lie between a few tenths of a second of arc and several seconds. Seldom do they exceed 20″ or 30″.

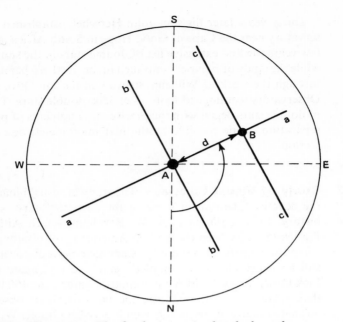

*The Filar Micrometer. The fixed wire as is placed along the
direction AB of the components to measure a position angle, θ.
The movable wires bb, cc are set on A and B to measure their
separation, d.*

The Danish astronomer, E. Hertzsprung, in the 1910's introduced modern photographic techniques for the measurement of double stars. Multiple-
image photographs with automatic cameras provide many measures that,
when combined, exceed the visual techniques in precision. An accuracy of
0".003 or better is thus possible. This type of observation, however, cannot
be made for the closest doubles, whose images are not well separated on
the plate.

In the next chapter we shall discuss the interferometer technique, which
is especially valuable for double stars of small separation, particularly
those with components of nearly equal brightness.

21-3   Orbits of Visual Binaries.   The orbits of visual binaries can be
drawn accurately only when many observations are available, and preferably when the stars have made at least one full revolution. Only a few
hundred pairs have made a complete circuit since the time of the Herschels
and the Struves (Fig. 21-3). A visual double star with a period as short as a
decade is rare. Many have periods of hundreds, or even thousands of
years, so that their orbits will not be precisely known for a long time.

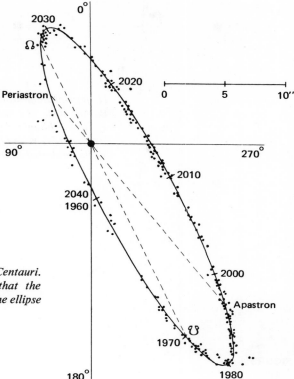

FIGURE 21-3

Observed orbit of the double star, α Centauri. The orbit is elliptical. The fact that the primary does not lie at the focus of the ellipse is an effect of projection.

When the necessary measures are available, one can plot the apparent orbit of the fainter star (the companion) relative to the brighter (the primary). Since we usually determine only the position of the fainter star relative to the brighter, we do not know the motion of each star independently. But this *apparent relative orbit* can still convey much information.

Figure 21-4 shows a schematic example of apparent relative orbit. Note, first of all, that the orbit is elliptical. However, the primary, instead of being at one focus of the ellipse, lies at the center. A little consideration will make the explanation evident. The true orbit must be a circle, with the primary actually at the center of the circle. The elliptical appearance of the orbit is merely an effect of projection (something like Saturn's rings). In fact, we can deduce the angle of projection from the degree of ellipticity. In other words, the ellipticity tells us the angle that the plane of the orbit makes with the observer's line of sight. Once we recognize the influence of projection of the actual orbit, we can proceed to show that the apparent motion must obey Kepler's second law (the law of areas) as well as the first law.

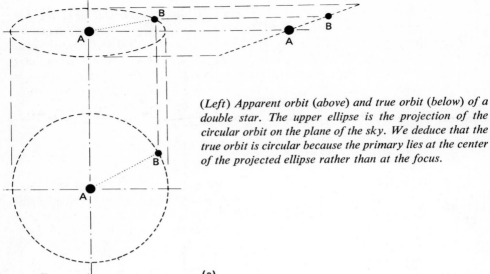

(*Left*) *Apparent orbit* (*above*) *and true orbit* (*below*) *of a double star. The upper ellipse is the projection of the circular orbit on the plane of the sky. We deduce that the true orbit is circular because the primary lies at the center of the projected ellipse rather than at the focus.*

(a)

**FIGURE
21-4**

(b)

(*Right*) *Relative and absolute orbits. The stars* A, B *move around their center of mass,* O, *in orbits of radii* r_A, r_B, *inversely proportional to their respective masses. The observed apparent orbit of* B *around* A *has the same shape but radius* $r_A + r_B$.

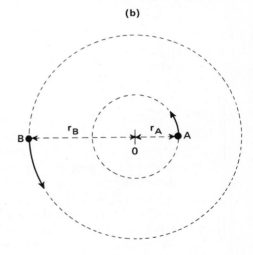

Figure 21-4(a) displays the orbit of the fainter star relative to the primary. Actually, both stars move in similar orbits around their center of mass, the so-called barycenter, with orbits whose radii are inversely proportional to their masses. For example, Figure 21-4(b) shows the circular orbits of two stars A and B around their barycenter O. A, having the larger mass (m_a), will lie closer to the barycenter at a distance r_a. Since the periods of revolution for the two bodies are identical, the linear velocities, v, will also be inversely proportional to the masses. Thus:

$$\frac{m_a}{m_b} = \frac{r_b}{r_a} = \frac{v_b}{v_a}.$$ (21-1)

These relationships indicate that one can determine the mass ratio and relative radii of the orbits by noting the period and measuring the velocities from the Doppler shifts of the two bodies. These relationships also show that the centripetal forces, as required by Newton's laws of motion, are equal and opposite. Hence

$$\frac{m_a v_a^2}{r_a} = \frac{m_b v_b^2}{r_b}.$$ (21-2)

In practice, visual double stars rarely have circular orbits. The orbits are elliptical, as in Fig. 21-3, where the larger and brighter primary lies far from the focus of the observed ellipse described by the secondary. Our example of the circular orbit suggests that here again we have to do with projection. A little experience will prove that, no matter how we tilt an ellipse, the position of its geometrical center remains unchanged. Hence, if we draw a line from the geometrical center of the apparent orbit through the position of the brighter star and out to cross the observed orbit, this line must be the projection of the major axis of the original orbit, even

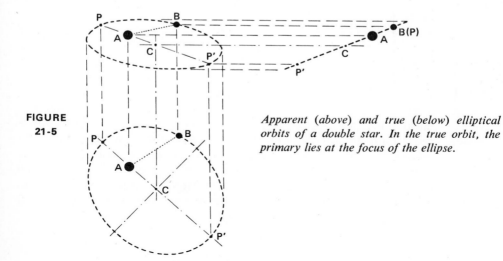

**FIGURE
21-5**

Apparent (above) and true (below) elliptical orbits of a double star. In the true orbit, the primary lies at the focus of the ellipse.

though it may be even shorter than the major axis of the observed ellipse or *apparent* orbit (Fig. 21-5). This fact enables us to carry out our analysis, and it shows again that Kepler's first law is obeyed. We can then verify the applicability of the second law.

Let us return for the moment to the simplest case of all, the circular orbit. We can measure the angular distance between the brighter and fainter stars at maximum separation, the angular radius of the orbit. If we know the distance of the double star (that is, its parallax), we can further calculate how many astronomical units separate one star from the other. Now we can make use of Kepler's third law (Sec. 3-3), which tells us that if this distance were one astronomical unit, and if the brighter star had the mass of the sun and the fainter star possessed negligible mass, the period of the binary would be exactly one year. For then we should have essentially duplicated the conditions under which the earth goes around the sun.

Thus, in general, if we can determine the apparent semimajor axis a'' of an elliptical orbit in seconds of arc, and if we know the parallax p'', we can calculate the actual separation, s, of the stars in astronomical units by the simple relation:

$$s = \frac{a''}{p''}.$$

(21-3)

We could employ the nomogram of Chapter 4 to calculate the mass. However, we note that the mass so determined is not the mass of either of the individual components, but their sum. And, as previously noted, we can determine the individual masses only if the absolute orbits have been determined by independent measurements—that is, by reference to other stars in the field.

The orbits of double stars of known parallax have proved to be extremely important for our understanding of the physical nature of the stars. The data are consistent with the idea that most stars have masses of the same order of magnitude as the sun, some of them smaller, some larger. In the absence of additional information, we should not make an enormous error if we were to assume that all stars have masses approximately equal to that of the sun.

Often we have no direct information about the parallax of a double star. Consequently, we cannot employ equation (21-3) to convert the measured separation in seconds of arc to the actual separation in astronomical units. However, we can put our data to another use. Even with a rough estimate of the masses, such as the assumption that the sum of the masses is equal to twice the mass of the sun, we can deduce the actual separation of the stars in astronomical units from a knowledge of the period. Then, knowing the apparent separation in seconds of arc, we can reverse equation (21-3) and calculate the parallax. Parallaxes so determined are called *dynamical parallaxes*.

The more information we get concerning the orbits of double stars, and the more complete orbits we measure precisely, the more direct information we shall have concerning the masses of stars. The *only* stars whose masses we can measure are members of binary systems. But, as we shall see, visual binaries are but one type of double star.

21-4   Spectroscopic Binary Stars.   Spectroscopic binary stars are double stars that appear single in the largest telescopes, but whose duplicity becomes apparent from periodic changes in their spectra. As previously mentioned, the brighter component of the double star Mizar, ζ Ursae Majoris, was the first star to be recognized as a spectroscopic binary. In 1889 E. C. Pickering observed that the spectral lines of this star occasionally appeared double, at other times single. On further study, the changes showed themselves to be periodic. The lines doubled at intervals of about ten days. Recognizing that the Doppler effect had caused the doubling, he interpreted the changes in terms of orbital revolution. When the double lines appeared, one star of the pair was receding, the other approaching. Since the stars had identical brightness and spectrum, this doubling occurred twice in one orbital revolution. Hence Mizar A (as the brighter of the two visible components is called) is a binary with a period of $20\frac{1}{2}$ days.

Mizar A looks like a single star in any telescope. Mizar B, at a present distance of $14\overset{''}{.}5$, takes at least 3000 years to revolve around A. But the spectrograph proves that Mizar A itself consists of two stars, about equal in brightness. Alcor, the faint naked-eye companion already mentioned (Sec. 21-1), distant $11'$, seems to share the proper motion of Mizar and so must be related to it. Here we have a group of at least four associated stars, and possibly five if Mizar B should also prove to be a spectroscopic binary. In rare instances double stars have been observed both as spectroscopic and as visual binaries, just resolved by the largest telescopes.

Accurate spectroscopic observations give a determination of the velocity changes of a spectroscopic binary. In Mizar A the spectra of both components are observable. We often observe similar changes where only one spectrum is visible. The star seems to recede for a time, then becomes stationary (all with respect to the average velocity of approach or recession), before the sense of motion is reversed, and these changes are repeated at regular intervals. Even though only one spectrum is visible, we infer that a second star is indeed moving about the center of gravity of the system, its spectrum invisible, perhaps, because the star is considerably fainter than its primary. We cannot determine the velocity or the mass of this invisible star. It might be a star of small mass, moving with high velocity in a large orbit, or (less probably) a massive star moving slowly in a small orbit.

Measures of the velocity, made at intervals during the orbital period, are used to graph the *velocity curve* (Fig. 21-6). The velocity curve defines

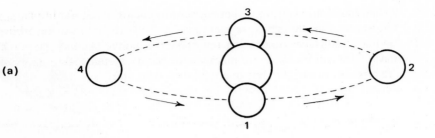

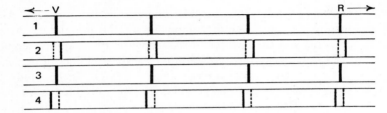

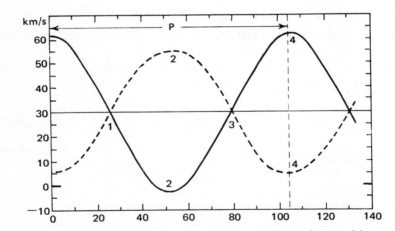

(a)

(b)

(c)

FIGURE 21-6 *Velocity curves of the components of a spectroscopic binary. (a) Orbital positions, (b) schematic spectrum. Note the doubled spectral lines at positions 2 and 4. (c) Velocity curves.*

the star's motion. Suppose for simplicity that both stars are visible, that their orbits are circular, and that they are oriented exactly edgewise. Then V_1, the maximum velocity of the first component, equals half the difference between the maximum and minimum observed speeds. For example, if the maximum speed (of recession) is $+26$ km/sec, and the minimum speed (of approach) is -14 km/sec, then $V_1 = 20$ km/sec, and the radial velocity of the center of gravity of the double system is $+6$ km/sec. Multiplying the velocity, V_1, by the period, we obtain the circumference of the orbit described by the first component; finally we get the radius a_1 of this orbit by dividing by 2π. We could treat the velocity V_2 of the second component in the same way, and thus find the radius a_2 of the orbit of the second component. Just as for a visual binary, these orbits are both described about the center of gravity of the system. The radius of the relative orbit would clearly be $(a_1 + a_2)$. If now this relative orbit is expressed in astronomical units, we can use the nomogram of Fig. 4-6 to read off the *sum of the masses*, since orbit radius and period are known.

We can also determine the individual masses, m_1 and m_2, by equation (21-1).

If, however, we can observe only one spectrum, we can find the radius only of the orbit of the visible star, and can determine neither the sum of the masses nor their ratio, but only a quantity known as the *mass function*, which involves the sum of the masses. By making plausible guesses as to the ratio of the masses, we can extract some information from even a single velocity curve.

Even spectroscopic binaries with both components visible present some unavoidable difficulties. In the simplified example just discussed, we supposed the orbits to be exactly edgewise, but such alignment happens only rarely. What is worse, the mere appearance of the velocity curve does not reveal whether, for example, the stars are moving at low velocities in an orbit whose plane lies directly in our line of vision, or whether they form a system whose true velocities are very much higher, but where orbital tilt has reduced the components of measured velocities on our line of sight. Unless we have direct and independent means for determining the orientation of the orbits, the uncertainty about their *inclination* will always enter into the determination of the masses. On the other hand, the eccentricity of the orbit can be deduced from the shape of the light curve. Even without knowing the inclination, one may obtain useful information about the mean masses for a number of stars by making plausible guesses about the average inclination of their orbits.

Despite these uncertainties, spectroscopic binaries have given us a vast amount of important data on binary systems. The distribution of orbital periods and eccentricities among binary stars of different spectral class, for example, contributes significantly to the study of the evolution of double stars.

For a limited group of spectroscopic binaries, described in the next section, the orbital inclination can be determined without ambiguity.

21-5 **Eclipsing Binaries.** Eclipsing binaries are, in a sense, a special group of spectroscopic binaries, whose orbital planes lie so near the observer's line of sight that the components pass in front of one another periodically.

The resulting changes of brightness enable us to recognize that an eclipse has occurred.

The first known eclipsing binary was the bright star Algol, or β Persei, which undergoes eclipse with a period of 2.86 days. The decrease in brightness is easily detectable, since it amounts to 1.1 magnitudes. The *light curve* of Algol appears in Fig. 21-7. One of the striking features of the Algol variation is the sharpness of the minimum, and the nearly constant

FIGURE 21-7 *The eclipsing system, β Persei, (Algol) and its light curve.*

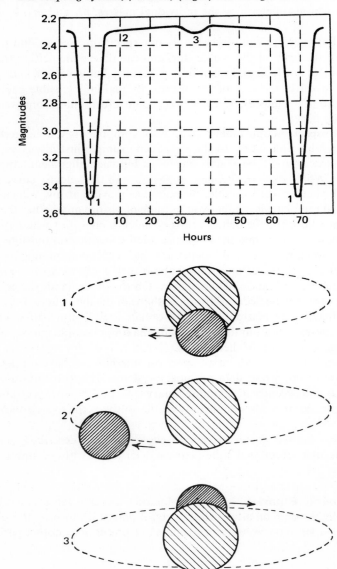

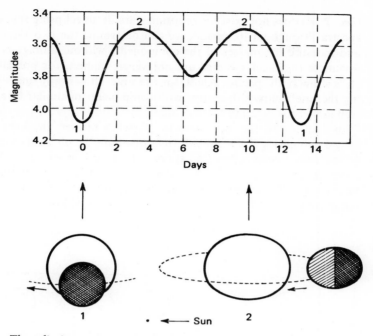

FIGURE
21-8 *The eclipsing system, β Lyrae and its light curve.*

brightness between eclipses. Another barely perceptible eclipse occurs between the deep minima. Algol differs in this respect from another naked-eye eclipsing star, β Lyrae, whose brightness varies continuously between eclipses (Fig. 21-8). The total variation of β Lyrae is only 0.7 mag.; its orbital period is 12.92 days, but eclipses occur at half this interval.

The contrast between Algol and β Lyrae is a result of the differences between their component stars. Algol consists of a large, faint star eclipsing a smaller bright one at intervals of nearly three days. About midway between the major eclipses, the brighter star eclipses the fainter one, but so little light is lost that the second eclipse is very shallow. β Lyrae consists of two stars more nearly equal in brightness, so that each one causes a perceptible eclipse as it passes in front of the other. The components of β Lyrae, moreover, are so close to one another that mutual gravitational attraction distorts their figures into lemon-shaped objects, whose cross section continually changes, with consequent variability of light intensity even outside of eclipse. The components of Algol lie farther apart and undergo only very slight distortion.

The fact that the components of such a system eclipse one another shows that we are very nearly in the plane of their orbits. A study of the details of the light variation determines the exact tilt of the orbit. The uncertainty caused by the unknown orientation of the orbit in computing stellar masses

(Sec. 21-4) does not exist for eclipsing stars. If an eclipsing star shows two spectra throughout its period, we can determine the two velocity curves and measure the masses of the components without ambiguity. It is unfortunate that such systems are not very common, for not all eclipsing stars show both spectra; Algol, for example, shows only one, except during the brief eclipses. But stars with components of similar brightness are well suited for determination of mass.

In addition to the measurement of stellar masses, the light curves of eclipsing binaries give much information about the physical characteristics of the stars that compose the system. To take a simple case (Fig. 21-9), suppose that the orbits are circular and that a large, faint star revolves about a smaller bright one. Then, during the time that the darker star completely covers the bright one, the brightness will be constant—that is,

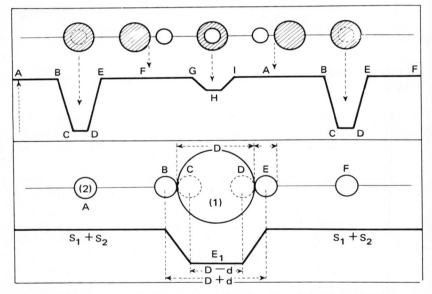

**FIGURE
21-9**

*Schematic light curves of eclipsing binary. In the upper diagram both components are visible for the intervals **AB**, **EFG**, and **IAB**. During the interval **BC**, the larger dark star is covering the smaller bright one. During totality **CD**, the light is constant, coming only from the darker component. The interval **GHI** represents the partial eclipse of the dark star by the smaller brighter one.*

*The lower diagram shows details of the primary eclipse, which begins when the smaller bright star is at **B**. Totality begins when the smaller star reaches the position **C** and ends when it reaches **D**. The durations of totality and of the partial phases measure the diameters of two stars. The depth of the eclipse measures the relative luminosities of the two stars.*

the eclipse is a total one. Another interval of equal length will occur half-way between the main eclipses, when the bright star passes in front of the faint star. This eclipse will be annular and the loss of light may be too small to be observed. A small variation of light during the annular phase gives information about the distribution of brightness over the surface of the larger star, which may, like the sun, show limb-darkening (see Sec. 9-1). Under such circumstances the primary eclipse shows no such effect. The whole duration of the eclipse from beginning to end depends on the sum of the radii of the stars; the length of time taken to go from maximum to minimum (the partial phase) depends on the difference of the radii. This interval is least when one star is very much smaller than the other. Thus the shape of the eclipse curve yields data about the dimensions of the two stars. The relative depths of the main (or primary) and intermediate (or secondary) eclipses are proportional to the surface brightnesses of the two components when the eclipses are total. Application of the Stefan-Boltzmann law therefore leads to a determination of the ratio of their surface temperatures.

The circular orbit that we have just described represents the most usual case for eclipsing stars. A few, however, have elliptical orbits. If the major axis of the ellipse is perpendicular to the line of sight, the large star will take longer to go from secondary minimum to primary minimum than from primary to secondary. The secondary minimum may thus occur asymmetrically between two successive primary minima. This asymmetry furnishes a measure of the shape of the ellipse. If the major axis of the ellipse is pointed toward the observer, then, by Kepler's second law, the fast-moving fainter star will pass through the primary eclipse more quickly than it will through the secondary. The eclipses will therefore have unequal durations. In most actual cases of elliptical orbits a combination of the two effects occurs as we encounter the different orientations of the ellipse in space. A few elliptical orbits turn regularly around in space, and the effects just described are then continually changing. But most eclipsing binaries have nearly circular orbits.

When the orbits are tilted so that the eclipses are no longer total, we observe continuous variation of brightness during eclipse, and the eclipse has no flat bottom, as in Fig. 21-7. Among stars that show continuous variation of light, the β-Lyrae type (Fig. 21-8), radiation from the surface of the brighter component may be absorbed or reflected by the surface of the other component. In consequence, when we see the full face of the latter star just before secondary minimum, its light may be slightly increased. This *reflection effect* is perceptible for Algol (Fig. 21-7).

The precise mathematical methods that we use today for analyzing an observed light curve and determining the size of the various effects listed above were developed mainly by H. N. Russell and H. Shapley in the 1910's. The data derived from precise light curves of eclipsing stars have provided us with our most direct information about the geometrical properties of stars. And, since many eclipsing binaries are also spectroscopic binaries, the additional information that we get from combining the two types of observations is extremely important.

21-6 Astrometric Binaries. Some stars have companions too close and/or too faint to be resolved optically with existing telescopes. The presence of such unseen companions can sometimes be detected by their gravitational action on the motion of the visible primary. The effect resembles somewhat the discovery of Neptune through the study of the perturbations of the motion of Uranus. The German astronomer F. W. Bessel noticed in 1834 that the proper motion of Sirius did not lie along a perfectly straight line, but was slightly wavy (Fig. 21-9). In 1840, he found a similar motion for Procyon. He concluded correctly in 1844 that the perturbation in the motions of these two bright stars resulted from the action of invisible companions. As for the earth-moon system (Chapter 5), the two components revolve about their common center of gravity. If the visible body deviates from a straight-line path, it must be describing a Keplerian orbit about the center of gravity formed by itself and an invisible companion. Bessel wrote: "I am convinced that Procyon and Sirius form true binary systems composed of one visible and one invisible star. There is no reason to suppose that luminosity is an essential characteristic of the heavenly bodies. The visibility of a multitude of stars is no evidence against the invisibility of a multitude of others."

In 1851 another German astronomer, C. A. J. Peters, proved that the observed motion of Sirius could be completely explained by orbital motion about the center of gravity of the system. In 1862, the American optician, Alvan Clark, while testing an 18-inch objective, then the largest astronomical lens in existence, discovered visually the "invisible" companion of Sirius. The powerful instrument disclosed a tiny satellite star (ten thousand times fainter than Sirius) in the exact position predicted by Peters' calculations.

History repeated itself almost exactly for Procyon, whose theoretical orbit was computed in 1862 by the German astronomer G. F. A. Auwers. J. M. Schaeberle, using the 36-inch refractor of Lick Observatory, discovered the companion visually in 1882 (in the predicted position). Since that time, perturbed motion has been detected in many double stars whose orbital paths appear slightly wavy when subjected to highly precise measurement and calculation.

The companions of Sirius and of Procyon have proved to be of outstanding significance, not only in astrometry but also and specially in astrophysics. Both stars proved to have abnormally low luminosities for their masses, which are about equal to that of the sun. They are the first two known examples of the so-called "white dwarfs" (sec. 20-9), stars that do not conform to the relation between mass and luminosity applicable to ordinary stars as discussed in the following section.

21-7   Mass-Luminosity Relation.   The mass-luminosity relation is the most important result of the study of double stars. The English astrophysicist Arthur Eddington, in his pioneering studies of the internal constitution of stars (see Chapter 23), concluded in 1924 that a fundamental relation must exist between the masses and luminosities of stars of similar structure. The more massive the star, the more luminous it should be. Eddington collected all the data on stellar masses and luminosities available at the time, concluding that they supported his contention. The data that have now been accumulated are shown in Fig. 21-10, which represents convincing evidence that such a relationship exists, at least for stars of the main sequence.

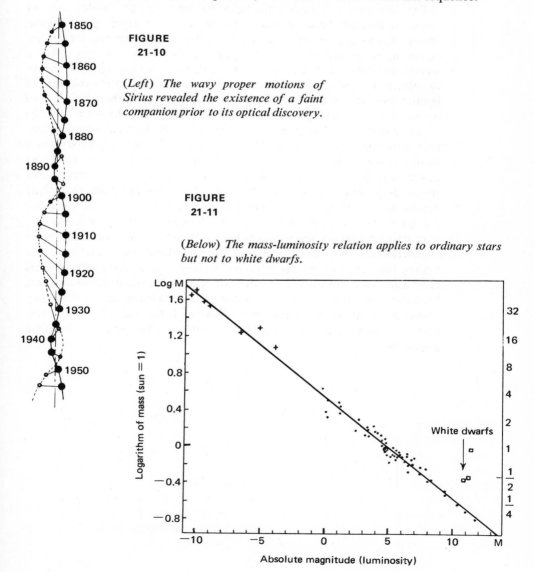

**FIGURE
21-10**

(*Left*) *The wavy proper motions of Sirius revealed the existence of a faint companion prior to its optical discovery.*

**FIGURE
21-11**

(*Below*) *The mass-luminosity relation applies to ordinary stars but not to white dwarfs.*

It is not surprising that such a relationship governs stars of similar structure and chemical composition, for the mass of a star will largely determine the temperature at its center (Chapter 23), and the central temperature in its turn may be expected to govern the rate of release of energy, which, of course, determines the star's luminosity. The nearly straight line shown in Fig. 21-11 may be represented by the simple equation

$$L = M^{3.82} \qquad \text{or} \qquad \log L = 3.82 \log M, \qquad (21\text{-}4)$$

where L is the luminosity and M the mass, expressed in units of the sun's luminosity and mass.

The data used to establish the equation, however, are limited to stars whose masses can be determined—double stars for which we know the orbital dimensions and inclinations. With very few exceptions, these stars lie on the main sequence. Strictly speaking, therefore, our data define a mass-luminosity relation for main-sequence stars, which we have reason to suppose possess similar structure. As more data on the masses of stars became available, a number of marked deviations from the relation were found. The most conspicuous discrepancies occur for the white dwarfs, which possess unusual properties, especially of chemical composition, such as scarcity of hydrogen. They certainly differ radically in structure from main-sequence stars. White dwarfs are far less luminous than our equation would predict. Another group of stars, the *subgiants* that lie between the giants and the main sequence, are more luminous than their masses would indicate. We cannot measure the masses of numerous other groups of stars, such as the intrinsic variables and all the nonbinaries.

In later chapters, when we study the evolution of stars, we shall find as a general rule that deviations from the mass-luminosity relation occur among the stars that are far advanced in their evolutionary history. This conclusion need not surprise us when we realize that the structure of a star probably undergoes large changes during its development, for the mass-luminosity relation was originally predicted only for stars of similar structure. It is well established only for the "rank-and-file" stars of the main sequence.

CHAPTER 22

The Diameters and Densities of Stars

22-1 The Diameters of Stars. Stellar diameters can be estimated from a knowledge of their total luminosities and surface temperatures—for the total light emitted by a star is simply the sum of the light emitted by each unit area of its surface. Suppose that we know a star's temperature from its spectral class, its color index, or the distribution of energy with wavelength. If the star radiates according to Planck's law—and in such determinations of temperature we have assumed that it does—then the brightness per unit area of the surface at any given wavelength depends simply upon the temperature, and can be calculated from the Planck formula [see Figs. 8-4 and 22-1(a)]. The total brightness of the star at a given wavelength is simply the product of this *surface brightness* (or luminance) and the star's surface area. If L is the star's luminosity, R its radius, and l the luminance,

$$L = 4\pi R^2 l. \tag{22-1}$$

If we know L from the absolute magnitude and l from the temperature, R can readily be calculated.

The relationship between stellar radius, color index, and absolute visual magnitude appears in the nomogram of Fig. 22-1(b).

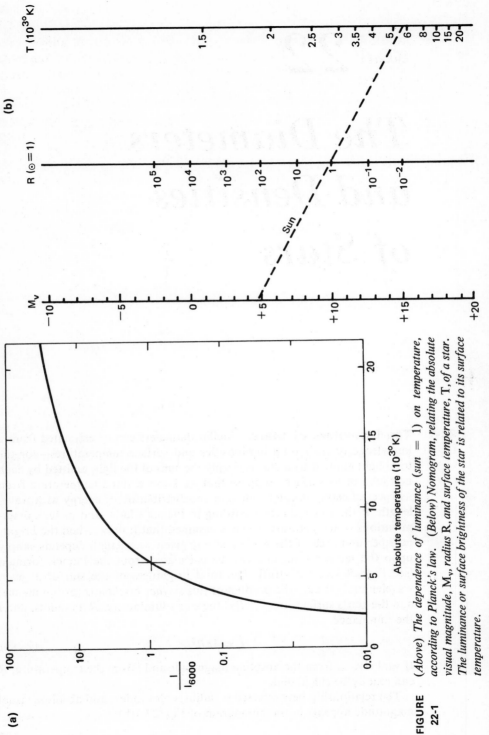

FIGURE 22-1 (*Above*) *The dependence of luminance (sun = 1) on temperature, according to Planck's law.* (*Below*) *Nomogram, relating the absolute visual magnitude, M_v, radius R, and surface temperature, T, of a star. The luminance or surface brightness of the star is related to its surface temperature.*

(a)

100
10
1
0.1
0.01

$\dfrac{I}{I_{6000}}$

0   5   10   15   20
Absolute temperature (10^{3}°K)

(b)

M_v

−10
−5
0
+5
+10
+15
+20

R (⊙ = 1)

10^{5}
10^{4}
10^{3}
10^{2}
10
1
10^{-1}
10^{-2}

Sun

T (10^{3}°K)

1.5
2
2.5
3
3.5
4
5
6
8
10
15
20

22-2 **Angular Diameters of Stars.** However, if we do not know the distance of the star, or its absolute magnitude, what information about stellar diameters can we derive from a knowledge of color index and apparent visual magnitude? The problem resembles that of finding the distance between the components of a double star, except that we are now observing a remote stellar disk instead of an orbit in space. For the double star we found that the distance between the components is given by the ratio of the angular separation to the parallax [equation (21-1)]. For the size of the star we find a similar relation: the actual diameter equals the ratio of the star's *angular diameter* to its parallax. As in the formula for double stars (21-1), the diameter is expressed in astronomical units.

For a stellar disk of apparent diameter d seconds of arc, at a distance $1/p$ parsecs, where p is the parallax in seconds of arc, the diameter D is given by the relation

$$D = \frac{d}{p},$$ (22-2)

where D is expressed in astronomical units,

or

$$D = \frac{107d}{p},$$ (22-3)

where D is expressed in terms of the sun's diameter.

The diameters of some typical stars can be read immediately from the nomogram of Fig. 22-1. Table 22-1 gives for some bright stars the name of the object, its apparent visual magnitude, absolute visual magnitude, spectral class, color index, linear diameter expressed in terms of the sun's diameter, apparent diameter in seconds of arc, density in terms of the sun's density, and mass in solar units.

Note that the line in the nomogram connecting the points representative of the sun's absolute visual magnitude, 4.8, and its color index, $+0.64$, interesects the diameter axis at $D = 1$, as required. With the scales shown, we cannot plot the sun's apparent magnitude, -27.8, on the magnitude scale. If, however, we extend this scale an appropriate distance, and draw the line between that point and that corresponding to the sun's color index, we shall obtain a value of 1920″ for the angular diameter of the sun, in agreement with the sun's measured apparent diameter of 32′.

For a given absolute magnitude and for stars of high luminosity, the diagram indicates that the larger the color index (the redder the star), the greater its diameter will be. Thus we find that red giant stars such as Antares, Betelgeuse, and Mira Ceti tend to have the largest diameters, which in some cases exceed that of the earth's orbit (Fig. 22-2). Such stars should also possess the greatest angular diameters, about 0″.05, but even the biggest existing telescope could not show them as perceptible disks.

TABLE 22-1 *Calculated diameters and densities of stars*

Star	m_v	M_v	S_p	C	$D,$ $\odot = 1$	$d,$ seconds	$\rho,$ $\odot = 1$	Mass, $\odot = 1$
Main Sequence								
α Vir	1.03	−3.3	B1 V	−0.35	8	0.001	0.05	(13)
α Lyr	0.00	0.6	A0 V	−0.13	2.4	0.003	0.22	(3.0)
α CMa A	+1.45	1.5	A1 V	−0.15	1.8	0.006	0.42	2.35
α Aql	0.74	2.39	A7 V	+0.09	1.4	0.003	0.62	(1.7)
α CMi	0.35	2.65	F5 IV	+0.31	1.9	0.005	0.35	1.74
α Cen A	−0.02	4.35	G2	+0.7	1.0	0.007	1.1	1.10
70 Oph A	4.05	5.7	K0 V	+0.84	1.0	0.002	0.9	0.89
61 Cyg A	5.25	7.65	K5 V	+1.05	0.7	0.003	1.7	0.58
Kruger 60A	9.9	11.9	M3		0.35	0.0008	6.3	0.27
Barnard	9.6	13.2	M5	+1.57	0.15	0.0007	(53)	(0.18)
Giants								
α Aur A	0.06	−0.5	G1 III	+0.74	17	0.012	5.10^{-4}	2.5
α Boo	−0.06	0.2	K2 III	+1.21	29	0.030	$1.4.10^{-4}$	(3.5)
α Tau	0.87	−0.6	K5 III	+1.59	85	0.040	6.10^{-6}	(3.7)
α Sco	0.98	−4.0	M1 In	+1.72	520	0.050	$1.4.10^{-7}$	(20)
α Ori	0.7	−4.1	M2 Ib	+1.74	580	0.060	$1.0.10^{-7}$	(20)
White Dwarfs								
α CMa B	8.6	11.4	wA5			0.022	9.10^4	0.99
40 Erid B	9.7	11.2	wB9			0.018	7.10^4	0.41
Van Maanen	12.3	14.2	wG	+0.35		0.007	4.10^5	(0.14)

Although, in principle, the 200-inch telescope should be able to detect stellar angular diameters of this order, optical aberrations and especially turbulence of the terrestrial atmosphere prevent the instrument from achieving its theoretical resolution. The problem can, however, be solved by means of the stellar interferometer.

22-3   The Interferometer. The interferometer has had many uses in astronomy. Its operation depends on the wave properties of light. When two waves of light are out of step, or out of phase, they "interfere" and cancel each other. When they are in step, they reinforce one another. In a plane where the relative phase of two waves varies, one observes a pattern of alternating dark and bright bands or fringes. The American physicist A. A. Michelson first applied this property to the detection of small angular separations, both of double stars and of the edges of stellar disks.

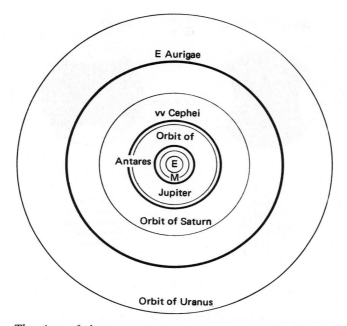

FIGURE
22-2 *The sizes of three supergiant stars compared with orbits of various planets of the solar system.*

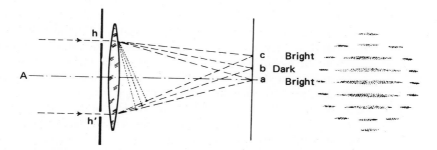

FIGURE
22-3
(a)

Principle of interferometer. Light beams from star A, *passing through two holes* h *and* h′, *equidistant from the center of the lens, reinforce one another at* a, *for which the distance* h′a = ha, *and a bright fringe appears; at* b, *for which* h′b − hb *is equal to a half wavelength of light, the two beams cancel one another and produce a dark fringe. When* h′c − hc *is a full wavelength, the beams again reinforce to produce a bright fringe. Alternate bright and dark fringes cover the diffraction image of the star.*

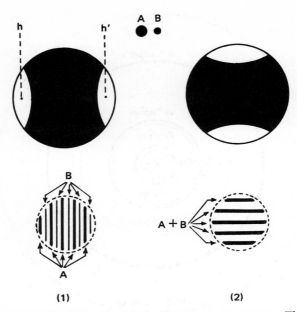

FIGURE 22-3 (b) *Application of interferometer to double-star measurements. The appearance of interference fringes depends on the angle between the line hh' through the centers of the apertures and the line joining the stars AB. The left-hand photograph (1) shows the two sets of fringes from the two stars. At the right (2), where hh' is perpendicular to AB, the two sets of fringes are superposed.*

As shown in Fig. 22-(3a), the objective of the telescope forms a diffraction image of a star *A* at the point *a*. The image, as magnified by an eyepiece, is circular, and to the eye it appears as a more or less uniform bright disk. If now we cover the lens of the telescope with a cap pierced by two circular holes, *h* and *h'*, not only will the light reaching the focus be greatly diminished in intensity, but the appearance of the image will be strongly modified. At the very center *a* of the diffraction image, now increased in diameter because the holes are smaller than the objective, the distance that light has traveled from the star and through the holes to the point *a* is exactly equal for both holes. The waves of light in the two beams will thus be in step—that is, in the same phase—and will add up to produce a bright spot, or rather a bright line perpendicular to *hh'*, in the center of the image. However, when we consider the point *b*, a small distance from *a*, we see that the path length *Ahb* through one hole will be somewhat less than *Ah'b*, through the other. We can find a position where the waves of light that have come through the two holes will be completely out of step (in opposite phase). The waves will therefore cancel one another and produce a dark streak. At a still more distant point, *c*, the waves will again be in step (their phases differing exactly by one complete cycle of vibration), and thus a bright streak will result, and so on. The diffraction disk of the star, which would have appeared almost uniformly bright before the cap was put on the telescope, will thus be covered by a series of alternate bright and dark bands, called *interference fringes* because they arise from the interference of two beams of light. The fringe spacing corresponds to the diffraction disk from an aperture whose diameter is equal to the separation of the holes. The introduction of the two holes has provided a powerful tool, for it has produced a pattern that depends critically on the precise direction of the star.

Let us now suppose that a second equally bright star B lies close to A, so placed that its bright fringes fall exactly upon the dark fringes in the image of the first star. The two patterns will add up to give a uniformly illuminated image. But the image will be uniform only if the spacing $h - h'$ is just right. If the distance is too small, the fringes will not cancel; if it is too large, they will not overlap exactly, and will not cancel either. Hence, if we make the holes movable, and adjust their separation until the interference fringes disappear, we can determine the angular distance between the two stars. This principle permits the measurement of close double stars with the interferometer, a topic mentioned in Sec. 21-2. The bright star Capella is a spectroscopic binary with components of nearly equal brightness, but too close together to be separated optically. P. W. Merrill, at Mount Wilson Observatory, used the interferometer to measure their angular separation, which never exceeds $0''.05$. The bright component of Mizar, which we discussed among the spectroscopic binaries, has also been shown, by means of the interferometer, to have an angular separation of about $0''.01$.

The interferometer can also measure the angular diameter of a single star, if the star is large enough. To understand the principle we may suppose that the two halves of the stellar disk act something like the two images of a binary. The interference fringes will disappear only when the centers of gravity of the two halves of the disk produce interference patterns that cancel one another, in the same way as for a double star. The distance between the holes can be adjusted, as before, until the fringes disappear. But in such an arrangement the distance between the holes is limited by the diameter of the telescope objective, and hence one cannot greatly increase the resolving power of the telescope by this means.

One can, however, effectively increase the diameter of the telescope, and therefore its resolving power, by using the arrangement shown in Fig. 22-4, where m and m' are small mirrors, inclined at 45° as shown. Instead of the two holes h and h', we have now introduced two more mirrors, k and k', each capable of being moved in and out along rigid tracks. The beams from the star, after striking the mirrors k and k', travel to the mirrors m and m', and thence into the telescope, where they come to a focus as before. We have, in effect, increased the diameter of the telescope so that it equals the spacing of the two mirrors k and k'.

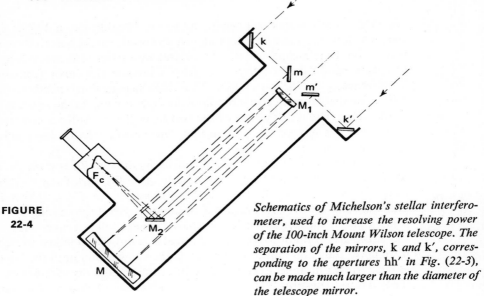

FIGURE
22-4

Schematics of Michelson's stellar interfero-meter, used to increase the resolving power of the 100-inch Mount Wilson telescope. The separation of the mirrors, k and k', corres-ponding to the apertures hh' in Fig. (22-3), can be made much larger than the diameter of the telescope mirror.

A heavy metal beam 20 feet long, attached to the 100-inch telescope at Mount Wilson Observatory, permitted the determination of the angular diameters of a number of red giant stars in the 1920's (Table 22-2). The red star Betelgeuse has, on two occasions, shown appreciably different dia-meters. The fact that Betelgeuse varies slightly in brightness lends some plausibility to the reported changes in diameter. These measures, made by F. Pease, were extremely difficult and have never been duplicated. However, other methods confirm the order of magnitude of the stellar diameters he measured.

TABLE
22-2

Diameters of red giants measured with the Michelson interferometer[a]

Star	m_v	Sp	Angular d''	π''	$D,$ $\odot = 1$
α Boo	−0.06	K2 IIIp	0.020	0.10	22
α Tau	0.87	K5 III	0.020	0.048	45
α Ori	0.7	M2 Ib	0.034–0.047	0.009	400–560
α Sco	0.98	M1 Ib	0.040	0.008	540
β Peg	2.52	M2 III	0.021	0.025	90
α Her	3.5	M8	0.030	0.003	(1000)
o Cet	1.9	M6e	0.056	0.013	460

[a] D calculated according to equation (22-2) with factor $\times 107$ to obtain D in terms of sun diameter (= 1/107 of one A.U.) as a unit.

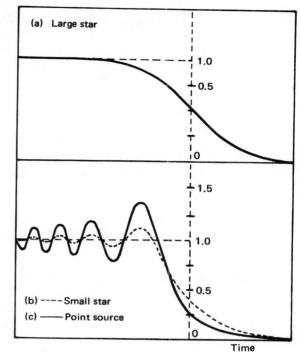

**FIGURE
22-5**

Measurement of stellar diameters by lunar occultations. (a) When a large star is occulted by the moon it fades gradually. (b) A small star disappears in a series of oscillations caused by diffraction effects at the edge of the moon. (c) A point source gives much more pronounced oscillations.

22-4 Lunar Occultations of Stars. Occultations of stars by the moon can be used to drive stellar diameters. The atmosphereless limb of the moon acts as a sharp edge, cutting across the light beam from the star to the telescope. If a star has a small but appreciable apparent diameter, its disk is not instantaneously occulted by the edge of the moon. During the time it takes for the lunar limb to move from one edge to the other of the stellar disc, the light fades progressively, though rapidly. This fading can be recorded by means of a photoelectric cell attached to the telescope and receiving the starlight through a small hole placed in the focal plane of the telescope. The photocurrent is displayed on the moving chart of a fast recorder or on the screen of an oscilloscope (cathode-ray tube) (Fig. 22-5a).

In practice, the apparent diameter of the star is usually so small that it is essentially a point source of light. Thus, the shadow of the edge of the moon projected by the star on the surface of the earth is outlined by *diffraction fringes*. Such fringes can be readily photographed at the edge of the shadow of any obstacle in a beam of star light. As the fringes sweep across the telescope aperture, the output of the photocell displays a series of maxima and minima (Fig. 22-5b) in rapid succession before the star vanishes from view. The relative intensities of these maxima and minima can be calculated exactly from diffraction theory. If, however, the apparent diameter of the star is not completely negligible—that is, if the star has an appreciable disk—the fringes produced by the various parts of its disk tend to overlap, and the observed diffraction pattern is somewhat blurred. The contrast between maxima and minima is not as strong as predicted by theory for a perfect point source (Fig. 22-5c). From the loss of contrast in the fringes the diameter of the star can be estimated.

The lunar occultation was first demonstrated by the French physicist A. Arnulf and by the American astronomer A. E. Whitford in the 1930's. It was further applied by the British astronomer D. S. Evans at Pretoria and the Cape, in South Africa, in the 1950's, to measure the diameter of Antares and other giant stars. His results are in good agreement with the values derived by Pease with the interferometer. Unfortunately, the lunar occultation method can be used only in a limited area of the sky, and the observer cannot choose the moment of observation—for example, to measure the variable diameters of pulsating stars.

22-5   The Intensity Interferometer.   The intensity interferometer for measuring stellar diameters is a sophisticated optical-electronic method developed and applied with great success by R. Hanbury Brown and R. Q. Twiss at Narrabri, Australia. Their method involves two large parabolic reflectors of diameter 6.5 m (256 in.), movable along a circular track of diameter 188 m (617 ft). In each reflector the light from the star passes through a light filter with a narrow spectral band (about 50 Å) and registers on a photocell. The amplified outputs of the two photocells are multiplied together electronically to measure the correlations of the fluctuations of light in the two telescopes down to times as short as a millionth of a second and averaged for 100-second intervals along with the total light-flux measures. In any one series of measures the separation of the reflectors is kept constant, and they move on the track so that the path difference of the light does not exceed 10 cm (corresponding to 3×10^{-9} second in light time).

For a star of small apparent diameter when the dishes are close together the inherent fluctuations in the light from the star register together in each receiver and the signal multiplication sum is a maximum. This corresponds to maximum interference-fringe visibility in the stellar optical interferometer. As the reflectors separate (up to 188 m), fringes could disappear in the use of the optical interferometer because light from different sides of the stellar disk becomes out of phase, owing to the different path lengths. In a difficult and complicated theory Hanbury Brown and Twiss show that the light fluctuations follow the fringe pattern in their correlation between the two reflectors. Lack of phase correlation reduces the signal multiplication sum of the fluctuations, so that stellar diameters can be determined if the star is apparently large enough to show an effect. Darkening of the

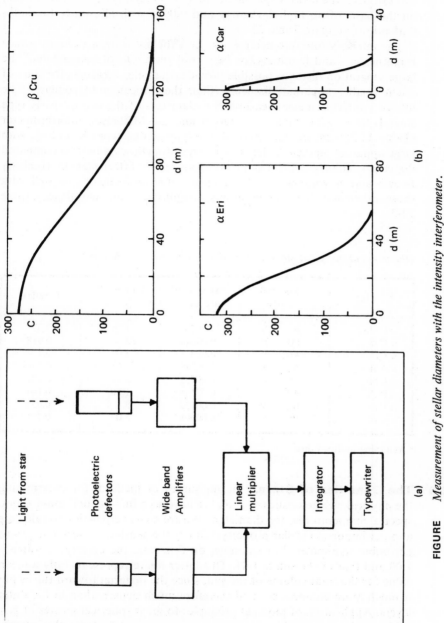

FIGURE 22-6 *Measurement of stellar diameters with the intensity interferometer.*
(a) Schematics of instrument.
(b) Correlation curves for several stars.

stellar disk from center to limb affects the result and must be allowed for. For example, the measured diameter of Vega (α Lyrae) is $0''.0033 \pm 0''.0002$, on the basis of no limb-darkening and $0''.0035$ with allowance for limb-darkening (compare Table 22-1).

The intensity interferometer works best for hot stars with high surface brightnesses—and is useless for very cool stars. On the other hand, the large separation distance possible provides diameter measures for stars of much smaller apparent diameters than the optical interferometer. The intensity interferometer measures the diameters of the nearest early-type stars B to F, while lunar occultation and the Michelson interferometer (Table 22-2) measure diameters of late-type giants, types K and M, with large apparent diameters. The two classes of method cannot be compared directly for the same star. Diameters measured for fifteen stars by Hanbury Brown and coworkers with the intensity interferometer agree well with those calculated from temperatures, magnitudes, and parallaxes (Table 22-3).

TABLE 22-3   *Diameters of hot giant stars measured with intensity interferometer*

Star	Spectral Type		Apparent arc sec[a]	Linear $\bigcirc = 1$	Parallax
α CMa	A1	V	$0''.00585$	1.76	$0''.375$
α Car	F0	Ib	0.00648	39.—	0.018
α Lyr	A0	V	0.00331	3.03	0.123
α Aql	A7	V	0.00279	1.65	0.198
α CMi	F5	IV	0.00531	2.17	0.288
α PsA	A3	V	0.00198	1.56	0.144
α Leo	B7	V	0.00133	3.8	0.039
α Gru	B5	V	0.00098	2.07	0.051

[a] Assuming uniform disk.

22-6   The Densities of Stars.   In our survey of methods for determining the physical characteristics of stars, we have seen how to find luminosities, spectra, temperatures, and diameters. We are now in a position to calculate another important stellar property, namely the density, expressed in grams per cubic centimeter. For example, on this scale the density of water is 1.00 and that of the sun is 1.41. This latter figure represents only a mean value for the total volume of the sun, since the material toward the center is much more compressed, and therefore much denser, than in the outer layers. Application of physical principles to great spherical masses of gas, held together by gravitational force, permits us to find out the internal characteristics of the sun and stars (see Sec. 25-2).

If we wish to calculate the average density of any star other than the sun, we simply multiply 1.41 by the ratio of the mass of the star to the cube of the star's radius, both expressed in terms of the sun as a unit. Hence, a star whose mass is five times the sun's, and whose diameter is ten solar diameters, would have a density $5/10^3$, or 0.005 times that of the sun. Thus the actual density of the star would be 0.007 times that of water.

Table 22-1 contains a list of the densities of representative stars, referred for convenience to the sun as a unit. Note the extreme range of densities, and the contrast between red supergiants such as Antares and the white dwarfs such as Sirius B and 40 Eridani B. Antares actually has a mean density only about 2/10,000 that of our atmosphere at sea level. How can we expect such a tenuous object to behave like a star, especially when allowance for greater concentration near the center implies even lower densities in the external layers? We shall answer this question in Chapter 23.

Table 22-1 shows that stars along the main sequence have a general tendency to increase in density from early to late spectral classes. This result conforms to our expectations, since the less luminous ones are usually smaller and more highly compressed. Stars have only a 100-to-1 range in mass. The range of diameters, however, exceeds 100,000 to 1, and the range of mean densities can be greater than 10^{13} to 1.

Difficult problems arise at very high densities. Under ordinary conditions, as we have already seen (Chapter 8), the atomic electrons arrange themselves in shells enormous compared with the tiny nucleus. The atom is mostly empty space. This electronic structure, however, can "buckle" under heavy pressure and collapse to form a superdense substance called "degenerate matter."

We have already noted this possibility in Sec. 13-3, remarking that Jupiter's radius is about maximum for a cold planet. Any cool star more massive than Jupiter would tend to collapse, as indicated in Fig. 13-3, to form a white dwarf. One of these objects has a density of 90,000 g/cm^3, about 3300 $lb/in.^3$. When we allow for the probable increase of density toward the center of this star, we infer that one teaspoonful of its core must weigh nearly a ton.

At some stage of evolutionary history, when internal sources of energy dwindle, a star may become a nova or supernova (see Chapter 27) and violently explode, ejecting a shell of gas into space. The remaining core may then collapse to form a white dwarf. Extreme compression can force the free electrons to unite with the protons to form neutrons. Such material is subject to still further compression, to form what is generally called a "neutron star." Such objects were theoretical curiosities until recently, when radio astronomers discovered a phenomenon known as "pulsars" (see Sec. 33-8). These appear to be neutron stars in extremely rapid rotation spewing out particle streams and generating radio waves in the process (Sec. 33-8).

CHAPTER *23*

Stellar Atmospheres

23-1 **The Stellar Photosphere.** The photosphere of a star, like that of the sun, comprises the layers that send us light and heat. Insofar as the sun and stars are gaseous throughout, the division of a star into atmosphere and interior appears somewhat artificial. But there is a simple way of looking at the problem, in which the distinction becomes reasonably obvious. The light energy originates in atomic transitions deep in the interior of the star and must work its way out through dense layers of absorbing gas. As the radiation pushes through gaseous layers of decreasing density, the temperature falls, for example, from a high value of about 14,000,000° K at the center of the sun to about 6000° K near the surface. There, the radiation suddenly finds itself free in atmospheric layers of low density. Even though more gas still lies above this level from which light can stream unimpeded into space, its absorbing power is much reduced. The layers where the sudden reduction occurs form the *photosphere*. Since the material that makes up the solar atmosphere may absorb differently in different spectral regions, the position of the photosphere may very well depend on wavelength. We may thus see deeper into the atmosphere of the sun or a star in one color than in another.

We now ask: "What are the physical processes that cause stellar absorption? Why are the outer layers opaque to radiation?" The principal ingredient of the atmosphere responsible for the absorption in the sun is the negative hydrogen ion, as we have noted in Sec. 9-1. Hydrogen atoms have each picked up an extra electron, presumably from some ionized metallic atom that has been unable to retain its full complement of electrons. Each time that a neutral hydrogen atom captures an extra electron to form a negative ion, it releases a quantum of light. Each time the quantum of light falls on a negative hydrogen ion, it removes the extra electron and its energy is absorbed. Radiation always tends to flow from the hotter regions toward the cooler ones. The radiation, finally escaping over a wide spectral range, produces the photosphere. The energy radiated by the photosphere forms the *continuous spectrum*.

Although the negative hydrogen ion is the chief source of opacity for stars whose temperatures lie within a few thousand degrees of the solar value, another important source of opacity takes over for the hotter stars, about class F0 and earlier. This source of opacity is the ionization of hydrogen, and subsequent recapture of the electron in some one of the various energy levels of the atom, as explained in Sec. 8-9.

We have seen that hydrogen has a number of different series, which result from atomic transitions from a given level upward (Fig. 8-6). The energy necessary to carry an electron from the ground energy level out to an infinite distance from the atom is called the *ionization energy* from that level. For hydrogen, this energy amounts to 13.54 electron volts, corresponding to a wavelength of 912 Å, the limit of the Lyman series, which lies so far in the ultraviolet that it does not penetrate the earth's atmosphere. At wavelengths shorter than 912 Å the atmosphere of any star containing hydrogen in appreciable abundance must be very opaque. Light quanta having such short wavelengths have more than enough energy to ionize the hydrogen atom. Radiation literally tears the electron away from the atom and puts the excess energy into motion of the electron.

Just as the Lyman series consists of lines and a continuous spectrum beyond the limit of array of lines, so, also, do the other series in turn. The Balmer series has its first members in the visible range and the limit occurs at 3646 Å, an easily observed wavelength in the ultraviolet.

Because hydrogen has greater atmospheric opacity at wavelengths shorter than the Balmer limit, we do not see down so deeply into the photosphere. The region that we observe, therefore, is somewhat cooler and accordingly radiates less light, so that we note a fairly sharp drop in the intensity of the spectrum beyond the Balmer limit. Through the visible range and in the infrared, the opacity results from many kinds of atomic transitions including the negative hydrogen ion.

23-2 The Stellar Reversing Layer.   Immediately above the stellar photosphere, which we have identified as the region that radiates a continuous spectrum, lies the stellar reversing layer. Sensibly transparent except in the various atomic and molecular lines, the layer consequently scatters and absorbs radiation within these lines; and, since it is cooler than the photosphere, any radiation it emits will be at a lower temperature. Thus dark "absorption" lines, the Fraunhofer lines, appear as gaps in the bright continuous spectrum from the photosphere. The phenomenon relates to the Kirchoff-Draper third law (Sec. 8-5). Thus the reversing layer derives its name from the fact that it produces a dark-line or "reversed" spectrum.

The upper reversing layer grades into the lower chromosphere, (Sec. 9-12), where the temperature may rise again and eventually attain values much greater than that of the photosphere. When viewed tangentially, as at the moment of totality during a solar eclipse, when the photospheric continuum vanishes, the upper reversing layer shows a bright-line spectrum.

In some stars the photosphere and the reversing layer tend to overlap. This condition occurs especially for the hotter stars, of spectral type A, for which hydrogen, the most abundant constituent, gives the strongest lines. Here the hydrogen lines may be so broad that their wings overlap and cut out an appreciable amount of energy between successive Balmer lines.

A somewhat similar condition obtains in many of the late-type stars, where the individual atomic lines and molecular bands are so numerous and so intense that the individual absorptions overlap, often over extended spectral regions. The apparent continuum may show sharp discontinuities resulting from such overlapping. The division into photosphere and reversing layer then becomes altogether artificial. Instead, we should say that the effective level of the photosphere is changing continuously throughout the entire spectrum, through the lines as well as through the less opaque regions in the wings of the lines.

23-3 Temperature Effect on Stellar Spectra. The effect of temperature on stellar spectra is a subject we have already discussed in considerable detail in Sec. 20-4, with relation to the spectral sequence. The importance of temperature in determining the spectral characteristics of a star is particularly evident along the main sequence. The hottest stars show the highest stages of ionization. Stars of intermediate temperature show lines of ionized metals. Stars of low temperature show neutral metals and even molecular bands. Thus, to a first approximation, we note that the temperature of a star defines its spectrum. However, we find that other factors contribute to the characteristics, the most important being the effect of gravitation.

23-4 Pressure Effect on Stellar Spectra. The effect of pressure on stellar spectra was generally overlooked by the early students of astrophysics. And yet the effect was immediately evident, once it had been pointed out. Many of the older textbooks printed spectra of the sun and Capella, one above the other, with a caption that emphasized the great resemblance between the two stars. They both were classified as G0. No one paid much attention to the various small differences that existed between the two spectra. Yet a few of these outstanding differences have played an important part in the development of astronomy.

We have already seen that Capella is a giant star, whereas the sun is a dwarf star on the main sequence, approximately five magnitudes fainter. Hence, if the two objects had the same temperature, we would ascribe the difference in magnitude to the sun's having about $\frac{1}{100}$ the surface area of Capella, or $\frac{1}{10}$ the diameter. A more precise calculation yields the ratio $\frac{1}{16}$, as given in Table 22-1.

Although Capella has a mass four times that of the sun, its mean density is much less, because its volume is so much greater. Furthermore, the effective gravity at the surface of the star will be 60 times smaller than that of the sun. Hence we conclude that the pressures in the atmosphere of Capella will also be lower than those in the atmosphere of the sun.

Pressure produces a well-known effect on the state of ionization in a gas (see Sec. 8-13). As the Indian physicist M. N. Saha showed in 1920, if two masses of gas have the same temperature but different pressures, the ionization will be greater in the gas at lower pressure. We readily see why the more tenuous gas possesses a higher degree of ionization. Absorption of radiation, the chief pocess tending to pull an electron away from an atom, is equally effective at all pressures. However, the reverse process, which requires that an ionized atom meet and combine with an electron, occurs much more readily in the denser gases. The denser the gas, the better the chances for an encounter between ion and electron. Since the recombination produces neutral atoms, there are indeed fewer ionized atoms in the denser gas. When we look at the spectra of the sun and Capella with this important principle in mind, we discover certain lines that are definitely strengthened in the spectrum of Capella. These lines all come from ionized atoms, such as ionized strontium, titanium, and many others. R. H. Fowler and E. A. Milne, and also A. Pannekoek further developed the theory of ionization. Detailed studies of the spectra of many stars confirm the theory completely.

The luminosity classification described in Sec. 20-7 is a direct application of the effect of pressure in stellar atmospheres. For a given temperature, the lower the density and the pressure, the higher the luminosity of the star. The spectroscopic characteristics, the relative intensities of certain lines from neutral and ionized atoms, used for determination of absolute magnitude, are closely related to pressure. Quantitative measures of ionized versus neutral lines allow for a finer subdivision among the various luminosity groups at a given spectral class and an improved procedure for the derivation of spectroscopic parallaxes (mentioned in Sec. 20-8). The measurement of pressure effects in stellar atmospheres has opened up a new field of analysis and has contributed a great deal to our knowledge of the distances and distribution of stars in the galaxy.

23-5 **Hydrogen in Stellar Spectra.** The appearance of hydrogen in stellar spectra is of great importance. We have already discussed the effect of the continuous spectrum of both atomic hydrogen and the negative hydrogen ion in the determination of the position of the photosphere.

Prior to observations made from rockets, the only lines of hydrogen directly observable were those from atomic levels 2 and higher, since the principal series, the Lyman series, lies in the far ultraviolet, which the earth's atmosphere heavily absorbs. Reference to Fig. 8-6 should refresh one's memory on a point made earlier: that the observable hydrogen lines come from atoms that have already undergone a very high excitation. For this reason, hydrogen is an atom fairly difficult to excite, whether in the star or in the laboratory. At the 4500° K temperature of the solar reversing layer, only one hydrogen atom in 10^{10} should be in a position to absorb the red Balmer line Hα. We are quite sure that some peculiar condition is acting to increase the amount of this absorption in the sun, because the Balmer lines are much more intense than they should be under average solar conditions. Our measures of solar prominences show how sensitive this atom is to regions of high excitation. In all probability, therefore, the hydrogen absorption as well as the hydrogen emission depends upon the high-temperature layers that exist in the chromosphere and prominences.

Whereas the hydrogen lines are more intense than we expect them to be in the solar spectrum, this type of anomaly increases for stars of still lower temperature, both main-sequence stars and giants, where the Balmer series may be surprisingly intense. Even though hydrogen is very abundant, the intense lines indicate the presence of peculiar excitation conditions, probably localized regions of high temperature. Thus our studies of the spectra of stars lead us to consider that stellar chromospheres and coronas may be fairly general.

The hydrogen lines reach their maximum in spectral classes A0 to A3, where, as previously mentioned, they may be extremely broad. These great line widths require something more than high abundance for their production. Theory and experiment have shown that an electric field will split the spectral lines of hydrogen into components, somewhat as a magnetic field splits atomic lines. This phenomenon is known as the *Stark effect*. Although we expect no large-scale, intense electric fields in stellar atmospheres, the electric charges of protons and free electrons may produce fairly intense, short-range fields. Such fields cause a broadening of the lines of hydrogen rather than a splitting, and thus the great widths of the Balmer lines are in accord with theory.

The number of Balmer lines that one can detect in the spectrum of a star is of some significance. If an atom of hydrogen is so closely surrounded by neighbors that its electron cannot attain some particular high energy value, then the spectral series will be cut off at the point determined by the space between the atoms. The higher the number of observable lines in a series, the greater is the overall free space surrounding the atom that produces them. Hence, the pressure in a stellar atmosphere will limit the number of observable Balmer lines. Giant stars and other stars with atmospheres of low pressure have their constituent atoms farther apart than denser stars, and consequently should show more members of the Balmer series than do main-sequence stars. Observation generally confirms this phenomenon, and the number of observable lines proves to be a useful index of stellar absolute magnitude (see Fig. 20-5).

In addition to its abundance, another reason why hydrogen appears in all spectral types, from early to late, is that it has but one electron to lose, whereas other atoms may have many. At the highest temperatures, therefore, any hydrogen nucleus, once it has picked up an electron, is ready to emit or absorb the spectrum of neutral hydrogen. By contrast, an atom of iron that has lost ten electrons will have to pick up ten in order to give us a spectrum of neutral iron. Neutral iron thus fades much more rapidly with increasing temperature than does hydrogen, because of the occurrence of multiple ionization.

23-6   Helium in Stellar Spectra.   Helium is also a very distinctive element. Although its absolute abundance is approximately one-tenth that of hydrogen, by volume, the lines do attain high intensity, especially in the early spectral types. Helium, with its two electrons, can exist in three stages of ionization: neutral, singly ionized, and doubly ionized. Doubly ionized helium, like singly ionized hydrogen, emits no spectrum whatever, since it has no electron attached.

Although helium lines are not particularly intense in spectral types later than B9, a few show up even in the spectrum of the sun. One of these is the infrared line at 10,800 Å. Occasionally the yellow line at 5875 Å also shows in absorption, but only over regions of high excitation, in the neighborhood of sunspots or flares. The appearance of helium in the sun, therefore, is something like the appearance of hydrogen in a late-type star, and clearly indicates the presence of regions of high excitation.

Singly ionized helium, with its one electron, has a structure and consequently a spectrum very closely resembling that of hydrogen. However, many of its spectral lines lie much farther out in the ultraviolet, although some fall in the visible region. The most intense and widely observed line of ionized helium occurs at 4686 Å. Also, in the visible range, we find the remarkable *Pickering series* of lines, the result of transitions from levels 5, 6, and higher to level 4. In many respects this series resembles the Balmer series, and every other member of it—from the even-numbered upper levels—coincides, within one or two angstroms, with a Balmer line. The Pickering series, however, contains twice as many lines as the Balmer series, with the extra lines set in regular progression between those that nearly coincide with the Balmer lines. Full development of this spectrum is found only at the highest stellar temperatures. The very luminous star ζ Puppis, spectral type O, shows it most conspicuously. Other stars of early type, including the W stars (see Sec. 20-6), have strong lines of ionized helium.

The occurrence of neutral helium in a stellar spectrum is a fairly good indication that the temperature is 18,000° K or more, whereas the presence of ionized helium points to an excitation temperature of at least 40,000° K.

23-7 Metals in Stellar Spectra. The metals in stellar spectra behave in a manner that a fairly simple diagram can summarize (Fig. 23-1). The lines of neutral metals and neutral nonmetals are generally strongest at low temperatures. However, to simplify the picture, we should divide spectral lines into two classes. Those lines coming from the lowest atomic levels, which need no prior excitation to a level from which the atom can absorb the line, we call *ultimate lines*. These ultimate lines from abundant, neutral atoms are generally very strong at low temperatures and decrease progres-

FIGURE 23-1 *The variation of line intensities of different atoms and molecules along the spectral sequence. The vertical scale schematically represents the logarithms of the numbers of atoms per square centimeter above the photosphere, active in producing the most intense lines in the visible range of the spectrum. The molecule CO is an exception, in that its strongest bands occur in the far infrared.*

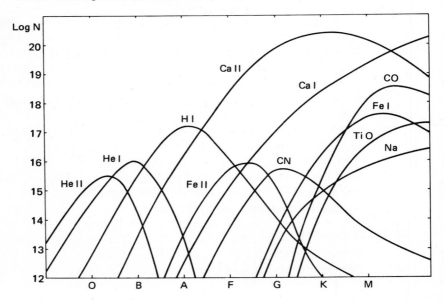

sively in intensity toward the earlier spectral types, finally disappearing altogether. The point at which these lines fade out provides a rough measurement of the abundance of the substance, lines from the most abundant elements of all persisting longest.

We also observe lines that require prior excitation for their production, since they come from higher atomic levels. Many neutral metallic lines of intermediate excitation occur in stellar spectra. These lines are usually weak at the lowest temperatures, gradually increasing in intensity at intermediate temperatures. Such lines finally reach maximum intensity, then fade at still higher temperatures as the atoms of the substance begin to be ionized. Such metals as iron, titanium, magnesium, and so on exhibit this behavior clearly. They have their maximum intensities between spectral types K5 and G0 according to the excitation potential and ionization potential.

Lines from ionized atoms behave much like those from neutral atoms requiring intermediate excitation, except that the temperatures are usually higher. Naturally, we do not expect to find much ionization at low temperatures, unless the atom itself happens, like sodium, calcium, and potassium, to require very little energy for removal of an electron. Of these, calcium has the strongest ultimate lines in the observable range, the H and K lines so intense in the solar spectrum. In stars like the sun or of somewhat later spectral type, practically all the calcium atoms in the atmosphere are ready to absorb the H and K lines. The contrast between these strong lines and the relatively weak Balmer lines illustrates the important part played by excitation in the intensities of lines. If we did not make allowance for this effect, we might erroneously conclude that calcium is by far the most abundant substance in the atmospheres of the sun and stars. The elements strontium and barium furnish other examples of ultimate lines in the visible range. The ultimate lines of neutral and ionized magnesium lie in the far ultraviolet, which can be observed only from rockets or satellites.

In the spectra of hotter stars the lines of ionized metals begin to weaken. For class A, all but a few have disappeared, the most persistent being the line of ionized magnesium at 4481 Å. For still hotter stars, the intensity of lines from atoms still more highly ionized rises and falls with increasing temperatures. Multiply ionized metals have few lines in the observable range, but a few atoms, including the relatively abundant silicon, have accessible lines in their doubly and trebly ionized states. Thus we can trace lines of silicon in four successive stages of ionization, throughout the spectral sequence from the M stars to the O stars.

23-8   Molecules in Stellar Atmospheres.   We have already seen that the solar atmosphere shows the spectra of at least two molecules, namely cyanogen and carbon hydride. These molecules have fairly intense and distinctive absorption bands, which indicate the presence of the corresponding chemical compounds. A compound can exist only where the temperature is low enough to permit the component atoms to enter into some sort of chemical union, sharing electrons, as discussed in Sec. 8-11.

In the spectra of sunspots we identify a few more molecules, some with less certainty than others. It is significant that most of these at present identified are hydrides, such as AlH and BH. Undoubtedly, the high abundance of hydrogen facilitates the formation of hydrides.

Molecules formed from other abundant atoms may also be present in the solar atmosphere, but many do not give spectra in the accessible range. Notable among these is the carbon monoxide molecule, CO, whose presence, predicted by H. N. Russell, was confirmed by O. Mohler, R. R. McMath, and L. Goldberg, who detected certain of its bands in the infrared spectrum of the sun. The bands are weak, partly because of their high excitation potential.

The most conspicuous compound shown by the spectra of late-type stars is titanium oxide, TiO, which appears in spectra from K5 and later. The intensity increases greatly toward later types. Its characteristically fluted bands are valuable criteria for the spectral classification of late K and M stars. The somewhat similar spectra of class S display bands of zirconium oxide, ZrO, rather than titanium oxide. The physical significance of this difference was mentioned in Sec. 20-5.

The spectra of the R and N stars are distinguished by the bands of the carbon molecule, C_2, and also by very intense bands of the cyanogen radical, CN. Two isotopes of carbon, with atomic weights 12 and 13, can form three different varieties of carbon molecules: C^{12}—C^{12}, C^{12}—C^{13}, and C^{13}—C^{13}. The spectra from these three compounds do not coincide exactly because of the different masses of the atomic nuclei. Spectra of all three substances occur in some of these cool stars. Why some cool stars show bands of TiO and ZrO, and others show carbon compounds, requires explanation. Temperature or excitation can scarcely account for the enormous differences. Astronomers are inclined to believe that the R and N stars are specially rich in carbon. As mentioned in Sec. 20-5 certain W stars, at the opposite (hot) end of the spectral sequence, also show anomalies apparently associated with high abundance of carbon.

The spectral sequence may well consist of several subsequences associated with different atomic abundances. For example, some distinctive groups of stars appear to be particularly rich in the rare-earth elements. For further discussion of these problems, see Chapter 25.

23-9   The Curve of Growth.   Astronomers have been able to express in quantitative terms certain of the qualitative ideas of the two earlier sections concerning the intensities of spectral lines. First, they have measured the total energy absorbed in each line in terms of the *equivalent width*, W (expressed usually in milliangstroms), of the line as defined in Fig. 9-3(a).

TABLE 23-1

The Abundances of the Elements, derived by Lawrence H. Allen (Interscience Monograph No. 7) from astrophysical, terrestrial, and meteoritic data.

Z	Elem.	log N	Z	Elem.	log N	Z	Elem.	log N
1	H	12.00	24	Cr	5.38	47	Ag	0.82
2	He	11.21	25	Mn	5.12	48	Cd	1.45
3	Li	3.50	26	Fe	6.57	49	In	0.75
4	Be	2.80	27	Co	4.75	50	Sn	1.57
5	B	2.88	28	Ni	5.95	51	Sb	0.95
6	C	8.60	29	Cu	4.00	52	Te	2.05
7	N	8.05	30	Zn	4.28	53	I	1.35
8	O	8.95	31	Ga	2.45	54	Xe	2.06
9	F	6.0	32	Ge	3.20	55	Cs	1.16
10	Ne	8.70	33	As	2.11	56	Ba	2.08
11	Na	6.30	34	Se	3.33	57	La	1.10
12	Mg	7.40	35	Br	2.65	58	Ce	1.29
13	Al	6.22	36	Kr	3.21	59	Pr	0.66
14	Si	7.50	37	Rb	2.35	60	Nd	1.36
15	P	5.40	38	Sr	2.70	61	Pm	—
16	S	7.35	39	Y	1.75	62	Sm	0.89
17	Cl	5.55	40	Zr	2.50	63	Eu	0.48
18	A	6.88	41	Nh	1.50	64	Gd	1.05
19	K	4.82	42	Mo	1.88	65	Tb	0.24
20	Ca	6.19	43	Tc	—	66	Dy	1.08
21	Sc	2.85	44	Ru	1.44	67	Ho	0.39
22	Ti	4.89	45	Rh	0.80	68	Er	0.84
23	V	3.82	46	Pd	1.26	69	Tm	0.08

Z	Elem.	log N
70	Yb	0.78
71	Lu	0.06
72	Hf	0.40
73	Ta	−0.75
74	W	0.60
75	Re	0.90
76	Os	1.40
77	Ir	1.20
78	Pt	1.70
79	Au	0.66
80	Hg	0.75
81	Tl	0.55
82	Pb	1.50
83	Bi	0.50
84	Po	−8.5
85	At	—
86	Rn	−10
87	Fr	—
88	Ra	−4.9
89	Ac	−8.2
90	Th	0.00
91	Pa	−5.0
92	U	−0.30

Suppose we have a number of lines from the same atom, for example, neutral iron, for which we have both the measured W's and the theoretical line intensities, properly corrected for the effects of excitation temperature. We plot W/λ, rather than W itself, against these adjusted theoretical intensities. This procedure, based on theoretical considerations, permits the combination of lines having widely different wavelengths. The resulting curve (Fig. 9-3b) shows how W/λ "grows" as the number of atoms active in producing the line increases. The diagram thus represents the *curve of growth* of a spectral line.

The curves for different elements of approximately the same atomic weight will be similar in shape, except for a horizontal shift. By measuring the shift for various atoms one can determine the chemical composition of the atmosphere. The results of such a study, combined with evidence from other data, appear in Table 23-1.

23-10 **Stellar Atmospheric Structure.** Atmospheric structure depends markedly on the physical processes that occur in the outer layers of the star. What processes, for example, govern the outward flow of heat and radiation? Three possibilities exist: *conduction, convection,* and *radiation.*

Conduction occurs most commonly in solid bodies, as when a spoon, placed in a bowl of hot soup, becomes warm. The rapidly moving molecules of the hot fluid transfer some of their energy to the metallic atoms of the spoon. These atoms, in turn, pass the energy to their neighbors, and so on, like an old-fashioned bucket brigade. Conduction, however, proves to be unimportant in most stars.

Convection occurs in the earth's atmosphere when masses of warm air rise and masses of cold air fall to take their place. Whether convection is effective or not depends on the rate of temperature change with altitude above the surface. As a bubble of warm air ascends, it expands and cools (see Fig. 23-2a). If, after this expansion, the gas inside the bubble is cooler than its surroundings, it will also be denser and will fall back to its original position. Such a condition inhibits convective flow. However, if the expanded gas (Fig. 23-2b) still is warmer than the surroundings, it will still continue to ascend. The cooler the upper layers, the stronger will be the convection.

If we could only impart heat to the rising bubble, it would become even warmer and ascend still more rapidly, like a hot-air balloon. Indeed, nature has a way of heating these expanding air bubbles, if they happen to contain quite a lot of water vapor. The water absorbs a certain amount of heat as it evaporates. Now, when the temperature of the expanding bubble falls low enough, the water will begin to condense out, forming fog or mist. At the same time it releases its original heat of evaporation, which becomes available for warming the air bubble. The gas, thus heated, ascends more rapidly. Vertical convection currents at cloud levels can be extremely violent, strong enough to buffet even a large passenger plane.

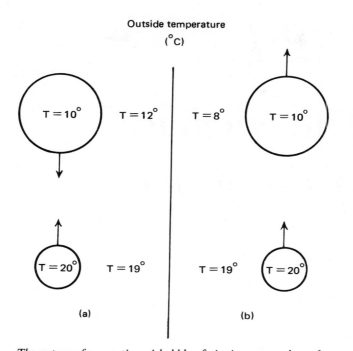

FIGURE
23-2   *The nature of convection. A bubble of air rises, expands, and
cools. (a) If, after expansion, the gas is cooler than its
surroundings, the bubble will be denser and will return to its
initial position.   (b) If, however, the gas is warmer than the
surroundings the bubble will continue to ascend like a hot-air
balloon. Convection results.*

Stellar atmospheres, of course, do not possess water vapor. However,
they have a similar form of stored energy in the ionized hydrogen. At some
depth below the photosphere of a star, hydrogen is largely ionized. As a
bubble of this gas starts to rise, some of the electrons recombine with the
protons. For the sun and, indeed, for many stars, this *hydrogen convection
zone* plays a very important role in the outward flow of heat, especially in
the last 10 percent or so of the stellar radius.

The third method of heat transfer, by radiation, is also important under
stellar conditions. As already noted, the gas below the photosphere has a
certain opacity for the radiation. In other words, the stellar material tends
to impede the radiation, by absorbing it and then reradiating it. Which
process, radiative transfer or convection, is the more important depends
upon the rate that the temperature increases inward. The steeper this rise,
the greater will be the tendency toward convection. When the temperature
increase is more gradual, convection ceases and radiative transfer becomes
the dominant process.

CHAPTER *24*

The Nucleus
of the Atom

24-1   Atomic Weight and Atomic Number.   Atomic weight and number
have already been briefly discussed in Chapter 8, where we confined our
attention mainly to the outside of an atom. There we noted that each atom
has a small core or nucleus, which contains most of the mass and all of the
positive charge, despite the smallness of the volume that it occupies. A
nucleus possesses an integral number of units of positive charge. We call
this figure the atomic number, because it specifies the chemical nature of
the atom. In order to become a neutral structure, such as we encounter
under ordinary terrestrial conditions, each nucleus must surround itself
with a number of negative electrons equal to the atomic number.

In general, atoms of high atomic number have heavier nuclei than atoms
of low atomic number. However, even though the atomic number uniquely
fixes the chemical nature of the atom, we find that some substances may
possess a variety of nuclei, having the same nuclear charge but different
atomic weights. We express our measured atomic weight on a relative scale,
so that the mass of a common carbon nucleus weighs exactly 12 units. On
this basis, the nuclear mass of the lightest atom, hydrogen, comes out
1.007825. This figure for the proton mass is almost but not quite an integer.
We shall find that the masses of many atomic nuclei, on this scale, approximate closely to integers—a fact known for more than a century. But the
small departure of the masses from being integral has a very deep significance in the modern interpretation of nuclear structure.

24-2 **The Structure of the Nucleus.** Atomic building-blocks p[...] considered are of two varieties: electrons and nuclei. In this chapt[...] shall have little concern for the electrons or for the outer structure t[...] determines the chemical characteristics of the atom; instead we are con[...] cerned with the detailed structure of the nuclei.

Originally, the atomic theory was developed to explain the *differences* between atoms. Later, the Russian chemist D. I. Mendeleev worked out his table (see Table 8-2) of the periodic structure of the elements, which exhibited the chemical kinship of elements in the same column of the table.

The fact that two atoms can resemble one another suggests the presence of some underlying regularities of structure. J. Dalton had suggested long ago (*c.* 1815) that the heavier atoms are all compounds rather than elements and built up from hydrogen. For example, sodium, with its atomic weight approximately 23 times that of hydrogen, was supposed to possess 23 atoms of hydrogen in its makeup. Although today we no longer accept this hypothesis in detail, it reappears in a slightly different form in the modern theory of nuclear structure.

24-3 **Fundamental Units.** Two basic units go into the makeup of nuclei: protons and neutrons, which we call by the general name *nucleons*. On the mass scale previously defined, neutrons have 1.008665 mass units. Thus protons and neutrons are very nearly identical in their atomic weights. They differ primarily in that the proton carries a unit of positive charge, whereas the neutrons, as its name suggests, is neutral. In other words, either the neutron carries no charge at all, or such positive charges as it may possess are balanced by the presence of an equal number of negative charges. The neutron, however, is unstable and decays with a half-life of 12 minutes into a proton and negative electron. Protons and neutrons are approximately 1840 times more massive than electrons. Since many atoms contain an almost equal number of protons and neutrons in their nuclei, and since the number of electrons must be equal to the number of protons, we get the rough rule of thumb that the outer electron structure of the atom weighs only $\frac{1}{3600}$ of the atom as a whole. Hydrogen, of course, is an exception.

A decided lack of symmetry exists in the properties of these fundamental particles—a condition particularly disturbing to those philosophers who like perfect order and balance. Certainly no one could have deduced from philosophical principles alone that the proton should be so much more massive than the electron. Indeed, nothing in physical law gives us the slightest hint as to why this imbalance occurs. Some philosophically minded physicists, therefore, have conjectured that regions of the universe may exist where the charged nucleon is negative and the surrounding electrons positive units. The English theorist P. A. M. Dirac even developed the

ich "antimatter," which has been occasionally called upon as a
hina to explain all sorts of apparently inexplicable cosmic events.
of cosmic rays and atom "smashers" have occasionally observed
short-lived particles whose mass is equal to, but whose charge
to, that of an electron. This positive electron or *positron*,
itself a fundamental unit of matter, is apparently significant as
dividual nucleons are concerned.

Protons, neutrons, positrons, and ordinary electrons possess one feature
in common. All appear to be rapidly rotating, like a miniature top—a
property that physicists usually designate as *spin*. All have the same
amount of angular momentum, equal to one-half a unit of spin. The
protons, positrons, and electrons also behave like tiny magnets, a fact in
accord with our expectations, since all three particles contain electric
charges that rotate. More surprising is the fact that neutrons also exert a
small magnetic effect, a property we interpret as signifying some imbalance
in the distribution of the internal electric charge.

Two other fundamental particles play an important part in atomic
reactions, the *neutrino* and *antineutrino*. Like the neutron, these particles
possess zero charge and a half-unit of spin. However, they have zero mass,
when at rest, and no magnetic field whatsoever.

Neutrinos and antineutrinos interact so weakly with atoms that, once
released from a nucleus deep inside the sun, they readily escape into space.
Such a particle, striking the earth, would have a high probability of passing
right through unimpeded and exiting on the far side. Sensitive laboratory
methods have been able to detect the neutrino and the antineutrino.

The neutrino differs from the antineutrino in only one respect. The
neutrino is "left-handed," a term used to indicate that, in the direction of
its motion, it turns like a left-handed screw. The antineutrino advances like
a conventional, right-handed screw.

Neutrino emission usually derives from the decay of nuclei having too
large a positive charge for their mass. The nucleus spontaneously ejects a
positron and a neutrino. Analogously, antineutrino emission occurs from
nuclei having an excess of neutrons. Such a nucleus ejects a negative
electron and antineutrino.

In passing, we mention several other particles, observed as part of the
debris resulting from fragmentation of an atomic nucleus by a cosmic ray
or by high-energy particles. These are the so-called mesons, short-lived
fragments intermediate in weight between an electron and a proton.

A number of these particles are known, some charged positively, some
negatively, and others neutral. Three kinds of π mesons exist (pions),
four varieties of K mesons (kayons), and two forms of μ mesons (muons).
The different kinds possess positive or negative charges or none.

We do not completely understand their roles, except that they seem to
bind together two or more particles of like charge into a stable nucleus.
Thus they are sometimes referred to popularly as nuclear glue. Some

heavier particles, called hyperons, of mass slightly greater than that of the proton, exist momentarily—of the order of 10^{-10} sec—as disintegration products. These seem to occur as the result of momentary association of a proton or a neutron with various kinds of pions. The hyperons have interesting properties, but they are not particularly important in problems of stellar energy.

Under certain conditions, which exist only in the atomic nucleus itself, various transitions occur, which are the key to basic changes in the nature of the nucleus. We denote the various fundamental particles by symbols, with the signs $+$ or $-$ to designate the charge. Absence of symbol means a neutral particle.

p^+ = proton e^+ = positron v = neutrino

n = neutron e^- = electron $\bar{v}$ = antineutrino

p^- = antiproton

Representative examples of various processes that occur inside the atomic nucleus follow:

Proton decay into neutron with positron and neutrino emission:

$$p^+ \rightarrow n + e^+ + v.$$

Neutron decay into proton with electron and antineutrino emission:

$$n \rightarrow p^+ + e^- + \bar{v}.$$

Electron capture by proton to form neutron with neutrino emission:

$$e^- + p^+ \rightarrow n + v.$$

Neutrino capture by neutron to form proton with electron emission:

$$v + n \rightarrow p^+ + e^-.$$

Positron capture by neutron to form proton with antineutrino emission:

$$e^+ + n \rightarrow p^+ + \bar{v}.$$

Antineutrino capture by proton to form neutron with positron emission:

$$\bar{v} + p^+ \rightarrow n + e^+.$$

Recent evidence, from both cosmic rays and atom-smashers, indicates that the antiproton can also exist momentarily, before it meets an ordinary nucleus and vanishes in a flash or radiation. But note that an antiproton and a positron would constitute a hydrogen atom of antimatter, which would have light-emitting properties identical with those of ordinary hydrogen. A few scientists have posed the question whether some of the stars or galaxies may consist of antimatter. The answer is not completely clear, but the atomic mixing that must occur in interstellar or intergalactic space would seem to preclude the possibility.

24-4   Equivalence of Matter and Energy.   The equivalence of matter and energy is the modern statement of what once were two major separate theorems of physics and chemistry: the laws of conservation of mass and conservation of energy. Where the old laws stated that neither matter nor energy could be destroyed or created, the new ruling permits the destruction of matter, for example, as long as it reappears in the form of energy— or vice versa.

Although A. Einstein originated the modern form of this "principle of equivalence," which appears as part of his famous theory of relativity, the basic concept is by no means new. Even Newton speculated on the possibility of converting "bodies into light or light into bodies." The real value of Einstein's contribution lay in the quantitative formulation of the problem. His law simply states that if a mass m of matter disappears, then the amount of energy ε that results from such conversion is numerically equal to that mass, m, times the square of the velocity of light, c. If we measure m in grams and c in centimeters per second, the energy so calculated is expressed in ergs. To reduce this energy to some more familiar (though usually less convenient) unit such as calories, or kilowatt hours, we should divide the number of ergs so calculated by appropriate conversion factors.

From Einstein's equation

$$\varepsilon = mc^2 \tag{24-1}$$

we thus derive

$$\varepsilon = 9 \times 10^{20} \times m \text{ ergs} = 2.15 \times 10^{13} \times m \text{ calories}$$
$$= 2.5 \times 10^7 \times m \text{ kilowatt hours.} \tag{24-2}$$

The enormous amount of energy that can become available as a result of mass conversion is evident immediately from the equation above. Only a few of the very best available fuels or explosives yield more than 2000 calories per gram of the materials required for the reaction. By contrast, complete conversion of a gram of matter into energy releases 2.15×10^{13} calories—ten billion times as much energy as a simple chemical process can yield. Although only about one per cent of all mass can be so converted into energy by currently known nuclear reactions, even so the energy released is enormous.

If we could start with a single gram of matter and cause only one percent of it to disappear into energy, we should have 250,000 kilowatt hours of energy available, worth $2500 at the nominal figure of 1 cent per kilowatt hour. By contrast, the amount of energy that would result from combustion of a full ton of coal and oxygen, mixed in the proper proportion for complete combustion, would be worth only $25. No wonder that industrial use of atomic energy for the production of electric power is being rapidly developed in many countries.

The foregoing equations found their first application in the field of astrophysics. We know that the earth has existed for several thousand million years (Sec. 10-10). Ordinary sources of energy, chemical or mechanical (heating by compression), are quite inadequate to keep the sun burning over more than a small fraction of geologic time. As soon as radioactivity was discovered, astrophysicists seized upon atomic energy as the only source of radiation that could possibly provide enough energy to bring the lifetime of the sun into accord with the estimated age of the earth. As nuclear physics and cosmic-ray research developed, physicists found increasing evidence in favor of these ideas, especially since laboratory experiments fully established the equivalence of mass and energy. And, insofar as the principle rests on the theory of relativity, the same experiments afford a verification for that theory as well.

24-5   Isotopes.   To illustrate the basic principle of nuclear construction, let us discuss the simplest atom of all, atom number 1, or hydrogen. By definition, such an atom must possess a positive nuclear charge of one unit. In other words, the nucleus must possess one and only one proton. Theory does not specify how many neutrons can enter the nucleus. Ordinary hydrogen has none, but we might expect it to be able to add one, two, three, or more nuetrons to produce heavier and heavier nuclei, all of the same atomic number and therefore all of them "hydrogen" in the chemical sense. The chemical nature of the atom depends primarily on the number of outer electrons and not upon the mass of the nucleus.

The type of hydrogen that contains one neutron in its nucleus weighs essentially twice as much as an ordinary hydrogen atom. The discovery of this substance, named *deuterium*, by the American chemist H. Urey in 1932 was a milestone in the sciences of chemistry and physics. In our atomic laboratories we have since been able to cause the nucleus of the deuterium atom, itself called the *deuteron*, to swallow and at least temporarily retain a second neutron, to form *tritium*, triple-weight hydrogen, an unstable isotope that decays, emitting a negative electron, with a half-life of 12.26 years. It becomes the rare but stable helium of weight three.

No one has yet been able to cause hydrogen to accept a third neutron. In fact, the indications are that such a nucleus would be completely unstable and would instantly and violently explode to form helium, with release of an enormous amount of energy. A hydrogen nucleus of mass 5, one proton and four neutrons, has been suspected but not confirmed.

We thus have three different kinds of hydrogen: three different nuclei, all of which occupy the same place in the periodic table of the elements because they have the same atomic number. We call such substances *isotopes*, and we designate the substance by the following notation. We use the ordinary chemical symbol to indicate the nature of the substance, repeating the atomic number associated with that letter as a preceding numerical subscript. The number of mass units—the number of protons plus neutrons—appears as a superscript following the letter. Hence we designate the three isotopes of hydrogen as $_1H^1$, $_1H^2$, $_1H^3$, and so on, if higher isotopes should exist in nature. Because of the importance of these particular isotopes of hydrogen, they are often distinguished by separate letters: H, D, T.

If we now go to atom number 2, helium, we enconter the following possibilities: $_2He^2$, $_2He^3$, $_2He^4$, $_2He^5$, $_2He^6$, ... Of these various atomic nuclei, the first one apparently does not exist in nature. Indeed we should not expect it to: two positive protons tend to repel one another, since there is nothing to hold them together. Apparently we need one or more neutrons in these complex atoms, to provide some degree of stability. Thus we cross off the first of these nuclei. The second exists as a stable but rare nuclide. The third, $_2He^4$, is the normal, abundant helium commonly found in nature. The fourth is so unstable that it effectively does not occur. The fifth exists, but it decays with a mean life of 0.81 sec. $_2He^8$ also occurs with a life of 0.3 sec.

Table 24-1 shows, for the first 14 members of the periodic table, the occurrence of the lighter nuclides, with the atomic number Z listed vertically and the number of nuclear neutrons N horizontally. The number following the chemical symbol represents the number of mass units, $Z + N$ The blanks indicate unobserved or nonexistent isotopes. For stable isotopes, the second number indicates the percentage that it contributes to the naturally occurring substance. For example, 99.985 per cent of hydrogen has a mass of 1; only 0.015 per cent consists of deuterium, H^2. When the isotope is unstable, the attached figure specifies the mean lifetime, expressed in seconds (s), minutes (m), days (d), or years (y). The times assigned to the very short-lived atoms are usually uncertain. A few of the nuclides can decay in two or more ways, but we need not consider this complication here.

Although the tabulation comprises but a minute fraction of the whole table of nuclei, it clearly exhibits one characteristic feature that persists, with some modification, throughout the entire table or over 100 natural and artificial elements. It shows the tendency of the stable nuclei to lie approximately along or close to the diagonal. These first few members, at any rate, show a tendency for the nuclei to possess approximately equal numbers of protons and neutrons. Hence, an atom that lies to the left of this diagonal has too many positive charges for its mass. One of the protons in the atom, therefore, tends to give up its positive charge as e^+ or positron radiation, so that the nucleus moves diagonally downward and to the right to the next row of smaller atomic number. In similar fashion, nuclei to the right of this diagonal have too few units of positive charge for their mass. Thus one of the neutrons in the atom will tend to emit a negatively charged electron (β radioactivity or e^- radiation), and the atom will move diagonally upward to the left, to the row of next higher atomic number.

Sometimes a slight rearrangement of the particles within the nucleus, or annihilation of a positron, causes the emission of a γ ray in addition. A γ ray is (usually) an X ray of very short wavelength. In this way we recognize the existence of nuclear energy levels that in many respects are similar to those of the external electron shell, which produces the optical spectrum.

The stable helium nucleus deserves special comment. Its two protons and two neutrons form one of the most stable of all atomic units. The helium atom with its four mass units possesses an effective atomic weight of 4.00260, whereas a single hydrogen atom has an atomic weight of 1.007825 units. Suppose that we could make four protons or hydrogen nuclei combine to yield a single nucleus of helium. This reaction would cause the disappearance of $1.007825 \times 4 - 4.00260 = 0.029$ unit of mass and their reappearance as various forms of energy. In other words, any given mass of hydrogen, built into helium, will release 0.007 (seven-tenths of one per cent) of its mass as energy. This figure is the basis for the rough estimate of 1 percent conversion of mass into energy, previously given. To manufacture energy inside a star, we require a reaction or a series of reactions resulting in the required conversion of mass into energy. Since one mass unit corresponds to 931.141 megaelectron-volts (Mev), the conversion of four protons into a single helium nucleus releases 27.09 Mev of energy. (One electron-volt is the kinetic energy acquired by an electron that has been accelerated by a potential difference of 1 volt in an electric field.)

24-6   Radioactivity and Stability of Nuclei. Scientists recognized the existence of unstable nuclei long before they learned how to manufacture a wide variety of these unstable types in their particle accelerators or in atomic piles. Uranium, thorium, radium, and many other substances are naturally radioactive, decaying spontaneously, some slowly and others rapidly. The decay products include both the negative electron e^- and the positron e^+ with occasional emission of γ radiation in addition. Or sometimes a heavier atom may eject an entire helium nucleus, the so-called *alpha particle*, as the primary product of its instability.

Z \ N	0	1	2	3	4	5	6	7	8	9	10	11	12	13	14	15	16	17	18
14 (Si)												Si 25, 0.23 s	Si 26, 2 s	Si 27, 4.2 s	Si 28, 92.92	Si 29, 4.70	Si 30, 3.09	Si 31, 2.62 h	Si 32, 700 y
13 (Al)											Al 23, 0.13 s	Al 24, 2.1 s	Al 25, 7.2 s	Al 26, 7.4×10^{5} y	Al 27, 100	Al 28, 2.30 m	Al 29, 6.6 m	Al 30, 3.3 s	
12 (Mg)										Mg 21, 0.12 s	Mg 22, 3.9 s	Mg 23, 12 s	Mg 24, 78.70	Mg 25, 10.13	Mg 26, 11.17	Mg 27, 9.5 m	Mg 28, 21.3 h		
11 (Na)										Na 20, 0.4 s	Na 21, 23 s	Na 22, 2.58 y	Na 23, 100	Na 24, 15 h	Na 25, 60 s	Na 26, 1.0 s			
10 (Ne)								Ne 17, 0.69 s	Ne 18, 1.46 s	Ne 19, 18 s	Ne 20, 90.92	Ne 21, 0.257	Ne 22, 8.82	Ne 23, 38 s	Ne 24, 3.38 m				
9 (F)								F 16, 10^{-19} s	F 17, 66 s	F 18, 110 m	F 19, 100	F 20, 11 s	F 21, 4.4 s						
8 (O)							O 14, 71 s	O 15, 24 s	O 16, 99.759	O 17, 0.037	O 18, 0.204	O 19, 29 s	O 20, 14 s						
7 (N)						N 12, 0.011 s	N 13, 9.96 m	N 14, 99.63	N 15, 0.37	N 16, 7.35 s	N 17, 4.14 s	N 18, 0.63 s							
6 (C)				C 9, 0.13 s	C 10, 19 s	C 11, 20.5 m	C 12, 98.89	C 13, 1.11	C 14, 5730 y	C 15, 2.25 s	C 16, 0.75 s								
5 (B)				B 8, 0.78 s	B 9, 3×10^{-19}	B 10, 19.78	B 11, 80.22	B 12, 0.020 s	B 13, 0.019 s										
4 (Be)			Be 6, 4×10^{-21} s	Be 7, 53 d	Be 8, 3×10^{-16} s	Be 9, 100	Be 10, 2.7×10^{-6} y	Be 11, 13.6 s	Be 12, 0.011 s										
3 (Li)			Li 5, 10^{-21}	Li 6, 7.42	Li 7, 92.58	Li 8, 0.85 s	Li 9, 0.17 s												
2 (He)		He 3, 0.00013	He 4, 1.00…	He 5, 2×10^{-21}	He 6, 0.81 s		He 8, 0.122 s												
1 (H)	H 1, 99.985	H 2, 0.015	H 3, 12.26 y																
0 (n)		n 1, 12 m																	

Consider the behavior of magnesium, atom number 12 in the periodic table, which has eight known isotopes, as listed in Table 24-1. These undergo successive radioactive decays as follows:

$$_{12}Mg^{21} \rightarrow {}_{11}Na^{21} + e^+,$$
$$_{11}Na^{21} \rightarrow {}_{10}Ne^{21} + e^+.$$

Atom $_{10}Ne^{21}$ is stable, and no further disintegrations occur. Similarly

$$_{12}Mg^{22} \rightarrow {}_{11}Na^{22} + e^+,$$
$$_{11}Na^{22} \rightarrow {}_{10}Ne^{22} + e^+ \quad \text{(stable)},$$
$$_{12}Mg^{23} \rightarrow {}_{11}Na^{23} + e^+ \quad \text{(stable)}.$$

$_{12}Mg^{24}$, $_{12}Mg^{25}$, and $_{12}Mg^{26}$ are all stable.

$$_{12}Mg^{27} \rightarrow {}_{13}Al^{27} + e^- \quad \text{(stable)},$$
$$_{12}Mg^{28} \rightarrow {}_{13}Al^{28} + e^-,$$
$$_{13}Al^{28} \rightarrow {}_{14}Si^{28} + e^- \quad \text{(stable)}.$$

The stable nuclei possess, on the average, more neutrons than protons. This required excess of neutrons increases toward the end of the periodic table. Thus, by the time we reach uranium, element 92, we find that the most abundant nuclide has mass number 238, so that it actually contains 54 more neutrons than protons. This neutron excess is necessary for stability. Consequently, if we were to break an atom of uranium into two equal pieces, each fragment would then contain too many neutrons for the element of half atomic weight. These neutrons would then escape from the nucleus, carrying energy with them. We shall see that the nucleus of uranium-235 rather than the more common U-238 exhibits a natural tendency toward "fission."

24-7   Atom Building and Atom Smashing.   In recent years *cyclotrons*, *betatrons*, *synchrotrons*, and *linear accelerators* have vied with one another for the production of high-energy particles capable of breaking down the barriers of the atomic nuclei and causing profound changes in them. As a result, physicists have discovered that many nuclear reactions can occur, chiefly between the lighter particles such as protons, deuterons, or neutrons and the heavier nuclei.

The reactions roughly resemble those that the chemist has analyzed for years in his laboratory. The primary difference is that the chemist deals with the outside of the atom, producing temporary rearrangement of the electronic shells, as one atom shares electrons with another in the bonds of chemical union. From the standpoint of chemistry, the atom itself is permanent and the nucleus unchangeable. The possibility of changing one element to another was the dream of the alchemist. The modern physicist has achieved that dream. With his special devices, he operates deep inside the atom to change the chemical structure permanently.

Vast numbers of nuclear reactions are possible and known—far too many to list here. From our standpoint, however, the most important reactions will be those that involve the proton, partly because the proton is so important as a source of energy and partly because hydrogen is by far the most abundant of all the atoms in the universe. Hence we expect that energy production in the centers of stars will rely largely upon conversion of the excess energy of a proton. Therefore, the reactions of protons with one another, with deuterons, and with such light elements as carbon, nitrogen, and so on are particularly important. We shall not detail them here, but the most significant ones will be given in Chapter 25, which deals specifically with the sources of stellar energy.

One further analogy should be pointed out between these atomic reactions and the reactions of ordinary chemistry. The chemist speaks of endothermic or exothermic processes, reactions that respectively absorb or give off energy. We find evidence of both processes occurring in the atom, but from the standpoint of the production of energy in the sun and stars, the exothermic are by far the more important. Hence the production of atomic energy is the physical analogue of those chemical reactions capable of producing heat. Almost literally, stars are "burning atoms" to produce radiation.

24-8 **Nuclear Fission and Fusion.** Enormous publicity has been given to nuclear fission and fusion since the first atomic bomb focussed the world's attention upon the power inherent in the atom. We have already mentioned (Sec. 24-6) the tendency of uranium-235 to split into two roughly equal masses. Left to itself, the nucleus will divide with a half-life of 71×10^6 years. Another kind of splitting, however, does not take place spontaneously. It occurs only when a wandering neutron happens to penetrate the nucleus. Whenever this sort of encounter occurs, the nucleus explodes, each of its fragments necessarily releasing excess neutrons in the process.

Usually the fragments are not exactly equal. Suppose that the atom splits so that the 92 units of positive charge divide into unequal parts, respectively 42 and 50, corresponding to the elements molybdenum Mo and tin Sn. At the same time, suppose that the 235 mass elements divide respectively in the proportion of 105 and 130. When we consult an ordinary table of the chemical elements, we find that the only stable isotope of molybdenum resulting from fission contains just 102 mass units; thus to balance matters the fission product, molybdenum, will have to lose three neutrons. In the same way, the most massive known isotope of tin is 126, so that the tin fragment must release four neutrons. These seven, plus the original neutron that produced the fission, give eight neutrons—releasing a large amount of energy as uranium-235 decays into tin and molybdenum.

Consider how this type of reaction may build up in a mass of pure U-235. If the mass is small, the eight released neutrons will fly out into space and be lost. But if the mass is large enough, each of these neutrons will be trapped by the nucleus of another U-235 atom. Each neutron, therefore, will release eight more neutrons, and hence the original eight give rise to 64. At the third step we have 512 neutrons, and so on, with all of the fission products associated with them. This is the so-called "chain reaction." Usually it does not proceed quite so efficiently, since some neutrons are lost or captured by impurities, but in a large enough mass it does release enough energy to produce the explosion of the atomic bomb. The minimum mass necessary to sustain a chain reaction is known as the *critical mass*.

Nuclear fission, as a process, appears to be relatively unimportant for stellar interiors. Uranium and other elements likely to enter into such reactions are rare. Although occasional U-235 atoms on the surface of the earth do absorb neutrons and explode, the abundance of this substance is so small (about 0.7 percent in natural uranium) that the neutrons from this reaction fritter away their energies in other processes and do not have a chance to build up the devastating chain reaction.

The most important process for stellar energy generation is nuclear *fusion*, the opposite of fission. While fission of heavy nuclei releases energy, it is the fusion of several light nuclei, and especially of hydrogen, that produces free energy as described in Sec. 24-5. High temperatures and high pressures in the deep interiors of the stars make this fusion possible. As the fusion processes proceed, the quantity of hydrogen slowly diminishes while the percentage of heavier elements increases. When hydrogen is exhausted, will other processes keep the star going? Also, we may ask whether all the heavier elements resulted from fusion. If so, are the current relative abundances of these elements consistent with what we know about nuclear evolution? These and other problems will be discussed in the next chapter.

CHAPTER *25*

Sources of Stellar Energy, Nuclear Reactions, and Atomic Evolution

25-1   Theories of Stellar Radiation.   The energy radiated by the stars was long a puzzle to astronomers. Where did it come from? Early views that the sun might be actually burning, producing heat by *combustion*, had to be abandoned when geologists overthrew the idea that the whole universe was only a few thousand years old. No conceivable fuel could keep the sun going by ordinary combusion. In 1854, the German physicist H. von Helmholtz suggested that gravitational contraction might be the source of solar energy. But the span of twenty million years that this process allows is still much too short even to encompass the recorded paleontological and geological history of the earth.

During the second decade of the twentieth century, the conviction grew among astronomers that the energy of the sun and stars came from the conversion of matter into radiation. At first this idea was mere speculation, since no one had suggested a probable mechanism for the conversion. Experimental research in the field of atomic physics, however, led to the discovery of actual nuclear processes that must certainly occur under the conditions found within the stars. Thus the vague speculations of the 1920's have developed into quantitative theories that account for the stars' outpouring of energy, in complete harmony with the observations and theories of nuclear physics and our knowledge of the conditions in stellar interiors.

25-2 Equilibrium of Stars. The internal equilibrium of a star depends on a balance between various forces. The stars are gaseous throughout, and gases tend to expand. But the stars are also very massive. Thus the dead weight of the overlying layers of matter will produce a downward force, and gas pressure from the inside will produce an upward thrust, which together will tend to compress a thin shell that is some distance below the surface of a star (Fig. 25-1). If the shell is to be stable, the pressure within it must be great enough to withstand the effect of these forces. We can picture the whole star as built up of a very large number of such shells. If the star is stable, each shell must be in equilibrium, so that at every point the outward push of gas pressure balances the inward push caused by its own weight.

Because the stars are wholly gaseous, we can confidently use the known law of perfect gases to find out conditions in their interiors. We know that increased pressure means higher temperature, or greater density, or both. Thus we are able not only to infer that both temperature and density grow higher, the deeper into a star we penetrate, but even to calculate exactly how the density and temperature change with depth.

Of considerable importance in the equilibrium of a star is the rate that energy leaks out from the interior. The *opacity* of the stellar material, its ability to allow its own radiation to pass through, controls the rate of leakage. Although the problem of the internal opacity of a star is very complicated, the laws of physics—of matter and radiation—lead us to a solution.

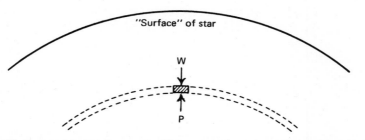

FIGURE 25-1 *The pressure of lower layers balances the weight of upper layers in all parts of a star in stable equilibrium.*

The opacity depends upon the chemical composition of the star, increasing with the average atomic weight of the stellar matter. As energy forces its way outward through the star, the high-energy radiation (γ rays and X rays) characteristic of the deep interior gradually changes into the low-energy radiation (ultraviolet, visual, or infrared light) that finally emerges from the star's surface. The opacity must control the flow of energy through the star, so that the rate of energy production in the interior is exactly equal to the rate of energy radiation by the star into space; otherwise, the star would not be in equilibrium. The outflowing radiation can exert a pressure on its own account, but for most stars its effect is minor. Early theories of stellar structure exaggerated the part played by radiation pressure in the equilibrium of stars, but it may profoundly affect some of the very hottest stellar interiors of the most massive stars.

If the mass, radius, and composition of a star are known, and if the opacity depends primarily on the interplay of radiation and electrons (either attached to atoms or "free"), the central temperature and density can be simply calculated. For example, the sun has a central temperature of about 14,000,000° K. We must see whether known nuclear reactions can proceed at this temperature, to provide a sufficient supply of energy to account for the observed output of the sun. Similar calculations can be made for a variety of other stars, some of which turn out to have even higher central temperatures.

25-3   Chemical and Nuclear Reactions.   In the previous chapter we noted that, from the standpoint of modern physical chemistry, nuclear reactions differ from chemical reactions only in degree. Here are three chemical reactions, somewhat simplified for purposes of demonstration:

$$Na + Cl \rightarrow NaCl + heat, \tag{25-1}$$

$$NaOH + HCl \rightarrow NaCl + HOH + heat, \tag{25-2}$$

$$C + O_2 \rightarrow CO_2 + heat. \tag{25-3}$$

The first of these equations illustrates the reaction between metallic sodium (Na) and chlorine (Cl), which results in the formation of common salt (NaCl) with the liberation of a certain amount of heat. In much the same way, the reaction between sodium hydroxide (NaOH) and hydrochloric acid (HCl) gives rise to sodium chloride (NaCl) and hydrogen hydroxide (HOH), which is another name for water (H_2O). This reaction too, produces heat. The third equation represents the familiar combustion process of carbon into carbon dioxide (CO_2). A chemical equation is really not complete unless we include the emitted heat on the right-hand side. We call such reactions *exothermic* because they lead to the production of heat. Some chemical reactions require added heat for their operation; these we call *endothermic*. For example, we may write, as a continuation of (25-3),

$$C + CO_2 + heat = 2\ CO. \tag{25-4}$$

Note that the heat produced in the step represented by (25-3) now induces the further reaction, leading to the formation of carbon monoxide (CO).

We have already noted that energy and mass are equivalent. Hence the energy in these equations must come from a conversion of mass into energy. For example, the CO_2 produced by the reaction of equation (25-3) must weigh less than the original ingredients. Likewise, the CO formed by the reaction (25-4) must weigh more than the initial ingredients. The energy of such chemical reactions is so minute, however, that balances are not sensitive enough to record the difference in mass between the initial and final products. For example, in the formation of a full ton of salt from its basic elements, only one ten-thousandth of a gram of mass disappears, in the form of released heat.

Nuclear reactions, unlike chemical reactions, often involve changes of mass that amount to nearly one percent. The vast amount of energy liberated in nuclear processes provides a source of energy for the sun, for an interval as great as the age of the galaxy, provided that physical conditions in stellar interiors permit such reactions to occur.

Our ideas about the part played by nuclear energy in the stars have changed considerably. Sir James Jeans once suggested that superradioactive atoms, left over from the formation of the universe, might be responsible for stellar energy. On this theory, the sun and stars would be something like enormous superuranium atomic bombs, with their atomic nuclei exploding spontaneously and breaking down into the ligher chemical elements. Although the idea initially seemed attractive, it proved to be unacceptable.

For a star to be stable, the rate of release of energy, far from being independent of external conditions, must vary markedly, increasing with both higher temperature and greater density. Suppose, for example, that a star should start to contract, perhaps as a result of gravitation. The resulting compression would then raise the internal temperature slightly. But the temperature increase would make the gaseous star expand. The star might explode except for the fact that the expansion cools the gas. The temperature falls and the rate of energy generation drops. In this way the atomic processes themselves provide a sort of safety valve.

Most stars are extremely stable. Some oscillate between narrow limits, perhaps from the very effects just described. We do witness occasional stellar explosions (Chapter 27), but for the most part these are localized, not detonations of a whole star. Therefore, we infer that the nuclear reactions responsible for the energy of the stars are highly sensitive to temperature changes.

25-4 Proton Reactions. Reactions with protons involve the most abundant of all atomic nuclei, hydrogen; hence we should expect them to dominate the production of stellar energy. If a nucleus of charge Z, containing a number M of mass units, captures a proton, the end product will be a nucleus of positive charge $Z + 1$ and mass units $M + 1$. Such a capture causes the atom to move vertically upward to the next row of Table 24-1.

Some of the nuclei formed by such a process are unstable, for reasons discussed in Chapter 24. The most stable of the lighter nuclei are those having about an equal number of protons and neutrons. If a previously stable nucleus were to capture a proton, it would then have one positive charge too many. Once unstable, such a nucleus would tend either to emit a positron, an electron of positive charge (see Sec. 24-5), or capture a negative electron. Positrons, like neutrons, are extremely short-lived outside the atomic nucleus. A positron can unite with a negative electron, so that both particles disappear, giving rise to two quanta of γ radiation moving in opposite directions with a total energy of 1.02 Mev. The unstable products of proton reactions decay with the production of energy—in other words, the reactions are exothermic.

The British astronomer R. d'E. Atkinson was the first person to suggest, about 1931, that stellar energy could result from successive captures of protons by an atom. He was primarily concerned with the reaction between helium atoms and protons, but he also discussed how heavy atoms might develop from lighter ones. H. Bethe in the United States, and C. von Weizsacker in Germany, presented the same principle with better evidence in 1939 and formulated the carbon cycle discussed in Sec. 25-8. Through the work of many other physicists and astrophysicists the theory of nuclear reactions and atomic synthesis as a source of stellar energy has gradually evolved, during the past three decades, as a coherent and increasingly detailed doctrine.

25-5   **Proton-Proton Reaction.**   The proton-proton reaction is fundamentally the simplest of all atomic processes of the kind termed *thermonuclear reactions.* As first pointed out by Menzel in 1932, the high abundance of the proton, or hydrogen nucleus, may make this reaction extremely important for the energy production in certain stars, including the sun. We note that apparently the simplest possible reaction—the direct union of two protons to form an isotope of helium, $_1H^1 + {}_1H^1 \rightarrow {}_2He^2 +$ energy— does not occur. There is no evidence for even the temporary existence of an isotope of helium of atomic weight 2. This isotope must be so extremely unstable that it decays spontaneously into a deuteron and a positron, before the interaction becomes complete. The *deuteron* is the nucleus of the hydrogen isotope of mass 2, commonly known as deuterium. It contains a proton and a neutron in combination.

Thus we have the following reaction:

$$_1H^1 + {}_1H^1 \rightarrow {}_1H^2 + e^+ + \nu + \gamma \quad (1.44 \text{ Mev}), \quad (25\text{-}5)$$

with the release of a positron, a neutrino ν, and nuclear energy in the form of a gamma ray. The reaction is exothermic, and the combined energies of the positron and the radiation released correspond to the mass difference between the two original protons and the deuterium nucleus. The energy released, expressed in Mev, is given in parentheses.

Even under the close-packed conditions in a stellar interior, the production of deuterium would be quite slow. However, deuterons can rapidly undergo other reactions, of which the most probable involves capture of another proton:

$$_1H^1 + {}_1H^2 \rightarrow {}_2He^3 + \gamma \quad (5.49 \text{ Mev}). \quad (25\text{-}6)$$

This reaction, which produces an isotope of helium of mass 3, proceeds very fast. The newly formed helium nuclei can now react with one another:

$$_2He^3 + {}_2He^3 \rightarrow {}_2He^4 + {}_1H^1 + {}_1H^1 \quad (12.86 \text{ Mev}). \quad (25\text{-}7)$$

The original union of two protons initiated a *chain reaction*. Finally, an ordinary helium nucleus of mass 4 and two protons result from six interacting protons. This chain reaction is known as the *proton-proton reaction* or *p-p chain*. Under conditions like those at the center of the sun, the first step would occupy an average time of 5×10^9 years; the second, about 4 seconds; the third, about 4×10^6 years. The rate at which the whole chain of reactions would go is determined by the first, because it is the slowest. Energy, of course, is a by-product of all these reactions. The total energy released by the conversion of four protons into $_2He^4$ equals twice the sum of the energies expressed in equations (25-5) and (25-6), for the production of the two interacting nuclides of (25-7). The total energy, then, is 26.72 Mev.

The proton-proton reaction is a major contributor to the production of stellar energy, because protons are a very abundant constituent of stellar matter. The reaction rate is much less sensitive to temperature than the rates of reaction of protons with some heavier nuclei.

The proton cycle can continue in other ways. For example:

$$_2He^3 + {}_2He^4 \rightarrow {}_4Be^7 + \gamma \quad (1.58 \text{ Mev}),$$
$$_4Be^7 + e^- \rightarrow {}_3Li^7 + \nu \quad (0.86 \text{ Mev}), \quad (25\text{-}8)$$
$$_3Li^7 + {}_1H^1 \rightarrow 2{}_2He^4 \quad (17.35 \text{ Mev}),$$

again resulting in the production of He and liberation of energy. As still another possibility, we have

$$_4Be^7 + {}_1H^1 \rightarrow {}_5B^8 + \gamma \quad (0.10 \text{ Mev}),$$
$$_5B^8 \rightarrow {}_4Be^8 + e^+ + \nu \quad (18.01 \text{ Mev}), \quad (25\text{-}9)$$
$$_4Be^8 \rightarrow 2{}_2He^4. \quad (0.10 \text{ Mev}).$$

Of special significance is the fact that the heavier atoms in the chain, Li, Be, and B, appear only temporarily. Eventually they too are consumed in the reaction with protons, giving stable $_2He^4$ as an end product. Note that the boron isotope, $_5B^8$, is unstable, decaying spontaneously to an isotope of beryllium, which is also unstable in its turn. The nuclide $_4Be^7$ is unstable with a mean life of 53 days. Hence, for the second set of reactions to be important the proton capture must occur within a comparable time interval.

25-6 **Proton-Deuteron and Proton-Tritium Reactions.** Reactions of protons with deuterons and with tritium have special interest, because of the part they play in the hydrogen thermonuclear bomb. As we have seen, the former [equation (25-6)] represents part of the proton-proton chain. This reaction is possible because a deuteron is a stable particle, surviving to undergo further reaction with another proton. The porton-deuteron reaction most important in stellar interiors [equation (25-7)] produces a stable helium nucleus of mass 3, which reacts with other similar nuclei to complete the proton-proton chain.

Tritium, the hydrogen isotope of mass 3, can result from the interaction of two deuterons:

$$_1H^2 + _1H^2 \rightarrow _1H^3 + _1H^1 \rightarrow _2He^4 \quad (23.84 \text{ Mev}). \quad (25\text{-}10)$$

Tritium, if it encounters another proton soon enough reacts to form an ordinary helium nucleus of mass 4 with release of an enormous amount of energy. However, unlike deuterium, tritium is unstable. Within about twelve years it disintegrates into the stable helium nucleus $_2He^3$ and a negative electron. The proton-tritium reaction is therefore not likely to be important in stellar interiors.

25-7 **Proton-Lithium Reaction.** The proton-lithium reaction is important because it can occur at fairly low temperatures, of the order of a million degrees. Perhaps it is the first reaction to start as an evolving star heats up internally (see Sec. 35.3). Lithium, then, could act as a sort of stellar kindling wood. We write the reaction as follows [see equations (25-8)]:

$$_3Li^7 + _1H^1 \rightarrow 2_2He^4 \quad (17.35 \text{ Mev}). \quad (25\text{-}11)$$

Lithium serves as a fuel in certain kinds of fusion bombs, loosely termed the "H-bomb." Note the high energy yield of 17.35 Mev.

As the central temperature of a developing star rises, all the lithium in its interior will be converted into helium. The apparently low cosmic abundance of lithium indicates that most of the $_3Li^7$ has disappeared in this way. When we find abundant lithium, as we do in the atmospheres of a few cool stars, we conclude that the element has always existed only in the outer regions and never in the interior.

25-8 **The Carbon Cycle.** Suggested by H. Bethe and C. F. von Weizsacker, the carbon cycle has been the most widely discussed of all the processes involved in the production of stellar energy. In its successive reactions the carbon nucleus $_6C^{12}$ captures protons to form increasingly more complex nuclei until finally, after four protons in turn have been captured, the product splits into a helium nucleus and a carbon nucleus like the one that started the chain. This carbon atom is ready to start the process over again. We therefore call the reaction a *cycle*, in which carbon plays the role of what a chemist would call a *catalyst*. The cycle consists of six successive steps, with times roughly given below for solar conditions. The reaction rates are, as noted earlier, very sensitive to temperature.

$$
\begin{aligned}
{}_6C^{12} + {}_1H^1 &\rightarrow {}_7N^{13} + \gamma & \text{(1.95 Mev)} & \quad (10^6 \text{ yr}), \\
{}_7N^{13} &\rightarrow {}_6C^{13} + e^+ + \nu & \text{(2.22 Mev)} & \quad (10 \text{ min}), \\
{}_6C^{13} + {}_1H^1 &\rightarrow {}_7N^{14} + \gamma & \text{(7.55 Mev)} & \quad (2 \times 10^5 \text{ yr}), \\
{}_7N^{14} + {}_1H^1 &\rightarrow {}_8O^{15} + \gamma & \text{(7.28 Mev)} & \quad (< 3 \times 10^7 \text{ yr}), \\
{}_8O^{15} &\rightarrow {}_7N^{15} + e^+ + \nu & \text{(2.76 Mev)} & \quad (2 \text{ min}), \\
{}_7N^{15} + {}_1H^1 &\rightarrow {}_6C^{12} + {}_2He^4 & \text{(4.97 Mev)} & \quad (10^4 \text{ yr}).
\end{aligned}
\tag{25-12}
$$

These equations, which represent the reactions that govern transformation of matter into energy, follow the pattern previously laid down. Each step of the reaction produces energy. The two intermediate nuclear products, $_7N^{13}$ and $_8O^{15}$, are unstable, because they have too many positive charges for their mass, and they decay spontaneously. Thus, each of them quickly expels a positron, the radioactive nitrogen nucleus decaying into a stable carbon nucleus within about ten minutes after its formation, and the radioactive oxygen nucleus reverting to a stable isotope of nitrogen within about two minutes. The stable products, as in the reactions previously studied, are then available to react with more protons. The combination of four protons into a single helium nucleus has released in all 0.0287 mass units, or 26.72 Mev in energy.

The only surprising step is the last, where we might expect the equation $_7N^{15} + {}_1H^1 \rightarrow {}_8O^{16}$. Although this reaction doubtless has a tendency to occur, we find that it produces an unstable excited oxygen nucleus, which splits immediately into a carbon atom and a helium atom, so that the end products are the same as those of the last equation (25-12).

Physicists have discovered a second cycle, which will occur about two thousand times less frequently than the cycle previously discussed.

$$
\begin{aligned}
{}_7N^{15} + {}_1H^1 &\rightarrow {}_8O^{16} + \gamma & \text{(12.13 Mev)}, \\
{}_8O^{16} + {}_1H^1 &\rightarrow {}_9F^{17} + \gamma & \text{(0.59 Mev)}, \\
{}_9F^{17} &\rightarrow {}_8O^{17} + e^+ + \nu & \text{(2.76 Mev)}, \\
{}_8O^{17} + {}_1H^1 &\rightarrow {}_7N^{14} + {}_2He^4 & \text{(1.21 Mev)},
\end{aligned}
\tag{25-13}
$$

which may then continue as before:

$$
{}_7N^{14} + {}_1H^1 \rightarrow {}_8O^{15} + \gamma,
$$

and so on. Since two coupled cycles operate, astronomers often refer to the combined chains as the "CNO bi-cycle."

The carbon cycle is much more sensitive to temperature than the proton-proton reaction, the energy production varying as about T^{17}, the temperature to the seventeenth power. A change of 1 per cent in temperature would

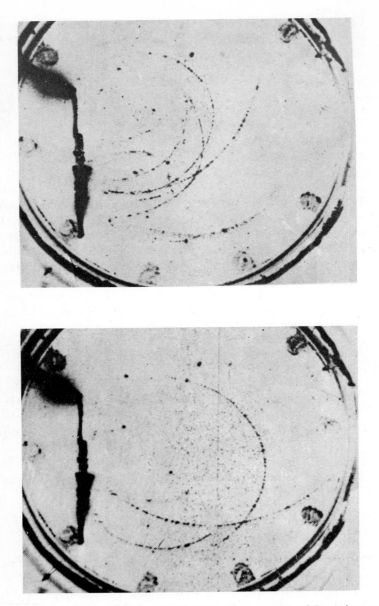

I

II

FIGURE 25-2   *Wilson cloud chamber photograph of a nuclear reaction of the carbon cycle. Carbon target at left has been bombarded by protons. The radioactive N^{13} nuclei formed in the target by reaction number 1 of the carbon cycle decays into C^{13} nuclei, with the emission of positrons, by reaction number 2 of the cycle. The photographs show the trajectories of the positrons, curved by a magnetic field. The lower photograph, taken 20 minutes after the first, shows reduced activity corresponding to decay lifetimes of N^{13}. (K. Anderson and S. H. Neddermeyer.)*

lead to an increase of almost 20 per cent in energy production. Hence the carbon cycle produces a greater variation along the stellar main sequence than does the proton-proton reaction, more in harmony with the observed one. For this reason, the carbon cycle is the principal source of energy for the hotter main-sequence stars, with the proton-proton reaction important for the cooler ones. We note that although the reactions involved are very different, the end product of both chains is the same: the helium nucleus.

Although the carbon cycle may be responsible for the energy of main-sequence stars, it does not seem to possess explosive properties. Many of the nuclear processes are unstable in times shorter than one-millionth of a second. Two of the intermediate steps of the carbon cycle take several minutes for the atoms to readjust themselves, and one of those in the proton-proton reaction requires four seconds. These delays exert a stabilizing influence. Thus, the carbon cycle could lead to an explosion only under very artificial circumstances, which might be difficult to produce. It probably has no applications in the construction of a nuclear bomb.

25-9   Nucleogenesis or Atomic Evolution. The nuclear reactions responsible for stellar radiation gradually alter the chemical composition of a star. Operation of the CNO bi-cycle, however, depends on the prior existence of the catalyst, C^{12}. At this time we may well raise the question: where did the heavier elements come from? And how did heavy, radioactive substances such as uranium come into being? Perhaps they were there from the beginning of the universe, but the high general abundance of hydrogen suggests that once upon a time the primitive world contained nothing but that element. If so, we must account in some way for the presence of the heavier nuclides. Many scientists have worked on this problem, including E. Öpik, E. Salpeter, G. and M. Burbidge, and W. Fowler in the United States and F. Hoyle in England.

The proton-proton reactions can, as we have seen, build up the stable, nuclei He^3, He^4, and Li^7, in addition to the radioactive atoms Be^7, Be^8, and B^8. But no natural process involving proton captures is evident for building stable atoms heavier than He^4 and Li^7. Similarly, the CNO bi-cycle comes to a dead end. It can build stable N^{14}, N^{15}, O^{16}, and O^{17}, but further proton captures produce unstable atoms that eject He^4 and revert to the original C^{12}. However, when a star has finally exhausted its original hydrogen, helium will be by far the most abundant residue. Hence, we should next look at reactions involving alpha particles.

If two helium nuclei unite, the resulting nuclide, $_4Be^8$, is unstable, breaking into two alpha particles in about 10^{-16} sec. However, if three helium nuclei collide simultaneously, the possibility exists of the reaction:

$$_2He^4 + {}_2He^4 + {}_2He^4 \rightarrow {}_6C^{12} \quad (7.26 \text{ Mev}). \quad (25\text{-}14)$$

Such a process, however, requires higher temperatures and densities than those existing in the sun.

When a star is old and running out of hydrogen, further collapse will raise the temperature and density sufficiently to initiate the helium reaction, which can continue to build up heavier atoms by processes such as the following:

$$
\begin{aligned}
{}_6C^{12} + {}_2He^4 &\rightarrow {}_8O^{16} \quad &(7.16 \text{ Mev}), \\
{}_8O^{16} + {}_2He^4 &\rightarrow {}_{10}Ne^{20} \quad &(4.72 \text{ Mev}), \\
{}_{10}Ne^{20} + {}_2He^4 &\rightarrow {}_{12}Mg^{24} \quad &(9.31 \text{ Mev}), \\
{}_{12}Mg^{24} + {}_2He^4 &\rightarrow {}_{14}Si^{28} \quad &(9.97 \text{ Mev}),
\end{aligned}
\quad (25\text{-}15)
$$

and so on through $_{20}Ca^{40}$, the most abundant isotope of calcium. When we try to add an alpha particle to this nuclide, however, a different situation arises, because the product is unstable:

$$
{}_{20}Ca^{40} + {}_2He^4 \rightarrow {}_{22}Ti^{44}.
$$

This substance decays, with a mean half-life of 47 years, into $_{21}Si^{44}$, which further decays in a few hours to another stable isotope of calcium, $_{20}Ca^{44}$. Successive captures of alpha particles produce $_{22}Ti^{48}$, $_{24}Cr^{52}$, and $_{26}Fe^{56}$, the most abundant isotopes of these three elements.

This process clearly favors the production of isotopes of even atomic number and even atomic weight, a result in agreement with what we find in nature, on earth, in meteorites, or in stellar, atmospheres. The even-numbered elements are more abundant than those with odd numbers. Another factor that undoubtedly enters is the tendency of the even nuclei to be more stable. Occasional captures of hydrogen residue will, of course, produce odd nuclei.

The picture looks reasonably convincing, but these reactions take place only at very high temperatures and densities. One might expect them to occur in a very old star (see Sec. 35-3). Current theories suggest that some process of this type is necessary to account for the existence of heavy atoms. The "cooking" temperature must exceed 5 billion degrees. Suddenly the star becomes unstable. It explodes, ejecting atoms into space, where they get mixed with other atoms, especially hydrogen. Then, after the passage of millennia, the material gravitates together and forms a new star. Thus the sun, like most stars, is not a "first-generation" star. Its complement of heavy atoms must have been retrieved from the remains of a star that exploded. We shall return to this question again in Chapter 27, when we discuss *supernovae*, and in Chapter 35, when we discuss stellar evolution.

One thing we should particularly note about these reactions involving the fusion of heavy atoms. They give off less and less energy per unit mass. The fusion of four protons yielded 26.72 Mev, or 7.64 Mev per mass unit. The fusion of three alpha particles to give C^{12} yielded only 0.45 Mev per mass unit, about seventeen times smaller.

Z	58	59	60	61	62	63	64	65	66	67
					Te 114 16 m	Te 115 6 m	Te 116 2.5 h	Te 117 61 m	Te 118 6.0 d	Te 119 4.7 d
51				Sb 112 0.9 m	Sb 113 7 m	Sb 114 3.4 m	Sb 115 31 m	Sb 116 60 m	Sb 117 2.8 h	Sb 118 5.1 h
50	Sn 108 9 m	Sn 109 81 m	Sn 110 4 h	Sn 111 35 m	Sn 112 0.96	Sn 113 118 d	Sn 114 0.66	Sn 115 0.35	Sn 116 14.30	Sn 117 7.61
49	In 107 32 m	In 108 58 m	In 109 4.3 h	In 110 4.9 h	In 111 2.81 d	In 112 14 m	In 113 4.28	In 114 50 d	In 115 95.12	In 116 54 m
48	Cd 106 1.22	Cd 107 6.5 h	Cd 108 0.88	Cd 109 1.3 y	Cd 110 12.39	Cd 111 12.75	Cd 112 24.07	Cd 113 12.26	Cd 114 28.86	Cd 115 43 d
47	Ag 105 40 d	Ag 106 8.3 d	Ag 107 51.82	Ag 108 5 y	Ag 109 48.18	Ag 110 260 d	Ag 111 7.5 d	Ag 112 3.2 h	Ag 113 5.3 h	Ag 114 5 s
46	Pd 104 10.97	Pd 105 22.23	Pd 106 27.33	Pd 107 7×10^6 y	Pd 108 26.71	Pd 109 13.5 h	Pd 110 11.81	Pd 111 5.5 h	Pd 112 21.0 h	Pd 113 1.5 m

Z	68	69	70	71	72	73	74	75	76	77	N
	Te 120 0.089	Te 121 150 d	Te 122 2.46	Te 123 0.87	Te 124 4.61	Te 125 6.99	Te 126 18.71	Te 127 105 d	Te 128 31.79	Te 129 33 d	
51	Sb 119 38 h	Sb 120 5.8 d	Sb 121 57.25	Sb 122 2.8 d	Sb 123 42.75	Sb 124 60.2 d	Sb 125 2.7 y	Sb 126 12.5 d	Sb 127 3.9 d	Sb 128 9.6 h	
50	Sn 118 24.03	Sn 119 8.58	Sn 120 32.85	Sn 121 25 y	Sn 122 4.72	Sn 123 125 d	Sn 124 5.94	Sn 125 9.4 d	Sn 126 10^5 y	Sn 127 2.1 h	
49	In 117 1.9 h	In 118 4.4 m	In 119 18 m	In 120 44 s	In 121 3.1 m	In 122 7.5 s	In 123 36 s	In 124 3.6 s			
48	Cd 116 7.58	Cd 117 3.2 h	Cd 118 50 m	Cd 119 9.5 m							
47	Ag 115 21 m	Ag 116 2.5 m	Ag 117 1.1 m								
46	Pd 114 2.4 m	Pd 115 45 s									
	68	69	70	71	72	73	74	75	76	77	N

i **the Heaviest Elements.** Capture of protons and of ʲ, astrophysicists believe, can account for most of the nuclides For still heavier elements, from iron to lead, other processes ʲrtant. At the high central temperatures of these special stars, of neutrons will come into play. Many nuclei readily capture process that increases N by one, in the table of nuclides. In s, neutron capture moves a nucleus one block to the right.

ʲe captures build up nuclei with a richer and richer complement or ʲ.... ʲs, until finally the nucleus can hold no more. It becomes unstable, ejecting a negative electron. As a consequence of this decay, the nucleus moves diagonally upward to the left, a step that increases Z by one, but leaves the number of mass units unchanged.

Table 25-1 represents part of the table of nuclides, showing the evolutionary path as a stable nucleus of cadmium, Cd^{106}, successively captures 21 neutrons. The heavy line with arrows indicates the evolutionary path, and the numbers at the bottom of each box give the mean lifetimes for unstable nuclides. Sometimes a given nuclide possesses two or even three varieties called *isomers*, each with a different lifetime, according to the state of excitation of the nucleus. We have here given the longest-lived isomer. The other numbers refer to the percentage occurrence of the natural isotope.

Follow the track. Cd^{107} is unstable and emits a positron, changing to stable silver Ag^{107}. Ag^{107} captures a negative electron in the nucleus and decays to palladium Pd^{108}, which is stable. The next neutron yields unstable Pd^{109}, which loses an electron and becomes stable Ag^{109}. The process, therefore, produces both of the abundant isotopes of silver. Another neutron capture, followed by electron emission, leads to stable Cd^{110}. The next four isotopes of cadmium are stable, but Cd^{115} decays through the abundant indium isotope In^{115}, unstable In^{116}, to stable tin, Sn^{118}. Note that it is very much more abundant than its neighbor, Sn^{115}, which cannot be reached by the "slow" or s-process. Presumably the tin nuclides in the range 108–115 originated by some form of proton capture, the " p-process."

The path then continues unbroken to Sn^{121}, which has two isomers of respective mean lifetimes 25 yr and 27 hr. In the "slow" or s-process (when the neutron flux is weak) Sn^{121} decays to stable antimony, Sb^{121}. Successive neutron captures continue along the path through tellurium until the next instability at Te^{127}.

In the presence of a strong flux of neutrons, Sn^{121} may be able to capture a neutron before it decays in 25 yr. When this so-called "rapid" or r-process occurs, Sn^{122} will be formed and the process continued via the other abundant isotope, Sb^{123}, while the slow process leads to Sb^{121}.

These tracks are, of course, only representative of many that can be traced through the periodic table all the way from iron up to lead and bismuth. For the last two elements the rapid process probably dominates and also accounts for the still heavier, neutron-rich, naturally radioactive substances, such as thorium and uranium.

Various nuclear reactions can lead to the evolution of neutrons necessary to sustain the foregoing reactions. For example,

$$He^3 + C^{13} \rightarrow O^{15} + n \qquad (7.1 \text{ Mev}),$$
$$He^3 + N^{15} \rightarrow F^{17} + n \qquad (5.0 \text{ Mev}),$$
$$O^{17} + He^4 \rightarrow Ne^{20} + n \qquad (0.61 \text{ Mev}).$$

In addition to these exothermic processes, a few endothermic reactions exist:

$$C^{12} + C^{12} \rightarrow Mg^{23} + n \qquad (-2.60 \text{ Mev}).$$

However, it seems doubtful that this process is effective as a neutron source.

25-11 **Very Luminous Stars.** The very luminous stars present a special problem. At its present rate of radiation, the sun could continue shining for about 10^{11} years, if it could consume all its internal hydrogen by the proton-proton chain or CNO bi-cycle reactions. However, S. Chandrasekhar has pointed out that a star can actually consume only about 10 per cent of its total hydrogen, because the nuclear processes can occur in only a very limited central volume at high temperature. Therefore, unless the hydrogen in the outer layers can reach the interior by some sort of convective process, which seems unlikely, the maximum age for the sun would be some 10^{10} years.

A main-sequence star of absolute magnitude zero, which radiates energy 100 times faster than the sun, could subsist for only 10^8 years. Similarly, a supergiant of absolute magnitude -5, which dissipates energy 10,000 times faster than the sun, should not last more than a million years. These luminous stars, therefore, must be newcomers to the celestial scene. We shall consider their birth and development in Chapter 35.

The red giant stars present another problem. They, too, are radiating energy at a far greater rate than the sun, and so they cannot have existed in their present state since the beginning of the galaxy. Moreover, their source of energy has presented a puzzle. If they are built like the stars of the main sequence, their large radii must cause their central temperatures to be very low—too low for the carbon cycle to operate. And if they are subsisting on the reactions between protons and lithium and boron, they will exhaust these rare substances rapidly and have an even shorter lifespan than main-sequence stars of similar luminosity—a time far too short to be likely, since red giants are not uncommon, and the frequency of stars of a certain type is a fair indication of their life span. For these reasons, astronomers now believe that the red giant stars do not have the same structure as main-sequence stars, but have central densities and temperatures at least as high as those in the interiors of such stars, and probably very much higher. What we see is only a tenuous, distended atmosphere of low density.

Today we infer, from our studies of globular clusters (Chapter 30), that red giant stars were once main-sequence stars. They are stars that have begun to adjust themselves to their decreasing supply of protons by changing their internal structure. This change causes them to expand superficially as they grow denser internally. The point at which they attain their maximum size marks the beginning of the stage when reactions between helium nuclei begin to take over the energy supply. Thereafter, they contract and the central temperatures rise. The stars run fairly rapidly through the cycle of neutron capture and synthesis of heavy nuclei. Eventually such a star becomes unstable, exploding as a supernova (Chapter 27), with the remnants collapsing to form a white dwarf.

All stars spend more of their lives on or near the main sequence than in any other stage except the last, when they have nearly exhausted all possible nuclear sources of energy. The more massive and luminous stars run their course most rapidly, and accordingly stay near the main sequence for a much shorter time than do stars like the sun. We shall return to these problems again when we take up the question of stellar populations and evolution (Chapter 35).

25-12   Technetium.   Element number 43 in the periodic table presents a special problem. Technetium has 14 isotopes, none of which is stable. In this respect Tc is unique among the earlier elements. Mean lifetimes of all but three of the isotopes are short, ranging from a few seconds up to four days. However, we do find three isotopes with relatively long lives: Tc^{97}, 2.6×10^6 yr; Tc^{98}, 1.5×10^6 yr; and Tc^{99}, 2.1×10^5 yr. Lifetimes of even a few million years are short compared with most stellar lifetimes; hence we should not expect to find lines of this element in the spectra of the sun and stars.

Nevertheless, as P. Merrill of Mount Wilson Observatory showed, Tc does appear in the spectra of certain late-type stars, of classes M, N, R, and S, many of them long-period variables. This fact indicates that the atmospheric shells of these stars must not be more than a few million years old. More significantly, it suggests that formation of Tc is still occurring in the core of these distended red giants and that the material somehow or other flows from the interior to the exterior. A. G. W. Cameron has suggested that Tc^{97} can result from the decay of a nucleus of molybdenum, Mo^{97}, a usually stable nuclide that becomes unstable when, at high temperatures, it absorbs an X-ray photon.

25-13 **Chemical Composition of Stars.** Chemical composition may well vary, according to the past history of the parent star that produced the various elements. In the universe we see many examples of anomalous abundances. These peculiarities apparently arose from special stellar conditions of density and pressure, which favored the production of this or that group of nuclides.

Many of the planetary nebulae and certain of the novae, especially Nova Persei 1901, appear to possess a high abundance of neon. We also find the helium stars, carbon stars, silicon stars, manganese stars, iron stars, strontium stars, zirconium stars, and rare-earth stars. The metal-poor stars of Population II, dwarfs or subdwarfs, contrast with the normal main-sequence stars of Population I. For further discussion of stellar populations see Chapter 30, 31, 32, and 35.

26

Variable Stars

26-1 What Is a Variable Star? Originally the term *variable star* was restricted to stars that vary conspicuously in brightness. In the larger sense, all stars are probably variable in one way or another. Our sun, for example, has slight variations in its spectrum during the sunspot cycle, and though its visual brightness varies inappreciably, its ultraviolet light and X rays sometimes change considerably, and its emission in the radio frequencies has striking fluctuations. However, the stars that we shall describe in this and the following chapter are much more variable than the sun.

Changes of brightness led to the discovery of the first recognized variable stars. Astronomers soon found that variations of color, of spectrum, or of radial velocity often accompanied such changes. For some stars, indeed, these changes are more conspicuous than those of brightness, and have led to the recognition of *spectrum variables* and *magnetic variables* (sec. 26-10).

There are almost twenty thousand recognized variable stars. Some, like the first-magnitude star Betelgeuse, show changes conspicuous to the naked eye (see Table 19-2). Many others were discovered visually through telescopes. But the great majority were found by intercomparison of photographs of the sky made on different dates. Photoelectric observations are necessary to detect the variability of stars whose light varies only slightly.

Individual variable stars within each constellation are named by letters and numbers that indicate the order of discovery. The first variable discovered within any constellation has the letter R assigned to it. Thus we speak of R Coronae Borealis. Subsequently discovered variables take the letters S, T, . . . , Z; then two-letter combinations are used, from RR, RS, . . . , SS, ST, . . . , and so on through ZZ. Thereafter we start at the beginning of the alphabet with AA, AB, . . . , AZ, BB, . . . , BZ, . . . , to QZ. This lettering system will take care of 334 stars in each constellation. When we have found more than this number in a given constellation, additional stars receive numbers from 335 on, preceded by the letter V, with names such as V444 Cygni and V805 Aquilae.

Visual methods for the observation of the brightness of variable stars are now largely obsolete. Photographic methods are valuable in discovering and observing large numbers of variables, for a single exposure can record hundreds or even thousands of these objects simultaneously. For highly precise measurements of brightness the photoelectric cell has no equal, even though only one variable at a time can be observed. Variable stars provide one of the most important fields of application of the photoelectric cell in present-day astronomy.

26-2   **Types of Variable Stars.**   Variable stars fall into two well-marked classes: the intrinsic and the extrinsic variables. Strictly speaking, the *extrinsic variables* are not true variable stars. Their changes of brightness are caused not by their own behavior, but by the intervention of some external action, or by the changes of aspect, as when an ellipsoidal star revolves or rotates. The *eclipsing variables*, for example, change in brightness because two stars periodically eclipse one another, either totally or partially (Sec. 21-5). Occasionally, we find stars that seem to vary slightly because obscuring material is drifting in front of them, and many of these stars show evidence of physical interaction with the interstellar matter (Sec. 26.13).

The remaining variable stars are known as *intrinsic variables*. Their changes of brightness indicate that something is happening to them. The cause must be truly physical, because changes of color, spectrum, and radial velocity accompany the changes in light. Some intrinsic variables display a more or less regular rhythm, or period, and are known as the *periodic variables* (Secs. 26-4 to 26-8). Others, only roughly periodic, are known as the cyclic or *semiregular variables* (Sec. 26-9), and these grade imperceptibly into stars whose variations show no obvious pattern, the *irregular variables* (Secs. 26-11, 26-12). The present chapter deals with the periodic, cyclic, and irregular variables. Far more spectacular are the changes shown by a group of variable stars that undergo some sort of explosion—the so-called cataclysmic variables discussed in Chapter 27. These include the *novae* or "new stars," a group of *dwarf novae*, and the *supernovae*, which undergo the largest changes and attain the greatest luminosities recorded for any variable or nonvariable stars.

The distinction between intrinsic and extrinsic variables is a useful one, but sometimes a star may fall in both categories, as when an eclipsing star, such as β Lyrae, displays true physical variations as well as the changes due to eclipses. Even the line between periodic, irregular, and cataclysmic variables is sometimes hard to draw. A few cataclysmic variables are, in addition, definitely recurrent if not strictly periodic.

26-3 Periodic Variables. The periodic variables are stars that exhibit more or less periodic changes of brightness. Most of the stars in the group are giants, and some of them are supergiants. The periods may be as short as an hour or so, or as long as three years. These periodic variable stars embrace a great variety of objects. In spectral class they range from A to M and N, and some closely allied stars have spectra of class B.

When we examine the properties of the periodic variables, we note at once that those of longest period have spectra of latest type, and that the short-period members tend to be A (or B) stars. As we have already seen, red giant stars are larger than bluer stars of similar luminosity, and therefore of far lower mean density. There is in fact a clear relationship between the periods and mean densities of many stars. Such a relationship suggests that the variations may indeed be intrinsic.

Simple arguments show that, if a star is *pulsating* or vibrating, the period of the pulsation will be inversely proportional to the square root of the star's mean density. Many of the periodic variables display a period-density relationship of this kind, and we can therefore regard them as pulsating stars. We must admit that we do not know the cause of the pulsation, but theoretical astrophysicists have found out a great deal about the physical changes that go with it.

Changes of radial velocity are observed for pulsating stars, with the same period as the changes of brightness. Therefore they represent the rise and fall of the star's atmosphere, rather than motion in an orbit, such as the eclipsing stars show. In addition, the color and spectrum change in time with the pulsations. The stars are bluest at their maximum brightness, reddest at minimum. In fact, the changes of temperature revealed by these variations of color and spectrum are the principal causes of the changes in brightness, for the star happens to be about the same size when it is brightest as when it is faintest. However, because of the difference of temperature, the surface brightness is very different on the two occasions.

Although all the periodic variables seem to share a common behavior, a physical pulsation, they do not form a continuously graded sequence. Detailed study shows that they fall into four main groups, which are found in different areas of the H-R diagram (Sec. 20-7). The stars of longest period, known as the *long-period variables* (Sec. 26-8), are red giant stars that populate quite a large area in the upper right corner of the diagram. Stars with periods between a day and fifty days or more, the *Cepheid variables* (Secs. 26-4 and 26-5), are supergiant stars with spectral types near F or G at maximum. Stars with periods less than a day, the *RR Lyrae variables* (Sec. 26-6), are confined to a small area of the diagram, with spectral types between A and F, and absolute magnitudes near zero. Between the long-period variables and the RR Lyrae stars occurs a less well-defined series of periodic variables, with giant or supergiant luminosities and spectral types between F and K (Sec. 26-7). The dwarf Cepheids (Sec. 26-10) are even fainter than the RR Lyrae stars.

26-4   Cepheid Variables.   The Cepheid variables may conveniently be studied first. They were so named because the naked-eye star δ Cephei is a typical example and was the first discovered of the group. All are giants or supergiants. In photographic and visual light they vary on the average by about a magnitude. Some have larger ranges than this, others smaller ones. The pole star, α Ursae Minoris, is a Cepheid variable with the extremely small range of about 0.1 magnitude in blue light. At maximum light, all these stars have spectral types F or G, depending on the period. All are reddest at minimum light, as we should expect, but never of spectrum later than K. Periods range all the way from a few days to several months.

The changes of brightness and of radial velocity of a Cepheid variable are illustrated in Fig. 26-1. Note that the radial velocity curve is a mirror image of the light curve. Since positive velocities imply recession, and negative velocities imply approach, we see that the surface is approaching us most rapidly when the star is at maximum light, and receding most rapidly when it is at minimum light.

Before we knew that the spectra and colors of Cepheids change continuously during the light cycle, astronomers tried to interpret the velocity curve as that of a binary system. But the results were inconsistent with the light variation, and indeed implied that one member of the supposed binary must be inside the other. When in 1914 H. Shapley advanced the hypothesis that Cepheids are pulsating stars, the binary idea was at once discarded.

The light variations of Cepheids are not all precisely similar. A definite tendency exists for stars of any one period to have light curves of similar shape. Figure 26-2(a) shows the progression in the average shape of the light curve for Cepheids in our stellar system. The definite humps at some periods and the unsymmetrical light curves at others appear to arise from peculiar features of the pulsation, such as shock waves. The velocity curves show similar humps and asymmetries; moreover, the ranges of brightness and of velocity are closely related.

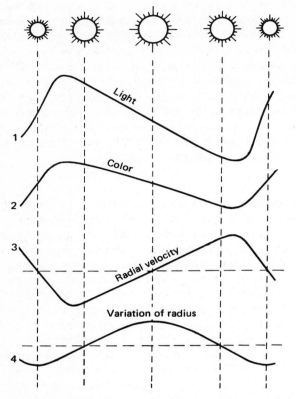

FIGURE
26-1

Variations in size: light (1), color (2), radial velocity (3), and radius (4) of a cepheid variable.

Cepheid variables are bright enough, as will be shown in the next section, to be observable in other stellar systems than our own, such as the great spiral galaxy in Andromeda. They are particularly numerous in the two nearest galaxies, the Large and the Small Magellanic Clouds. These two systems, visible to the naked eye in the southern hemisphere as two faint patches of luminosity, are close enough, about 170,000 light years, to be resolved into individual stars (see Chapter 32). Figure 26-2(b) shows the progression of the shapes of light curves with period for Cepheids in the Magellanic Clouds. It may be compared with Fig. 26-2(a) for galactic Cepheids. Clearly the general patterns are very similar, and we are justified in saying that these variable stars must be the same kind of object in the two systems. However, the differences in detail are great enough to restrain us from asserting that they are identical. In Fig. 26-2(a) and (b) the light curves are all reduced to the same horizontal (time) scale, for convenience of comparison.

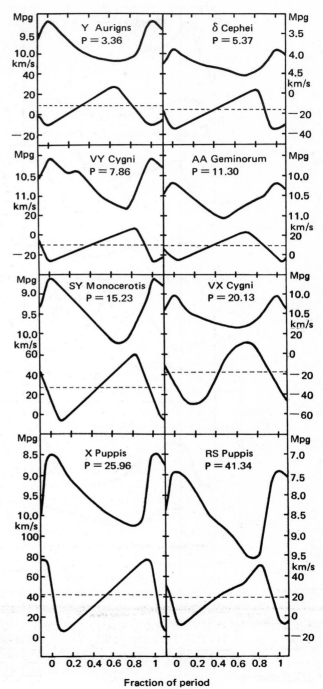

FIGURE
26-2
(a)

Shapes of light and velocity curves of galactic Cepheids. (A. H. Joy, Mount Wilson and Palomar Observatories.)

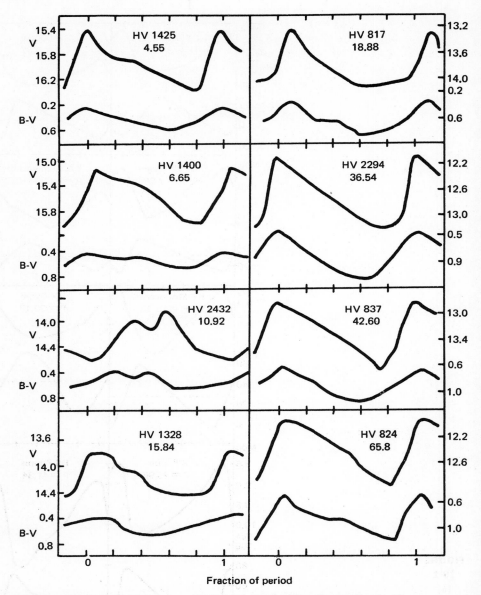

FIGURE 26-2 (b) *Shapes of light and color curves of Cepheids in the Magellanic Clouds. The stars are designated by their Harvard Variable Number. (S.C.B. Gascoigne, Mount Stromlo Observatory.)*

26-5 **Period-Luminosity Relation.** The period-luminosity relation is an important property of the Cepheid variables. The longer the period of a Cepheid, the more luminous it is. The importance of this relation in astronomy has far transcended its immediate physical implications. As we shall see in Chapter 32, it has provided one of our most powerful tools for exploring distant stellar systems in which we can see individual stars.

The two Magellanic Clouds present unusual opportunities for the study of stellar characteristics. Each contains thousands of observable variable stars, all practically at the same distance from us, except for the thickness of these systems in the line of sight. But apart from this minor uncertainty, the apparent magnitudes of the stars in each of the Magellanic Clouds differ from their absolute magnitudes only by a constant that depends on the distance.

Several thousand variable stars are known in the Magellanic Clouds, and their light curves identify most of them as Cepheids. Figure 26-3 shows the apparent magnitudes of the Cepheids of the Magellanic Clouds, plotted against the logarithms of their periods. The logarithmic scale is a useful device when, as in this case, the numbers cover so large a range. Each point represents the average brightness of one star.

FIGURE 26-3 *A period-luminosity diagram. Average apparent magnitude,* m, *of Magellanic Cepheids plotted against the logarithm of the period,* P *(in days). Note that longer periods correspond to brighter stars. The right-hand vertical scale represents the absolute magnitude,* M. *(S.C.B. Gascoigne, Mount Stromlo Observatory.)*

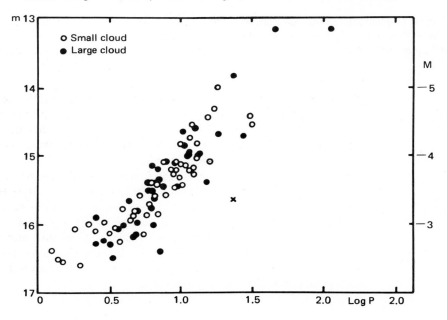

A plot of this kind was first made by Miss H. Leavitt, at Harvard Observatory, in 1910. The distribution of the points clearly defines a relationship and must express some physical property of the stars. We can draw a smooth curve through the average of the points, as shown in Fig. 26-4, which defines the relationship between the period of a Cepheid variable and its luminosity. We call it the *period-luminosity curve*. Such a relationship could not have been discovered so readily from a study of the Cepheids of our galaxy, because they are all at different distances.

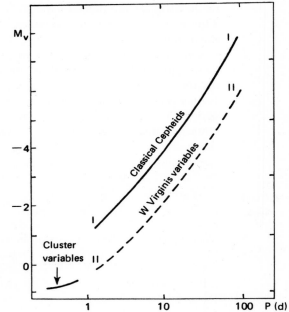

**FIGURE
26-4**

Period-luminosity curves for cluster variables, classical Cepheids, and W Virginis variables.

When the curves were first obtained, astronomers had little idea of the distance of either of the Magellanic Clouds. However, the fact that the curves for the two clouds practically coincided implied that the systems were approximately at the same distance. The problem, then, was to find some way of determining these distances.

The Magellanic Clouds are much too far away to measure their distances in any direct way, as by determination of trigonometric parallax. However the similarity between the Cepheids in the Clouds and in our galaxy (Fig. 26-2) leaves little doubt that the two groups of stars are physically similar. If we could determine the luminosities of galactic Cepheids, and then assume that those of the Magellanic Cepheids were the same, period for period, the difference between absolute and apparent magnitude for the Clouds would give their distance at once.

Few Cepheids in our Galaxy are near enough for their trigonometric parallaxes to be measured. It is possible, however, to use their proper motions for an indirect determination of their average distance (Sec. 19-3). Such methods have shown that Cepheids are giant and supergiant stars. The early determinations were beset with difficulties, and the luminosities first determined in the 1920's and 1930's were considerably revised upward in the 1950's. The best method of calibration rests on the fact that a few Cepheids are members of galactic clusters whose distances can be reliably estimated independently (Sec. 30-3). The most recent results give values of 18.7 and 19.0 for the distance moduli—or difference between apparent and absolute magnitude, respectively—for the Large and Small Magellanic Clouds. Utilizing the nomogram of Fig. 19-6, we see that the distance of the Clouds is 170,000 light years. The scale of absolute magnitudes is shown on the right-hand edge of Fig. 26-3.

Several other stellar systems, such as Messier 31 (the Andromeda nebula) and Messier 33, which are spiral galaxies, and IC 1613, an irregular system, contain Cepheids that seem to conform to the period-luminosity relation, and to have light curves like those of the galactic and Magellanic Cepheids. We can therefore use the scale of absolute magnitudes just described to determine the difference between apparent and absolute magnitude for all these systems, so that their distances can be determined. The method can be applied for any nearby galaxy containing Cepheids (Chapter 32).

The points in Fig. 26-3 possess an appreciable scatter about the mean curve. Various factors contribute to this scatter. Minor errors may exist in some of the periods or apparent magnitudes. The Clouds contain irregular dust clouds that absorb light, making some of the variables appear too faint. Also the two systems have appreciable thickness, and hence all the stars are not at exactly the same distance. Finally, the stars may themselves differ in physical structure or chemical composition. A few stars shown in the diagram have periods greater than fifty days. The galactic Cepheid having the longest recorded period is RS Puppis, with a period of 45 days, but others with longer periods may exist. No abrupt upper limit occurs for the period of a Cepheid, for Messier 31 contains some with periods well over a hundred days. Some 1150 are known in the Small Cloud and 1110 in the Large.

The Cepheids of the Magellanic Clouds, and similar stars in our galaxy, of which over six hundred are known, are often called *classical Cepheids*. We shall see that these highly luminous pulsating stars tend to be associated with dust-filled regions of space. There are other periodic variables, with periods in the same range, that do not resemble the classical Cepheids in light curve; their spectra and velocity curves also show significant differences. Unlike the Cepheids, these stars tend to be of high velocity, and are associated, as a group, with dust-free regions of interstellar space. They have proved to be less luminous, period for period, than the classical Cepheids; we discuss them in Sec. 26-7.

26-6 **RR Lyrae Stars.** The RR Lyrae stars are extremely numerous in our own galaxy. They are too faint to be seen in any but the nearest external galaxies, such as the dwarf system in Sculptor. A few have been found in the Magellanic Clouds, at about the observable limit of brightness. In Messier 31 and other more distant spirals they are below the threshold of detection. Thus, information about them depends mainly on specimens in our own galactic system. As a group, they possess a high-velocity motion that associates them with objects in the nucleus of our own galaxy and other dust-free regions of space.

Stars of this type were first discovered in large numbers in globular clusters, which themselves are objects of high velocity (Chapter 30). Hence, they were first known as *cluster-type variables*, but this name is misleading (because other types of variable stars also occur in globular clusters), and they are now known as RR Lyrae stars, since RR Lyrae is one of the brightest and best known among them.

The galactic RR Lyrae stars have absolute magnitudes near zero and spectral types at maximum near class A5. Their periods are shorter than those of classical Cepheids, usually being less than one day. Only a slight dependence of period on luminosity exists, so we commonly regard them all as having about the same luminosity. It is unfortunate that they are intrinsically too faint to be detected in the nearer spiral galaxies, since a direct measure of the difference of luminosity between them and the classical Cepheids must depend on the few faint specimens in the Magellanic Clouds.

The periods of RR Lyrae stars are related to the shapes of their light curves. Stars with periods between about half a day and a day exhibit a very steep rise of brightness of about a magnitude. For periods less than about half a day, the forms of the curves show a sharp transition to symmetry with small amplitude. The two types of light curve represent physically distinct modes of pulsation. The variables with asymmetrical and symmetrical light curves actually follow two parallel period-luminosity curves that conform to the theoretical period-density relation described above for the Cepheids. The RR Lyrae variables, then, are pulsating stars of very short period.

On account of their high velocities, RR Lyrae stars have motions far easier to measure than those of Cepheids, a fact that permits us to determine their luminosities with greater certainty. Those in our neighborhood have absolute magnitudes near zero or perhaps $+1$. When we discuss globular clusters (Chapter 30), we shall find that the absolute magnitudes of the RR Lyrae stars probably differ from one cluster to another, and that the differences have an evolutionary significance.

26-7 **The W Virginis and RV Tauri Stars.** In Sec. 26-4 we mentioned that some stars having periods in the same range as the Cepheids do not resemble the classical Cepheids in light curve or in spectrosopic details.

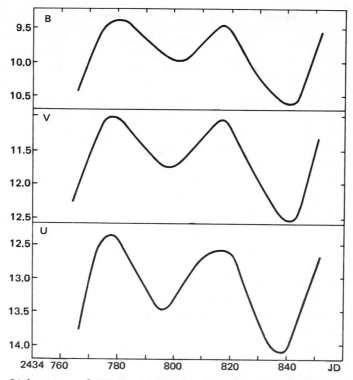

Light curves of RV Tauri in three colors. (F. Kameny, Harvard
College Observatory.)

Since the periods of these stars extend from a little more than a day to
over a hundred days, they span the range between the RR Lyrae and the
long-period variables. The group of longest period is often called the *RV
Tauri stars*, after a typical specimen (Fig. 26-5). Those with periods between ten and thirty days are similarly known as the *W Virginis stars*.
Sometimes we call the whole group *type II Cepheids*.

We have a clue about stars of these types when we find that they all
occur in globular clusters, usually in company with RR Lyrae stars; evidently they all belong to the same family. Finding them associated with RR
Lyrae stars in space, we can find their absolute magnitudes relative to those
of the RR Lyrae stars. The latter are fairly well known (Sec. 26-6), so that
we can form a period-luminosity relation for the type II Cepheids. The
curve lies parallel to the period-luminosity curve for classical or type I
Cepheids (Sec. 26-5), but the luminosity at a given period is about 1.5
magnitudes fainter (Fig. 26-4).

The type II Cepheids that occur outside globular clusters are often very
like those in the clusters, but no sharp distinction exists between these stars

and the classical Cepheids. The two groups grade into one another, as studies of their light curves, their spectra, and their motions clearly show.

A. H. Joy and R. F. Sanford of Mount Wilson and G. Wallerstein of Lick Observatory have contributed much of our knowledge concerning the spectra of type II Cepheids. These stars resemble the Cepheids in having spectra and colors characteristic of class F to G. However, they differ in displaying strong bright lines of hydrogen. Moreover, their radial-velocity curves, instead of showing a continuous variation like those of the Cepheids, are discontinuous, and for part of the period display two different sets of lines. Thus, while these stars are evidently executing some sort of pulsation, the motions of their atmospheres are far more complex than those of Cepheids. The phenomena are related to the great extent of the atmospheric envelope and the violence of its motions. The differences are not completely understood, but probably they result from differences in evolutionary stage, atmospheric structure, and perhaps also composition.

All the type II Cepheids tend to be more irregular in period, and to have light curves that repeat less precisely than those of the classical (type I) Cepheids. They show another marked difference. Whereas all the classical Cepheids lie near the galactic plane, within the layer of dust and gas (Sec. 26-5), the type II Cepheids occur at large distances from the galactic plane, and form a more nearly spherical system, like that filled by the globular clusters.

26-8   Long-Period Variables.   The long-period variables are easily recognized by their large ranges of light, between three and six magnitudes in visual and photographic brightness. Their periods have a large spread, from ninety days up to six or seven hundred. Their spectra are those of the coolest stars, classes M, S, R, and N.

Because their ranges are so large, long-period variables are easily discovered, and several of them, including Mira Ceti, "the Wonderful," have been known for several centuries (Fig. 26-6). A great many have also been found from their peculiar spectra, which display a variety of bright lines, notably the Balmer series. It is remarkable to find bright lines of hydrogen, whose production requires high temperature or strong excitation, associated with molecular bands, which occur only at low temperature.

Long-period variables are very common in our galaxy, perhaps about as common as RR Lyrae stars, of which our stellar system may contain perhaps 100,000. However, unlike the RR Lyrae stars, they are rarely or never found in globular clusters.

The physical nature of the long-period variables is not as clearly defined as that of the types of variable stars hitherto described. Though they exhibit definite periodicity, their variations are not punctual, and the forms of their light curves differ from one cycle to another. The changes of radial velocity throughout their cycles are extremely small and do not appear to

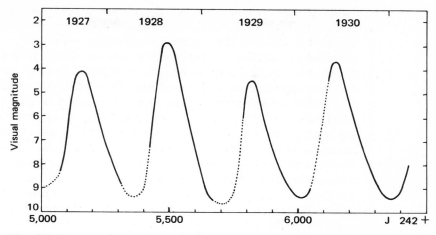

FIGURE
26-6 *Visual light curve of Mira Ceti. The lower scale represents Julian days. The dotted portions are interpolations when the star is unobservable.*

be related to the changes of light in the same way as for the Cepheids (See. 26-4). Further, although they do appear to display a relation between period and luminosity, it runs the opposite way to that for the Cepheids, at least for galactic long-period variables: long-period variables of shortest period have the brightest absolute visual magnitudes.

This apparent anomaly may be clarified in two ways. First, consider the observed variations of brightness, with ranges of several magnitudes. The variations of spectrum show that the stars are much cooler at minimum (less than 2000° K) than at maximum (about 3000° K) and therefore, by Wien's law, the maximum of spectral intensity is displaced toward longer wavelengths at minimum. Calculation shows that the maximum of the curve of spectral energy distribution vs. wavelength lies near 0.96 μ at maximum, and near 1.44 μ at minimum. Hence most of the star's radiation lies in the infrared regions of the spectrum. The cooler the star, the farther its energy curve moves into the infrared. We observe visually only a very small fraction of the total energy, much smaller when the star is faint than when it is bright. Most of the star's large change in visual magnitude results from the shift of its energy curve into the infrared. Actually, the total ranges of energy output of long-period variables are not much greater than those of Cepheids, as a bolometer or thermocouple easily demonstrates (Chapter 6). Consider, in addition, that the long-period variables show a marked relation between period and spectrum—those of shortest period have the earliest spectral class (about M0) and, therefore, the highest surface temperatures. Even at these temperatures, most of the energy is in the infrared, and the visual magnitude represents only a small fraction of the whole. For stars of longest periods, with spectra about M7, the temperature is even lower and the fraction of the total light seen visually is very much smaller still. Thus, the absolute *bolometric* magnitudes of long-period variables show a period-luminosity relation.

Second, the relation between period and luminosity for long-period variables cannot be determined by a study of such stars in a remote system, as was done for the Cepheids, for we know of no distant system that contains long-period variable stars. It is true that a few luminous stars of comparable period have been found in the Magellanic Clouds, but their ranges are rather small and their spectra, when observable, have been mostly of classes F and G, though at least one is an M star. The absolute magnitudes of long-period variables must thus be determined by indirect methods, from proper motions and radial velocities. When we examine the observations that have gone into these determinations, we notice that long-period variables of the longest periods tend to occupy the same regions as the galactic Cepheids, within the gaseous, dusty regions of the galactic plane, whereas those of the shortest periods are distributed like the RR Lyrae stars and are evidently associated with the dust-free interstellar regions. Thus long-period variables probably represent not a single family of stars but a mixtute of two and perhaps more.

Although the relation between light, spectrum, and radial velocity is not so clear-cut as for Cepheids, the long-period variables too are probably executing pulsations of some kind. But their spectral peculiarities indicate that they possess a very complex atmospheric structure.

We have previously noted that the presence of hydrogen emission suggests high-temperature excitation, whereas the molecular bands point to a low-temperature atmosphere. It happens that the H line of ionized calcium, a particularly strong absorption line in these spectra, coincides almost exactly with the Balmer line Hε. Whereas most of the Balmer lines appear in emission, Hε is always absent or very weak indeed over most of the cycle. Clearly, the hydrogen emission line is absorbed by the calcium gas. Other missing Balmer lines farther to the violet are similarly found to coincide with strong absorption lines of iron. We infer that the hydrogen emission lies deep in the atmosphere, and that the low-temperature metallic lines arise in the outer parts of the atmosphere. Here, then, is evidence that long-period variables possess extensive atmospheres, which are more highly excited in the deeper layers. Doubtless, the pulsations of such an atmosphere would be complicated phenomena, perhaps to the degree that the effects of several successive pulsations can be seen at one time.

The type example of long-period variables, Mira Ceti, is a double star. The spectrum of a faint companion can be studied only when the red star is at minimum. The companion is a hot star of low luminosity, lying perhaps five magnitudes below the main sequence in the H-R diagram. However, when the large star is at minimum, the small, hot star may become visible and produce the observed spectrum. Further discussion will appear in Chapter 28, including the occasional association of such stars with planetary nebulae.

As mentioned, long-period variables may have spectra of class M, S, R, and N. All of them show similar behavior, and the bright-line phenomena do not depend on the composition of the stellar envelopes. It seems, then, that composition (at least insofar as it involves carbon and oxygen content) has nothing to do with whether a star can be a long-period variable. The same conclusion can be drawn for the semiregular and irregular red variables.

26-9 **The Semiregular Red Variables** form a group of stars that extends continuously into the long-period variables. They have similar periods and spectra, but their ranges of brightness and their spectral changes are much smaller. Some show weak bright lines; others have small ranges and no obvious emission. Many red semiregular variables have luminosities very like those of long-period variables. A well-defined group contains a number of supergiant M stars, such as Betelgeuse and Antares, which have absolute visual magnitudes near -4, and are large enough in angular diameter to be measured with the interferometer (Sec. 22-3).

All these stars may be considered cyclic rather than periodic. The lengths of individual cycles and the forms of individual light variations are much more irregular than for the long-period variables, which in turn are less regular than the Cepheids. Probably these semiregular red variables are stars that are, so to speak, on the verge of being long-period variables.

The semiregular red variables, again, grade imperceptibly into the *irregular red variables* of spectral classes M, S, R, and N. These have very small ranges and erratic variations. The American astronomers J. Stebbins and A. E. Whitford established in the 1930's, by very accurate photoelectric studies, that all giant stars of these spectra vary slightly, and that the range of variation is greatest for stars of latest spectral class and of highest luminosity.

26-10 **Other Periodic Intrinsic Variables.** The variable stars hitherto discussed are all of spectral class A or later. An interesting group of periodic variables occurs near spectral class B2. They bear the name of *β Canis Majoris stars* after a typical example. Here, the most striking variations are those of radial velocity, but small changes of brightness are also present. Their periods are very small, from four to six hours, and their magnitude ranges are almost always less than 0.1. The few known members form a series roughly parallel to the main sequence. They show a period-luminosity relation.

The striking feature of the β Canis Majoris stars is the coexistence of two or more periods, which are nearly alike and cause periodic amplitude

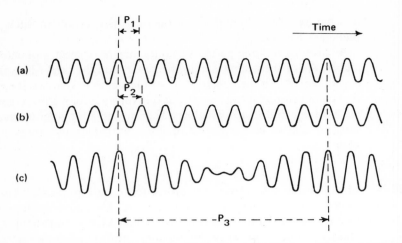

**FIGURE
26-7** *Illustration of the mechanism by which two oscillations combine to produce beats. Two regular (harmonic) oscillations* a *and* b *of different periods* P_1 *and* P_2 *combine as in* c *to produce a modulated oscillation with alternating high and low peaks with the beat period* P_3. *The high peaks of* c *occur when* a *and* b *are in additive or "constructive" interference; at the low peaks, when* a *and* b *are in subtractive or "destructive" interference.*

fluctuations in the velocity curve. The mechanism by which "beats" occur between two oscillations on different frequencies is illustrated in Fig. 26-7. Most of the stars show small changes in spectrum that can be related to the changes of velocity.

We owe most of our knowledge of these stars to W. F. Meyer and to O. Struve and his collaborators. Several theories have been advanced to interpret the observations. P. Ledoux thought that forced oscillations by a very small, dense companion might be responsible. D. H. Menzel and O. Struve have suggested that the effects result from stellar magnetic fields; pulsations of the star would then be affected, because the periods at the star's equator and pole would differ slightly. This theory holds considerable promise, but no magnetic effects are observed in the spectra of the β Canis Majoris stars. On the whole it seems probable the the β Canis Majoris stars are pulsating. Their period-luminosity relation makes such activity plausible, but we do not know exactly what causes the pulsation and produces the two nearly equal periods. Chandrasekhar and Lebovitz endeavor to explain the instability in terms of a complex coupling between two vibrational patterns having nearly the same frequency.

Another class of stars, the so-called *dwarf Cepheids*, also shows evidence of two periods that combine to produce "beats." These have periods usually near a tenth of a day, and ranges of magnitude about 0.25 on the average. In luminosity they lie near to, perhaps slightly below, the main sequence. Their spectral types are from A to F. Their velocity curves and their light curves are closely related and show simultaneous fluctuations of amplitude ("beats") that can be analyzed into the two periods. The star of shortest known period, $1^h 19^m = 0.055^d$, SX Phoenicis, belongs to this class of stars. Probably the dwarf Cepheids are closely allied to the classical Cepheids, but at present we know only half a dozen of them, and we need more information before their relationships can be determined with certainty. Some stars, like SX Phe, appear not to show the "beat" phenomenon.

Another important type of star is the *magnetic variable*. These stars were first discovered because their spectra showed large periodic variations in lines sensitive to the zeeman effect (Sec. 9.5). One of the most striking is Cor Caroli (α Canum Venaticorum), extensively studied by O. Struve. Horace Babcock at Mount Wilson Observatory found that many of these spectrum variables possess strong variable magnetic fields. Most of them are also slightly variable in brightness, but with amplitudes of the order of only 0.01 magnitude.

This group of stars is still very incompletely studied. A. Deutsch, at Mount Wilson Observatory, has found that a number of them are periodic. The spectrum, the magnetic field, and the brightness all seem to show the same period. The observations of magnetic field and light variations are very exacting, and it is difficult to say definitely whether all the stars actually vary in all three properties. However, it seems likely that the three types of variation are associated with the same cause. Most of the known periods lie between half a day and ten days. The spectra appear to be class A.

Some investigators believe that these variables are *oblique rotators*— rotating stars that possess a magnetic axis not coincident with the axis of rotation. We may remark that the earth is in fact such a body, since the magnetic poles do not coincide with the geographic poles. Others attribute the variations to a sort of pulsation of the stellar magnetic fields themselves.

With reference to magnetic stars, RR Lyrae itself has been found by H. Babcock to possess a large magnetic field, variable but not with the same period as that of the variations in brightness, radial velocity, and spectrum. Whether other stars of the RR Lyrae class possess such fields is not known, because most of them are too faint to permit this type of observation.

26-11   Irregular Variables.   Truly irregular variables, like R Coronae Borealis, suffer brightness changes in abrupt and unpredictable fashion (Fig. 26-8). Such a star may continue at constant brightness or, after tiny fluctuations for months or years, drop suddenly in brightness by perhaps six magnitudes in days or weeks. It may then execute erratic variations and finally return gradually to maximum brightness, sometimes after an interval of several years.

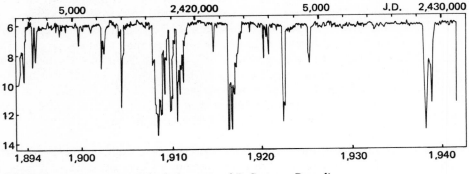

FIGURE 26-8   *Thirty years of the light curve of R Coronae Borealis.*

The spectra of R Coronae stars show them to be rich in carbon, as L. Berman of Lick Observatory first discovered for R Coronae Borealis itself. This property is universal for such stars. Strong carbon bands led to the actual discovery of one of these stars. G. Herbig, at Lick Observatory, found that R Coronae Borealis shows a bright-line spectrum at minimum brightness. The R Coronae Borealis stars are very luminous. Their spectra show the marks of high luminosity, and several of them are bright enough to be observed in the Large Magellanic Cloud at absolute photographic magnitude − 5.

26-12   Flare Stars. Flare stars are radically different from any intrinsic variables hitherto considered, all of which, except the dwarf Cepheids, were giant stars. Flare stars are main-sequence stars. Occasionally, a red K or M dwarf star may brighten abruptly by several magnitudes for a very short time, then rapidly revert to its usual brightness in a matter of minutes (Fig. 26-9a). These stellar flares occur erratically; no periodicity is detectable. In order to estimate the average time interval between successive flares, a systematic surveillance or photometric monitoring of typical flare stars is preferable to statistical considerations based on the fraction of unplanned observations that happened to catch various stars during a flare. The amplitudes of the flares must be considered also; some produce a very large increase in the star luminosity (especially in the ultraviolet), while others, more frequent, cause only a modest increase ("microflares"). In three typical stars, AD Leonis, YX Canis Minoris, and Wolf 359, monitored by W. Kunkel at McDonald Observatory in 1965, microflares occurred as often as several times each night, while the spectacular larger flares were observed with frequencies of once a week to once a month (Fig. 26-10). It may be noted that our nearest neighbor among the stars, Proxima Centauri (Table 19-1), is a flare star.

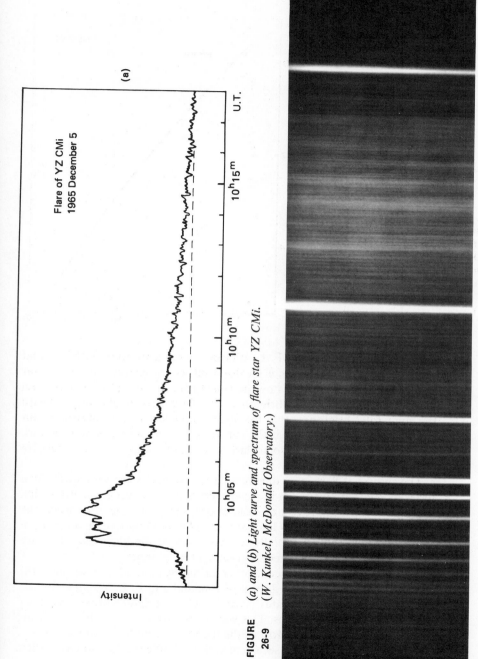

FIGURE 26-9 *(a) and (b) Light curve and spectrum of flare star YZ CMi. (W. Kunkel, McDonald Observatory.)*

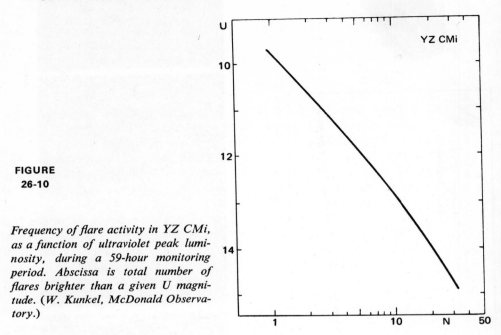

**FIGURE
26-10**

Frequency of flare activity in YZ CMi, as a function of ultraviolet peak luminosity, during a 59-hour monitoring period. Abscissa is total number of flares brighter than a given U magnitude. (W. Kunkel, McDonald Observatory.)

Following G. Kron and F. L. Whipple, most astronomers believe that the phenomenon represents a localized release of energy within the star atmosphere similar to flares on the sun. From this point of view the sun itself is a flare star, but the brightness of a flare is so small compared with the sun's total brightness that the solar luminosity is not appreciably increased during a flare. But if the sun were as faint as Proxima Centauri, an eruption as bright as a solar flare would cause a much greater relative change in total brightness.

The spectra of flare stars during a flare have been observed initially in a series of photographs taken in rapid succession, and more recently with a moving plate to improve time-resolution. Such spectrograms reveal the presence of bright emission lines and of enhanced continuum emission in the ultraviolet (Fig. 26-9b). Both facts confirm the view that the flare phenomena occur in localized, high-temperature regions.

From continued simultaneous observations by radio (125-m wavelength) and photography, A. C. B. Lovell of England and F. L. Whipple and L. Solomon of the United States proved that the brighter optical flares in UV Ceti are also radio flares. This was the first radio noise observed from a star other than the sun. Most of the flare-star radio flares appear similar to type I solar radio flares (see Chapter 9), but, like the optical flares, they are orders of magnitude more intense than those on the sun. Also UV Ceti shows some evidence of a 5-year cyclic change in flare frequency, perhaps analogous to the solar 11-year cycle. There seems to be little question that the flare phenomena are similar in physical character to both the flare and the sun.

26-13   T Tauri Stars. The T Tauri stars exhibit erratic variations that may be partly extrinsic. These objects, of which a great number are now known, have luminosities comparable to that of the sun. They tend to occur in groups, which are always associated with regions full of interstellar nebulosity and dust. Their spectra are peculiar, with bright lines of hydrogen and the metals, and continua of unusual energy distribution.

The studies of A. H. Joy, O. Struve, G. H. Herbig, and others have shown without question that the T Tauri stars actually lie within the nebulous masses, and that their behavior represents interaction between the stars and the nebulosity. A. H. Joy pointed out, and O. Struve confirmed, that they occur in abnormally rich concentrations of stars; hence it is possible that they are actually being born within, and out of, the nebular material. G. H. Herbig has actually observed the appearance of previously unobserved stars in such regions. The Armenian astronomer V. A. Ambartsumian sees in these compact groups of variable stars embedded in nebulosity, which he styles "T associations," a special class of stars, possibly stars in the condensing stage (see Chapter 27). Evidence in support of this interpretation is found in the fact, noted by M. Walker at Lick Observatory, that T Tauri stars are observed in extremely young galactic clusters (Chapter 30). Herbig finds that FU Orionis shows all the characteristics expected of a very young star. It does not, however, exhibit the characteristic spectrum of a T Tauri star.

26-14   Stellar Evolution and Variability. We may surmise the evolutionary stages of variable stars from the physical properties described in this chapter. The evidence from star clusters is especially important. In brief, the T Tauri stars seem to occur during an early stage of stellar evolution. While near the main sequence, stars do not tend toward variability, except for local flares. After evolution removes them from the main sequence, stars may become Cepheids. Long-period variables may mark the extreme point of distention of a star, while it is changing from proton reactions to the helium reactions as a source of energy. During such evolutionary stages, as C. Hoffmeister and R. Kippenhahn have shown, massive stars may oscillate in one or more patterns, analogous to the "fundamental" or "first harmonic" modes in a musical instrument. The more helium the star contains, the more likely will be the excitation of the first harmonic. The semiregular red variables may mark the edges of the domain in which stars incline toward long-period variability. During its contraction, after the helium reactions come into play, the star reaches another point at which pulsation is initiated, and becomes an RR Lyrae star. This part of the evolution is not yet understood. The W Virginis and RV Tauri stars might be RR Lyrae stars, with some exceptional properties.

This scheme does not include the β Canis Majoris stars or the magnetic variables, and we do not know why these types of variation are confined to such a small range of spectral class and luminosity.

TABLE 26-1

Main types of intrinsic variables

Type	Period range	Character-istic mag. range	Character-istic spectrum	Mean Abs.Mag.	Space Distribution
Classical cepheids	$2-80^d$	1 mag.	F, G superg.	-3	Dust-filled regions, galactic plane
RR Lyrae (or cluster)	$0.1-1^d$	1 mag.	A, F giants	0	Dust-free regions, nucleus of Galaxy
W Vir, RV Tau (or Type II cepheids)	$1-100^d$	1 mag.	F–G, G–K	-2	High galactic latitude, halo
Long period	$90-600^d$	3–6 mag.	M, S, R, N, emission	$-1, 0$	Dust-free regions, galactic plane
Semiregular	$\sim 100^d$	1 mag.	M, S, R, N	-2	Dust-filled regions, nucleus of Galaxy
Irregular	—	0.1 mag.?	M, S, R, N	-2	Dust-filled regions, nucleus of Galaxy
β Canis Majoris	$3-6^h$	0.1 mag.	B	-3	Dust-filled regions
Dwarf cepheids	$1-3^h$	0.2–1 mag.	A–F	$+2$	Dust-filled regions
Magnetic or spectrum	$0.5-10^d$	0.1 mag.	A	0	
R Cor Bor	irreg.	6 mag.	G, K, R emission	-3	Low galactic latitude, carbon
Flare stars	irreg.	6 mag.	K, M emission	$+10$	Lower main sequence
T Tau	irreg.	1–3 mag.	G, K–M	$+5, +2$	Dark clouds of dust and gas

Table 26-1 summarizes all the types of variable stars described in this chapter, giving their range in period and in magnitude, their mean absolute magnitudes, and their characteristic spectra. Figure 26-11 is the H-R diagram for all types of intrinsic variables and illustrates the characteristics summarized in the table.

We turn now to the explosive or cataclysmic variables.

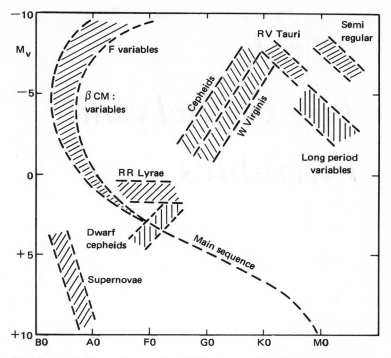

FIGURE
26-11 *Position of different classes of variable stars on H-R diagram.*

CHAPTER 27

The Cataclysmic Variables

27-1   Explosive Variables.   For all the intrinsic variable stars hitherto considered, the changes in brightness result from moderate variations of surface temperature and radius. Cataclysmic variables, on the other hand, apparently undergo some sort of explosion. The brightness rises abruptly, quickly reaches maximum, and then fades away more gradually. For example, the luminosity of an ordinary nova may increase by a factor of about 100,000 (12.5 mag.) within only a day or two.

If a mere increase of surface brightness were responsible, the Stefan-Boltzmann law (Sec. 8-7) requires a more than twentyfold rise in temperature. Or, at the other extreme, if the star were to expand, keeping its surface brightness constant, the radius of the star would have to increase by a factor of about 400. For a *supernova*, which brightens far more than does an ordinary nova, the corresponding figures would be even more extreme.

Changes in both temperature and radius occur, but the latter appears to be the more important. We observe high radial velocities of approach. Therefore, we are indeed dealing with rapid radial expansion of the star.

As for the temperature, the spectral changes occurring during the outburst suggest that the star grows cooler as it nears maximum. The temperature is probably appreciably lower than before the outburst, a fact we infer from measures of the star's color prior to and during the outburst.

Hence we must conclude that the sudden brightening of novae results from a rapid swelling of their photospheres. Indeed, we may truthfully say that the star is undergoing an explosion.

The spectrum of a nova after maximum light bears out the idea of an explosion. Bright lines appear and grow progressively more conspicuous, as if matter had left the surface of the star, surrounding it with a sort of extensive chromosphere. As the process continues, the spectrum gives evidence of decreasing density. Bright lines characteristic of the most diffuse astronomical objects, the gaseous nebulae, finally appear. We infer, therefore, that the envelope of the star is expanding into the surrounding space. In fact, the nova process clearly requires the explosive ejection of part of the star. The details differ for novae of various kinds, and indeed no two explosions are exactly alike.

27-2 Classification of Cataclysmic Variables. The classification of such variables relies on the intensity of the outburst. We recognize three groups of such stars. The first contains the *dwarf novae*, usually known as the *SS Cygni* or *U Geminorum* stars. These "repeating variables" have ranges up to about six magnitudes; intrinsically faint at minimum, at maximum they may attain about zero absolute magnitude. They repeat their outbursts at quasi-periodic intervals of a few weeks or months. The second group contains the ordinary *novae*, which have ranges of about twelve magnitudes. They, too, are of low luminosity at minimum. Most normal novae have, in our experience, brightened only once. However, a closely allied group, the *recurrent novae*, repeat their explosions at intervals of a few decades. The third group includes the *supernovae*, which have very large ranges and attain enormous luminosities. The supernovae also subdivide into two classes that differ in both luminosity and spectral characteristics.

TABLE 27-1 *Main types of cataclysmic variables*

Type	Period range	Magnitude range	Abs. Mag. (max.)	Spectra
Dwarf novae (SS Cyg, U Gem)	A few months	6	0	Emission at min., absorption at max.
Recurrent novae	Several decades	8 to 10	-4 to -8	Coronal lines
Novae		10 to 12	-6 to -10	B, A, F (Ia); absorption before max., emission at max.
Supernovae		12 to 14	-16 to -20	Broad emission bands

We shall now discuss each of these classes in more detail. A summary of their properties appears in Table 27-1.

27-3 **Dwarf Novae.** The dwarf novae go through a well-defined series of spectral changes. Their spectra at minimum ordinarily show bright lines, those of hydrogen and helium being by far the most conspicuous. These lines are fairly broad. The light may fluctuate slightly during minimum, but no major change occurs until the star suddenly brightens. The rapid change may often increase the star's brightness six- to tenfold in less than a day. The actual increase can differ from one maximum to the next, and the details of the change of light and shape and duration of the maximum are not the same on each occasion. The times of brightening are not exactly periodic, and the interval from one cycle to the next often differs greatly from the average cycle for the star. A series of sample maxima of the typical example SS Cygni appears in Fig. 27-1.

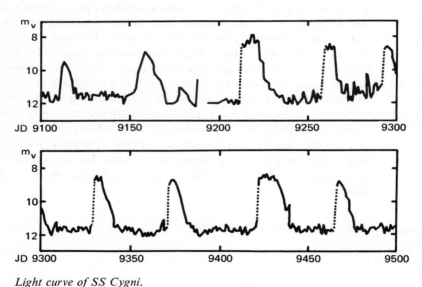

FIGURE 27-1 *Light curve of SS Cygni.*

As the star brightens, the emission lines that characterize the minimum disappear. An intensely blue continuous spectrum develops, indicative of very high temperature. Broad, shallow absorption lines replace the bright lines. At some maxima they are so broad and shallow as to be difficult to see. Hydrogen lines are present, and the high temperature is further attested by the presence of ionized helium. As the star begins to decline after maximum, weak emission lines appear at the centers or the broad absorptions, gradually replacing them as the strong continuum fades away and minimum is approached.

As we shall see when we examine the spectra of ordinary novae, the dwarf novae differ from them in several respects. They show no negative radial velocity (no perceptible rise of their surface layers), and the bright lines are not displaced relative to the absorption lines; in the novae proper, the absorption lines lie at the violet edges of the bright lines. Moreover, although they have bright-line spectra at minimum, the dwarf novae never show the bright lines characteristic of gaseous nebulae. The very broad absorption lines suggest violent mass motions in the star's envelope, or a large Stark effect (Sec. 23-5). Perhaps these stars have envelopes that swell abruptly and then contract again, rather than blowing off explosively into space.

At least three of the dwarf novae are binary systems of very short period. SS Cygni consists of a K dwarf (or subdwarf) and a blue star that undergoes the outbursts; they revolve round one another in 0.7 day. The outbursts recur at much longer intervals than the orbital period, but it is possible that they are caused by the close proximity of the K star. The distended envelope at its largest is about the same size as the K star, and about fills a critical volume governed by the mutual gravitation of the two. Possibly the envelope of the blue component swells until it fills this critical volume. Very likely all the dwarf novae are binary systems similar to SS Cygni, and their outbursts are in some way triggered by the proximity of their companions. At least one nova is also a close binary. Hence, a blue star with a nearby companion can also undergo a far more violent explosion, with the ejection of material into space.

27-4   Recurrent Novae.   The Recurrent Novae are stars that have gone through novalike outbursts more than once during the past century. Their ranges of light, between eight and ten magnitudes, are much larger than those of the dwarf novae, and their outbursts occur at irregular intervals of several decades. The spectral changes are much more like those of

TABLE 27-2   *Recurrent novae*

Star	Max.	Min.	Years of Maxima
T Cor B	2.0	10.6	1866, 1946
U Sco	8.9	17.6	1863, 1906, 1936
T Pyx	7.2	13.9	1890, 1902, 1920, 1944
RS Oph	4.3	11.7	1898, 1933, 1958
V1017 Sgr	7.2	14.0	1901, 1919
WZ Sge	7.3	15.2	1913, 1946
VZ Aqr	8.0	<16	1907, 1962

novae proper than of dwarf novae. Because of the short span of our observations, only a few have been observed (Table 27-2), but it appears possible that many of the recently observed novae may repeat their outbursts at intervals of centuries or longer.

The fact that a star can become a nova more than once has great significance. The nova process, violent though it appears to be, can be only a passing incident in the history of the star, which returns to very nearly the same condition that existed before the outburst. Later on, the star repeats the same process again (Fig. 27-2).

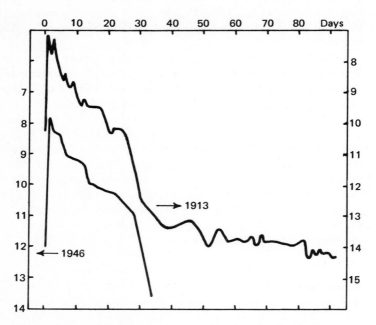

FIGURE 27-2   *Two outbursts of a recurrent nova, T. Sagittae (1913, 1946).*

At least one of the recurrent novae, T Coronae Borealis, is a member of a binary system. The star that undergoes the explosions is a faint blue star of high temperature, and its companion is a red giant star. Possibly other recurrent novae and perhaps all novae are also binary systems. The rapid periodic variations of the remnants of nova DQ Hercules 1934 (Fig. 27-3), first observed by M. Walker at Lick Observatory, are consistent with an ultrarapid orbital motion of close, dense stars.

The stars of this class that have been bright enough for spectroscopic study have displayed some remarkable features—namely, the development of intense emission in the red and green lines characteristic of the spectrum of the solar corona. These lines are emitted respectively by ionized atoms

**FIGURE
27-3** *Nova DQ Herculis 1934. (Left) Near maximum (March 10, 1935). (Right) Near intermediate minimum (May 6, 1935). Compare with light curve in Fig. 27-4.*

of Fe X and Fe XIV, which implies that certain regions of the envelope after the outburst possess temperatures of the order of a million degrees. The increase of brightness and the other features of the spectrum suggest temperatures of perhaps 50,000° K. These very intense coronal lines are characteristic of the recurrent novae, but the same lines have been detected, though they were much fainter, in the spectra of a few of the more violent ordinary novae.

The recurrent nova T Coronae Borealis had major outbursts in 1866 and in 1946, but it is not quiescent when it is faint. For several months before the brightening of 1946 the blue component of the binary was varying slightly, and its spectrum gave evidence of considerable activity. Since its decline from its bright outburst the star has continued to vary in brightness, slightly but spasmodically. Perhaps the binary nature of the system predisposes it to instability.

27-5 The Normal Novae. The novae proper occur rather frequently in our own stellar system. More than a hundred have been observed during the past one hundred years. Many others have doubtlessly been missed, some because they are distant and faint, others because they may have appeared in the part of the sky temporarily concealed by the sun. A fair estimate is that 25 novae, brighter at maximum than the ninth magnitude, occur every year in our galaxy. Novae visible with the naked eye are not uncommon, and some have been as bright as Sirius when at maximum. Novae are designated by constellation and year of appearance; Nova Aquilae 1918 was the brightest seen in this century. Table 27-3 gives a list of bright galactic novae.

**TABLE
27-3**   *Selected bright galactic novae*

Star		Max.	Min.	M (max.)	Type
μ Car	1843	−0.8	8	(−11)	v. slow
V841 Oph	1848	5.0	12.6	−7.8	?
T Cor B	1866	2.0	10.6	−8.4	v. fast
Q Cyg	1876	3.0	14.8	−8.3	v. fast
T Aur	1891	4.2	14.8	(−5.3)	slow
V1059 Sgr	1898	4.9	16.5	−8.3	v. fast
GK Per	1901	0.2	13.5	−8.3	v. fast
RS Oph	1901	4.3	11.7	−8.3	v. fast
DM Gem	1903	5.0	16.5	−8.3	v. fast
DI Lac	1910	4.6	14.0	−7.3	fast
DN Gem	1912	3.5	14.8	−7.7	fast
V603 Aql	1918	−1.1	10.5	−8.4	v. fast
GI Mon	1918	5.6	15.1	−8.1	fast
V476 Cyg	1920	2.0	16.1	−8.3	v. fast
RR Pic	1925	1.2	12.7	−6.2	slow
DQ Her	1934	1.4	14	(−5.5)	slow
CP Lac	1936	2.1	15.3	−8.3	v. fast
V630 Sgr	1936	4.5	15	−8.4	v. fast
Mon	1939	4.3	16	−7	fast
CP Pup	1942	0.2	17.0	(−11)	v. fast
DK Lac	1950	5.4	13.4	−8.0	fast
Her	1963	3.9	>8	−8.5	fast

We know little about the prenova stage. There is no way of predicting when or whether a particular star will become a nova; only after the event can we search collections of photographs to discover what the star was like before the explosion. About twenty novae have been traced on photographs made before their explosion. All were faint stars with a slight tendency to variability. Only one prenova spectrum has been recorded, that of Nova Aquilae 1918. This star was near the eleventh magnitude and was recorded at Harvard on several objective-prism plates. The spectrum was described by Miss Cannon as resembling an A star in energy distribution, so it was probably of early type, but no spectral lines were visible, perhaps because the spectra were all of poor quality. We do not know whether other novae had similar spectra before their outbursts, but many astronomers regard this similarity as probable.

The brightness of the prenova cannot always be determined, for many are too faint at minimum to be observed at all. Those that are well observed have an average range of about thirteen magnitudes, corresponding to an increase in brightness by a factor of 160,000. Usually the rise in brightness requires only a day or two. The absolute visual or photographic luminosity at maximum is very high, ranging up to a million times that of the sun. Some novae decline rapidly from their peak of luminosity. Others fade

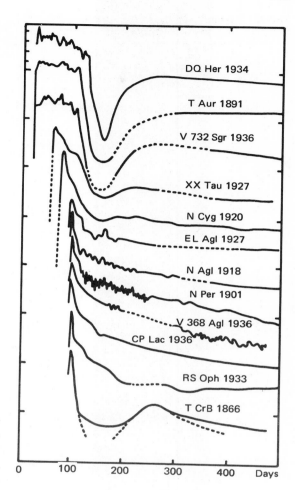

DQ Her 1934

T Aur 1891

V 732 Sgr 1936

XX Tau 1927

N Cyg 1920

EL Aql 1927

N Aql 1918

N Per 1901

V 368 Aql 1936

CP Lac 1936

RS Oph 1933

T CrB 1866

0   100   200   300   400   Days

**FIGURE
27-4**

Typical light curves of novae, arranged in systematic order. (L. Campbell, Harvard College Observatory.)

more slowly, often with irregular oscillations (Fig. 27-4). The novae that run their course most rapidly, the fast novae, attain absolute magnitudes −9 to −10 at maximum. The slow novae may be several magnitudes fainter, but all are extremely luminous, even compared with supergiant stars and high-luminosity variables such as the Cepheids. With an average range of 12 to 14 magnitudes, the prenova must have an absolute magnitude about +4 to +6, comparable to that of the sun. However, all prenovae are almost certainly much bluer than the sun, and therefore much smaller and denser. When placed on the H-R diagram, they fall several magnitudes below the main sequence, around spectral type A (see Fig. 26-8).

In the short observable interval before maximum, the spectrum of a nova shows absorption line strongly displaced to the violet, indicative of the rapid expansion of the star's envelope that accompanies the explosion.

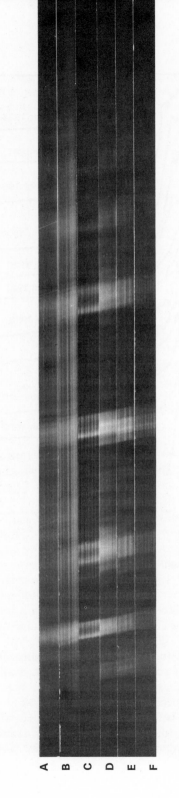

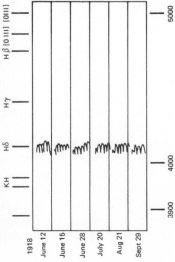

**FIGURE
27-5**

(*Above*) *Spectra of Nova Aquilae, 1918.* (*a*) *June 12, 1918;*
(*b*) *June 15, 1918;* (*c*) *June 28, 1918;* (*d*) *July 26, 1918;* (*e*)
August 21, 1918; (*f*) *September 24, 1918.* (*Left*) *Note the
gradual emergence of the emission lines, and the doubling of
the lines followed by intense emission centers.* (*Harvard
College Observatory.*)

Most novae at this stage have spectra like those of supergiant B, A, or F stars. At one time astronomers thought that the whole star might expand and finally burst. However, we now believe that the explosion is confined to a fairly thin layer of the star's envelope, which expands and makes the whole star appear to swell. However, the star itself has undergone little change, and after the bubble has blown off, it reappears unaltered in either brightness or color.

During the early stages of the outburst, the shell of gas is thick enough to hide the star that lies within it. We see only a little way into the expanding envelope. The limiting depth to which we can see defines the photosphere, which rises as the envelope expands, and the regions above it provide the reversing layer where the absorption spectrum is formed.

The temperature at maximum is probably not very different from that of the star at minimum. If anything, the surface is cooler. Thus we must ascribe the great change in brightness to a change in surface area. If the surface brightness stays constant, the increase in area by a factor of 160,000 corresponds to an increase in diameter by a factor of 400. If the surface brightness has fallen, the increase of diameter must be even greater. Here is a figure that we can check roughly. If the prenova had a radius of 10^5 km, the expansion would increase it by a factor of 400, to 4×10^7 km. The surface would therefore move through 400×10^5 km in a couple of days, about 2×10^5 sec, and the speed of expansion would be about 200 km/sec. Many of the observed expansion velocities have been greater than this, up to about 1000 km/sec. As pointed out above, the expansion factor is probably even greater. Also, the rise of the photosphere will not be identical with the outward velocities of the atoms, because the density of the photosphere decreases as it expands, and the photospheric level will therefore not move outward as fast as the atoms. When these factors are considered, the agreement in order of magnitude is sufficient to suggest that the hypothesis of an expansion fits the facts.

Actually the ejected material probably does not form a sphere. It may be more like a doughnut thrown off from the star's equator. Or, possibly, matter ejected from opposite poles may build up a dumbbell-shaped envelope. Magnetic fields of the star may participate in focussing the ionized gas. The differences between novae may in part result from viewing the phenomenon from different angles.

Almost immediately after maximum brightness the spectrum undergoes a striking transformation (Fig. 27-5). Prior to maximum it has consisted principally of absorption lines. Emission lines now appear on the redward edges, particularly of the Balmer lines and the lines of ionized metals. These lines rapidly become the dominant features of the spectrum. Meanwhile, the absorption spectrum grows complex, suggesting successive shells of gas leaving the star's surface with different speeds. These newly appearing spectra are always associated with higher outward velocities than the original spectrum. The lines tend to be broad and diffuse, with emission on the red wings. The analysis of the nova spectrum becomes exceedingly complicated and the details are difficult to unravel. The degree of excitation increases progressively, but the absorption spectrum at this stage does not resemble anything in the normal spectral sequence.

As the density of the expanding envelope dwindles, the brightness fades. The continuous spectrum weakens, and the absorption lines diminish in intensity and finally disappear. Apparently the effective photospheric envelope has grown too rarefied to emit a continuous spectrum. It slowly becomes transparent as the expanding gas evolves into a nebula. The bright lines become relatively more conspicuous. At the beginning of the nebular stage we often observe strong and very broad emission lines of nitrogen (N III). Presumably they are broadened by Doppler effect, but the associated velocities are much higher than any others observed in the spectrum. In some novae these lines attained an overall width of 100 angstroms, corresponding to the enormous outward velocity—if velocity it is—of about 3200 km/sec. Why nitrogen in particular should show this extreme broadening and intensification is not known.

Finally, all the normal emission lines begin to weaken, except those of hydrogen and helium. Other lines characteristic of the spectra of gaseous nebulae make their appearance. We shall have more to say about these lines in later chapters. They tend to appear in a very rarefied gas at low pressure (see Sec. 28-3).

Some years after the original explosion, an expanding shell of gas can sometimes actually be photographed around the star (Fig. 27-6). Some of these shells appear more or less uniform, but many show irregularities that indicate emissions from localized knots of nebulosity. For a few novae we have been able to make a spatial reconstruction of the expanding nebulous cloud, which clearly is far from uniform.

Some of the excitation of the nebulous cloud doubtlessly comes from the ultraviolet radiation of the star. Some may result from high-speed particles ejected from the stellar surface. And in some instances the rapidly expanding shell may be interacting with the relatively stationary gas clouds of interstellar space, forming what aerodynamicists call a "shock wave."

Not all novae have shown observable nebulous envelopes, but most of the bright and well-observed ones have done so, and there is no doubt that all novae eject material into space. The total amount of matter lost by a nova in the course of an explosion, however, is small, perhaps a thousandth of its mass or less. It is not surprising, therefore, that the explosion passes away and leaves the star essentially unaltered.

The foregoing description of the spectroscopic behavior of novae is very general. Actually, each nova is an individual, so characteristic that an expert can usually identify it from a glance at one or two of its spectra. Some, such as Nova Persei 1901, seem to have attained very high excitation for the forbidden lines of Ne V were prominent features in its spectrum, though absent from those of many other novae. Nova Pictoris 1925, which

(a)

**FIGURE
27-6**

*Expanding shell of Nova Aquilae, 1918.
(Mount Wilson and Palomar Observatories.) (a) July 20, 1922; (b) Sept. 3,
1926; (c) Aug. 14, 1931.*

(b)

(c)

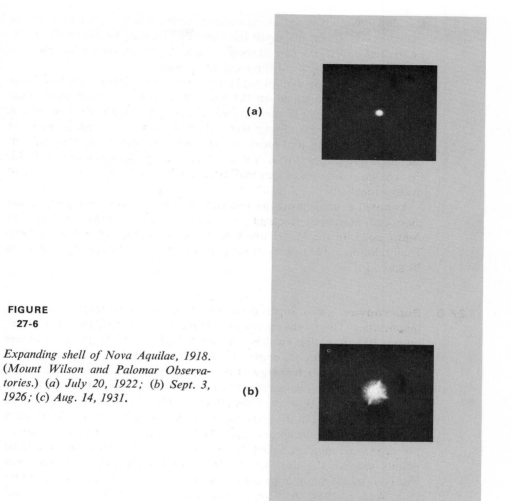

is a slow nova, appeared at first to display much lower excitation, but in the course of time it displayed the forbidden lines of Fe II, Fe III, Fe V, Fe VI, and Fe VII. We should recall the even higher excitation indicated by Fe X and Fe XIV lines of the recurrent novae.

Perhaps excitation cannot explain all the effects. The features of Nova Persei could be explained on the basis of a temperature of more than 200,000° K, but Nova Pictoris is more of a problem. The forbidden lines of oxygen, normally very strong, were relatively weak in its spectrum; those of neon were strong, and those of S II were conspicuous, as were those of the successive stages of iron. There is a possibility, discussed in Sec. 25-13, that chemical differences may exist between different novae, as the result of nuclear reactions.

As noted earlier, novae are relatively common in our own galaxy, and they occur even more frequently in some others, such as the great spiral in Andromeda. In the Magellanic Clouds, however, they are relatively rare. Their relation to the type of stellar population will be discussed in Chapters 30 and 32.

27-6 Supernovae. The Supernovae are exploding stars reaching extreme luminosities. Their light curves are generally similar to those of ordinary novae, but their luminosities are much higher, the brightest attaining absolute magnitudes of about -18. At maximum, such a star radiates in a single second as much energy as the sun gives out in several years.

Supernovae are comparatively rare; none has been observed in our galaxy since the invention of the telescope. (See Table 27-4). The famous "new star" of 1572 in the constellation of Cassiopeia, known as "Tycho's Nova" because of the special study Tycho Brahe made of it, was a supernova. At maximum it rivalled the planet Venus in brightness and could be plainly seen in the daytime. In 1604 another galactic supernova was observed by Kepler and his contemporaries in the constellation of Ophiuchus. And, in 1054, the Chinese and Japanese chronicles record a "guest star," which we identify with the supernova whose remnant still remains visible as an expanding nebulosity in Taurus.

Over a hundred supernovae have been recorded in other galaxies. One of the best known supernovae appeared in the Andromeda galaxy, Messier 31, in the year 1885. It was of the sixth magnitude at maximum and thus approached the luminosity (fourth magnitude) of the entire system in which it appeared. Over a hundred ordinary novae observed in the same system are fainter than the fifteenth magnitude at maximum, hence the maximum luminosity of the supernova was some 10,000 times that of ordinary novae.

The spectra of supernovae show what appear to be extremely broad emission lines, whose chemical source has not yet been identified. The great width of the lines, presumably caused by Doppler effect, makes analyses of

TABLE
27-4

Galactic and extragalactic supernovae

Galaxy	Type	Year	Mag. at Max. App.	Abs.	Type
Galaxy	Sb	1054[a]	−6	−16.5	I
		1572[b]	−4		I
		1604[c]	−2		I
NGC 224	Sb	1885	7.2	−17.6	I
NGC 4424	Sa	1895	11.1	−19.4	
NGC 5253	Imp	1895	8.0	−20	
NGC 4321	Sbc	1901	11.9	−18.4	
NGC 5457	Scd	1909	(12.1)	−16.4	
NGC 4486	E0p	1919	11.5	−18.5	
NGC 2608	Sb	1920	11.0	−20	
NGC 3184	Scd	1921	11.0	−19	
IC 4182	Sm	1937	8.2	−20	I
NGC 1003	Scd	1937	12.8	−17.4	I
NGC 4621	E5	1939	11.8	−19	
NGC 5195	I0p	1945	11.2	−17.5	I
NGC 4214	Im	1954	9.0	−19.4	(I)
NGC 5668	Sd	1954	12.0	−19.5	I
NGC 3992	Sbc	1956	12.5	−18.1	I
NGC 4496	Sm	1960	11.0	−19	
NGC 4564	E6	1961	11.0	−19	I
NGC 1313	Sd	1962	10.0	−18.5	(II)
NGC 3913	S	1963	12.5		I

[a] Crab Nebula, supernova remnant.

[b] Tycho's star in Cassiopeia.

[c] Kepler's nova in Ophiuchus.

FIGURE
27-7
(a)

Supernova in spiral galaxy NGC 3389. Photograph, May 3, 1967. (G. deVaucouleurs, McDonald Observatory.)

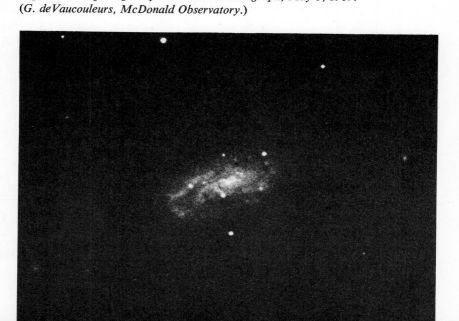

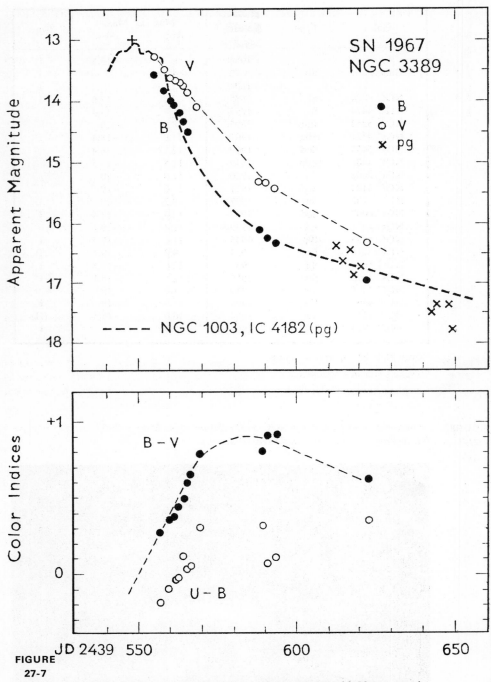

FIGURE
27-7

(b) *Light and color curves. (G. deVaucouleurs, McDonald Observatory.)*

these spectra very difficult, especially since supernovae in external systems are seldom brighter than the tenth apparent magnitude, and are often much fainter (Fig. 27-7).

The observations of W. Baade, F. Zwicky, and R. Minkowski at Mount Wilson Observatory in the 1930's have indicated the existence of at least two groups of supernovae: type I, to which belong SN 1885 in Messier 31 and also Tycho's nova, and type II, which is perhaps ten times fainter at maximum, and with light curves of different shapes (Fig. 27-8a,b). In 1940, F. L. Whipple and C. P. Gaposchkin produced a reasonable replica of the spectrum of a type II supernova by superposing broadened lines from the spectrum of a normal nova. Perhaps this second type of supernova represents extreme specimens of ordinary novae. In the last decade or so, F. Zwicky at Mount Palomar Observatory, and collaborators around the world, have made extensive search for supernovae and have found that as their number increases, additional types may be recognized. Five types have been determined according to the various shapes of the light curves and spectral characteristics.

Within our galaxy, the remnant of the supernova of 1054 is the remarkable "Crab Nebula" in Taurus (Fig. 27-9). Photographs and spectra reveal the nebula expanding away from a central star with a velocity of about 1300 km/sec. Recent studies have shown that the light of this nebula

FIGURE 27-8 *Light curves of three supernovae. Tycho's star, B Cassiopeia; Kepler's nova in Ophiuchus; and the brightest known modern supernova in I.C. 4182. The magnitude scales appear on the left for the first two novae and on the right for the third. (W. Baade and F. Zwicky, Mount Wilson and Palomar Observatories.)*

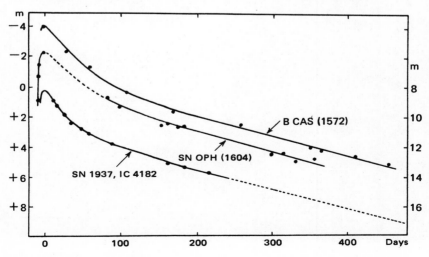

FIGURE
27-9   *The Crab Nebula in Hα light. (W. Baade, Mount Wilson and Palomar Observatories.)*

is polarized, and that light "ripples" are passing through its structure. These effects probably result from the ejection of high-velocity particles from the central star, and their interaction with a magnetic field. Thus, although the outburst took place over 900 years ago, the star that gave rise to it is still far from quiescent. The Crab Nebula itself is a strong source of cosmic radio waves emitted by electrons moving at speeds close to the velocity of light in a magnetic field (see Chapter 33). Detailed studies of the strength and direction of polarization of visible light (Fig. 27-10) enable us to map approximately the structure of the magnetic field. Such objects also are sources of intense X-ray emission. Sources of radio emission and faint wisps of nebulosity also occur near the positions of the supernovae of 1572 and 1604.

After many centuries, the expanding nebulosity of a supernova dissolves in interstellar space and ceases to be visible optically. However, the radio emission from the electrons in the ionized gas cloud may still be detected. Several large radio sources in the galaxy may well be the remnants of long-vanished supernovae. Of late, astronomers have taken a renewed interest in old medieval chronicles and still older Chinese or Japanese annals in the hope of identifying such sources with forgotten records of unusually bright "guest stars."

27-7 Superdense Stars. Superdense stars are the result of supernovae. A complete collapse of the stellar core accompanies the violent explosion of the outer layers, which go to form the nebular shell.

The white dwarfs (see Sec. 20-9) represent one class of dense stars, which have moved off the main sequence during their evolution. A typical white dwarf would have shed a large fraction of its mass into space, until its mass was, say, about 0.6 times that of the sun. One suspects that planetary nebulae (see Secs. 28-3 and 28-4), arise in such a manner. Its collapsed radius would be about one per cent that of the sun, or about equal to that of the earth. Its mean density would be of the order of 5×10^5 g/cm^3. The atoms have been stripped of their orbital electrons because of the high internal pressure. But the electrons, themselves, still exert an outward pressure, so that the star resists further collapse and a stable configuration results. The properties of such a compressed electron gas are well known, so that one can calculate the characteristics of such a star. The atomic nuclei retain their identity. Nuclear reactions, which include mainly the burning of helium and gravitational energy of contraction, continue to furnish energy to keep the white dwarf feebly shining.

A star appreciably more massive than the sun, having largely depleted its internal energy, may also become denser and denser, until it becomes unstable and violently collapses. The core contracts until it reaches a density of some 10^{11} g/cm^3. At such a density the electrons no longer possess any resilience. Nor do the atomic nuclei preserve their individuality.

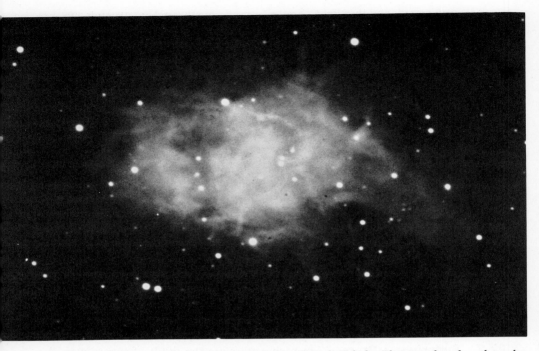

FIGURE 27-10 (a) *Polarization of continuum emission of Crab Nebula. Photographs taken through a polarizing filter and a yellow color filter transmitting only the continuous spectrum. Light vibration vertical.*

**FIGURE
27-10
(b)**   *Light vibration horizontal.*

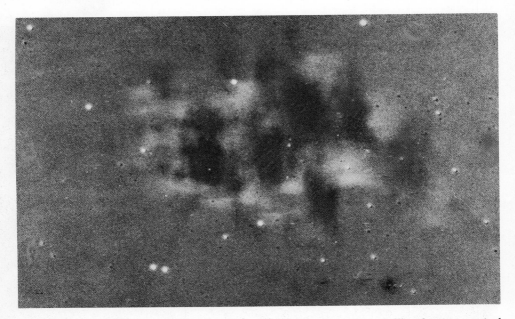

**FIGURE
27-10
(c)**   *Composite print of a positive of (a) and a negative of (b), showing vertical polarization in white, horizontal polarization in black, and unpolarized light in gray. (F. Zwicky, Mount Wilson and Palomar Observatories.)*

The electrons start to combine with the freed protons, as indicated in one of the reactions discussed in Sec. 24-3, to form neutrons with release of neutrinos. This phenomenon is termed "electron crush." The collapse continues to densities of from 10^{14} to 10^{15} g/cm^3, when the star has a radius of some 10 km.

In the early stages the neutrinos readily penetrate the outer layers and escape. But as the compression continues, the shell absorbs the neutrino energy, and heats up. This heating has two consequences: (1) it produces a violent shock wave, which makes the outer atmosphere explode; (2) the shock reacts against the core, causing it to implode and become superdense. The explosion furnishes the energy for the supernova. A quivering, pulsing neutron star remains.

We simply don't know enough about such highly compressed matter to predict exactly what will finally happen. One's instinct suggests that, at such high densities, the neutrons will resist further compression. And the discovery of a star rapidly varying in both radio and light emission in the Crab Nebula tends to substantiate our hypothesis (see Sec. 33-8). Other "pulsars" are probably the cores of old supernovae.

Some theoretical difficulties still remain, however. As we shall see later (Sec. 34-6), a spatial distortion occurs in the vicinity of a large mass like the sun or star. The enormous gravitational forces arising from a neutron star could cause the object to collapse completely and vanish from the observable universe. One should further note that such supernova explosions may very well be the primary source of the energetic cosmic rays.

CHAPTER 28

Stars with Atmospheric Shells

28-1 Stellar Atmospheric Forces. The forces in a stellar atmosphere can arise in a number of ways. For a completely static atmosphere, the outward push of the gas pressure must counterbalance the pull of gravity toward the center of the star. In other words, the gas pressure must equal the weight of an overlying column of gas. Since the supported mass increases toward the center of the star, the pressure must increase inwards. An early analysis by E. A. Milne gave some hope that the distension of the solar chromosphere and the support of gases in prominences might arise from the pressure of radiation flowing outward. Later studies showed, however, that radiation pressure was generally ineffective.

O. Struve and others have proved that the centrifugal force of rotation acts to distend stars, causing them to bulge at the equator or, in some extreme cases, to take the form of a flattened egg, no two axes alike. In such stars, the centrifugal force so nearly counterbalances the gravitational, that they tend to vibrate and even split in two. Very close binary pairs, like the Beta Lyrae stars, may originate in this fashion.

Although certain stars rotate very rapidly, the rotational forces are by no means large enough to cause the extremely high distension that exists in many stellar atmospheres. We have to look for still other forces.

We know that the earth's atmosphere is not completely static. It possesses a rotational bulge like that of the earth's surface, and the winds flowing parallel to the surface experience a deflecting force, termed the force of Coriolis (Sec. 4-5). These forces produce cyclones and hurricanes.

In some respects, our sun is a star with a distended atmosphere, though we should hardly recognize it as such, if it were as far away as even the nearest star. The distended portion of the solar atmosphere consists of the corona and prominences, and perhaps some of the chromosphere. But the total light from these regions is so small that only our proximity to the sun enables us to detect them, except in the rocket ultraviolet, where strong emission lines from the hot corona dominate the picture.

The solar atmosphere displays considerable dynamic circulation. Jets and surges transport material explosively to high levels. Violent shock waves cross the solar surface. The observations suggest that the major force behind such motions must be some sort of explosive action. Solar flares, as we have seen, often explode violently. We have traced the pumping action to the presence of intense magnetic fields and associated electric currents in the sunspots. Magnetic forces may also play an important role in the atmospheres of certain types of stars. The fields would provide major support for the equatorial regions.

Magnetic forces within a stellar atmosphere can be considerable. However, it should be emphasized that magnetic forces between stars are entirely negligible compared with gravitation. Also, because stellar matter is so highly conductive, no appreciable electric charges can build up in stellar atmospheres. Thus electrostatic forces are negligible.

28-2 Stars with Distended Atmospheres. Although stars with extensive shells of gas around them are probably not particularly plentiful in space, the fact that they often display bright lines in their spectra calls our attention to them, and they have been intensively studied. Many different types exist—so many that anything like a complete classification is difficult. They cover such a wide range of objects that one may well question whether they are all related.

Most tenuous of all are the planetary nebulae, where the central or nuclear star is surrounded by truly enormous shell of tenuous gas. Next, we encounter what P. Merrill has termed the "symbiotic stars," stars that appear to possess simultaneously the spectral characteristics of a cool star and a gaseous nebula. "Symbiosis" is a biological term, used to designate the union of two organisms that depend on one another for their existence. Perhaps the most familiar example is the "lichen," which is a sort of partnership between a type of fungus and an alga. There is no question but that the luminosity of the gaseous nebular spectrum results from its close association with the star, from which it gains its excitation. One might question whether the nebulosity is necessary for the existence of the star.

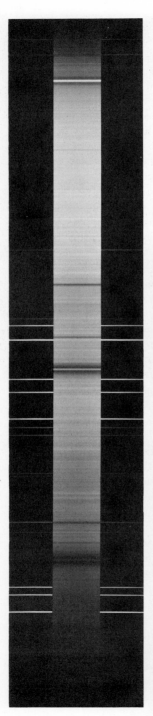

**FIGURE
28-1**

(*Left*) *Note the hydrogen emission cores, characteristic of the shell.* (*Lick Observatory.*) *Spectrum of Shell Star, 11 Camelopardalis.*

**FIGURE
28-2**

(*Below*) *"The Ring." Planetary Nebula M57 in Lyra.* (*Mount Wilson and Palomar Observatories.*)

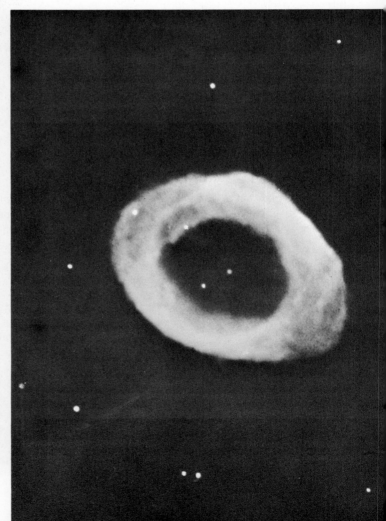

FIGURE 28-3   *"The Dumbbell." Planetary Nebula M27 in Vulpecula. (Mount Wilson and Palomar Observatories.)*

However, the union may be somewhat closer than we have thought. In some ways we might consider these symbiotic objects as a sort of missing link between the planetary nebulae, which we shall discuss later on, and ordinary distended stars. We shall continue to use the term symbiotic star to designate this type of object.

The long-period variables, which we have previously discussed, seem to be stars with distended atmospheres. Indeed, as we have already seen, the hydrogen emission associated with these stars probably comes from fairly deep levels in the atmosphere, where temperatures are high, but densities still low. By no stretch of the imagination could we expect the atmospheres of such stars and of other bloated objects, which include the red and yellow giants and possibly even some of the Cepheids, to be in simple, quiet, hydrostatic equilibrium. We also note the Wolf-Rayet stars, which appear to be continually ejecting gas into space.

FIGURE 28-4   *Planetary, NGC 7662. (Yerkes Observatory.)*

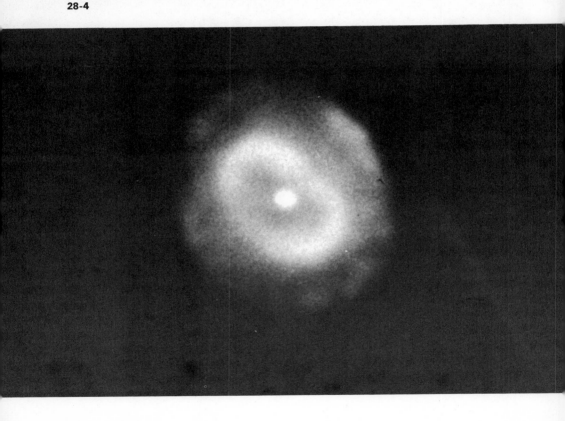

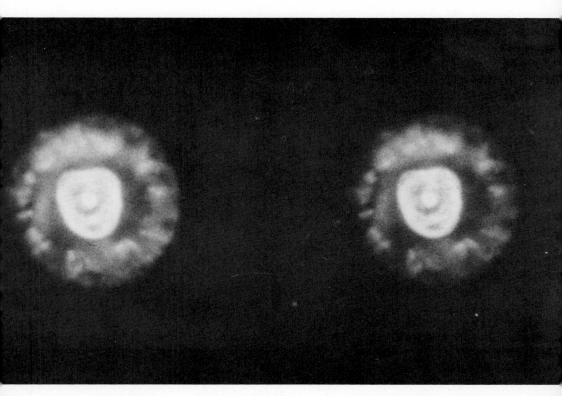

FIGURE   *Two different exposures of "The Eskimo" planetary. (I. S. Bowen, Mount Palomar*
28-5   *Observatory 200-inch telescope.)*

Many early-type stars possess shells of gas apparently forming an almost detached or semidetached atmosphere. Such stars, of which Pleione in the Pleiades is a leading example, comprise what O. Struve terms shell stars. Radiation pressure would not in general be sufficient to cause the distension of their shells. A combination of rotation, convective, and magnetic forces seems to be necessary. Shock waves probably play a major role. Practically none of these shell stars can be regarded as having an atmosphere in stable, static equilibrium. Rather, the equilibrium is dynamical, with a continual replacement of material from below and probably a circulation along some very definite path. The variability of many of these stars may result from some form of atmospheric circulation (Fig. 28-1).

28-3   Planetary Nebulae.   A planetary nebula, in its simplest form, appears to be a roughly spherical or ellipsoidal shell of gas, with a nuclear star in or close to the center. These objects have received their name, not because of any possible generic relationship to planets, but because early observers, searching for planets with the telescopes of their day, sometimes came across these disklike objects that at first glimpse looked like planets. They proved not to be planets because they did not move among the stars. Many of these objects have received distinctive names, suggestive of some terrestrial object. Thus we have "The Ring," "The Owl," "The Dumbbell," and many others (Figs. 28-2–28-5 on pages 582–585).

Approximately 300 planetaries are known, the great majority of them having been discovered by R. Minkowski. Their spectra are so distinctive with bright emission lines, that they are conspicuous on an objective-prism plate.

The nuclear stars of these objects very plainly provide the main source of luminosity for the nebula. W. H. Wright has shown that the nuclei are of two main types. Some of them are primarily continuous O-type spectra; others are W stars. Analyses show that the temperatures of these nuclear stars are always high. Studies by H. Zanstra, B. Vorontsov-Velyaminov, and D. H. Menzel have shown that they range between 30,000° and 200,000° K and possibly even higher. The interesting fact is that the nuclei of planetary nebulae, despite their high temperature and high surface brightness, are extremely faint, not much brighter than the white dwarfs.

These hot stars radiate by far the greatest portion of their energy in the far ultraviolet, beyond the Lyman limit of hydrogen. The abundant hydrogen in the outer shell of the planetary nebula absorbs this far-ultraviolet radiation, causing the atoms to become ionized. Other hydrogen nuclei, wandering through the nebula, capture these free electrons on the second, third, and higher levels, in addition to the first. The electrons then fall to lower levels, emitting radiation in the Lyman, Balmer, and other series. Thus the original ultraviolet radiation of the star gradually degrades

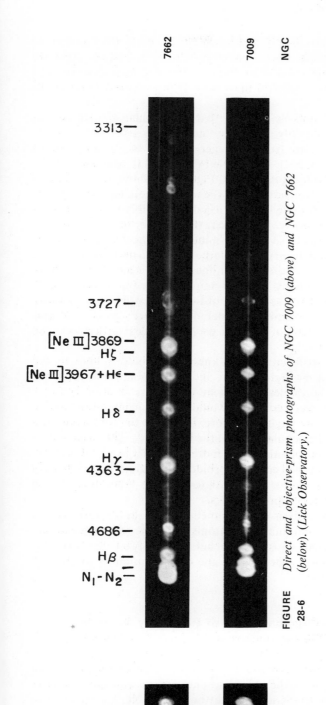

3313 —

3727 —

[Ne III]3869
Hζ —

[Ne III]3967+Hε —

Hδ —

Hγ
4363 —

4686 —

Hβ —
N₁-N₂ —

7662

7009

NGC

**FIGURE
28-6** *Direct and objective-prism photographs of NGC 7009 (above) and NGC 7662
(below). (Lick Observatory.)*

**FIGURE
28-7** *Slit spectrum of NGC 7662. (G. deVaucouleurs, McDonald Observatory.) Compare
with Fig. 28-6.*

into energy of longer wavelength and Lyman α radiation. Most planetary nebulae show strongly the lines of neutral helium. A few, those whose excitation is high enough, also show intensely the leading lines of ionized helium. We also observe spectral lines from highly ionized oxygen and nitrogen.

Of particular significance is the fact that the nebulae radiate certain characteristic lines, commonly called the "nebular lines." These lines are the "forbidden" radiations from low-energy levels of various ionized atoms, chiefly O II, O III, N II, Ne III, Ne IV, Ne V, S II, and A III, A IV. I. S. Bowen identified many of these lines. Figure 28-6 shows spectra of two bright planetaries. The strongest lines to the right are nebular lines of O III.

As noted in Chapter 8, these lines are usually negligibly weak in laboratory spectra, for two main reasons. First, the chance that an atom will emit such radiation is small, so that the minute volumes of gas we ordinarily deal with in the laboratory are not large enough to produce these lines with appreciable intensity. Second, other lines that have a high probability of occurring are many millions of times more intense. But in the gaseous nebula, far from this central star, the ordinary lines are tremendously weakened, whereas the intensities of the nebular lines are scarcely changed. Thus, the enormous volume of gas that we have to deal with makes the nebular lines appear plainly.

Many textbooks assert that the weakness of forbidden lines in the laboratory results from the high pressure and the dominance of collisions with other atoms and molecules, which tend to deexcite an atom before it has a chance to radiate. Such collisions would be infrequent in the rarefied gaseous nebulae. And one might, therefore, infer that the forbidden lines should be stronger in the nebulae than in the laboratory. This argument is wrong. Collisional excitations also occur more frequently in the laboratory. Hence a given mass of gas will emit the forbidden radiations more intensely at high than at low pressures. The weakness in the laboratory is due entirely to the low transition probabilities.

A physical-chemical analysis of the planetaries has shown how rarefied they actually are. The total number of atoms per cubic centimeter ranges between 1000 and 10,000. Thus, these nebulae are only a trifle more concentrated than interstellar space, where the density amounts to about 1 atom per cubic centimeter.

28-4   The Excitation and Physical State of Planetary Nebulae. Spectroscopic analysis of internal motions of the planetary nebulae by W. W. Campbell, J. H. Moore, O. C. Wilson, and others indicates that the objects are expanding. The idealized form of a complete spherical shell rarely exists. In many cases, the objects seem to possess an axial symmetry of some sort, for which magnetic fields may be responsible. The shell can be incomplete in various positions, thus giving the forms referred to earlier.

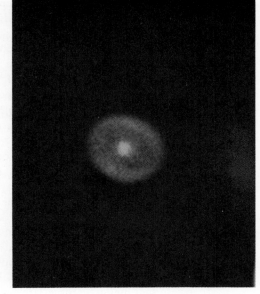

(a)

**FIGURE
28-8**

(*a*) *The ring nebula, IC 418 in Hα light.
(G. deVaucouleurs, McDonald Observatory.)   (b) The doubling of spectral lines
indicates an expanding envelope. (O. C.
Wilson, Mount Palomar Observatory
200-inch telescope.)*

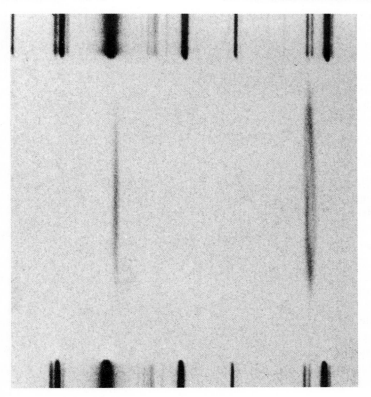

(b)

The nebulae are generally transparent to their own radiations in the visible region, and we actually see both sides of the expanding shell. The spectral lines are double: the component displayed to the violet represents the near side with motion toward us; the component displaced to the red comes from the receding portion on the side away from us (Fig. 28-8). The velocities of expansion, as observed, are generally fairly low. Most of them range from a few kilometers a second to 30 km/sec, with an average around 20 km/sec. Because the objects are expanding, they are probably evanescent and will disappear as a result of expansion in something like 30,000 years.

Although, as previously pointed out, many novae develop nebular shells some few years after the outburst, their velocities of expansion are very much greater, perhaps as much as 1000 km/sec. However, a few of the objects initially classified as planetaries are actually remnants of old novae or supernovae. The Crab Nebula (Sec. 26-6) is one example.

Planetary nebulae show many knots and condensations and filamentary structure. The "Helical" nebula in Aquarius resolves into thousands of comet-shaped objects, with nuclei pointed toward the central star and long, fan-like tails spreading away from it (Fig. 28-9). It seems reasonable to suppose that magnetic forces are acting to hold together the gas filaments, which otherwise would disperse into interstellar space.

Detailed spectra of some planetaries, taken by O. C. Wilson, have shown that whereas the lines of O III and Ne III share in the velocity of expansion, the lines of Ne V are relatively stationary. This important result has not been explained. But it does indicate that the motions within the shell can be extremely complicated.

As for the physical significance of the planetaries, one can conclude only that they result from gas ejection by "slow novae." Some indirect evidence exists, as we shall see, to link them with the long-period variables.

28-5 Symbiotic Stars. The symbiotic stars, discussed by P. Merrill and O. Struve, are in a sense the missing link between the planetaries on the one hand and truly stellar objects on the other. Only a relatively few stars of this type are known. But they are extremely important for us. These stars are anomalous at first sight, since they combine the characteristics that we normally associate with the hottest stars and the coolest stars. At some stages, most of them have nebular lines and other intense emission lines in their spectra; at other times they show strong titanium oxide absorption and have a distinct red color. This multiple character once led astronomers to conclude that these objects were double stars consisting of a hot blue component and a cool red one. The long-period variable star Mira Ceti provided some indirect evidence in favor of this view. But it now seems more likely that in many cases a single star is involved: a small, hot, blue star, surrounded by an extended, variable envelope, which is rarely complete.

FIGURE 28-9 *The "Helical Nebula" NGC 7293, in Aquarius, largest example of planetary nebula displays fine jetlike structure. (W. Baade, Mount Palomar Observatory 200-inch telescope.)*

These symbiotic stars are variable in both light and spectrum. From time to time they exhibit strong absorption lines, displaced to the violet, and in this sense resemble the recurrent novae. Many details are missing, but here we seem to find explosions occasionally occurring over only portions of the star, explosions that lead to the formation of a temporary dynamic shell, which cools as it expands and acts like a stellar photosphere. Since it does not cover the star completely, enough ultraviolet radiation escapes in the open zones to excite the nebula that lies beyond. In fact, the nebula itself may very well be renewed by atoms ejected in these recurrent outbursts.

For us, the significant facts are the hot central core, the explosive variability, and the formation of partial photospheres. Such a star will be far from symmetrical. It may possess enormous irregular blobs protruding for many diameters beyond either pole. Perhaps some of the material may tend to settle back over the equator, where it receives some partial support from the presence of magnetic fields.

One remarkable star, which seems to fall just within this category, is R Aquarii, definitely a long-period variable, despite the fact that a planetary nebula surrounds it. Here again, we are confronted with the apparent anomaly of an extremely cool, distended red star that is probably the source of excitation of a gaseous nebula whose luminosity, we know, must depend on intense ultraviolet radiation. Again the double-star possibility suggests itself, but by analogy with Z Andromedae and others in this class, we assume that we are also dealing with a partial photosphere, a star whose distended atmosphere covers only part of the surface so that ultraviolet radiation can leak out from certain unprotected zones and thus continue to excite the nebulosity. This association suggests that the long-period variables are related to the repeating novae, and that the primary distinction between the two is the amount of material ejected.

28-6 **Long-Period Variables and Red Giants.** The long-period variables and the red giants by virtue of the argument presented in the previous section, may thus belong to a class of highly bloated stars, in which the distension results from dynamic action, a sort of deep-seated convection in the presence of magnetic fields.

Although astrophysicists have tried to represent these red giants as stars in static equilibrium, none of the models appears really convincing. These stars, like the ordinary red giants, are old stars. Having exhausted most of their hydrogen, they are now temporarily employing the helium-helium reaction to provide their energy. Such stars will not live long. They radiate energy at a very high rate, and their store of energy is extemely limited. The turbulence in their atmospheres attains some velocity. The violent convection leads to further distension of the outer envelopes and escape of material into space.

28-7 K-Type Giants and Supergiants. K-type giants and supergiants exist in considerable abundance. Some of these stars are eclipsing variables, a fact that makes it possible for us to study in great detail the atmospheric structure of the component with a distended atmosphere. Consider the system ζ Aurigae, which consists of a supergiant K star and a main-sequence B, revolving around one another. In the visible portion of the spectrum, the two stars—the large cool one and the small hot one—send out about the same amount of light. In the violet and ultraviolet, the early-type star generally dominates. But as this blue star goes into eclipse behind the giant red star, we can see the absorption effects that the extended atmosphere produces. Strong absorptions of variable amount suggest the presence of irregularities more like prominences than an absolutely uniform atmosphere. In any event, the K star does not seem to have a sharp outline like the sun.

Whether this shell model can be extended to include the Cepheids and other supergiants is not known. Perhaps the main sequence, after all, is the only basic sequence. Some of its stars, however, may have mechanical, vibrational, or magnetic properties that lead to the formation of a distended atmosphere. In such cases, since the stars are bigger, if the amount of energy radiated remains the same, the surface temperatures of the effective photospheres are lower. Hence the star assumes a spectral type somewhat later than that of the normal star on the main sequence.

28-8 Wolf-Rayet Stars. The Wolf-Rayet stars, or W stars, have already been discussed from the standpoint of spectral classification. We have seen that they comprise two separate sequences, carbon and nitrogen stars, whose spectral characteristics indicate different chemical compositions. In this respect perhaps we are dealing with two independent groups, something like the M type vs. N or R type. L. F. Smith in 1966 listed 127 W stars down to magnitude 12. The brightest is gamma Velorum, of visual magnitude 1.74.

The Wolf-Rayet stars are blue and very hot. Their spectra display wide emission lines. About 25 per cent are spectroscopic binaries, with O or B companions. A number of them have proved to be eclipsing variables. Their light curves clearly indicate the general accuracy of the shell model of the W component, with a small, highly compressed core and a very distended atmosphere, evidently in dynamic equilibrium.

Early studies by J. S. Plaskett suggested that the W stars were perhaps two magnitudes fainter than O stars. More recent work, mainly by M. Roberts and L. F. Smith, has shown, however, that these objects are enormously concentrated toward the galactic equator. Forty per cent of them lie within 1 degree of the equator, 70 per cent within 2 degrees, and 95 per cent within 5 degrees. Two that lie farthest from the galactic equator, at $10°.1$ and $7°.7$, are among the brightest of the group. This high concentration implies that the W stars are highly luminous. However, they are short-lived. They are, therefore, potentially interesting for locating and tracing the spiral pattern of our galaxy, since they do not have time to move far from the spiral arms where they originated.

It has not yet been decided precisely where W stars fit into the H-R diagram. Their high temperatures and high luminosities indicate that they should come before the O's. On the other hand, their tenuous atmospheric envelopes suggest a relationship with the giant M's or the symbiotic objects.

CHAPTER 29

Interstellar Matter

29-1   Dust Clouds in Space. Early astronomers assumed that the immense spaces between the stars were empty—that interstellar space is effectively a perfect vacuum. Some observations by William and John Herschel, who noticed strange vacant areas in the dense star fields of the Milky Way, did not seem to disturb this belief. The Italian astronomer A. Secchi was possibly the first to argue, over a century ago, that these vacant spaces are not regions devoid of stars but areas where vast clouds of absorbing matter obscure or hide the stars. The assumption of a genuine vacancy would require the implausible hypothesis that space contains long narrow tunnels devoid of stars and pointing directly at us from many directions. Yet so strong was the dogma of a perfectly empty space that the existence of dust clouds did not find acceptance until the first half of this century.

The beautiful wide-angle photographs of the Milky Way taken at the turn of the century by E. E. Barnard at Yerkes Observatory and by M. Wolf at Heidelberg, in Germany, disclosed the complex and interwoven network of bright nebulosities and dark regions covering the great star clouds along the galactic equator. Both Barnard and Wolf—unaware of Secchi's early argument—suspected that the dark regions were obscuring

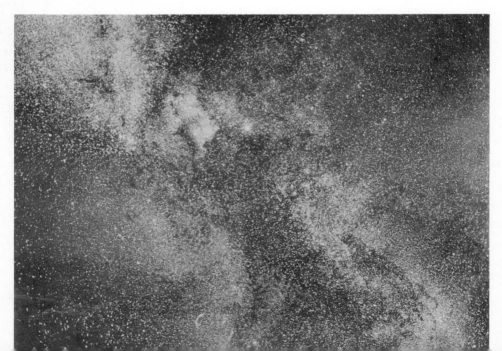

clouds that blocked out the distant stars. A few of the dark areas are readily visible to the naked eye. A classical example is the "Coal Sack" visible near the Southern Cross (Fig. 29-1). Another almost equally remarkable lies in the constellation of Cygnus. The "great Rift" of the Milky Way, which splits its luminous path from Cygnus to Sagittarius (Fig. 29-2), results from a succession of large and overlapping dark clouds in the equatorial plane of our star system (see Chapter 31).

FIGURE 29-3

Dark clouds in silhouette against southern Milky Way.
(Boyden Station, Harvard College Observatory.)

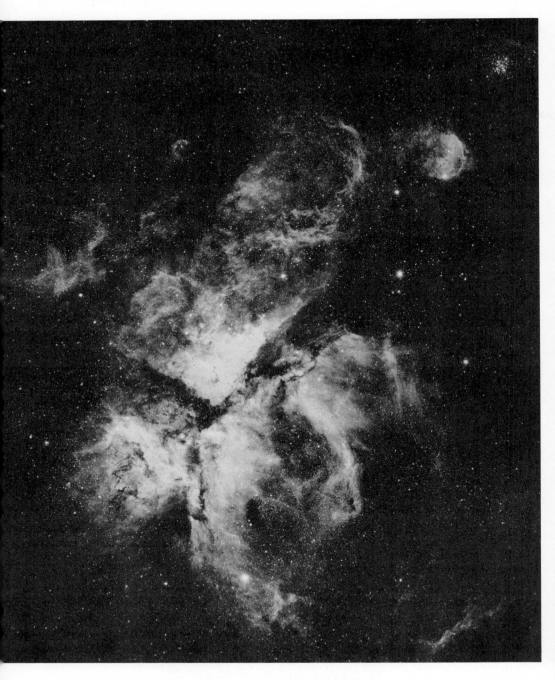

FIGURE   *The η Carinae Nebula contains both bright and dark nebulosity. (Boyden Station,*
29-4   *Harvard College Observatory.)*

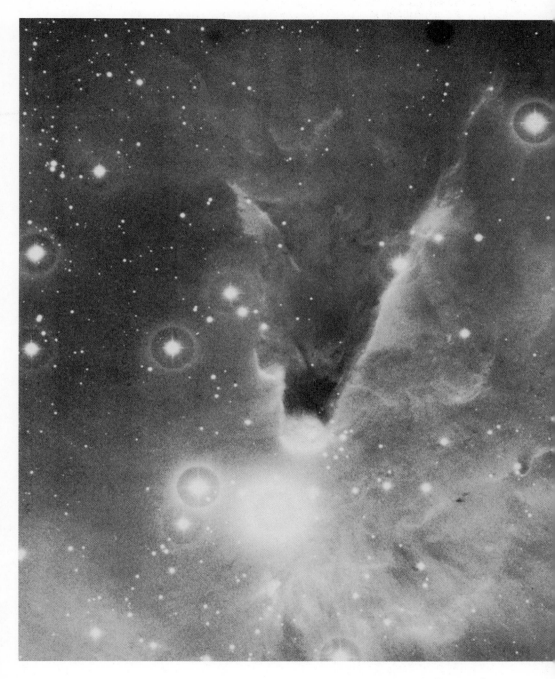

FIGURE   *The phenomenon of "elephant trunks," a dark region extending radially away from*
29-5   *a bright star. (Mount Wilson and Palomar Observatories.)*

FIGURE 29-6 *Globules seen in projection against nebulosity M8 Sagittarii. (Mount Wilson and Palomar Observatories.)*

Telescopic photographs disclose many smaller dark clouds, often curiously elongated and twisted. Some are completely black and stand out only as silhouettes against a rich star field (Fig. 29-3). Others are associated with and projected against a bright diffuse nebulosity (Fig. 29-4). Of the latter, some point straight at a bright star and are sometimes called "elephant trunks" (Fig. 29-5). These various shapes indicate the presence of dust clouds associated with a diffuse gaseous medium and subject to various forces. Some minute, more or less circular dark patches, called "globules" by B. J. Bok, may represent an early stage of condensation of the interstellar medium leading to the formation of a star or star cluster (Fig. 29-6).

The conclusion that obscuring interstellar clouds must consist of small grains of matter—variously described as "dust" or "smoke"—is not immediately obvious. However, a quantitative calculation of the amount of diffuse matter necessary to produce the observed absorption confirms the dust hypothesis. Briefly, large chunks of dark matter such as planets or meteorites are extremely inefficient absorbers in relation to their mass. As matter breaks up into smaller and smaller parcels, its absorbing efficiency increases, because the absorption area or cross section of a given total mass increases while the diameter of the grains decreases. This fact results because for a given total mass M divided into N identical grains, the number N varies as the inverse cube of their diameter D, while the absorbing cross section of each grain varies as the square of the diameter. Thus the total screening area varies proportionately to $ND^2 \propto D^2/D^3 = 1/D$, which increases rapidly when $D \rightarrow 0$. Hence, clouds of dust or smoke particles of diameters of the order of 1 micron (0.001 mm) are enormously more efficient absorbers of starlight than swarms of cold planets or meteorites. In other words, a much smaller total mass of matter will produce the observed absorption. This reasoning, however, fails when the particles become small compared with the wavelength of light. It would be wrong to conclude that atoms or electrons are even more efficient absorbers than dust. Although tenuous gas clouds in space are indeed present, they reveal themselves by the discrete absorption lines that they produce in the spectra of distant stars or by their own emissions. They will not, however, produce the complete obscuration of all wavelengths of the type we see in the dark clouds.

The effects of absorption will become more obvious as we continue our exploration of the rarefied interstellar medium.

29-2   Diffuse Obscuration in Interstellar Space.   In addition to the discrete dark clouds seen in projection against the bright star fields of the Milky Way, a more subtle, diffuse obscuring medium pervades all interstellar space. This elusive general absorption was not generally recognized until about 1930, mainly through the work of R. J. Trumpler at Lick

Observatory. Trumpler, who was studying galactic star clusters (see Sec. 30-3), had devised two methods of measuring their distances. One involved the apparent luminosities of stars of known absolute luminosity and an application of the inverse-square law of photometry (see Chapter 19). The second method was based on the assumption that the apparent diameter of the clusters, defined in a consistent manner, would vary in inverse proportion to the distance. As long as space was believed to be empty and perfectly transparent, astronomers had no reason to expect that the two methods of measuring distances would disagree.

When Trumpler discovered that the distances derived from the luminosity criterion were systematically larger than those derived from apparent diameter, he correctly concluded that starlight dims as it travels through space. Thus the stars appear fainter and more distant than they really are. From a comparison of the true and apparent distances he made a first estimate of the general absorption in interstellar space, about 1 magnitude per kiloparsec, which is equivalent to a reduction in the ratio of 10 to 1 after a travel of 8000 light years. Such a widespread obscuring haze between the stars should limit severely our exploration of the stellar system by optical means. In fact it is very difficult to see out beyond 10,000 light years near the Milky Way. Before Trumpler's discovery, models of the stellar system based on the assumption of a perfectly transparent space were grossly in error.

Another proof of the presence of a widespread obscuring medium in our system came to light about the same time as Trumpler's work. As we shall see later (Chapter 32), the existence of stellar systems outside our own was firmly established, mainly by E. P. Hubble at Mount Wilson Observatory during the 1920's. Hubble proceeded to make sample counts of these "extragalactic nebulae," as they were then called, over a large area of the sky. The counts show that these external systems become more difficult to see and appear fewer per unit area of the sky near the plane of the Milky Way. The effect resembles the dimming of starlight by the earth's atmosphere toward the horizon, and it follows a similar law, namely that the dimming expressed in magnitudes is proportional to the path length of the rays through the absorbing layer.

By simple geometry (Fig. 29-7) Hubble derived, from the counts of the number of visible galaxies, the total absorption through the Milky Way perpendicular to the dust layer, about 0.5 magnitude. More recent estimates from more extensive counts at Lick Observatory indicate a higher value, avout 0.8 magnitude, corresponding to a dimming factor of 2 to 1 for a ray crossing the absorbing layer at right angles to its equatorial plane. Absorption is much larger in the direction of the plane and, in fact, external systems are totally obscured in a belt or "zone of avoidance" of variable width along the path of the Milky Way. In only a few directions do outer systems show dimly through narrow "windows" or gaps between

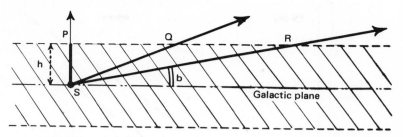

FIGURE
29-7 *The interstellar absorption increases toward the galactic plane. The line of sight crosses minimum thickness, SP = h, of the absorbing layer in the direction of the galactic pole. At lower galactic latitude, b, the path through the layer increases as in SQ or SR.*

the dust clouds. This fact and the appearance of the discrete dark nebulae show that interstellar absorption is spotty and the widespread dust far from uniformly distributed.

29-3 Star Counts and Absorption. We notice the presence of a dark absorbing cloud projected against a rich star field because it produces a deficiency in the star density or the number of stars per unit area of the photographic plate. When the absorption is large, say one magnitude or more, the deficiency is obvious. Individual obscuring clouds that reduce light by one-half or less are not so evident, but we can still recognize them by means of star counts. The German astronomer Max Wolf at Heidelberg first employed this method of studying dark nebulae nearly half a century ago. The most important tool he introduced is a graph of star counts versus magnitude, known as a "Wolf diagram," derived as follows.

Suppose we count all stars brighter than successive magnitudes, say 10, 11, 12, etc., in a certain area of the sky. If the stars were uniformly distributed in space, the numbers should increase for each magnitude step by a factor of about 4.0. The inverse-square law of brightness requires that two groups of stars differing by one magnitude should have average distances proportional to the square root of their relative brightnesses, or $\sqrt{2.512}$. Then the volume of space probed by the counts should increase in proportion to the cube of the distance, or $(\sqrt{2.512})^3 = 3.98$, which for a transparent, uniformly populated space should equal the ratio of the numbers of counted stars per magnitude.

The Wolf diagram is a plot of the logarithms of the numbers counted to successive magnitude limits against the magnitude (Fig. 29-8). Since the logarithm of 3.98 is 0.6, the points should lie on a straight line of slope 0.6 if space were transparent. If, however, a thin obscuring cloud absorbing Δm magnitudes occurs at a distance corresponding to apparent magnitude

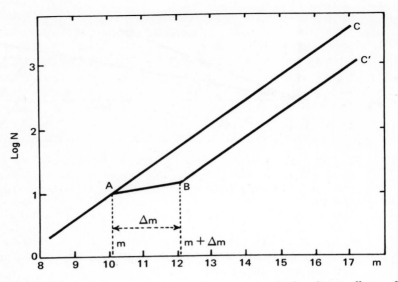

FIGURE 29-8   *Wolf diagram, a plot of star members vs. magnitude, shows effects of absorbing cloud on star counts. In an unobscured region the counts increase along* AC, *while in an obscured region the counts increase as in* ABC′.

m (if one assumes for simplicity that the stars observed have all the same absolute magnitude), the magnitudes of all stars fainter than m will be increased by Δm, and the curve breaks into two segments separated by a horizontal step AB between m and $m + \Delta m$ (Fig. 29-8). The magnitude at point A fixes the distance of the cloud, and the length of the segment $AB = \Delta m$ gives us the absorption in the cloud.

In practice the dark nebula is rarely, if ever, like a thin black curtain. Also the stars have a large range in absolute magnitude and therefore in distance. Actually the appreciable thickness of the cloud and the spread in absolute magnitudes of stars wash out the sharp step to produce a smooth transition, as indicated by the dotted line in the figure. The step becomes somewhat steeper if we restrict counts to stars of only one spectral type, say A-type stars, which have little spread in absolute luminosities. But then the amount of work involved greatly increases, since one must determine the spectra of all counted stars. A further complication results from the fact that, as we have seen, space is not quite transparent between dark nebulae.

If several such absorbing clouds of different depths and opacities lie at different distances along the same line of sight, the Wolf diagram then looks more like an irregular, rising curve than the simple step expected in principle and its interpretation is much less definite. Nevertheless, Wolf-diagram analyses, made during the 1920's and 1930's for most of the conspicuous dark nebulae, gave fairly satisfactory results. For example, the great Coal Sack of the Southern Milky Way (Fig. 29-1) results from a dark cloud some 40 light years across, absorbing somewhat more than one magnitude and located at a distance of some 500 light years. Few single clouds absorb more than three magnitudes, but the accumulation in depth of many individual clouds in the vast cloud complexes of the Great Rift of the Milky Way can produce in places almost total obscuration.

29-4   Interstellar Reddening.   Up to this point we have discussed interstellar absorption without reference to the wavelength of light and with little indication of the mechanism of this "absorption." In reality most absorption mechanisms depend more or less on wavelength and may or may not produce true absorption of luminous energy (that is conversion into heat or kinetic energy in the absorber). For instance, gas molecules of the air cause very little true absorption. Most of the intercepted starlight is immediately reradiated in all directions by the molecules. The absorption is only apparent. The missing energy is simply scattered from the light beam, but not transformed into some other form of energy by the molecules.

This scattering process—first elucidated by the English physicist Lord Rayleigh—is much more efficient for the shorter light waves (blue, violet, ultraviolet) than for the longer waves (yellow, red, infrared). Thus scattering removes from a beam of white light a greater proportion of the shorter wavelengths. The scattered light is bluer than the original light, whereas the transmitted light is redder. For Rayleigh scattering by atoms or molecules, this selective scattering varies inversely with the fourth power of wavelength (as λ^{-4}), and the change of color is conspicuous. The daytime blue sky and red glare of sunset owe their striking colors to this phenomenon.

Small particles, such as cigarette smoke, also scatter blue light more effectively as long as the particles are small compared with the wavelength of light. Larger particles scatter all wavelengths more uniformly. Cigarette smoke from the lighted tip is bluish. However, if one keeps cigarette smoke a few moments in the mouth, the exhaled smoke is whitish rather than bluish, because water droplets condensed on the carbon particles possess diameters greater than the small range of wavelengths of visible light.

From a precise study of the wavelength dependence of scattering and/or absorption of light by a cloud of small particles, we can gain some idea of the sizes of the particles, as the German physicist Mie first showed over fifty years ago (see Sec. 6-16). Astronomers have thus applied the Mie theory of scattering to learn something of the nature and sizes of the dust particles producing the dark nebulae and the general interstellar absorption. This is why astronomers have devoted much effort to measure the *reddening* of light of distant stars. The existence of interstellar reddening proves immediately that the dust particles must be small compared with the wavelengths of visible light (0.4 to 0.8 μ), because interstellar dust scatters blue light more effectively than red light. However, the particles cannot be very much smaller, since the selectivity is less than that for Rayleigh scattering.

More precisely, detailed measurements show that interstellar absorption (in magnitudes) varies approximately in inverse proportion to wavelength (as λ^{-1}). Hence atoms or molecules do not contribute appreciably, because they would require a λ^{-4} law. Nor can particles larger than, say, 1 micron, which would cause little or no selectivity, play an important role.

C. S. Schalén of Sweden was the first to make extensive theoretical studies on interstellar particles in the 1930's. He concluded that the observations were compatible with metallic particles about 0.1 micron in diameter. Sir Arthur Eddington, however, pointed out that we should not expect many metallic particles in interstellar space. The overwhelming abundance of hydrogen and the large abundances of carbon, nitrogen, and oxygen in the universe as compared with the metals should lead to a great preponderance of interstellar *ices*, particularly ordinary water-ice crystals, in space. The temperatures between the stars should be only a few degrees above absolute zero, so that ices could form and persist.

J. Oort and his colleagues at Leiden have shown how interstellar particles can grow from interstellar gas through atomic collisions and then be broken up again as they collide with one another in the turbulent gas and dust clouds of space. Over long periods of time the two processes must reach a stable statistical equilibrium of particle sizes and composition. Thus the particles may be described as grains or smoke if we emphasize the building-up by accretion and condensation, or as dust if we consider the destruction by collision and scattering.

In order to understand better the physics and chemistry of interstellar matter, we need even more refined studies of its effects on starlight passing through it. We derive the law of reddening from the *color excess* of stars strongly obscured by dark clouds as compared with the normal color index of nearby stars of the same spectral type. We recall (Chapter 19) that the color index measures the difference of magnitudes of a star measured at two different wavelengths—for instance, in blue light B and in yellow light corresponding to ordinary visual magnitudes V. Thus $C = B - V$. As we know (Chapter 20), a close relationship exists between color index and spectral type. Hence, if we measure the color index C of a distant star and compare it with the color index C_0 of a nearby star of the same spectral type and unaffected by absorption, the color excess

$$E = C - C_0 = (B - V) - (B - V)_0$$

measures the differential absorption between blue and yellow (visual) suffered by light of the distant star in its journey through space. If, further, we can reliably derive the distance of the star—for example, through membership in a star cluster (see Chapter 30)—we can also derive not only the differential absorption between B and V, but also the total absorption A of, say, its visual apparent magnitude. Thus we find the ratio A/E of total to differential absorption.

This ratio has a mean value of 3 and apparently varies little from region to region. Since this ratio depends sensitively on the sizes of the scattering particles, its near constancy tells us that the statistical properties of interstellar grains are remarkably similar in different parts of our system, at least as far as we can explore it optically. Studies of the galaxies where dark clouds are observed indicate that about the same value of A/E obtains in other systems and thus that the basic properties of interstellar dust are pretty much the same throughout the universe.

The range of wavelengths between blue and yellow—about 0.45 to 0.55 μ—is really very small compared with the many octaves covered by electromagnetic radiation. Atmospheric absorption limits our studies with ground-based telescopes to a range from 0.3 to about 13 μ. Since the end of the Second World War, photoelectric photometry has greatly extended the range of wavelengths over which the absorption curve of interstellar matter can be determined. This curve turns out to be S-shaped (Fig. 29-9), starting from essentially negligible absorption in the infrared and increasing steadily toward the ultraviolet. The λ^{-1} law is only an approximation,

FIGURE 29-9 (a)   *Interstellar absorption increases rapidly toward shorter wavelengths.*

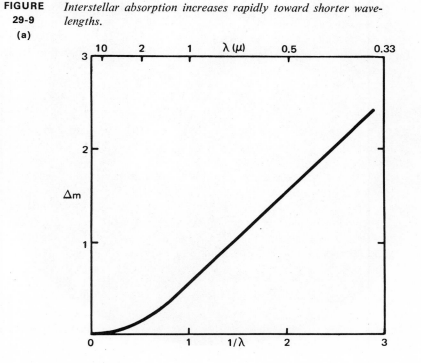

valid somewhere near the middle of the visible range. Thus we can conclude that true absorption is negligible compared with apparent absorption caused by scattering. This result indicates that interstellar grains are of dielectric composition, possibly containing ice crystals, and not as strongly absorbing as metallic grains would be.

29-5 Interstellar Polarization. Particles with spherical symmetry would be expected to scatter light uniformly around the direction of propagation of a light beam. In other words, transmitted light would be unpolarized; no direction of vibration would be favored. But polarization *is* observed! Asymmetric or elongated or flat particles, such as the small ice crystals sometimes postulated to explain interstellar reddening, would be expected

**FIGURE
29-9
(b)**

Star cloud near galactic center, obscured in blue light.

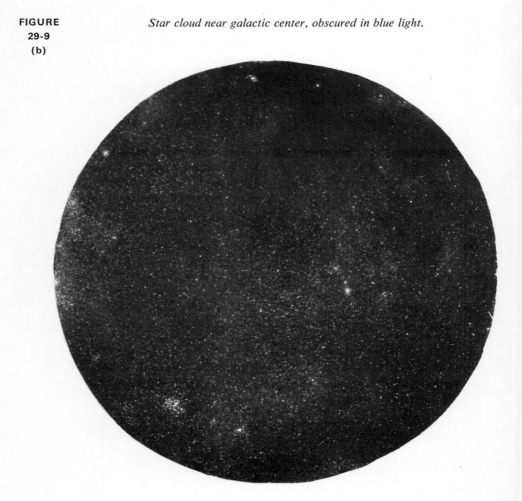

to show some preference individually, but this preference would cancel out in a cloud containing billions of randomly oriented particles. However, suppose that the cloud is immersed in or contains a homogeneous magnetic field—that is, one having parallel lines of force. Then the particles may tend to orient themselves in a systematic fashion with respect to the magnetic field. If this happens, even to a small degree, natural (unpolarized) starlight passing through such a cloud will emerge partly polarized (the vibration will be stronger in one direction and weaker at 90° from it.

This concept provides a probable explanation for the phenomenon discovered in 1948 by J. S. Hall then at the U.S. Naval Observatory, and W. A. Hiltner of Yerkes Observatory. They found that unobscured nearby stars showed no polarization, whereas distant stars, strongly affected by interstellar absorption and reddening, showed small but measurable polar-

FIGURE 29-9 (c)

Cloud appears clearly on infrared photograph.
(J. Dufay, J. Bigay, and J. Texereau, Haute-Provence Observatory.)

ization. Also, stars seen through the same obscuring cloud showed very nearly the same degree and direction of polarization. This result furnished additional proof that the polarization originates not in the stars but in large volumes of interstellar space.

Detailed maps of polarization over the whole sky correlate well with the distribution of dark clouds near the plane of the Milky Way. Nevertheless, in some regions stars with large color excesses show little or no polarization, either because grains are not oriented in the absorbing cloud or clouds in these particular directions or because starlight crosses several distant dust clouds whose different polarizing properties more or less cancel out.

The theory of interstellar polarization invoking a systematic orientation of elongated grains by magnetic fields was put forward by L. Davis and J. Greenstein. Many details are still obscure. For instance, are the grains pure paramagnetic ices or do absorbed metallic impurities make them ferromagnetic? Why should they be elongated? Are they perhaps flakes of graphite, as in the concept of an interstellar soot advanced by E. Schatzman? How strong are the postulated interstellar magnetic fields? Field strengths of the order of 10^{-4} gauss may be required to orient the grains, while the generally accepted theory of cosmic-ray acceleration in the Galaxy, first proposed by E. Fermi, considers fields only of 10^{-6} gauss. Whatever the details, the introduction of magnetism into the large-scale phenomena of interstellar space was a major advance of astronomy in the 1950's.

29-6   Interstellar Absorption Lines.   Dust particles are not the only form of obscuring matter in interstellar space. In 1904, J. Hartmann at Potsdam discovered the first indication that gas pervades space between the stars. He observed that the H and K lines of ionized calcium, Ca II, in the spectrum of the spectroscopic double star δ Orionis remain stationary, while the stellar lines shift back and forth by the radial motions of the star during each revolution (Fig. 29-10). From his further observation that the Ca II lines are very sharp compared with the diffuse stellar lines, he concluded that ionized calcium atoms must lie in interstellar space rather than in the star's atmosphere. The interstellar D lines of sodium were detected some twenty years later at Lick Observatory, while Mount Wilson observers discovered most of the other known interstellar lines during the 1920's and 1930's. A. McKellar of the Dominion Astrophysical Observatory first proved the existence of molecules in space. Later, absorption bands from radicals such as ionized CH were identified. The origins of several diffuse bands, first noted by P. Merrill at Mount Wilson, still remain unsolved. Particularly challenging is a strong band, some 40 angstroms wide, centered at λ 4430 Å. Table 29-1 gives, after J. Greenstein, a listing of all the identified interstellar lines and bands. See also Figs. 29-10 and 29-11.

D LINES OF SODIUM I

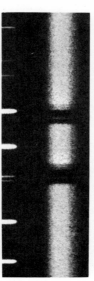

Five components are also visible in each of the interstellar D lines in the spectrum of Epsilon Orionis. They yield the same radial velocities as the calcium lines.

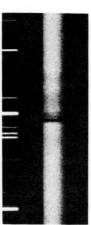

Each D line in the spectrum of 6 Cassiopeia is made up of two groups of lines, each group arising in separate clouds of sodium atoms in two different arms of our Galaxy whose radial velocities relative to the sun differ by about 30 kilometers per second.

In the spectrum of HD 14134 is visible the same complex structure of the interstellar D lines that was shown in the spectrum of 6 Cassiopeia. This indicates that this star also lies in or beyond a second spiral arm of our Galaxy.

H LINE OF CALCIUM II

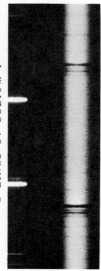

ments of these lines indicate heliocentric velocities for the five absorbing clouds of +3.9, +11.3, +17.6, +24.8, and +27.6 kilometers per second, respectively.

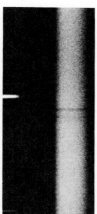

calcium lines as shown above. The broad faint lines adjacent to the H line originate in the atmosphere of the star.

**FIGURE
29-10**

The stationary lines of calcium, Ca II and sodium Na I in different stars. (*Hale Observatories.*)

K LINE OF CALCIUM II

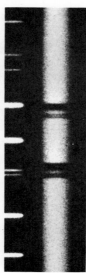

Five components are visible in the interstellar H and K lines in the spectrum of Epsilon Orionis. The displace-

The star HD 172,987 is about 20,000 light years distant, and shows in its spectrum unusually strong, complex, interstellar

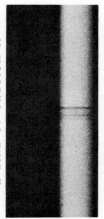

The star HD 47240 shows in its spectrum three weak, highly displaced interstellar K lines. This star lies in the direction of, and beyond the gaseous nebula NGC 2237, in the constellation of Monoceros.

TABLE 29-1 *Interstellar lines and bands*

Source	Wavelength
A. Optical Spectrum[a]	
Na I	3302*, 3303*, *5890, 5896* Å
Ca I	4227 Å
K I	7665, 7699 Å
Fe I	3720, 3860 Å
Ca II	*3933, 3968* Å
Ti II	3073, 3229, 3242, 3384 Å
CH	3143, 3879, 3886, 3890, 4300* Å
CN	3580, 3874, 3875, 3876 Å
CH⁺	3579, 3745, 3958*, 4233* Å
Unidentified	⎰ 4440 (40), 4760 (80), 4890 (50), 5780* (8), 5797 (4) Å
Bands	⎱ 6180 (80), 6203 (7), 6270 (5), 6284* (10), 6614 (4) Å
B. Radio Spectrum	
H, spin reversal	21.1 cm
H, high-level transitions	3.4, 5.2, 6.0, 17.5, 17.9 cm
OH	4.982, 4.971, 4.968, 4.956; 18.593; 6.380, 6.356 cm
H_2O	1.3481 cm
NH_3	1.2652, 1.2637 cm
HCHO	6.207 cm

[a] Strongest line italicized, stronger lines marked by asterisk, bandwidth in parentheses.

The absence of interstellar absorption lines from such abundant elements as hydrogen and helium does not indicate that these elements are absent in space, but merely that conditions are not favorable for their observation in the optical spectrum or that atomic levels leading to observable transitions are not populated. Thus, while the presence of hydrogen had been inferred from the presence of hydrides, its direct observation was not possible until the early 1950's, when its emission at a wavelength of 21 cm was detected by radio-astronomical techniques (see Chapters 6, 8, 31). The reason why very few atomic transitions are observed in absorption requires some explanation. It rests on the very peculiar conditions that obtain in interstellar space: mainly, extremely low densities and temperatures, first analyzed by Eddington in the 1930's.

Starlight in space is so feeble that an atom may absorb an ionizing or exciting photon only once in several years. The particle density is so low that an ionized atom will collide with another atom not more than once a week, whereas neutral atoms will suffer encounters only at intervals of a year. Since excited atoms ordinarily fall back to lower energy states and radiate after some 10^{-8} second and stay in most metastable states not more than a few seconds, the chance that an incoming photon will find an interstellar atom in an excited state is negligibly small. Consequently, interstellar line absorption usually takes place from the very lowest energy level in the atom or molecule. Molecules that produce extremely complicated band systems in the laboratory may show only a few absorption lines in interstellar space.

Since most atoms have their ultimate lines arising through transitions from the lowest energy states in the far ultraviolet, we can find only a few in interstellar space that absorb in the region of the spectrum that can be photographed from the earth's surface. From a satellite observatory we could expect to observe many more lines in the far ultraviolet and thereby greatly increase our information about the composition of interstellar matter. Nevertheless, through a combination of existing observations and physical theory, astronomers have already learned a great deal about conditions in interstellar space.

The level of ionization in space is fairly high, corresponding roughly to a temperature of 10,000° K. Thus calculations predict that the ratios of Ca III to Ca II ions to Ca I atoms are 10 : 1 : 0.1. Such high temperatures may seem inconsistent with our statement in a previous section that the temperature of interstellar dust is only a few degrees above absolute zero. Consideration of both the *quantity* and the *quality* of the radiation in interstellar space resolves this difficulty. The quantity of radiation or heat is negligible because of the enormous dilution factor corresponding to the average distances of several light years between the stars. Hence a solid in radiative equilibrium in space must, by Stefan's law (see Chapter 8), cool down to a few degrees above absolute zero, where its reduced emission of radiation can balance the meager absorption of radiant energy. But the energy per quantum of radiation in space is high, more or less averaging that of an A-type star. Since the gas pressure and density are correspondingly as low as the radiation density, when compared with conditions in a stellar atmosphere, the atoms and molecules can respond to the high-energy photons of the weak radiation and show strong ionization. In other words, the few high-energy quanta in space are frequent enough to ionize the few atoms because recombination is so slow, but the total amount of radiation is not enough to heat even a minute dust particle appreciably.

Although hydrogen and many other elements cannot produce observable interstellar absorption lines, those substances that we do observe show relative abundances similar to those in stellar atmospheres. In an average cubic meter of interstellar space we find on the average about one atom of sodium, one of calcium, and a few of iron. We should have to search a volume equal to that of a small room in order to find a single atom of titanium or potassium.

C. S. Beals, in Canada, discovered in 1936 that some of the interstellar lines are multiple in structure; he concluded that each component represents a separate gas cloud moving with its own specific radial velocity. W. S. Adams of Mount Wilson Observatory later studied the spectra of some 300 distant bright stars in the Milky Way plane to determine the intensities and complex structures of the interstellar lines. Out of 300 stars he found 95 with double lines, 17 with triple, and 4 with quadruple. The average radial velocities of the stronger lines, with the galactic rotation subtracted, amount to some 8 km/sec. However, 21 of the stars show components moving faster than 30 km/sec and 7 more than 50 km/sec. Adams' results proved not only that the bulk of interstellar gas revolves about the galactic center together with the majority of the stars (see Chapter 31), but also that some of the smaller gas clouds move turbulently or irregularly at speeds greater than most stars.

More recently G. Münch, who followed up Adam's work at Mount Wilson, showed that most discrete components of interstellar lines are associated with one or several distinct spiral arms in our galaxy (Chapter 31).

29-7   Reflection Nebulae.   In most places interstellar dust clouds are cold and dark. Occasionally, however, they become luminous when a nearby bright star causes them to shine by reflected light. In 1912 V. M. Slipher at Lowell Observatory first recognized reflection nebulae when he discovered that the diffuse nebulosities in the Pleiades did not possess the line emission spectra characteristic of gaseous nebulae such as Orion (see following section), but continuous spectra showing absorption lines identical with those from the bright stars of the cluster. Clearly the nebulosity of the Pleiades reflects the light of the embedded stars. In the early 1920's, E. P. Hubble gave further proof of the scattering mechanism, when he measured the color index of these reflection nebulae and found them about one-quarter of a magnitude bluer than the illuminating stars. This fact is consistent with scattering by small particles of the order of size of the wavelength of light. As we explained earlier (Sec. 29-4), molecules would scatter more blue light, while large particles would scatter all colors equally. The blueing of scattered light in reflection nebulae is the complementary phenomenon of the reddening of transmitted light by interstellar dust clouds.

We should not conclude that reflection nebulae are merely dark clouds that happen to be near a bright star. Some actual physical relationship, not a temporary chance encounter, may exist between the reflection nebulae and the stars that illuminate them. Hubble observed that stars of type B1 or later produce reflection nebulae, while stars of type B0 and earlier are associated with emission nebulosities. Here the difference is primarily one of temperature and degree of excitation. However, the hotter stars may blow away or evaporate the icy particles of interstellar dust.

**FIGURE
29-11**

Variable nebula NGC 2261 and associated variable star R Monocerotis. (Yerkes Observatory.)

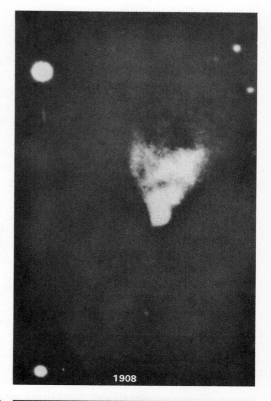

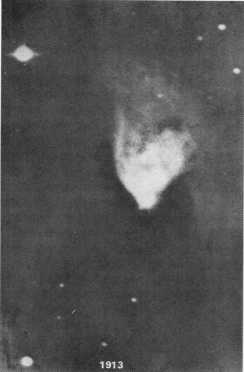

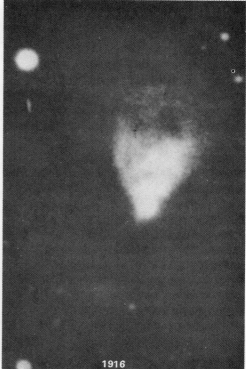

A more definite physical relation between star and nebula occurs for the variable star R Monocerotis, which illuminates the cometary or fan-shaped nebula NGC 2261 (Fig. 29-11). Both the star and the nebula display irregular variations in light. On many occasions a dark shadow appears to sweep across the entire nebula in a few days. These intensity variations apparently move with the velocity of light. There seems little doubt that the illuminating star at the tip of the "fan" is passing through dense dark clouds that irregularly shield the nebula from its radiation. It is also probably more than a coincidence that R Monocerotis is a T Tauri variable (see Chapter 26).

These peculiar variables studied by A. Joy and J. Greenstein at Mount Wilson and especially by G. Herbig at Lick Observatory are faint dwarf stars of type G, K, and M, and they seem to be always immersed in dense regions of interstellar gas and dust. In fact Herbig has suggested that they are stars still in the process of condensation from the interstellar medium (see Chapter 35). They display rapid "flares" in brightness and bursts of line emission in their spectra, probably produced by the irregular influx of interstellar dust and gas, which causes unusual activity in their photospheres and chromospheres.

29-8   Emission Nebulae. Emission nebulae—or nebulosities—are huge masses of gas that absorb ultraviolet radiation from nearby hot stars and reradiate it as bright-line emission, often by metastable or "forbidden" transitions (see Chapter 8). The classical example is the great nebula Messier 42 in Orion (Fig. 29-12); another fine example is the Eta Carinae nebular complex in the southern sky (Fig. 29-4) and the spectacular North America nebula in Cygnus (Fig. 29-14). The largest emission nebulosities are almost always associated with O and B0 stars and in most cases with dense groups of extremely hot and luminous stars. Furthermore, the associated stars show large color excesses, which indicate the presence of large amounts of interstellar dust in their neighborhood.

Dark dust clouds clearly appear on photographs of M 42 or Eta Carinae. Whether because of high luminosity, radiation pressure, a "stellar wind," or some other cause such as vaporizing of the icy dust, the central stars appear to have cleared the dust away from their immediate surroundings. Baade and Minkowski showed that the central stars in the Orion nebula delineate a hole or dust-free bubble, inside a rather dusty region of space. On the other hand, the gas densities in these great nebulae are considerably higher than the average in interstellar space. In the Orion nebula, for example, the density may be as high as 300 ions/cm^3, according to J. Greenstein.

O. Struve observed that emission nebulae show somewhat less obscuration by interstellar dust than do the reflection nebulae. Since the emission-

FIGURE   *The Great Nebula, M42, in Orion.*
29-12   *(Mount Wilson and Palomar Observatories.)*

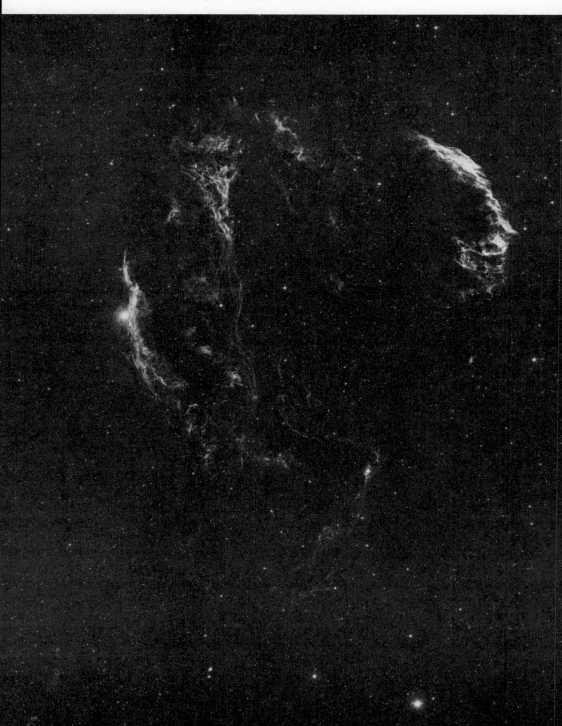

line spectrum should be added to the reflection when gas and dust are both present, one might expect to find the surface brightness of emission nebulae relatively greater than those of reflection type. Hubble's photometric observations of the diffuse nebulae failed to detect such a relationship, a fact supporting the view that the hotter stars eliminate the dust immediately around them.

In 1939 B. Strömgren showed how an ionized region in the interstellar gas must form about a hot star. Since the gas is almost all hydrogen, the emissions from hydrogen will dominate the far-ultraviolet region of the spectrum. For a considerable distance from a hot star practically no hydrogen can remain neutral. The gas is a rarefied plasma of protons and free electrons. Eventually, at still greater distances, the rate of recombination of the protons and electrons, which depends on density and remains constant in a uniform gas cloud, surpasses the ionization rate determined by the intensity of ultraviolet radiation, which decreases as the square of the distance from the star. At a certain distance the continuous absorption at the limit of the Lyman series ($\lambda < 912$ Å) begins to cut off the ultraviolet light of the stars, and the ionization of hydrogen stops abruptly. Thus a roughly spherical volume of ionized hydrogen, an *H II region*, forms about the hot star. Other regions of space, called H I regions, contain only neutral hydrogen. The transition between the two regions where hydrogen is partially ionized is very sharp.

Hydrogen can be ionized not only by absorption of ultraviolet light but also by collisions with other atoms, if the impacts are sufficiently violent, i.e., if the speeds of the colliding atoms are large enough. This completely different excitation mechanism is responsible for the luminescence of another type of emission nebulosity exemplified by the Network nebula, NGC 6992-6995 in Cygnus (Fig. 29-13). Here only narrow filaments near the rim of a wide arc appear luminous. No hot star or cluster of hot stars can be found to account for the emission. Precise measurements of bright knots in the filaments show that the Network nebula is expanding outwards at a rate of some 0".03 per year, and since the Cygnus loop has an overall diameter of the order of 3°, an expansion age of some 180,000 years would follow on the assumption that the expansion velocity is constant. The maximum radial velocity is 116 km/sec, which, combined with the angular expansion rate, indicates a distance of 770 parsecs and an overall diameter of about 40 parsecs.

J. Oort suggested in 1946 that this type of nebulosity can best be explained as the remnant of the high-velocity gas expelled in a supernova explosion. Its luminescence results from collision between the expanding cloud and the low-density gas of the interstellar medium. At the interface a violent shock wave occurs, as it does about a jet plane moving faster than the velocity of sound. Excitation of hydrogen luminescence by shock waves has been produced in the laboratory. Precise measurements of intensity ratios of hydrogen Balmer lines, which depend on the excitation mechanism, confirm that atomic collisions produce the luminescence of the Loop nebula. Since several other nebulosities of the same type are known, this phenomenon must be relatively frequent in space. Most of these objects have also been observed as extended radio sources (see Chapter 33), and the spectrum of the radiation is also consistent with the old supernova-shell interpretation.

In addition to individual, well-localized H II regions, enormous volumes of interstellar space show a weak hydrogen-line emission without definite center. O. Struve and C. T. Elvey in the late 1930's were the first to observe systematically the diffuse background of emission in the Milky Way, using a special wide-field nebular spectrograph. They detected emission lines over large areas of the Milky Way in regions where no nebulosity had previously been observed. Only low-excitation lines appear, such as those of the hydrogen series, mainly Hα, and in the ultraviolet the forbidden line λ 3727 of [O II]. Systematic spectroscopic surveys of the complete galactic belt made in Australia, Europe, and the United States, in the 1940's and 1950's, demonstrate the presence of this weak, widespread emission over a large portion of the sky. In recent years, special techniques such as narrowband interference filters and composite printing have permitted the photography of some of these areas. The regions display a complex filamentary structure, which extends in places to great distances from the Milky Way, especially where obscuring dark clouds are present. One such region, twenty degrees across, lies near the south celestial pole.

Strömgren's theory applies to the diffuse background of line emission in the Milky Way, as excited by the general ultraviolet radiation from all stars. The observed intensities can be explained if only 2 or 3 atoms of hydrogen per cubic centimeter occur in these regions of weak emission. This amount is less than 1 per cent of the density in the classical emission nebulosities, but still somewhat higher than the overall average density of neutral hydrogen near the galactic plane, as derived from radio observations.

29-9 Neutral Hydrogen in Space. The ubiquitous presence of the recombination spectrum of hydrogen in emission and the visibility of absorption lines of hydrides of several light elements (Sec. 29-6) had long been taken as a strong presumption that neutral hydrogen atoms are present in interstellar space. In fact Eddington had already concluded from his early analysis of physical conditions in interstellar space that hydrogen must be by far the most abundant element in space, as indeed it is in the stars. However, neutral hydrogen in its ground state emits and absorbs radiation only in the extreme ultraviolet, a region unobservable by optical techniques from the earth's surface. Hence there was no way of verifying Eddington's conclusion about the high abundance of hydrogen. The advent of radio astronomy changed this situation.

In 1944 H. van de Hulst of Leiden Observatory predicted that a transition between sublevels in the fine structure of the lowest energy level (ground state) of the hydrogen atom should produce an emission or absorption line at wavelength 21 cm, in the microwave range, and that this transition, then unobservable in the laboratory, should be observable in the radio spectrum of interstellar space. This transition, as we explained earlier (Chapter 8), arises when the spin or rotation vector of the electron reverses itself spontaneously or by absorption of radiation from parallel to antiparallel (and vice versa) with respect to the spin of the proton. The energy of the hydrogen atom is slightly different in these two conditions, and the difference corresponds to the energy $E = h\nu$ of the quantum emitted or absorbed in the transition. This difference being very small, the frequency of the radiation $\nu = 1420$ MHz is correspondingly small (compared with the optical frequencies of 10^{14} to 10^{15} Hz). The wavelength $\lambda = c/\nu = 3 \times 10^{10}/1.42 \times 10^9 = 21.1$ cm lies in a region of the radio spectrum to which our atmosphere is transparent. Sensitive radio telescopes and receivers should be able to detect the radiation, despite the very low probability for the transition, corresponding to the very long lifetime of 3.5×10^{14} seconds. In other words, a transition between the two states would occur only once every 11 million years in a given hydrogen atom. In the depths of interstellar space, however, where hydrogen atoms can remain neutral for very long intervals of time, the total mass of hydrogen in the line of sight, especially near the galactic plane, is large enough to produce measurable emission or absorption of the 21-cm line.

Radio telescope receivers capable of detecting the 21-cm radiation from space were not available in war-torn Holland in 1944, but such instruments were rapidly developed after the war in several countries. The 21-cm line of emission of neutral hydrogen was successfully detected in 1951, first by H. I. Ewen and E. M. Purcell at Harvard. C. A. Muller in Holland and F. J. Kerr in Australia quickly confirmed the result. All reported a significant increase in intensity of the sky radiation at the predicted wavelength when the Milky Way was crossing the cone of reception (or "antenna beam" in the radio astronomer's parlance).

Within a few years extensive programs for the detailed study of the 21-cm line in Australia, England, Holland, and the United States provided a wealth of new information. Soon the line was observed not only in emission but also in absorption against the brighter or "hotter" background of strong, distant radio sources (cf. Chapter 33). Figure 29-15 shows the complex structure of the 21-cm line in absorption in the spectrum of the radio source Cassiopeia A; the several components correspond, as in the optical case, to the presence in the line of sight of several hydrogen clouds moving with different radial velocities. A strong, narrow absorption core also appears in the H I emission observed in the direction of the center of the galaxy (see Chapter 31), and recently a weaker double line of the OH molecule was detected in the same direction.

FIGURE   *The North America Nebula. (Lick Observatory.)*
29-14

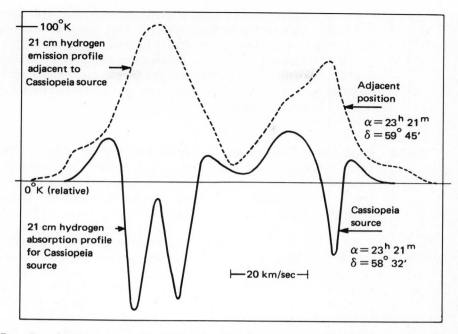

— 100°K

21 cm hydrogen
emission profile
adjacent to
Cassiopeia source

Adjacent
position

$\alpha = 23^h \, 21^m$
$\delta = 59° \, 45'$

0°K (relative)

21 cm hydrogen
absorption profile
for Cassiopeia
source

Cassiopeia
source

$\alpha = 23^h \, 21^m$
$\delta = 58° \, 32'$

├─20 km/sec─┤

**FIGURE
29-15**

Complex structure of hydrogen 21-cm absorption line (solid curve) observed against a bright emission background from "hotter" gas (dashed curve).

As we shall explain more fully later (Chapter 31), the detection of the 21-cm line of emission of neutral hydrogen was not only an outstanding achievement of theoretical spectroscopy and electronics engineering, but also a major breakthrough in the study of the large scale structure of our galaxy.

Thus for the first time the bulk of the interstellar medium became accessible to direct observation. A detailed theory of neutral-hydrogen emission led to a determination of the density and temperature of neutral hydrogen. Typical densities were of the order of 0.05 to 5 atoms per cubic centimeter. The derived temperatures from 10° to 100° K agreed well with the theoretical expectations for the average H I regions near the galactic plane. The typical density of matter in interstellar space is about $\rho_1 \approx 2 \times 10^{-25}$ g cm^{-3}, which corresponds to 0.1 hydrogen atom per cubic centimeter. Other atoms, molecular substances, and dust particles contribute not more than a few per cent to the total density. For comparison, the average density of the matter condensed in stars in the neighborhood of the sun is about 0.14 solar mass per cubic parsec, or $\rho_s = 2.8 \times 10^{32}/(3.08 \times 10^{18})^3 = 10 \times 10^{-24}$ g cm^{-3}. We conclude that interstellar matter, although of great astrophysical and cosmogonic interest, contributes no more than about 2 per cent to the total mass density of our region of space.

The question of the possible presence of *molecular* hydrogen H_2 in interstellar space cannot be answered at present, because the critical emission (or absorption) lines are in the far-infrared region of the spectrum—unobservable from the earth's surface because of atmospheric absorption. Conceivably hydrogen molecules could be much more abundant in space than free hydrogen atoms, but their detection must await future observations from space telescopes.

29-10   High-Level Transitions in Hydrogen and Helium.   In 1952 J. P. Wild of Australia suggested that transitions between very high-energy levels of hydrogen could produce radiations in the range of radio frequencies. N. S. Kardashev of the Soviet Union elaborated this idea in 1959. Shortly thereafter, many radio astronomers detected a number of lines in the radio region.

As noted earlier, the energy-level diagram (Fig. 8-6) of hydrogen resembles a stepladder, with the spaces between successive steps diminishing with height. Each step bears a label, the quantum number n, which ranges from 1 to infinity. Electronic transitions between relatively low levels produce the familiar Balmer series, $H\alpha$, $H\beta$, $H\gamma$, and so on, as discussed in Sec. 8.9. The Greek letters α, β, γ, δ, ε, and so on indicate transitions in which the quantum number n changes by 1, 2, 3, 4, 5, . . . units, respectively.

The rapid convergence makes it impossible to represent the higher levels in Fig. 8-6. Figure 29-16 shows only the upper levels, on a scale 50,000 times greater than that used for Fig. 8-6. On this scale, $n = \infty$ lies about five times higher than the range of levels 100–112. As indicated in Sec. 8-9, we usually label these high-level transitions by the lower quantum number, and the index, α, β, γ, . . . , as explained above. Lines and wavelengths include the following: 90α—3.4 cm, 104α—5.2 cm, 109α—6.0 cm, 156α—17.5 cm, 158α—17.9 cm. The β lines and γ lines are much fainter than the corresponding α's, but several such lines have been detected.

Emission of high-level hydrogen radiation comes from the H II regions. Such radiation occurs when electrons, captured in high levels, cascade in steps to lower levels. Observed lines include those from capture of electrons by singly ionized helium.

29-11   Free Radicals and Molecules in Space.   The incomplete molecule or radical, OH, exists abundantly in various regions of space. The lines appear sometimes in absorption and other times in emission. The complete water molecule, H_2O, has also been detected, with such a high temperature that some astronomers refer to the material as "steam." Interstellar ammonia, NH_3, and formaldehyde, $HCHO$, also exist.

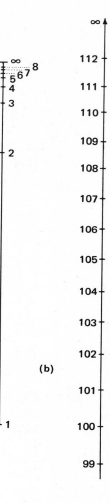

**FIGURE
29-16**

*The higher energy levels of hydrogen, responsible
for radio emission.   (a) At left, location of
energy levels is designated by quantum numbers
1 through 8. Note how the upper levels converge
toward $n = \infty$. Compare Fig. 8-6.   (b) The
center line represents a 10,000-fold enlargement
of an upper section of the left-hand scale, to show
the arrangement of a few high-energy levels of
hydrogen.*

*On this enlarged scale, the point labled ∞ lies
on a vertical extension of the line above $n = 112$,
equal to five times the distance, from 1 to ∞, in
(a).*

(a)   (b)

CHAPTER *30*

Star Clusters and Associations

30-1   Galactic Clusters.   Stars do not occur in space at completely arbitrary and independent places. Some, such as our sun, are single, but others are members of pairs or form multiple-star systems. Still others form clusters of various types, sizes, and populations.

Two main types of star clusters occur: the small and sparse open or *galactic clusters*, and the large and dense *globular clusters*. Large-scale groupings of some stellar types are called *associations*. We shall discuss first the galactic clusters, of which several hundred known examples exist in regions of the Galaxy accessible to direct optical exploration.

Several of the nearer galactic clusters can easily be seen with the naked eye. The best known are the brighter stars of Ursa Major, of Coma Berenices, the Hyades, and the Pleiades in Taurus (Fig. 30-1). The last furnishes a good test of keen eyesight; normal vision will detect 6 to 8 stars in it. Dozens of clusters become visible with binoculars or small telescopes— for example, the double cluster h and χ in Perseus, which to the unaided eye appears as a fuzzy double star roughly halfway between α Persei and Cassiopeia. Praesepe in Cancer is another example. Several hundred galactic clusters have been catalogued and thousands of them must exist in our Milky Way, most of them too far away or too obscured by interstellar dust for detection.

(a)

FIGURE
30-1 *The Pleiades, M 45, a typical galactic cluster. (a) Small-scale photograph. (Mount Wilson and Palomar Observatories.) (b) Large-scale photograph showing reflection nebulosities. (Lick Observatory.)*

(b)

Many star clusters are free of nebulosity. The Hyades and Praesepe are examples. Other clusters are imbedded in bright nebulosity often mixed with dark absorbing clouds. The Pleiades furnishes a fine example of bright nebulosity reflecting the light of the cluster stars (Fig. 30-1). Here tenuous veils of interstellar dust shine by reflected starlight, much as a thin cloud of dust or smoke, reflecting the sunshine, will appear bright against a darker backdrop. In many cases, however, the interstellar gas becomes self-luminous not merely by reflection but by fluorescence, if the associated cluster includes some high-temperature supergiant stars of types O or B whose light is rich in ultraviolet radiation. We have discussed this emission mechanism in Chapter 29 in connection with the luminescence of H II regions.

30-2 **The Motions of Nearby Galactic Clusters.** Star clusters provide one of the basic methods of measuring stellar distances both near and far. Consider a nearby cluster such as the Hyades or the Ursa Major cluster, covering a fairly large area of the celestial sphere. Except for small peculiar motions, all stars that are members of the cluster move in space along essentially parallel paths, much like meteors in a meteor stream. If, then, we measure with high accuracy the proper motions of the cluster stars with respect to the fixed background of the distant stars, we find that the individual proper motions seem to converge to or diverge from a common point on the sky (Fig. 30-2), in the same fashion that meteors in showers appear to diverge from their radiants (see Chapter 15). The point

FIGURE 30-2   *Proper motions (arrows) of stars in the Hyades converge toward a point, C, on the celestial sphere.*

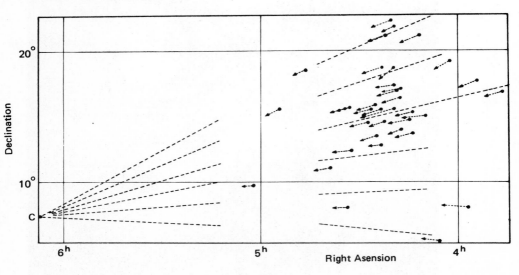

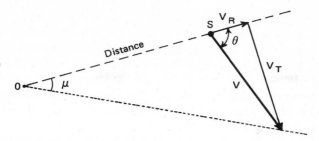

FIGURE
30-3 *Relation between proper motion, μ, radial velocity, V_R, and distance when convergent is known.*

of divergence marks the direction in space toward which the sun is moving with respect to the cluster. The point of convergence of the proper motions (diametrically opposite to the point of divergence on the celestial sphere) marks the direction toward which the cluster is moving with respect to the sun. All motions are relative, of course. Now if we knew the actual space velocity V, say in kilometers per second, of each cluster star with respect to the sun, we could calculate the distance of each star by comparing the linear motion and the apparent (angular) proper motion. We can in fact, measure directly the line-of-sight component of the motion, the radial velocity V_R of the star. Then, let V_T be the tangential velocity, the component of the velocity across the line of sight, and since we know the direction of the space velocity V from the position of the convergent, we know the angle θ between V_R and V in the right-angle triangle SRV (Fig. 30-3); thus

$$V = \frac{V_R}{\cos\theta}, \tag{30-1}$$

and

$$V_T = V\sin\theta = \frac{V_R\sin\theta}{\cos\theta} = V_R\tan\theta. \tag{30-2}$$

We know already that the apparent proper motion μ is related to the tangential velocity V_T (in km sec^{-1}) through equation (19-3),

$$V_T = 4.74\frac{\mu}{p}, \tag{30-3}$$

where μ is the annual proper motion and p the parallax (in seconds of arc). Equating the two values of V_T we derive for the parallax

$$p = 4.74\frac{\mu}{V_R\tan\theta}, \tag{30-4}$$

and consequently the sought distance $R = 1/p$ of the cluster (in parsecs).

By this *moving-cluster* method B. Boss found in 1915 that the distance to the center of the Hyades is 135 light years, a value in good agreement with but more reliable than the distance derived from trigonometric parallaxes of individual stars in the cluster. The mean parallax is only $p = 0''.024$, a figure not much larger than the accidental and systematic errors of an average trigonometric parallax. Hence a direct determination of the distance of the Hyades would be difficult, and the result would be of low relative accuracy. Thus, the few nearby moving clusters, mainly Ursa Major and the Hyades, provide us with the first stepping-stone to the great galactic distances that lie beyond the range of direct triangulation.

30-3 The Distance of Galactic Clusters. The next step involves a consideration of the color-luminosity diagrams of stars in galactic clusters. Having derived the distances to the Hyades and other nearby clusters by fundamental geometric methods, we can immediately calibrate the apparent magnitudes m of the stars in terms of absolute magnitudes M, through equation (19-6):

$$M = m - 5 \log r + 5. \tag{30-5}$$

Now if we plot, as in Fig. 30-4 for the Pleiades, the color-luminosity diagram of a cluster, a graph of absolute magnitude versus color index (similar to the Hertzsprung-Russell diagram, see Chapter 20), we notice that the stars do not scatter randomly in the M-C plane, but fall along narrow, well-defined bands or sequences. In particular we recognize the main sequence of ordinary dwarf stars going through the known solar luminosity and color at $M_v = +4.8$, $B - V = +0.6$. Suppose now that we plot a similar diagram for a galatic cluster of *a priori* unknown distance. We plot simply the observed apparent magnitude versus measured color index. The vertical displacement $m - M$ that brings into coincidence the main sequence of the observed cluster with the standard main sequence provides, through equation (30-5), a measure of the distance or *distance modulus* of the cluster.

R. J. Trumpler of Lick Observatory, in 1930, first applied this simple and powerful method of measuring the distances of galactic clusters. In fact he used the Hertzsprung-Russell diagram in its original form. He obtained low-dispersion spectra of the brightest stars in many clusters and plotted apparent magnitudes against estimated spectral types. He then read the distance moduli from the graphs as explained above.

The method, however, involves some complications, which led Trumpler to the major discovery of interstellar obscuration. We have already discussed this topic in Chapter 29, but let us retrace the steps leading to Trumpler's discovery, because they provide a striking example in scientific methodology.

From the apparent diameters of the clusters and their distances derived from the Hertzsprung-Russell diagram, Trumpler computed the linear

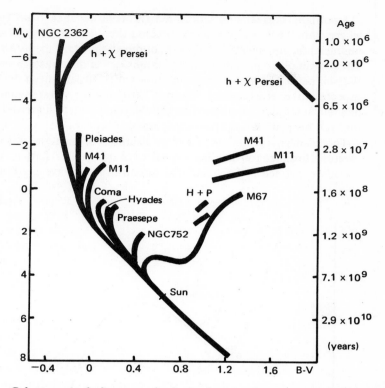

**FIGURE
30-4**

*Color-magnitude diagrams of galactic clusters. The ages of the clusters correspond-
ing to the positions of the breaks from the main sequence are shown by the scale on
right. Compare Fig. 35-6.*

diameters of about one hundred clusters. He then found that he could
classify the clusters according to their richness (number of stars) and their
concentration towards the center. These quantities correlated well with
the measured linear diameters of the nearer clusters. Thus Trumpler decided that he could calculate the true (linear) diameters of distant clusters
merely from their appearance (richness and concentration) on photographs.

Although the method checked well with diameters derived through the
H-R diagram modulus for relatively nearby clusters, it gave significant
discrepancies when applied to the more distant clusters. The diameters
calculated from the appearance were smaller than those calculated from
the H-R diagram, and the greater the distance the greater was the relative
error. The discrepancies are real and do not depend upon the various uncertainties in the measuring or observational procedures. Trumpler correctly interpreted these results as indicating the presence of an absorbing
medium in interstellar space. He judged the diameter-by-appearance to be

substantially correct and calculated the true distance from that figure, whereas the H-R diagram overestimated the distance because interstellar obscuration, not mere remoteness, dimmed the stars in distant clusters. Thus, Trumpler concluded that the brightness of distant stars in the Milky Way is reduced by about one magnitude (or to 40 per cent of the geometric expectation) over a distance of some 5000 light years. All subsequent studies have substantially confirmed Trumpler's conclusion, first arrived at in 1930 through his penetrating studies of galactic clusters.

The modern method of deriving distances of galactic clusters makes careful allowance for the dimming effect of interstellar matter. As we have already explained in Sec. 29-4, the obscuration is wavelength-dependent.

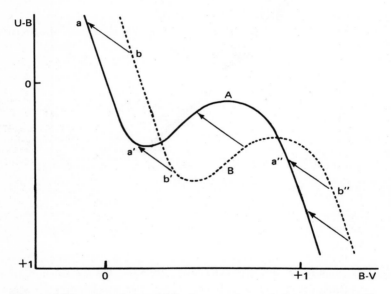

FIGURE 30-5 *Obscured (B) and unobscured (A) clusters. Arrows indicate corresponding points.*

Obscured stars also appear redder than nearby stars of the same type. This reddening is best derived from a color-color diagram of the cluster stars as in Fig. 30-5, a plot of the $U - B$ color index (ultraviolet magnitude minus blue magnitude) against the $B - V$ index (blue minus yellow). This graph has a peculiar and unmistakable sinuous shape (caused by the strong absorption of the near ultraviolet by hydrogen in A-type stars). A sliding fit of the cluster main sequence to the standard main sequence unambiguously determines the color excess or differential absorption. We can then derive the total obscuration in any given color, as explained in Sec. 29-4. Alternatively, one may measure the colors of cluster stars of known spectral types and hence of predictable true color (cf. Fig. 20-7); the difference between the observed and true colors again fixes the color excess. Then the distance follows from equation (30-5), where the apparent magnitudes are first corrected for the calculated absorption A; thus

$$\log r = 0.2(m - A - M) + 1. \tag{30-6}$$

The true distances of several hundred galactic clusters have been so derived. Some include so few stars that we can recognize them as clusters only if they happen to be very near. The Ursa Major moving cluster is an example. Others contain several hundred stars, down to the limit of faintness where cluster-membership can be established by one or more criteria (common proper motion, photometric parallax, color-color relation). Some very rich galactic clusters may total several thousand members strongly condensed toward a common center. The stellar density, or number of stars per unit volume, in the center of a galactic cluster varies over a wide range, from less than 1 per cubic parsec in poor clusters up to perhaps 100 per cubic parsec in the densest clusters. Thus the star density runs well above the average value, about 0.1 per cubic parsec, in our solar neighborhood. But for very poor and loose clusters such as Ursa Major, no central condensation can be detected and the average density may be much lower, perhaps as low as 0.0001 star per cubic parsec.

Nearly all galactic clusters lie very close to the plane of the Milky Way and hence are lost in the rich stellar background at great distances. Only a few galactic clusters can be identified beyond 10,000 light years from us, and those that have been detected suffer heavy obscuration and reddening by interstellar absorption. The few clusters that appear at high galactic latitudes, such as the Coma Berenices cluster, are quite close, and their actual distances from the galactic plane are not large, a few hundred parsecs at the most (see Fig. 31-4).

The color-luminosity diagrams of galactic clusters have proved to be extremely useful tools—not only to explore our galactic neighborhood out to fairly large distances (see Chapter 31), but even more importantly to serve as crucial indicators of stellar evolution and of the ages of the clusters, a field of study developed mainly after 1950 (see Chapter 35). The next section, which treats the stability of galactic clusters and of the remarkable stellar associations, will indicate the nature of these studies.

30-4 Stellar Associations. The low stellar density of most galactic clusters suggests that they cannot be permanent and must possess relatively short lives on a galactic time scale. Even if it were completely isolated in space, a cluster would slowly lose stars. Close approaches or "encounters" between stars of the same cluster will lead to exchange of energy and occasionally give some star a velocity greater than the small velocity of escape from the cluster. This evaporation process operates only very slowly.

The clusters, however, are not isolated in space but are part of the galactic system, continuously moving with reference to the field stars and star clouds of the Galaxy. Encounters with these stars will occasionally eject a star from the cluster. Conversely, of course, stars from the field can occasionally be captured by the cluster. The second process, however, is much less efficient than the first, and the clusters suffer a net loss of stars. Finally the clusters are subject to the effect of differential rotation in the general gravitational field of the Galaxy (cf. Chapter 31). The galactic perturbation acts somewhat like a tide-raising force; this tidal force, in fact, far exceeds the others in disrupting clusters.

The speed of "evaporation" of the cluster stars depends greatly on their density in the cluster. A dense, compact cluster will be much more stable than a cluster of low density. Thus, calculations indicate that in our neighborhood a cluster will be stable for a long period of time if its average density exceeds 0.1 solar mass per cubic parsec. Hence the Hyades cluster is probably safe for about a thousand million years, whereas the Pleiades and Praesepe may endure perhaps ten times longer. Nearer the galactic center the disruptive forces increase and the lifetimes become much shorter. We therefore conclude that galactic clusters, especially the loose clusters, have very short lives on a galactic time scale. Their ages may be less than one-tenth, and in some cases less than one-hundredth, of the age of the Galaxy (see Chapter 31). Why, then, do we still see so many around us, even very loose ones? The answer must be that galactic clusters form continually in condensations of the interstellar medium.

This conclusion was strongly reinforced through the discovery and investigation of certain very loose galactic clusters or *stellar associations*, whose nature and significance was first pointed out in the late 1940's by the Armenian astronomer V. A. Ambartsumian of Burakan Observatory. He noted that these associations are nearly spherical in shape (circular outline), although their space density is very low. The differential revolution about the galactic center should have elongated these groups in a few hundred million years, as the portion nearer the center revolves more rapidly in pseudo-Keplerian motion. The lack of elongation, explained Ambartsumian, indicates that the stellar associations were formed recently and must be expanding rapidly at velocities of the order of 5 to 10 km/sec. If so, their ages cannot exceed a few millions or tens of millions of years. Conspicuous examples of such stellar groupings appear in Orion, where the great nebula and its central cluster form the nucleus of a large expanding association. Other associations exist, in Perseus around α Persei, in the vast Scorpio-Centaurus association, and so on. Scores of such associations have been identified around the galactic belt. Many of them include well-known galactic clusters and emission nebulosities.

Some striking confirmations of the instability and youth of stellar associations were quickly discovered. Most significant, perhaps, was the discovery of "runaway" stars with large proper motions at great distances from the Orion nebula, such as AE Aurigae and μ Columbae. These stars are moving directly away from the Orion Center. Their present distances from the center and their annual proper motions enable us to calculate how long ago they must have escaped from the core of the association. Such studies lead to remarkably short ages, of the order of a million years. Another confirmation of Ambartsumian's view came from a study of the correlation between the shape and age of the associations. While young associations are roughly spherical, or circular in outline, as noted by Ambartsumian, older associations are distinctly elongated, as would be expected from the effect of galactic differential rotation. An example is the Scorpio-Centaurus association (Fig. 30-6).

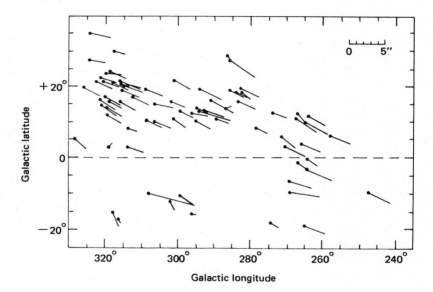

FIGURE 30-6   *The Scorpio-Centaurus association.*

We have referred to young and old associations. This designation finds justification not only in the "expansion ages" derived as explained above, but also in the evolutionary ages estimated from the H-R diagrams of stars in the association, as we shall explain more fully in Chapter 35. In brief, theories of stellar evolution enable us to predict the rate at which stars of various masses burn hydrogen to maintain their energy output. As the hydrogen core of the stars becomes depleted, the structure and luminosity slowly adjust themselves to the changing rate of energy production. The mass, radius, and surface temperature of the star change in a characteristic way, which manifests itself in the H-R diagram by a displacement of the representative point away from the main sequence and toward the upper right-hand part of the diagram. This evolution takes place very rapidly, in a few million years, for the OB supergiants, which use up their energy resources at a fantastic rate (up to a million times faster than the sun). The rate is slower for late B and A stars, which are less prodigal of their energy (luminosities 100 to 1000 times that of the sun), and slower still for the sedate stars of F and G types, which spend hundreds of millions of years on the main sequence before making any perceptible departure from it.

The mass of a star measures both the amount of hydrogen fuel available and the rate of its consumption. The mass thus determines the length of time that a star must spend on the main sequence. Therefore, the position of the "break-off" point in the H-R diagram—the point where evolving stars begin to move off the main sequence—gives a fairly good indication of the age of a galactic cluster or of an association, at least of its age since the formation of a stable main sequence.

Some clusters are in fact, so young, perhaps less than a few hundred thousand years old, that only their most massive stars have had time to condense out of the interstellar medium and reach the hydrogen-burning stage on the main sequence, while they have not yet burned their hydrogen supply and still lie on the upper part of the main sequence. Stars of lesser mass, however, which condense more slowly, have not yet settled on the

FIGURE 30-7 *Color-magnitude diagram of young cluster, NGC 2264. (M. Walker, Lick Observatory.) The cluster stars lie in hatched strip which departs from the standard main sequence (dashed).*

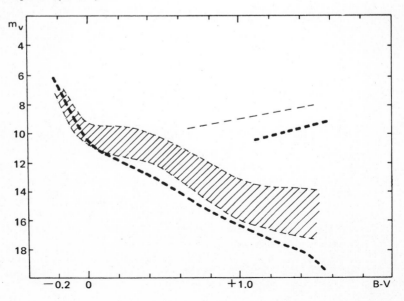

main sequence and form a scattered band to the right of it (Fig. 30-7). Such very young clusters with many stars still in the contracting stages of stellar evolution are usually imbedded in bright and dark nebulosity, and many of their stars are variable. One may even wonder whether in some cases stars may not evolve so rapidly that they could emerge from invisibility in only a few years. G. Herbig at the Lick Observatory has reported the case of a "star" becoming visible in the Orion association where a few years earlier none appeared on photographs. We have come a long way since the ancients affirmed their belief in the immutability of stars and starlight.

FIGURE 30-8 *The globular cluster ω Centauri. (G. de Vaucouleurs, Mount Stromlo Observatory.)*

It is now obvious that stellar associations are not permanent units of the Galaxy. They form out of the interstellar medium, only to scatter their stars in the general field after a relatively short lapse of time, some 10^7 or 10^8 years.

30-5   Globular Clusters.   The large globular clusters of stars are among the most beautiful celestial objects. The largest are easily visible to the naked eye as fuzzy patches of light. The two brightest are ω Centauri and 47 Tucanae, visible in the southern hemisphere as fourth-magnitude nebulous "stars" (Fig. 30-8). The third brightest is the great cluster M 13 in Hercules (Fig. 30-9). Thousands of stars, generally fainter than the eleventh magnitude, become visible in each cluster through large telescopes, and the unresolved light of hundreds of thousands of fainter stars forms a faint background nebulosity. About 100 globular clusters are known, most of which were discovered in the eighteenth and early nineteenth century by Messier and the Herschels. For example, M 2, M 3, M 4, M 5, M 9, M 13, M 15, M 22, M 62, M 71, M 92 are in the Messier catalogue.

The distribution of globular clusters over the sky (Fig. 30-10) shows a strong concentration in the general direction of the galactic center in Sagittarius, thus establishing immediately the close association of the system of globular clusters with the galactic system. However, the globular clusters are much less concentrated to the galactic plane than the galactic clusters. They form a roughly spherical system of "satellites" of the Galaxy, whereas the galactic clusters form an extremely flat system. The apparent absence of globular clusters close to the galactic equator is the result of strong absorption by interstellar dust clouds, which are heavily concentrated in the galactic plane and obscure the distant clusters. Infrared photography can sometimes penetrate the haze and reveal an occasional cluster near the galactic plane. The hemisphere of the sky opposite to the galactic center is nearly devoid of clusters (Fig. 30-10). This distribution results not from obscuration but from our outlying location in the Galaxy. Most clusters lie nearer to the galactic center than the sun, and only a very few are scattered at larger distances. One of the most distant globular clusters in the direction of the anticenter is NGC 2419, 180,000 light years away.

30-6   Distances and Distribution of Globular Clusters.   The modern era in the study of globular clusters and of their distribution with respect to the galactic system began with the epoch-making work of H. Shapley. To determine the distances of the globular clusters, Shapley employed the period-luminosity relationship of the Cepheid variables discovered by Miss Leavitt in 1911 (see Chapter 26). Many variable stars having light curves similar to the Cepheids, but of periods generally shorter than one day, had

FIGURE 30-9 *The globular cluster M 13 in Hercules. (Mount Wilson and Palomar Observatories.)*

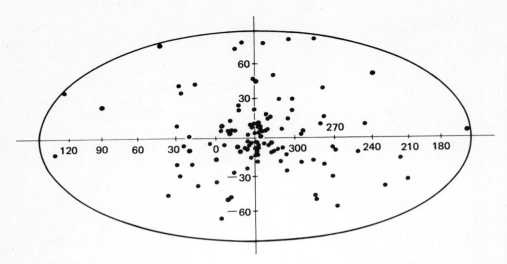

FIGURE
30-10

Apparent distribution of globular clusters on sky shows high concentration toward galactic center. Coordinates are galactic longitude and latitude.

been discovered by S. I. Bailey at Harvard Observatory around 1900. Shapley and other discovered many of these "cluster variables" by the "blinking" technique, wherein two photographs are optically superimposed in a "blink microscope" that shows one plate and then the other in rapid succession (Fig. 19-6). If a star has varied in brightness between the two plates, its apparent pulsation renders detection easy.

The cluster-type variables of periods shorter than one day have a remarkably small scatter in apparent magnitude and consequently in absolute luminosity, since all stars in a given cluster lie effectively at the same distance from us. If we can establish the mean absolute magnitude of the cluster variables by independent calibration (see below), the recognition of a cluster variable and measurement of the period of variation immediately gives its absolute magnitude. Comparison with the observed apparent magnitude then yields the distance, through equations (30-5) or (30-6). For many years a mean value $M = 0.0$ was used for the mean absolute magnitude of the cluster variables, but recent studies indicate that $M = +0.7$ (visual) or $+1.0$ (photographic) is more nearly correct. Thus in M 3, cluster variables have a mean photographic apparent magnitude $m = 16.0$, and the distance is $m - M = 16.0 - 1.0 = 15.0$, corresponding to a distance $r = 10,000$ parsecs.

This method works well for all clusters in which variables can be detected. For the same clusters Shapley found that the absolute magnitudes of the brightest stars correlate well with those of the cluster-type variables. Excluding the five brightest stars on the grounds that some of them might be foreground stars, he found that the next 25 brighter stars average about 1.3 mag. brighter than the cluster-type variables, thus providing a second brightness criterion applicable out to significantly greater distances. As an even simpler though rough method for estimating the distances of very distant clusters, Shapley found that the apparent diameter of the cluster itself gives an indication of its distance. Globular clusters of known distances turn out to have reasonably similar diameters. Hence the distance of a cluster is approximately inversely proportional to its apparent diameter.

In 1918, Shapley employed these estimated distances of most known globular clusters to show that the clusters comprise a nearly spherical system centered in the direction of Sagittarius and having an overall diameter initially estimated at 300,000 light years. This result gave the first indication of the great size of the galactic system and of the outlying position of our sun (Fig. 30-11). The actual value for the diameter has been revised downward twice since the 1920's, first through the discovery of interstellar absorption in the 1930's and then through the revision of the absolute magnitude of the cluster variables in the 1950's. Interstellar obscuration dims the light of distant clusters and makes them appear fainter and seemingly more distant than they actually are. The apparent magnitudes require correction for this effect, equation (30-6), as explained in Chapter 29, before one applies equation (30-5) to compute the distance.

The calibration of the mean absolute magnitude of the cluster variables is difficult, because no cluster is near enough for direct measures of parallax

FIGURE 30-11   *Space distribution of globular clusters and the galactic system in schematic edge-on view.*

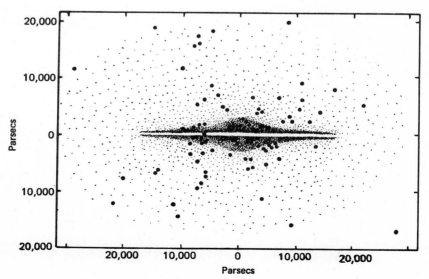

or proper motion. We must employ an indirect method of calibration. Variables of the RR Lyrae type, though not in clusters, seem to be the same kind of stars as the cluster variables. Their light curves, periods, colors, and other properties are very similar if not identical. Fortunately, as O. Eggen showed in 1959, several RR Lyrae variables are members of loose groups of stars having some common velocity components. If we can gauge the distance of a group from its radial velocity and proper motion (as explained in Sec. 30-2), the absolute luminosity of the variables follows immediately. In this manner, O. Eggen determined values ranging from $M = +0.5$ to $M = +1.0$ for several of the nearer RR Lyrae-type variables Further work will certainly lead to a definitive value, and the distance scale of the globular clusters will thus be finally settled.

30-7 **The Stars in Globular Clusters.** Detailed studies of the magnitudes and colors of stars in globular clusters by A. R. Sandage, H. L. Johnson, and others have led to a better understanding of the composition of the clusters and of stellar evolution. Figure 30-12 shows a color-magnitude diagram of the globular cluster M 3; comparison with similar diagrams for galactic clusters (Fig. 30-4) reveals some important differences. Most obvious is the absence of blue giant and supergiant stars (the upper left corner of the diagram is empty)—that is, superluminous stars of short

FIGURE 30-12 *Color-magnitude diagram of stars in globular cluster M3. (H. L. Johnson and A. R. Sandage.)*

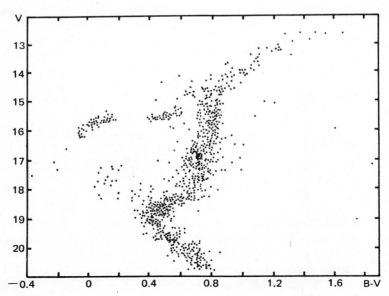

lifetimes are absent in globular clusters. If ever they existed, these stars must have vanished long ago. From this fact alone we infer that globular clusters must be very old. Stars of the main sequence appear only near the bottom of the graph, where the photographic limit of detection prevents the recording of still fainter objects. Dwarf stars of low luminosity, which have very long lifetimes on the main sequence (see Chapter 35), are present, but only the brightest can be photographed and measured. Many more must be present but too faint to be recorded. We see a characteristic sinuous belt heavily populated by subgiant and giant red stars extending toward the upper right-hand corner of the plot. The brightest stars used by Shapley as distance indicators lie near the top of this giant branch and have absolute visual magnitudes around $M_v = -3$. This branch characterizes stars of *population type II*, a term introduced by W. Baade in 1944 to distinguish the stars found in globular clusters from those found in the galactic plane and in galactic clusters. The latter stars he called *population type I*.

Finally, a scattering of stars forms a horizontal branch across the diagram, including the cluster variables in the narrow range of color index between 0.0 and +0.2. Some clusters, relatively poor in variables have an incomplete horizontal branch in this color range, but sufficient to estimate the mean magnitude of the (missing) cluster variables. We shall discuss later (Chapter 35) the implications of the globular-cluster, color-luminosity diagram for the theories of stellar evolution. However, we can already conclude that it labels the globular clusters as old and essentially permanent members of the galactic system, unlike the temporary open clusters.

The main reason for the long-term stability of the globular clusters is their compactness and large stellar population. Except for an occasional loss of a few stars whose velocities exceed the velocity of escape from the cluster, the cluster members form a permanent self-gravitating system, or at least one capable of maintaining cohesion for long periods of time against the disruptive galactic tidal force.

Only a few clusters are elongated, probably because of a slow motion of rotation. One of the most elongated clusters is ω Centauri, whose general rotation was first detected in 1964 from radial velocities of individual stars. Most clusters are essentially spherical, and stars must move more or less at random in all directions in elongated orbits approximating radial oscillations from the center of mass. By measuring the radial velocities of a sufficient number of individual stars in a globular cluster, we can estimate the average random motion or velocity dispersion in the cluster. Since the stars are bound together within a limiting radius by the aggregate attraction of all the stars in the cluster, measures of the velocity dispersion and the radius enable us to calculate the total mass. This method can be applied to only a few of the brighter stars in each of the nearer clusters, but the results are in agreement to indicate total masses of the order of 100,000 solar masses.

For instance, O. C. Wilson, using the 200-inch telescope and a fast spectrograph, observed 15 red giant stars in M 92 and derived a velocity dispersion of 46 km/sec and a total mass of 300,000 solar masses. From the total luminosity of the cluster, also expressed in solar units, one deduces that in M 92, two solar masses are present on the average where one unit of solar light output is observed. In other words the mass-luminosity ratio $M/L = 2$. Now for average dwarf stars in the solar neighborhood—and, as we shall see later, also for galaxies as a whole (Chapter 32)—this mass-luminosity ratio is usually much higher than 2, often of the order of 5 to 10 or even more. Through the mass-luminosity relation (Chapter 21) this result indicates that in our galaxy and other galaxies most of the mass resides in dwarf stars much fainter than the sun. In the globular clusters, the average stars must be only slightly fainter than the sun, and very faint K and M dwarf stars must be less abundant than in the galactic plane near the sun. We conclude that the cosmic environment in which globular clusters formed long ago was quite different from that currently prevailing near the galactic plane or that the small stars have largely escaped from the clusters.

30-8 Motions of Globular Clusters. The motions of globular clusters in space also give significant clues to the general properties of the galactic system. The clusters are too far away to show perceptible motions, but their radial velocities can be measured spectroscopically, either as the average of the velocities of several of the brighter stars, or more expeditiously from spectra of the integrated light of all the stars falling in the slit of a small spectrograph. About 70 per cent of all known globular clusters have been so observed, and they show a remarkable distribution of velocities in relation to the direction over the celestial sphere. The average velocity of the clusters greatly exceeds that of the stars, about 100 to 200 km/sec instead of 10 to 20, and it depends strongly on direction. Clusters in the hemisphere centered in Cygnus tend to approach with velocities up to 200 to 250 km/sec, while clusters in the opposite hemisphere tend to recede with about the same velocities. Since the distribution of globular clusters in space is spheroidal and little flattened, one does not expect the system to be rapidly rotating. In fact, the motions of the globular clusters about the galactic center must be like radial oscillations in and out, rather than circular orbits. The peculiar asymmetry in the clusters' radial velocities results from the motion of the sun in space as it partakes in the general rotation of the galactic system (or rather the stars in the "disk" or flat component of the system). The sun is moving toward Cygnus with a velocity of some 200 to 300 km/sec, according to various methods of estimation, and the apparent velocities of the globular clusters mainly reflect this motion.

When we subtract the solar motion around the Milky Way (see Sec. 31-4) from the observed motions, the globular clusters seem to move more or less at random about the galactic center, very much as individual stars in a cluster seem to have mainly incoherent motions about the cluster center (except perhaps for a slow general rotation of the system). The same argument of balance between attraction and random agitation, previously used for the stars in clusters, can be applied to the globular clusters and the Galaxy, to give an estimate of the total mass of the galactic system. It turns out to be of the order of several times 10^{11} solar masses, in general agreement with the estimates derived from analyses of the galactic rotation (Chapter 31).

CHAPTER *31*

The Milky Way

31-1 **Survey of the Milky Way.** A naked-eye view of the Milky Way gives us the first impression of the huge clouds of stars in which our solar system is imbedded. A good time to observe the Milky Way in the northern hemisphere is early on a summer evening when Cygnus, the Swan (or Northern Cross), is overhead. If the sky is free from clouds, moon, or city lights and the southern horizon unobstructed, we can see the Milky Way a few degrees east of the red giant star Antares, in the constellation Scorpius. Then as our eyes sweep northward towards the zenith they pass by the great star clouds of Sagittarius (Fig. 31-1) through the Scutum cloud, by Altair in Aquila, and through Deneb in Cygnus overhead (Fig. 29-2). We can still follow it towards the northeast through Cassiopeia, by Perseus and possibly even down to Auriga near the northern horizon.

To observers in the southern hemisphere, where it is winter, the view of the luminous belt culminating near the zenith in the brilliant star clouds of the galactic center through Sagittarius, Scorpius, Lupus, Centaurus, and the Southern Cross (Fig. 29-1) is a much more spectacular display and an unforgettable experience.

FIGURE 31-1   *The great star clouds in Sagittarius mark the direction of the galactic center. (G. de Vaucouleurs, Mount Stromlo Observatory.)*

The Milky Way, however, is not a smooth luminous band. It is distinctly patchy. A dark rift divides it from the northern part of Aquila down to Sagittarius and beyond it through Lupus. This *Great Rift* (Fig. 29-2) looks like a long dark cloud on a brighter background. A telescope or photograph resolves the smooth texture of the Milky Way seen with the naked eye into complex clouds of innumerable faint stars. We have proof (Sec. 29-1) that the great rift is truly a dark cloud of dust obscuring distant star clouds from our view. The cloud in Scutum is the most conspicuous of all the bright clouds, its brilliance enhanced by the nearer dark dust clouds that surround it.

FIGURE 31-2 *A sparser star field in Canis Major and Puppis, a portion of the Milky Way far from the galactic center, Sirius is near top of picture. (Mount Stromlo Observatory.)*

FIGURE *Path of Milky Way around the sky. (Lund Observatory. Courtesy Sky and Telescope.)*
31-3

In winter the northern Milky Way is much less brilliant but still conspicuous (Fig. 31-2). When Sirius appears in the southeastern sky, the Milky Way is slightly east of it and swings up northwesterly between Orion and Gemini and by Capella in Auriga. Again great irregular dark lanes and bright clouds are prominent. We find that many of the latter arise not only from stars but also from glowing masses of gas, excited by enormously bright stars in their neighborhoods (see Chapter 29).

A casual inspection of the sky with the eye over the course of a year, or a study of large-scale photographs, shows that the star clouds are denser and brighter toward Sagittarius than they are in the opposite direction toward Taurus and Gemini. Later on we shall find still better evidence that the center of our stellar system or Galaxy lies in the direction of Sagittarius.

The fact that the Milky Way encircles the whole sky (Fig. 31-3) indicates that the Galaxy is a flat system, and that our sun lies within it, near its equatorial plane. The *galactic plane* is the plane of the great circle around the sky that best approximates the center line of the Milky Way. It is tilted about 63° to the earth's equator, so that the galactic poles lie in Coma Berenices (north) and Sculptor (south). The galactic equator crosses the celestial equator in Orion and Aquila. With respect to this equator we can define a galactic longitude l and a galactic latitude b, measured north or south of the galactic plane, similar to longitude and latitude on the earth. At first the origin of galactic longitudes was placed at the ascending node of the equator in Aquila. However, after radio-astronomical studies pinpointed the direction of the center of the Galaxy, the origin was shifted to this center in Sagittarius, at RA = $17^h 42^m 4$, $D = -28° 55'$ (1950). Galactic coordinates are independent of precession.

31-2 The Concentration of Stars Toward the Milky Way. Counts of stars show how the apparent density of stars (number per unit area) increases toward the plane of the Milky Way. Table 31-1 lists the number of stars per square degree brighter than the magnitude limits given in the left-hand column. The tabulation includes values near the galactic plane ($b = 0°$), at galactic latitude $b = 40°$ and at the galactic pole. Even among

TABLE 31-1 *Number of stars per square degree brighter than certain magnitudes*

Magnitude limits	In the Milky Way	40° from Milky Way	At the pole
5th	0.045	0.018	0.013
9th	2.80	1.07	0.72
13th	146	42	21
17th	4,780	744	288
21st	73,600	5,000	1,670

the naked-eye stars brighter than the fifth magnitude we find more than three times as many in the plane of the Milky Way as at its pole. As we go to fainter and fainter magnitudes, this ratio increases from about 17 at the seventeenth magnitude to about 44 at the twenty-first magnitude. Not only are there more stars of all magnitudes towards the galactic plane, but as we go to fainter limits the ratio increases compared with other regions of the sky.

William Herschel argued that these star counts, or *star gauges* as he called his visual counts, indicate that the system of stars extends farther in all directions of the plane of the Milky Way than towards its poles. One might, of course, argue that the stars are just closer together near the galactic plane, or perhaps envision that obscuring matter cuts out the light from distant stars near the poles. Again one might speculate that stars are intrinsically brighter towards the plane, but no more distant or more concentrated in space.

Of these possibilities we can immediately discard the presence of much obscuration near the galactic poles. *Extragalactic nebulae* acquired this name from the fact that they appear only in regions well away from the galactic plane and are clearly distant star systems well outside our galaxy (see Chapter 32). In fact they are totally absent from a "zone of avoidance" closely following the galactic belt. It would be a coincidence beyond belief if distant external systems were to be so distributed throughout all space in such a fashion as to avoid the plane of our own system. In fact, as we know, obscuration exists near the plane of the Milky Way, not away from it.

By analogy with other galaxies we may expect that the stars should extend to greater distances and also be more crowded together near the equatorial plane of the Milky Way. Astronomers have long known that certain types of objects are more strongly concentrated toward the Milky Way than others. The high-temperature O and B stars and associated emission nebulosities almost always appear near it. The naked-eye O and B stars, however, congregate in the so-called Gould belt, which is inclined by some 15 to 20° to the main galactic plane, a mere local structure among the nearby stars. Cepheid variables and novae also occur much more frequently near the Milky Way than elsewhere. Likewise the galactic clusters show a strong galactic concentration (Fig. 31-4) while the globular clusters, as we have already noted, show only moderate concentration (see Chapter 30). Almost every individual type of object that we can identify shows some degree of preference for the plane of the Milky Way. The only exceptions are external galaxies and radio sources far beyond the borders of our system.

At the turn of the century, J. C. Kapteyn of Groningen, Holland, planned a large-scale program of determining stellar magnitudes and colors, classifying spectra, measuring radial velocities, counting stars, and mapping the Milky Way to determine its structure. To avoid the impossible task of carrying out these operations for *all* stars, he adopted the principle

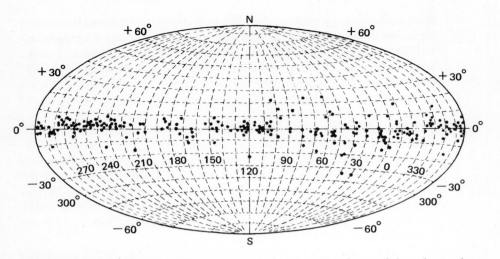

FIGURE 31-4   (*Above*) *The open clusters are strongly concentrated toward the galactic plane.*

FIGURE 31-5   (*Below*) *The relative space densities of various objects show different degrees of concentration to the galactic plane, but most are within one kiloparsec from the plane. (1) B-type stars; (2) Cepheid variables, A-type stars, open clusters; (3) average of all stars, F-type stars; (4) K-type giants; (5) cluster variables, long-period variables.*

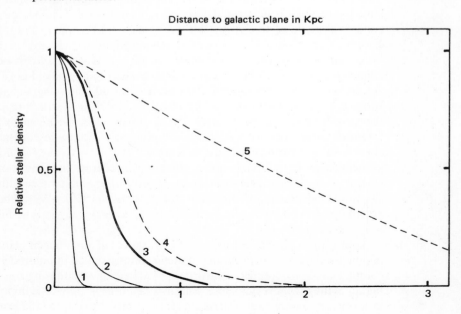

of concentrating efforts on 206 small regions, or "Selected Areas," distributed over the entire sky. Cooperative work by astronomers from many nations produced a fund of information that is still useful, although the massive application of stellar statistics to the data, in 1922, led to a model of the stellar system or "Kapteyn Universe" that was much too small (10,000 light years in diameter) and improbably heliocentric. The reason for this error is instructive.

The fact that intrinsically bright stars such as the O and B stars, novae, and Cepheid variables hug the galactic plane, and also that many appear to be faint and therefore distant, proves that our system must be strongly flattened (Fig. 31-5). We can, of course, calculate the distance of a star of known absolute magnitude, M, from its apparent magnitude, m, by the inverse-square law. We have already derived the relation in equation (19-6). If we measure distance, r, in parsecs, $r = 1/p''$ (p'' is the stellar parallax), we find

$$\log r = -\log p'' = 0.2(m - M) + 1. \qquad (31\text{-}1)$$

An O-type star of absolute magnitude $M = -5$, showing an apparent magnitude $m = +10$, will lie at a distance of 10,000 parsecs by our equation. Unfortunately we cannot accept this result at face value, because the obscuration of light by interstellar matter will reduce the apparent brightness of the star and make us overestimate the distance. We must measure the obscuration independently, by the method outlined in Chapter 29, and then apply a correction to the apparent magnitude before we can be certain of the distances. Thus equation (31-1) should be modified to read

$$\log r = 0.2(m - A - M) + 1, \qquad (31\text{-}2)$$

where A is the amount of absorption in magnitudes.

Kapteyn was correct in his conclusion that our galaxy is a highly flattened system extending for thousands of parsecs in its plane, but he was led astray in estimating its size and the position of the sun by neglect of the correcting term A, whose importance had not yet been realized. In fact, the effect is so large that most stars are obscured beyond a few thousand light years from us, thus creating the illusion that we are in the center of the star system.

The correct order of size of the galactic system and the true outlying position of the sun were discovered in 1918 by H. Shapley by a much simpler but more revealing method. Shapley estimated the distances of globular clusters from the apparent luminosity of their RR Lyrae variables and from the apparent diameters of the clusters themselves. He found that the system of the globular clusters is roughly spherical and that the center of the system does not lie near the sun, but near a distant point in the direction of Sagittarius. He boldly concluded that this point is also the center of the galactic system, whose diameter he estimated to be some 300,000 light years (Sec. 30-6). Discovery of interstellar absorption in the 1930's reduced this estimate to about 100,000 light years. The sun lies in the outer parts of the disk some 30,000 light years from the center. This model of the Galaxy was the key to all subsequent progress.

31-3 The Sun's Motion. Halley's discovery of the proper motions of stars (see Chapter 19) implied that the sun itself must be moving in space. In discussing the motion of the sun we must define our reference system, since all motion is relative to some standard of rest. Only two methods are available to us for measuring motions through space: (a) proper motion across the line of sight (see Chapter 19), and (b) radial velocity along the line of sight, from the Doppler effect on spectral lines (see Chapter 8).

If we wish to measure the sun's motion with respect to the nearby stars, we must assume, at least as a first approximation, that these stars themselves move at random, like molecules in a gas. If so, the stars in the general area of the apex (the point towards which the sun's motion is directed) should show systematic or averaged motions away from the apex, as the trees ahead appear to spread apart as one drives through a forest. Conversely, the average proper motions should close up towards the antapex (the corresponding point on the other side of the sky). Or, stated differently, the average proper motions of groups of stars in small areas of the sky should be systematically directed along the great circle from the apex towards the antapex (Fig. 31-6).

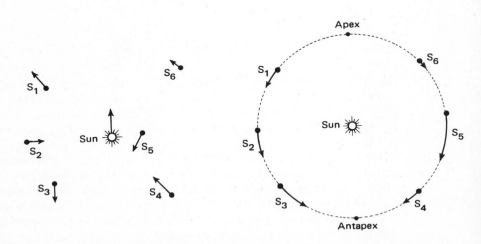

FIGURE
31-6

Solar motion in space (left) is reflected by the tendency of apparent proper motions (right) of stars on celestial sphere to diverge from apex and converge toward antapex.

From the proper motions of only seven stars (later 13 and still later 36),
William Herschel concluded in 1787 that the sun is moving in the direction
of the constellation Hercules. Modern investigations agree with Herschel's
result within a few degrees and place the apex of the sun's motion (relative
to nearby stars) in Hercules (right ascension 18^h, declination $+30°$). The
antapex is in the constellation of Columba. In the direction towards which
the sun is moving, the stars are systematically approaching—that is, the
averaged radial velocities are negative. In the direction of the antapex,
from which the sun is moving away, the radial velocities are systematically
positive, indicating recession. At right angles the radial velocities should
average zero.

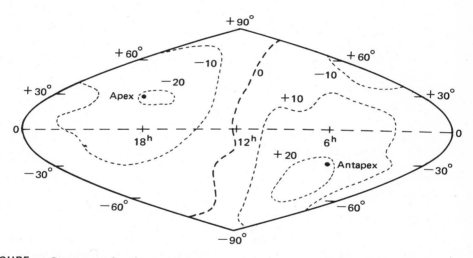

FIGURE   *Systematic distribution of radial velocities of stars reflects solar motion in space.*
31-7

Herschel, of course, had no means of determining the sun's *linear*
velocity with respect to the nearby stars. W. W. Campbell of the Lick
Observatory carried out the first extensive analysis of the radial velocities
of 2149 stars. He placed the direction of the apex at 18.1 hours, $+29°$,
and the velocity at 19.6 km/sec, or rather less than the earth's orbital speed
(30 km/sec). Figure 31-7 shows the systematic nature of the averaged
radial velocities over areas of the sky. The sun's motion with respect to the
brighter or nearby stars is thus well established by both methods.

Studies of the sun's motion with respect to fainter stars, however, and
particularly with regard to those stars moving with relatively high velocities
show significant departures from the direction of the sun's motion deter-

mined from the nearby and brighter stars. For example, P. van de Kamp and A. N. Vyssotsky, from an analysis of the proper motions of 18,000 faint stars, derived an apex at right ascension 19.0 hours and declination +36°, in Lyra. Now most of the stars in the solar neighborhood are moving with their own peculiar velocities in space at much the same speed as the sun. G. Stromberg of Mount Wilson analyzed all the radial velocities of stars available in 1925. He found that those of increasingly greater random velocities indicated a solar motion of increasingly greater velocity. Also the direction of the sun's way tended to shift towards increasing right ascension and increasing declination. The globular clusters furnished the most extreme case, increasing the solar motion to 286 km/sec towards right ascension 20.4 hours and declination +62°. We shall shortly return to this problem.

Kapteyn had earlier, in 1905, discovered another remarkable effect. The "random" motions of the stars are by no means random in direction. After subtracting the solar motion, he found that the intrinsic motions of the stars with respect to their average tend to be greater in the plane of the Milky Way than perpendicular to it. Also at least two preferential directions exist in space along which the stars tend to *stream*. This effect, called *star streaming*, indicated large-scale systematic motions in our region of the galactic system.

31-4 Galactic Rotation. In 1926 B. Lindblad, of Stockholm, cleared up the questions of the star streaming and the high-velocity objects in one sweeping dynamical explanation. He argued simply that the Milky Way must be in rotation, because otherwise it would possess a spherical instead of the recognized flattened form. V. M. Slipher had already discovered the rotation of some external galaxies, as indicated by the tilt of their spectrum lines (see Chapter 32). If the Milky Way is rotating, then stars must be moving about the center in orbits that are roughly ellipses, at least at great distances from the center. We can scarcely expect all of the stars to be moving in exactly circular orbits.

Stars in the solar neighborhood may lie anywhere between their maximum and minimum distances from the galactic center. Stars that move in elliptical orbits and lie near their minimum distance to the center (like the perihelion for a planet in the solar system) should be revolving faster than those that move in nearly circular orbits. Stars with smaller elliptical orbits and near their maximum distance (like aphelion) should be moving more slowly. Stars in between might be moving somewhat slower or faster and also inward and outward with respect to circular motion. Most of the star streaming comes from these in-and-out components of the motions in elliptical orbits.

High-velocity stars tend to move in more eccentric orbits, spending a major fraction of their time near their "aphelion" points. Average stars like the sun and most of its near neighbors are very well behaved, moving in nearly circular orbits. Relatively very few stars occur near "perihelion" in the sun's neighborhood. We conclude that the stars with apparent high velocity are really the laggards, not the speedsters in the Milky Way.

After Lindblad's explanation of star streaming and the high-velocity stars, Oort devised a method for measuring the actual rotational speed and the mass of the Milky Way. The fact that practically no stars near us are near "perihelion" indicates that the sun lies in a region of the Galaxy remote from the center, as indeed Shapley had concluded from the distribution of globular clusters. Hence almost all of the mass of the Galaxy falls inside the sun's orbit, and we should expect the gravitational effect of the entire Galaxy to be somewhat like that of a point mass at the center. Therefore the velocity of revolution of stars about the center of the Galaxy will decrease with increasing distance, as is true for the planetary orbits in the solar system, and we can expect to observe effects of *differential rotation*.

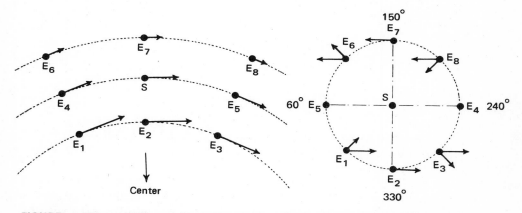

**FIGURE
31-8**   *Effect of differential rotation of galaxy (left) is reflected in systematic distribution
of stellar radial velocities (right).*

Assume for simplicity that orbits are practically circular. Figure 31-8 shows that stars moving in circular orbits located on the line joining the sun and Milky Way center will have zero radial velocity with respect to the sun. Correspondingly, at distances small compared with the radius of the orbit, stars on a line tangent to the sun's orbit will also have zero radial velocity. But stars in the galactic plane between these four cardinal directions will have systematic radial velocities towards or away from the sun. Thus we should expect a double sine wave of average radial velocities as we proceed through 360 degrees of galactic longitude *l* (counted from the direction of the center; cf. Sec. 31-1) around the sun. Radial velocities of Cepheid variables, whose distances can be estimated through the period-

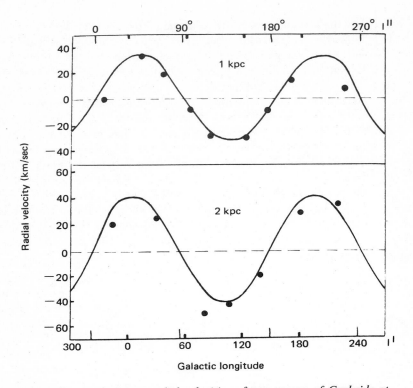

**FIGURE
31-9** *Oort effect in the mean radial velocities of two groups of Cepheids at average distances of 1 and 2 kiloparsecs is explained by differential rotation of galaxy.*

luminosity relation (see Sec. 26-5), are especially suited to show the Oort effect and how it increases with increasing distance from the sun (Fig. 31-9). The systematic proper motions of stars also obey a double sine wave like that in Fig. 31-9, except for a 45° displacement in the longitudes of the maxima of the waves.

The direction of the galactic center indicated by these Oort curves lies in Sagittarius, in agreement with all other indications from the distribution of stars and clusters and from the great brightness and breadth of the Milky Way in that direction (Fig. 31-2).

All other methods so far devised for checking on the rotation of the Galaxy confirm the preceding conclusions. The motion, incidentally, is clockwise as seen from the north pole of the Milky Way. Spectral lines from interstellar gas (Chapter 29) show Doppler shifts consistent with those of stars and indicate that interstellar clouds share in the general galactic rotation. The velocities obtained by radio astronomy from the 21-cm line of hydrogen (Chapter 29) also prove that great clouds of neutral hydrogen move in a similar fashion (see Sec. 31-6). As we have noted (Sec. 31-1), the system of globular clusters is roughly spherical, indicating that their motions are more nearly radial than circular (a situation similar to long-period comets in the solar system). Hence the motion of the sun with respect to the globular clusters or, even better, with respect to the distant galaxies, will mainly reflect its rotation about the galactic center. The orbital velocity appears to be of the order of 250 km sec^{-1}. The sun, therefore, completes one revolution about the galactic center in some 200 million years. The sun was last on this side of the Milky Way center with respect to the universe of galaxies at about the time that small dinosaurs were beginning to develop on the earth's surface; it has moved through an angle of about 120° of its orbit since the last great dinosaur disappeared. Altogether our earth and sun have completed only about 20 to 25 revolutions around the Milky Way since their formation nearly 5 billion years ago.

31-5 The Mass of the Galaxy. In principle, if all the mass of the Galaxy were concentrated in its center, we could employ the inverse-square law of attraction exactly as we did in determining the sun's mass from the earth's orbital motion (Chapter 7). In practice, however, we must allow for the fact that all the mass of the galaxy is not concentrated at the center but is spread out over a thick disk. M. Schmidt, L. Perek, and others have developed methods of solving this problem for various plausible density distributions in flattened ellipsoids more or less concentrated near the center.

The basic principle is not altered. In fact, the more complicated solutions do not appreciably change the total mass derived from the simple point-mass approximation, because the sun lies so far out near the edge of our system. All we need to know, then, is the orbital velocity of the sun and the orbital radius—that is, the distance to the galactic center. As noted above, the orbital velocity is not precisely known, but is probably in the range of 200 to 300 km/sec—let us say 250 km sec^{-1} as a plausible compromise. Measurements of distances in the Milky Way and especially to the center are difficult and always somewhat uncertain because of obscuration by interstellar dust. Nevertheless, the best results, based on measures of the distances of globular clusters, or of cluster-type Cepheid variables near the center of the Milky Way, indicate a value of the order of 10 kiloparsecs or 30,000 light years.

With these figures for the velocity and orbital radius, the point-mass approximation gives for the total mass of the Milky Way system

$$M = \frac{V^2 R_1}{G} = \frac{(2.5 \times 10^7)^2 \times (3 \times 10^{22})}{6.67 \times 10^{-8}}$$
$$= 2.8 \times 10^{44} \text{ g} = 1.4 \times 10^{11} \text{ solar masses.}$$

More refined analyses lead essentially to the same value.

Thousands of man-years of observation, analysis, and theoretical developments, and the use of the greatest instrumental equipment of our time have gone into this simple numerical result.

31-6 **Spiral Structure.** Many of the distant galaxies possess a beautiful vortical or spiral structure (see Chapter 32). For more than a century astronomers often surmised that our Milky Way is also a spiral galaxy, but this was more easily said than proved. Because we are located inside it and surrounded in all directions by an apparently chaotic distribution of stars, clusters, nebulae, and absorbing clouds, the overall plan of our system eluded us for a long time. The strenuous efforts of optical astronomers during the early 1950's first gave us a glimpse of spiral-arm structure in our galactic neighborhood. Soon afterward, the penetrating methods of radio astronomy revealed the general picture of the spiral pattern of our system, although much detail remains to be filled in.

Let us first outline the methods by which W. W. Morgan and his associates at Yerkes Observatory, in 1951 detected fragments of spiral arms in the vicinity of the sun. Observations of nearby galaxies, and especially the great spiral M 31 in Andromeda (see Chapter 32) by W. Baade with the 200-inch telescope, had shown that the bright emission nebulosities and the star clusters rich in O and B stars associated with them follow closely the paths of spiral arms. These easily recognizable objects can therefore be used as tracers or markers of spiral arms. Morgan undertook to map the distribution in the galactic plane of the more conspicuous OB associations. Since the directions are known, only the distance had to be determined. This part of the problem proved difficult, involving as it does the calibration of absolute magnitudes of distant, superluminous stars and precise correction for interstellar absorption (see Chapter 29). Morgan carried out his studies, initially for a score of associations, by methods described in several earlier chapters (19, 26, 30). A plot on the galactic plane (Fig. 31-10) disclosed several strings of associations that Morgan identified as marking the path of sections of several spiral arms. The sun appears to lie about 1000 light years from the central part of a spiral arm that includes the Orion nebula, the Coal Sack, and the North American nebula. An outer spiral arm including the double cluster in Perseus passes about 6000 light years beyond us, while the η Carinae nebula probably lies in an inner arm. The direction of galactic rotation is toward the right in Fig. 31-10, so that the tilt of the arms to the radius vector from the center indicates that the spiral arms are trailing in the rotation as in a vortex.

Studies of the 21-cm line emission of interstellar hydrogen soon confirmed and greatly extended Morgan's analysis. As we noted earlier (Chapter 29), the 21-cm line is not only broadened by random motions of atoms in a gas cloud but it is shifted by Doppler effect in proportion to the radial velocity of the cloud as a whole. By inverse application of the theory of

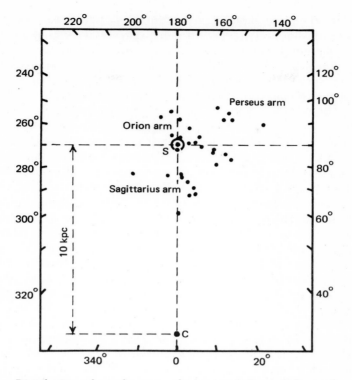

FIGURE
31-10
Distribution of nearby open clusters, emission nebulae and associations in galactic plane were the first indication of the spiral structure of the galaxy. Sun at S, galactic center at C. (W. W. Morgan, H. Sharpless, and D. Osterbrock, Yerkes Observatory.)

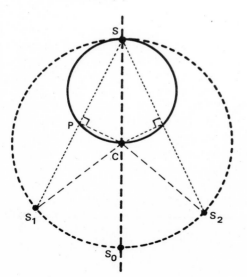

FIGURE
31-11

Circles in galactic plane where radial velocities with respect to sun (S) are zero (large circle), or maximum (small circle). Galactic center at C.

differential rotation of the galaxy (Sec. 31-4), we can estimate the distance corresponding to the observed radial velocity of a cloud, provided some model of the galactic rotation curve establishes for us the relation between rotational velocity and distance to the center. The latter we derive in part from optical studies and analysis of Oort's double wave for stars at different distances. Inside the sun's orbit (Fig. 31-11), the maximum radial velocities of interstellar hydrogen in different directions correspond to points where the line of sight is tangent to the orbital paths (assumed circular). The rotation curve of the Galaxy derived by such methods appears in Fig. 31-12. It then becomes a simple matter, at least in principle, to map in the galactic plane the positions of maximum concentration of interstellar hydrogen. The great advantage of radio astronomy lies in the fact that whereas optical studies cannot reach beyond a radius of a few kiloparsecs because of the strong absorption of light waves by interstellar dust clouds in all directions of the galactic plane, the dust is transparent to radio waves. These waves (including the 21-cm H I radiation) can traverse the whole galactic system from end to end virtually unobstructed.

Extensive observing programs with large radio telescopes in Holland and in Australia between 1951 and 1961 led to the combined map in Fig. 31-13.

FIGURE 31-12   *Rotation curve of galaxy derived from radial velocities of 21-cm hydrogen line. Orbital velocity, V, as a function of distance, R, in kiloparsecs from the galactic center, for northern (N) and southern (S) halves of Milky Way. Galactic longitudes of corresponding points are shown in upper scale. (F. J. Kerr, Radio Physics Laboratory, Sydney.)*

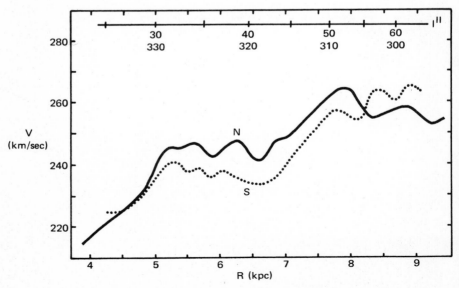

FIGURE *Spiral structure of neutral hydrogen in our galaxy from radio observations.*
31-13 *(F. J. Kerr and G. Westerhout.)*

Neutral hydrogen is clearly concentrated in a complex spiral pattern, and several of the nearby concentrations coincide more or less closely with the sections of spiral arms mapped by Morgan. It is a moot question whether the arms marked by the stars correspond precisely or only approximately with the spiral pattern of the gas, but the general correspondence is beyond doubt. A fair degree of correlation between condensations of gas and concentrations of dust occurs, but here again the correspondence is not always one-to-one.

31-7 Central Regions of the Galaxy. During the first few years of the exploration and mapping of the Galaxy by the 21-cm line, circular motions were adopted as the simplest hypothesis and a good enough approximation for a preliminary reconnaissance. In recent years, however, increasing attention has been paid to departures from circular motions involving not merely elliptical orbits but systematic gas streaming in the radial direction, and away from the galactic nucleus.

FIGURE 31-14

Possible barred structure of inner regions of galactic system according to (a) G. de Vaucouleurs (1963), and (b) F. J. Kerr (1967). Length of bar is about 5 kiloparsecs.

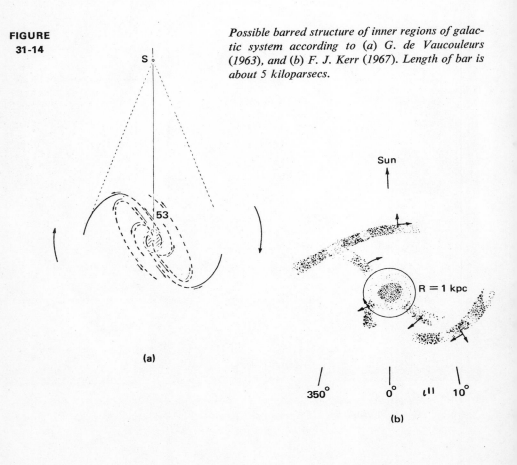

(a)

(b)

FIGURE *Spiral structure of NGC 4303 may resemble that of inner regions of our galaxy.*
31-15 *(G. de Vaucouleurs, McDonald Observatory).*

Such motions are directly in evidence when we observe the 21-cm line in the direction of the galactic center, where the effect of differential rotation should be exactly zero according to Oort's theory. Any nonzero velocity must represent noncircular motion. The hydrogen line appears in absorption in the direction of the galactic center, where an intense source of radio emission is located. The major absorption component appears at zero velocity in accord with Oort's model, but a weaker component appears Doppler-shifted by some −50 km/sec. In other words, a mass of neutral

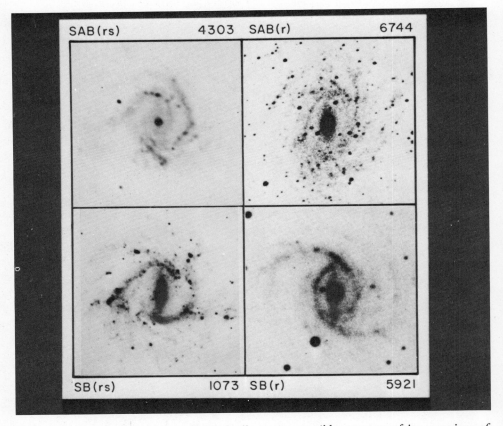

FIGURE 31-16 *Several types of barred spirals illustrating possible structure of inner regions of galactic system (G. de Vaucouleurs, 1969).*

hydrogen between us and the galactic center is approaching us and hence moving away from the galactic nucleus with a line-of-sight velocity component of 50 km/sec. This feature ties in with a structure seen in emission on both sides of the direction of the galactic center, presumably an inner spiral arm, at an estimated distance of 3 kiloparsecs from the center, a feature that Oort and coworkers interpret as indicating an expanding arm. Alternatively de Vaucouleurs proposed in 1963 that the central regions of the Galaxy possess a weak bar structure (Fig. 31–14), and that gas is streaming away from the center along the bar on both sides of the nucleus, a phenomenon first observed in other galaxies (Fig. 31-15). Obviously much more work will be needed before the structure of the central regions of our galaxy is clarified by radio methods. Optical observations, nevertheless, help indirectly by giving us examples of the various kinds of structure that occur in the central regions of other galaxies (Fig. 31-16).

31-8 **The Nucleus of the Galaxy.** As we noted earlier (Secs. 29-1 and 31-1) the nucleus of the Galaxy is hidden from our direct view by obscuring interstellar dust clouds. Because interstellar absorption is selective, decreasing toward the longer wavelengths (Fig. 29-9), some infrared waves from the distant galactic nucleus should penetrate the obscuring haze and reach us. Indeed the nucleus is prominent in the still longer radio waves (see Chapter 33). As soon as sensitive detectors of infrared radiation became available in the late 1940's, attempts to detect the hidden nucleus were made by direct photography (Fig. 29-9) and by systematic photoelectric scanning of the sky in the expected location of the nucleus. However, the early attempts failed because the wavelengths accessible were not yet quite long enough to penetrate the haze in measurable intensity. The first positive results, reported by Stebbins and Whitford observing (at $\lambda \simeq 1\ \mu$) from Mt. Wilson Observatory in 1945 and 1946 and by Whitford using the, then new, lead sulfide photoconductive detectors ($\lambda \simeq 2\ \mu$) at Lick Observatory in 1947, were sufficient to detect a strong enhancement of radiation from the general direction of the galactic center, but did not have sufficient resolution to pinpoint the nucleus itself. Success was only achieved in 1968 when E. E. Becklin and G. Neugebauer at Mt. Wilson and Palomar Observatories were able to observe at greater wavelengths ranging from 1.6 to 19.5 microns with a variety of improved photoconductive detectors refrigerated by liquid nitrogen or hydrogen to reduce thermal noise in the system. Shortly afterwards the nucleus was detected at still longer wavelengths ranging all the way from 5 to 100 microns by F. J. Low and collaborators working from high-altitude stations in Arizona and with a small telescope carried to 50,000 feet elevation by specially equipped jet aircraft. The resulting spectrum or energy distribution curve is most peculiar (Fig. 31-15); it shows a great excess of energy in the infrared compared with what could be expected from the optical ($v > 10^{14}$ Hz) and radio ($v < 10^{11}$ Hz) observations. The cause of this peculiar behavior is not yet fully understood but it does show that several extremely powerful excitation mechanisms are at work in the galactic nucleus. The total power emitted by the nucleus, a region not more than 10 pc in diameter, adds up to the total luminosity of some 10 million suns. Similar powerful sources have been discovered in some other galaxies (see Chapter 32).

CHAPTER 32

Galaxies

32-1 Discovery of Other Galaxies. The discovery of galaxies outside our own was a slow process. The philosophical concept of stellar systems similar to our Milky Way and far beyond its limits was first discussed by philosophers and scientists in the eighteenth century. E. Swedenborg in Sweden, T. Wright in England, I. Kant in Prussia, and J. Lambert of Alsace all discussed various aspects of a hierarchical order of nature, and during the nineteenth century von Humboldt popularized this concept of "island universes." Nevertheless, not until the early 1900's was the existence of other galaxies far beyond our Milky Way system established beyond doubt, mainly through the work of H. D. Curtis at Lick Observatory and especially of E. P. Hubble at Mount Wilson Observatory.

The brightest and nearest galaxies, the Clouds of Magellan (Fig. 32-1), are visible to the naked eye in the southern hemisphere and were known to the Arabs of the eleventh century, who had also noticed the brightest galaxy of the northern sky, the Andromeda nebula (Fig. 32-2). The telescopic rediscovery of this object by S. Marius of Germany in 1612 opened the way for the scientific study of diffuse celestial objects or "nebulae." Systematic searches by C. Messier of France in the eighteenth century, and especially by W. Herschel in England and later by J. Herschel in South Africa, greatly added to the numbers of such objects. The Messier catalogue numbers are still in use; for example, the Andromeda nebula is M 31. Several thousands were already recorded in the "General Catalogue" of J. Herschel published in 1864. A "New General Catalogue,"

FIGURE
32-1
(a) *The Clouds of Magellan. Large cloud.*
(G. de Vaucouleurs, Mount Stromlo Observatory.)

FIGURE
32-1
(b) *Small cloud. (G. de Vaucouleurs, Mount Stromlo Observatory.)*

FIGURE
32-2
The great spiral nebula M 31 in Andromeda.
(G. de Vaucouleurs, Lick Observatory.)

FIGURE 32-3   *The great spiral nebula M 33 in Triangulum. (G. de Vaucouleurs, Lick Observatory.)*

published in 1888 by J. L. E. Dreyer in Ireland, is still the standard reference work. Thus the Andromeda nebula also bears the label NGC 224. Recently, great catalogues recording tens of thousands of galaxies found on the Palomar National Geographic Sky Survey with the 48-inch Schmidt camera have been published by F. Zwicky and collaborators in the U.S. and by B. Vorontsov-Velyaminov in the U.S.S.R.

Application of the spectroscope to the study of nebulae in the second half of the nineteenth century led to the recognition of two classes of "nebulae": the "green nebulae," such as the Orion nebula, which showed emission lines—including a strong green line—in their spectra, and hence were clearly gaseous in nature; and the "white nebulae," such as the Andromeda nebula, which showed continuous spectra with dark absorption lines, and could therefore be composed of stars or include stars and diffuse matter.

The occasional apparition of a "new star" or nova (actually a supernova), such as S Andromedae, observed in M 31 in 1885, was additional evidence for the stellar composition of most white nebulae, although at the time the evidence seemed inconclusive. About 1850, Lord Rosse and his assistants, using what was then the largest telescope in the world—a 72-inch reflector erected on his estate of Parsonstown in Ireland—had visually discovered the beautiful vortex structure of several spiral nebulae. The great abundance of this type became evident when photography was applied to the study of nebulae toward the end of the century (Fig. 32-3).

As long as the distance was unknown, the scale was undefined, and the vortex structure was mistakenly regarded as an illustration and example of the Laplace nebula—the then popular theory of the origin of the solar system (cf. Chapter 17). The continuous spectrum was then supposed to arise from light reflected by the nebula from a central star.

The true nature of the white nebulae was finally established by the large reflectors of Mount Wilson Observatory, which resolved into individual stars the spiral arms of the Andromeda nebula M 31 and of the Triangulum nebula M 33. Among these stars Hubble recognized in 1923 several variable stars with the characteristic light curves of Cepheids. Through the period-luminosity relation (Chapter 26) he could estimate the absolute magnitude and hence the distance of the stars. This technique gave the final proof that the distances of the brighter, and presumably nearer, "white nebulae" were vastly greater than the maximum dimensions of our Milky Way. This decisive step opened new vistas for further exploration, and progress followed quickly. In little more than a decade, from 1924 to 1936, Hubble pushed his reconnaissance of extragalactic space from the nearer galaxies to the limit of penetration of the 100-inch telescope. In 1929 he discovered that the recession velocities of galaxies increase in proportion to their distances as estimated from their magnitudes. This phenomenon is commonly interpreted as evidence for an expansion of the universe, and it forms the observational basis for modern cosmological speculations (Chapter 34).

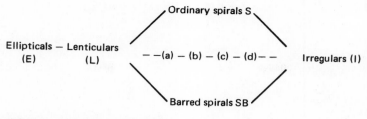

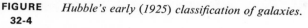

Elliptials — Lenticulars
(E)   (L)   — —(a) — (b) — (c) — (d)— —   Irregulars (I)

FIGURE   *Hubble's early (1925) classification of galaxies.*
32-4

32-2   Types of Galaxies.   A system of classification of galaxies introduced by Hubble in 1925 is, with some revisions, still in general use. It recognizes three main classes: ellipticals (E), ordinary spirals (S) and barred spirals (SB), and irregulars (I). Among spirals three stages, Sa, Sb, Sc, are distinguished according to the relative size of the nuclear or central bulge (decreasing from Sa to Sc) and the relative strength of the arms (increasing from Sa to Sc), as illustrated by Fig. 32-4. Elliptical galaxies have a smooth structure from a bright center to indefinite edges; they differ only in ellipticity from round (E0) to a 3:1 axis ratio (E7). Spiral galaxies show their typical spiral arms or whorls emerging either directly from a bright round nucleus (ordinary spirals) or at the ends of a diametral bar (barred spirals). Irregular galaxies are either of the Magellanic Cloud type, or chaotic and impossible to classify in the Hubble scheme (Fig. 32-5).

This classification system was later revised and refined to include new types or subtypes identified since 1925. In particular, Hubble himself described the S0 or lenticular type, which shares the smooth structure of the ellipticals but has a definite nucleus, occasional dark interstellar clouds, and a general luminosity distribution akin to spirals (Fig. 32-6). The current classification, codified by A. R. Sandage and G. de Vaucouleurs, also includes two varieties of spirals in each of the ordinary and barred families (denoted SA and SB), namely the classical S-shaped spirals (s) and the ringed type (r), in which the arms start at the rim of an inner ring (Fig. 32-7). Some spirals and lenticulars show a faint outer ring surrounding the whole structure (Fig. 32-7). Transition types exist between all the main types. Additional spiral stages Sd, Sm have been added, forming the transition to the Magellanic irregulars Im. This system can be illustrated by a three-dimensional diagram. A cross-section of the classification volume at stage Sb illustrates the continuous transition between the various families and varieties (Fig. 32-8).

In addition to morphological types, a spectral classification introduced by W. W. Morgan, of Yerkes Observatory, gives some indication of the stellar types that contribute most of the light of the nuclear regions of galaxies (the outer regions are usually too faint to produce readable spectra). Thus, in the blue-violet region of the spectrum, galaxies of types E, S0, and Sa show the absorption lines characteristic of stars of spectral

NGC 7097, type E5

NGC 1553, type SA(r)0

NGC 7689, type SA(s)bc

NGC 7793, type SA(s)d

NGC 1365, type SB(s)b

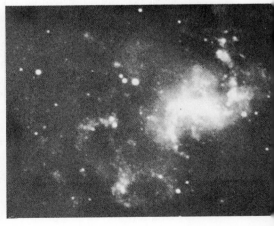

NGC 1313, type SB(s)d

FIGURE
32-5
Examples of elliptical, lenticular, and spiral galaxies.
(G. de Vaucouleurs, Mount Stromlo Observatory.)

FIGURE   *Example of lenticular galaxy seen edge-on, NGC 5866.*
32-6   *(Mount Wilson and Palomar Observatories.)*

(a)

FIGURE 32-7 *Examples of ordinary spiral galaxies seen edge-on show interstellar absorption by dust in equatorial plane. (a) (Above) NGC 4594, Sa. (b) (Below) NGC 891, Sb. (Mount Wilson and Palomar Observatories.)*

(b)

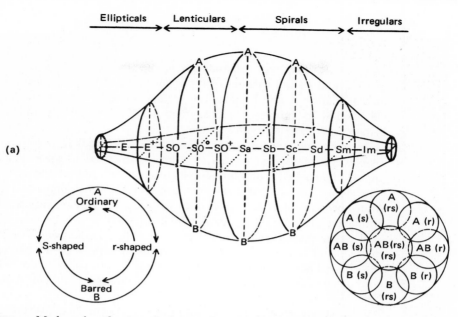

FIGURE 32-8   *Modern classification of galaxies takes into account different families and varieties.* (*a*) (*Above*) *classification volume.*   (*b*) (*Below*) *Cross-section at Sb.*

type K, galaxies of types Sb and Sc have spectra similar to stars of spectral types G and F, and galaxies of types Sd, Sm, and Im have spectral types approximating A and B stars. For instance, the Andromeda nebula was classified Sb in Hubble's original system; it is denoted SA(s)b in the revised system, and kS in Morgan's system.

An intriguing type (or types), to which F. Zwicky called attention in the 1960's, comprises the "compact" galaxies, so-called because they often appear almost stellar on photographs (except for a few that happen to be very close). The nearest is M 32, the bright elliptical companion of the Andromeda nebula. Many have a K-type spectrum and are very similar to ordinary elliptical galaxies, except for their small size, high density, and semi-stellar nucleus. Others, however, are bluer and have A- or B-type spectra, and some show only continuous spectra with no perceptible absorption lines. Perhaps such objects represent a transition form between normal galaxies and the quasi-stellar objects discussed in the next chapter (Sec. 33-6).

Other strange objects that do not fit in any current classification scheme are simply called "peculiar," a label that certainly covers a variety of rare (or perhaps short-lived) but not necessarily abnormal evolutionary stages. Some astronomers, in particular H. C. Arp in the U.S. and B. Vorontsov-Velyaminov in the U.S.S.R., have assiduously hunted for such objects and collected their catch in photographic atlases. The peculiar physical conditions in some of these objects are also betrayed by an intense flux of radio waves and X rays (see Chapter 33), others merely show mildly abnormal spectral emission lines (Sec. 32-6), and a few do not exhibit any other abnormality than their odd-looking shapes (see Figs. 35-10 and 35-11).

In the future a quantitative system of classification based on measured characteristic parameters, such as colors, line strengths, and so on, will probably replace or supplement the present qualitative classifications.

32-3 **Composition of Galaxies.** The different types correspond to fundamental differences in the stellar population content of galaxies. The detailed study of the various types of stars populating different galaxies or parts of the same galaxy is possible for the nearest systems that can be resolved into individual stars on photographs taken with large telescopes. Initially, photographs in blue light could resolve only the brightest blue supergiant stars of types O and B in some irregular galaxies, such as the Magellanic Clouds or NGC 6822, and also in the spiral arms of galaxies of types Sb, Sc, and Sd. The nuclear regions of spirals Sa, lenticulars S0, and elliptical galaxies could not be resolved and, hence, consisted presumably of fainter stars. The central parts of the Andromeda nebula and its elliptical companions M 32, NGC 147, 185, and 205 were finally resolved in 1943 by W. Baade with the 100-inch telescope of Mt. Wilson Observatory on photographs taken through a red filter. This crucial observation led Baade to distinguish two fundamental types of stellar populations:

(a) *Type I*, found in irregular galaxies and in the arms of spirals, is characterized by the presence of blue giant and supergiant stars of spectral types O and B, and also by some red supergiants of type K and M. Associated with this population is a great abundance of interstellar gas and dust in which the ultraviolet radiation of the OB stars causes the formation of bright regions of ionized hydrogen—that is, emission nebulosities of the Orion nebula type. The main-sequence stars found in the spiral arm of our galaxy, including the sun, are also part of this type I population, but are too faint to be resolved in other galaxies.

(b) *Type II*, found in elliptical and lenticular galaxies, in the nuclear regions of spirals Sa and Sb, and in globular clusters, is characterized by red giant stars of spectral types G5 to K5, by sub-giants, and probably by stars populating the lower part of the main sequence observed in globular clusters (which again are too faint to be detected in other galaxies).

The two population types are best separated by the corresponding Hertzsprung-Russell or color-luminosity diagrams (Chapter 20). Both types are usually mixed in various proportions in spiral and irregular galaxies, but it is possible that type II is by far the dominant population in elliptical and lenticular galaxies. If so, the ratio type I/type II increases more or less regularly along the classification sequence.

Although useful, Baade's two-population concept is probably an oversimplification. Detailed analyses of stellar types in a number of nearby galaxies indicate that age and chemical composition are probably the fundamental parameters (cf. Chapter 35) and that an evolved population I may be easily mistaken for a population II. The latter is now considered to be characteristic only of globular clusters, which are themselves far from being strictly homogeneous in composition (Chapter 30). Detailed studies of colors and spectra of galaxies and clusters will eventually lead to finer distinctions between the various population types found in stellar systems of different sizes, ages, compositions, and origins.

32-4 Distances of Galaxies. In principle several methods are available to derive distances of the nearer galaxies. We merely extend the techniques that have worked so well in our galaxy and its various star clusters. For instance, if we observe in another galaxy a variable star, such as a nova or a Cepheid, which can be unequivocally identified as having the same properties (period, color, light curve) as the prototypes in our galaxy, the distance modulus $m - M$ follows immediately from a comparison of the observed apparent magnitude, say, at maximum, with the standard value of the absolute magnitude of the prototype previously determined in our galaxy. Because of the difficulties involved in the absolute calibration of the novae at maximum, and especially of the period-luminosity relation of Cepheids (Chapter 26), the extragalatic distance scale initially set up by Hubble in the 1920's was seriously in error. A first revision advocated by Baade in 1952 was followed by renewed efforts of many astronomers to establish a definitive calibration of the primary distance indicators (novae, Cepheids, brightest OB stars) and to set up a reliable distance scale beyond the confines of our galaxy.

Although further progress is likely, it is clear that earlier estimates of extragalactic distances were too small by factors ranging from 3 to 6. At present the following distances of some nearby galaxies are generally believed to be reliable within 20 per cent:

TABLE 32-1

	Apparent modulus	Distance (megaparsecs)
Magellanic Clouds	19.0	0.05
Andromeda Group (M 31, M 32, M 33)	24.5	0.7
Ursa Major Group (M 51, M 81, M 101)	27.5	2.5
Virgo Cluster	30.5–31.0	12–16

FIGURE 32-9 *The depths of space. Very distant galaxies appear as diffuse specks on a photographic plate. (Mount Wilson and Palomar Observatories, 200-inch reflector.)*

Beyond a few megaparsecs the primary indicators lose their value, because the stars are too faint for individual studies and the H II regions are too small for accurate measurement. Secondary and tertiary indicators calibrated by means of the nearer galaxies come into play, mainly the apparent sizes of the H II regions and the total magnitudes and apparent diameters of the galaxies themselves. Thus, if the nearer galaxies of type Sb have an average absolute magnitude $M = -19$, we can presume that a more distant galaxy of the same type (or preferably the mean of several such galaxies in a group or cluster) has the same intrinsic luminosity. Therefore, if its apparent magnitude is, say, $m = 14$, its distance modulus is about $m - M \simeq 33$ or its distance $\Delta \simeq 40$ megaparsecs. If all galaxies of the same type had the same intrinsic luminosity, this method would work well. Unfortunately, a large range of absolute magnitudes exists among galaxies, even when restricted to a given type—perhaps as much as ± 2 or 3 magnitudes from the mean—and distances so estimated are obviously much less reliable than those derived from individual stars. Only by taking the mean of several estimates, derived from many galaxies of different types in a group or cluster of galaxies, can we hope to obtain a fair approximation. We are fortunate, however, that so far the various less direct methods of measuring these distances appear to be consistent, and show no evidence of intergalactic dust or absorption of radiation.

The difficulties are compounded as we consider more and more distant galaxies, whose individual types can no longer be determined as their photographic images shrink to mere specks of silver grains at the greatest depths of space that the largest telescopes can probe but not fathom (Fig. 32-9). We can only guess that at the indefinite horizon of the explored universe, galaxies several billion light years distant are recorded on our photographic plates. The light that reaches us today may have emerged from the stars of these distant systems before the birth of the solar system (cf. Chapter 35).

32-5 Luminosities and Dimensions of Galaxies. We can measure the luminosity or surface-brightness distribution in galaxies by photographic or photoelectric photometry. This luminosity per unit area measures the total light output of stars in this area and therefore is a measure of star density or, more exactly, of the stellar luminosity function through the thickness of the system (possibly modified by internal absorption).

By studying the variation of the luminosity as a function of distance to the nucleus we gain important information on the detailed structure of galaxies. Because of their regular and symmetrical structure, elliptical galaxies have been a favorite subject for such studies. They seem to obey a common luminosity law, indicating that, except for scaling factors, they are built on the same model (Fig. 32-10a). Lenticular galaxies display more variety but are characterized by three main regions: the nucleus, the

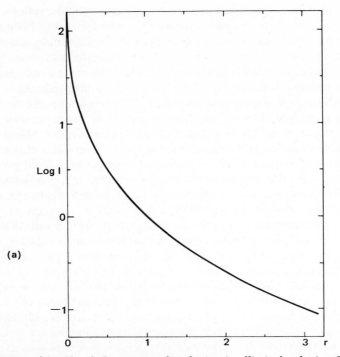

FIGURE
32-10

(a) (*Above*) *Luminosity distribution in elliptical galaxies. Surface brightness, I, is plotted vs. distance to center, r, both in reduced units. (I = 1 at radius r$_e$ within which half of total light is emitted.)* (b) (*Below*) *Luminosity distribution along major and minor axes of M 33 is typical of many spiral galaxies. Here B is in magnitudes per square second of arc, distance x is in minutes of arc. (G. de Vaucouleurs, Lowell Observatory.)*

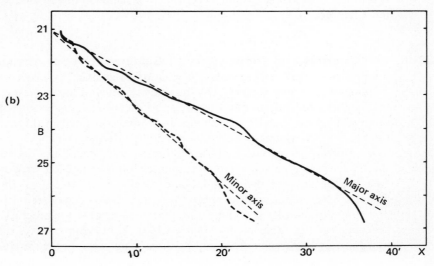

lens, and the envelope. There is a scarcity of detailed observations of this galaxy type, which comprises about one-quarter of the galaxy population.

The spiral galaxies have a variety of luminosity distribution types corresponding to the nuclear region, the spiral arms, and the faint outer fringes. At least for ordinary spirals and the Magellanic irregulars the average radial luminosity distribution approximates an exponential decrease (see the Mathematical Appendix). The spiral arms themselves are much less conspicuous on photometric luminosity profiles than visual inspection of photographs suggests. The interarm regions are far from empty but are occupied by the general population of dwarf stars pervading the whole galaxy (Fig. 32-10b). It is mainly the concentration of blue supergiant stars in the arms that makes them so conspicuous on photographs in blue light; they become much less prominent on photographs in red and infrared light (Fig. 32-11).

The total magnitudes of galaxies can be measured by summing the brightness distribution or by measuring with a photoelectric photometer

FIGURE 32-11   *Composite photographs of M 51 emphasizing the blue (a) and red (b) components of the stellar population. (F. Zwicky, Mount Wilson and Palomar Observatories.)*

FIGURE
32-12

Luminosity functions of galaxies.
(a) Selected by apparent magnitude,
(b) In a large volume of space. (After T. Kiang.)

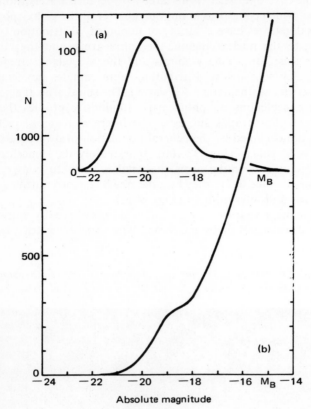

the amount of light received through a series of field apertures of increasing diameters. This amount (corrected for night sky light) tends to a finite limit, which measures the total or integrated luminosity of the galaxy.

Once the apparent total magnitude m of a galaxy has been measured, and if its distance modulus $m - M$ is known, its absolute magnitude M follows immediately. The frequency distribution $N(M)$ of the absolute magnitudes of galaxies, or luminosity function, is an important characteristic of the population of the universe, but is hard to determine accurately, since giant galaxies tend to dominate our catalogues and surveys.

If we select galaxies according to apparent magnitudes—for example, all galaxies brighter than $m = 13$—the luminosity function approximates a Gauss or "normal" error curve, with a mean photographic absolute magnitude $\overline{M}_m = -18.0$ and an apparent dispersion of about 1.0 magnitude from this mean. In other words, most "bright galaxies" are in the range $-15 < M < -21$ (Fig. 32-12).

However, if we attempt to observe all galaxies in a given volume of space—for instance, all the members of a group of galaxies—we discover a large number of "dwarf" galaxies that either are small or have a low surface brightness or both. Such dwarf galaxies are difficult to detect, because they are faint and do not attract the attention as readily as the giants, but certainly their true abundance in space must be many times that of the more conspicuous giants. It is possible that if we could count all such faint systems, the general luminosity function of galaxies would turn out to increase rapidly with the inclusion of fainter and fainter galaxies. How small and faint can a stellar system be and still be counted as a galaxy? No one knows for certain, but we already know of some dwarf elliptical galaxies that are fainter than the brightest globular clusters (that is, $M > -10$). It is an intriguing thought that perhaps millions of very faint stellar systems will always escape detection in the apparently empty space between giant glaxies. Nevertheless, we do know that dwarf galaxies exist only as elliptical or Magellanic irregular systems; apparently spirals occur only among giant galaxies.

When we speak of "giant" and "dwarf" galaxies we mean not only differences in absolute luminosity but also differences in size. However, the apparent "dimensions" of a galaxy are somewhat indefinite because the stellar density and luminosity decrease continuously outwards without any sharp edge. In principle one can define an "effective" or "half-power" diameter in terms of the circle (or ellipse) that contains half the total luminosity. However, this definition requires a detailed and tedious study of the luminosity distribution, and this information is currently available for less than one hundred galaxies. In practice we must for the present content ourselves with measuring the part of a galaxy that appears under some standard conditions on well-exposed photographs, either by simple visual inspection or perhaps on microphotometer tracings of the image density. Such studies limit the dimensions to regions brighter than 1 to 10 per cent of the luminosity of the night sky.

The diameters of galaxies so defined range normally from a few kiloparsecs to 20 or 30 kiloparsecs. For example, the maximum diameter of the Andromeda nebula detected on microphotometer tracings of long-exposure photographs is about 4 degrees, corresponding to some 25 kiloparsecs. This value is comparable to the diameter of our own Galaxy.

32-6 Colors and Spectra of Galaxies. We have noted that the nuclear regions of galaxies of different morphological types have different spectral types. This fact is reflected in their colors. The color index of a galaxy, measured as usual by the difference between the integrated "photographic" (or blue-light) magnitude B and the integrated "visual" (or yellow-light) magnitude V, varies along the classification sequence as follows:

**TABLE
32-2**

Type	E	S0	Sa	Sb	Sc	Sd	Sm	Im
$B - V$	+1.0	+0.95	+0.9	+0.8	+0.7	+0.6	+0.5	+0.4

Hence the color varies from reddish in the E to Sa types, to yellowish at Sb, Sc, substantially "white" at type Sd (the color index of the sun is +0.64), and slightly bluish at Sm, Im. The color range is much smaller than for stars (ranging from −0.3 to +1.6 or more), because galaxies always contain a mixture of many different stellar types of various colors.

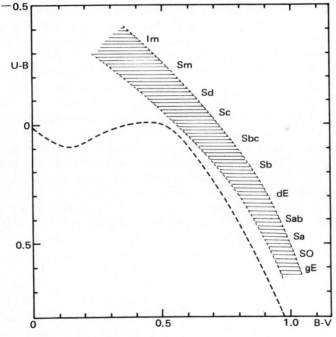

**FIGURE
32-13** *Diagram of U-B, B-V for galaxies.*

This mixing effect also shows up in a color-color diagram (Fig. 32-13), where we plot the $B - V$ color index against the ultraviolet minus blue ($U - B$) color index. Galaxies populate a sequence parallel to but not identical with the corresponding locus for stars (dashed line). A similar effect appears in color-color plots of double and multiple stars and star clusters, where we see a regular progression from the reddest galaxies (E at +1.0 and +0.6) to the bluest galaxies (Im near +0.4 and −0.2).

Occasionally the representative point of a galaxy stands above the sequence, usually because of an excess of ultraviolet light from emission lines. The most common is the forbidden doublet of singly ionized oxygen [O II] at λ 3726 and 3729 Å, arising from emission nebulosities. Another frequent emission line is H-alpha, and the forbidden nitrogen lines [N II] at λ 6548 and 6583 Å (Fig. 32-14). In about one or two per cent of the spirals the very strong emission lines observed in the nucleus are broad and diffuse, in particular the Balmer lines of hydrogen; this phenomenon, first investigated in detail by C. Seyfert at Mt. Wilson Observatory in 1942, demonstrates the presence of large random velocities up to several thousand kilometers per second in the nuclei of some galaxies (several are also variable in intensity). The presence of emission lines in some elliptical and lenticular galaxies indicates that these systems are not entirely devoid of interstellar matter, but in the absence of OB stars the light emission must be excited by some mechanism other than the absorption of ultraviolet quanta from hot stars. A plausible mechanism is excitation by collisions in a violently turbulent gas, but the origin of this violent motion is unknown.

FIGURE   *Typical galaxy spectra.   Nucleus of NGC 1291, type K,*
32-14   *λ 3900 to λ 4900 Å.*
(a)

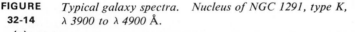

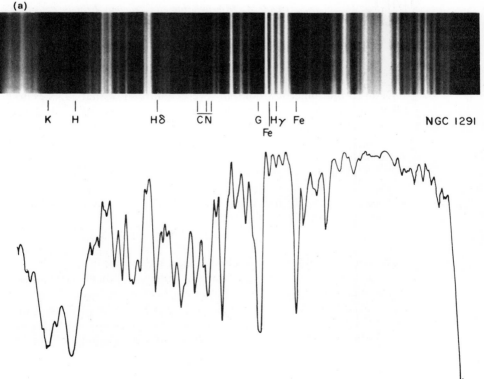

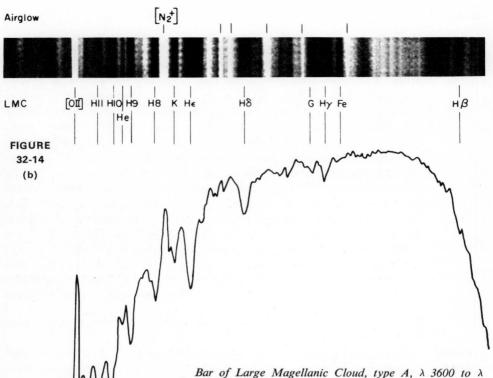

Airglow

$[N_2^+]$

LMC

[OII] HII HIO H9 H8 K Hε   Hδ   G Hγ Fe   Hβ

He

FIGURE
32-14
(b)

*Bar of Large Magellanic Cloud, type A, λ 3600 to λ
4900 Å. Microphotometer tracings help in detection of
faint lines. Note greater strength of metallic lines and
molecular bands (CN) in K type. Compare with Fig. 20-2.
(G. de Vaucouleurs, Mount Stromlo Observatory.)*

Detailed photometric studies in several colors show that galaxies do not
have a uniform color. The color index varies systematically with distance
to the nucleus. In general, the color index is highest (the color redder) in
the nuclear regions. It decreases outward, a little only in E and S0 galaxies,
much more so in spirals, whose arms are definitely bluer than the nuclei.
In the Magellanic irregulars, however, which lack a definite nucleus, the
color is bluer in the center and redder in the outer parts. This color
variation is related to the segregation of stellar types in different parts of
a galaxy, but as yet we have no detailed theory.

32-7 Rotation and Masses of Galaxies. The beautiful shapes of spiral
galaxies, the smooth elliptical outlines of others, and the marked flattening
of galaxies of all types when seen on edge clearly indicate that they are
rapidly rotating. The angular rotation is too small to be detected by direct
comparison of photographs taken a few decades apart, as some astronomers had initially hoped, but it is large enough to produce measurable

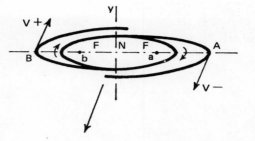

**FIGURE
32-15**

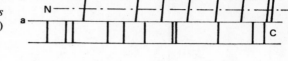

*Inclination of spectral lines indicates
rotation in edge-on galaxy. (Above)
Geometry. (Below) Spectrum.*

Doppler shifts of the spectral lines as in spectroscopic binaries (Chapter
21).

If we consider a galaxy seen nearly edge-on (Fig. 32-15a), the rotation
causes one side to move toward the observer (relative to the nucleus) while
the other side moves away from the observer. As a consequence of the
Doppler effect, the spectral lines are slightly displaced toward the violet
on one side and toward the red on the other side (relative to the position
of the same lines in the nucleus). In the central regions the linear velocity
of rotation is often proportional to distance to the nucleus, and the spectral
lines remain straight but inclined with respect to the lines of a terrestrial
comparison source (Fig. 32-15b).

If the distance Δ of the galaxy is known, we can immediately derive by
simple geometry the rotation period P of the galaxy from the formula

$$P = 29.7 \, \frac{\Delta r}{V_r}, \tag{32-1}$$

where V_r is the rotation velocity (in kilometers per second) at a distance r
(in minutes of arc) from the nucleus. In this formula P is in millions of
years if Δ is in megaparsecs. The rotation periods of the central regions
of many galaxies determined by N. U. Mayall at Lick Observatory between
1940 and 1960 seem to be loosely correlated with galaxy type as follows:

**TABLE
32-3**

Type	E, S0	Sa	Sb	Sc	Sd, Sm
Period*	5 to 10	10 to 20	20 to 40	40 to 100	100 to 300

* In millions of years.

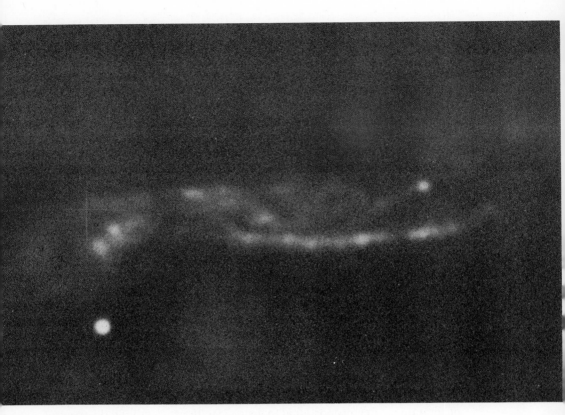

FIGURE   *Rotation-velocity curve of a galaxy NGC 1421.*
32-16   *(G. and A. de Vaucouleurs, McDonald Observatory.) Compare with 32-15.*

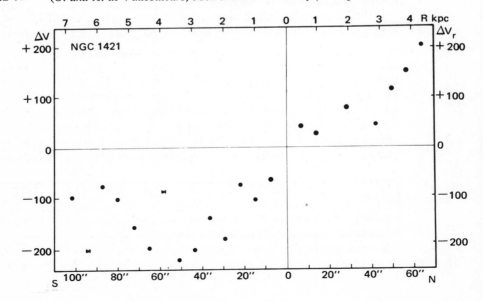

If the rotation period were strictly constant throughout a galaxy (the inclined spectral lines rigorously straight), the galaxy would rotate like a wheel—that is, like a solid body. Detailed measurements prove that this is not really the case; the linear rotation velocity does not increase indefinitely with r, but increases more and more slowly up to a maximum, beyond which a slow decrease takes over (Fig. 32-16). In other words, the angular velocity $\omega = 2\pi/P$ decreases and the rotation period $P = 2\pi r/V_r$ increases in the outer regions. For example, we have already noted that the sun, which is rather far from the galactic center, completes its revolution in about 200 million years, while the central regions of the Galaxy rotate much faster (Chapter 31). This relation also holds for other galaxies.

These observations, based initially on optical spectra, were confirmed and extended after 1953 by radio observations of the 21-cm line emission of interstellar neutral hydrogen in our and other galaxies (Chapter 29). Except for possible minor differences, such as gas streaming away from the nucleus in the central regions of barred galaxies, these observations prove that stars and interstellar gas rotate in the same direction and with about the same velocities, as in a gigantic whirlpool.

If the spatial orientation of a spiral galaxy is known—that is, if the side nearer to the observer can be identified (for instance, by the obscuring effect of dark matter in spiral arms projected against the bright central bulge) —the spectroscopic rotation determines the direction of rotation of the spiral arms. In all clear-cut cases the arms are *trailing* in the rotation, as in all natural vortices.

The mathematical analysis of the rotation curve of a galaxy showing the variation of velocity with distance to the nucleus provides the most direct method of estimating its mass. We merely extend to these huge clusters of stars the method that works so well for double stars (Chapter 21). In other words, we apply Newton's law to the rotation of a whole galaxy. We generally proceed by comparing the observed rotation curve with theoretical curves computed for various models of mass distribution in galaxies. The simplest "model" merely assumes that the motion of the external regions simulates that of a "planet" revolving in the gravitational field of a central mass in the nucleus of the galaxy. The mass then follows from the formula:

$$M = \frac{1}{G} r V_r^2, \tag{32-2}$$

where G is the constant of gravitation, and V_r the rotation velocity at distance r from the center. It is remarkable that more realistic models requiring much more complicated equations give results that differ but little from this simple approximation.

Masses of galaxies (first estimated in this manner by H. D. Babcock, N. U. Mayall, and A. B. Wyse at Lick Observatory in 1939–1940) range over a large interval reflecting the spread between giant and dwarf systems from less than 10^9 solar masses to more than 10^{11} solar masses. Table 32-1 lists typical estimates for some bright galaxies as well as the absolute photographic luminosities L and mass-luminosity ratios $f = M/L$ (in solar units).

A loose correlation exists between mass-luminosity ratio and galaxy type. The ratio decreases along the classification sequence from about 50 for giant elliptical and lenticular galaxies to about 5 or less for Magellanic irregulars. This variation is related to that of other properties such as color and spectral type and reflects the changing stellar population and luminosity function in galaxies of different types. It is interesting to note here that mass estimates furnish the only clue to the abundance of faint dwarf stars in galaxies. As we have already observed, the relatively rare but enormously bright giant and supergiant stars dominate the light from galaxies. Because of the mass-luminosity relation (Chapter 21) these stars emit much more light per gram than do the much more numerous, but inconspicuous, dwarf stars of the main sequence. The latter contribute very little to the visible properties (light, color, spectrum) of the galaxy, but they make up the bulk of its mass. The mass-luminosity ratio, then, tells us that faint dwarf stars are relatively much more abundant in E and S0 or Sa galaxies than in the more spectacular spirals of type Sb and later.

We can also compute the ratio $h = M(H)/M$ of the neutral-hydrogen mass to the total mass of the galaxy. We determine the mass of neutral hydrogen $M(H)$ by summing up the intensity of the 21-cm emission line over the whole area of the galaxy, and by using formulas giving the average amount of 21-cm energy emitted by each hydrogen atom. The hydrogen ratio increases from essentially 0 percent for types E to Sa, to 5 per cent or more for the Magellanic irregulars, in a fashion consistent with the other indices of abundance of interstellar matter.

32-8   **Pairs and Groups of Galaxies.**   Galaxies are often members of pairs, triplets, and groups of increasing multiplicity. Less than half (possibly one-quarter) of all galaxies appear to be single. Thus the Large and Small Magellanic Clouds form a close pair (Fig. 32-1) associated with our own Galaxy in a loose triplet. The Andromeda nebula is the major component of a triplet including its two elliptical satellites M 32 and NGC 205 (Fig. 32-2) and of a loose group including the spiral M 33 and the smaller ellipticals NGC 147 and 185 as more distant members. Both the Galaxy-Magellanic Clouds triplet and the Andromeda group are in turn associated in a larger grouping called the "Local Group" of galaxies, including a score or members, the more important of which are listed in Table 32-2.

Beyond the Local Group, other nearby groups of galaxies have been isolated in Sculptor, Ursa Major, Leo, Fornax, and so on (Fig. 32-17).

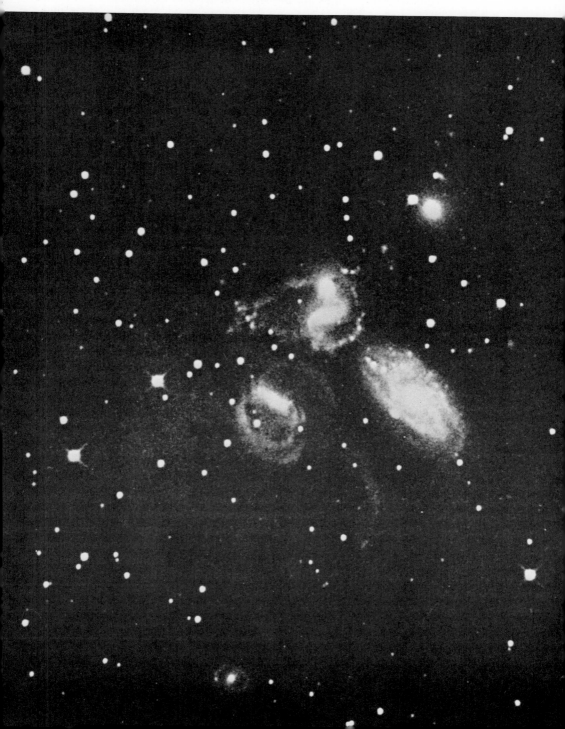

FIGURE 32-17 *A dense group of galaxies, the Stephan Quintet. (Mount Wilson and Palomar Observatories.)*

The tendency of galaxies to occur mainly in groups helps us greatly to ascertain their distances, since individual departures from mean luminosities and diameters average out over the group. Table 32-3 lists the known groups within 10 megaparsecs ($m - M < 30$).

Galaxies associated in groups reveal a remarkable similarity of morphological types. In dense and rich groups, such as the Fornax I "cluster," elliptical and lenticular galaxies are dominant with few or no spirals. In irregular, looser groups, such as the Local Group itself, spirals and Magellanic irregulars are most frequent, with few or no giant ellipticals or lenticulars (although dwarf ellipticals are present as satellites). This segregation of types undoubtedly indicates the existence of different physical conditions at the time the galaxies began to form, but we do not know what physical differences were responsible for the variations.

We can estimate the average masses of galaxies in *pairs* from their differential radial velocities, on the plausible assumption that the two members of a physical pair are in relative elliptical motion, like the components of a binary star. This method, introduced by E. Holmberg of Sweden in 1937, and applied by T. Page in the United States since 1950, is not precise, because it requires statistical consideration of many pairs (the orbit of each pair is unknown and the measured differential velocities are not much in excess of the accidental errors of measurement), but the average masses of various types of galaxies agree generally with masses derived from internal rotation, and it provides the best information on masses of elliptical galaxies, in which rotation and internal motions are difficult to observe.

However, if we consider *groups* rather than pairs, and derive the total mass from the velocity dispersion within the group, we find an apparently significant contradiction. A classical theorem of statistical mechanics, the "virial" theorem, requires that in an assembly of particles in stable statistical equilibrium under its own gravitation, the potential energy of the system due to the mutual attraction of its members must equal twice the total kinetic energy of the particles due to their random motions. We can measure the velocities of individual galaxies in a group and compute their residuals with respect to the mean for the group. If, then, the residual velocities are distributed at random in all directions, we can in principle calculate the total mass of the group and, after counting the membership, derive the average mass of the member galaxies. Another method sometimes used assumes merely that the largest residual velocity is close to the velocity of escape (Sec. 7-14) from the system, and this too leads to an estimate of the mass [equation (7-14)]. Both methods give average galaxy masses that are from one to two *orders* of magnitude (10 to 100 times) greater than the average masses derived from rotational motions.

This contradiction has led to much discussion since 1950. Some astronomers believe that both methods—rotation and velocity dispersion— are substantially correct but do not measure the same mass. The first method measures only the mass of matter in individual galaxies, while the second includes also the total mass of low-density matter dispersed in the vast volumes between galaxies. Whether this matter is mainly in the form of gas or of dwarf stars is not clear. Other astronomers, following a suggestion first made in 1955 by the Armenian astronomer V. A. Ambartsumian, of Burakan Observatory, conclude that the virial theorem does not apply because the groups are not in stable statistical equilibrium. The residual velocities exceed the velocity of escape, and the groups are actually flying apart in a relatively short time, perhaps in less than a billion years. If so, most galaxies in groups have formed relatively recently and are rapidly dispersing throughout space. Ambartsumian's suggestion merely extends to extragalactic space his concept of expanding stellar associations that has proved so successful in our Galaxy (Sec. 30-4). Whether this grandiose application of the concept will also prove to be correct is not yet decided. Nevertheless, a growing body of evidence points to the existence of very large-scale explosive phenomena throughout the universe, first in novae and supernovae (Chapter 27), then in the nuclei of some galaxies (see the next section) and (perhaps) in groups and clusters of galaxies, and finally in the universe as a whole—conceived as an expanding universe (Chapter 34).

32-9 **Interacting, Colliding, and Exploding Galaxies.** In close pairs and groups of galaxies, spectacular interactions often occur between neighbor systems. The manifestations of the interaction vary greatly according to the separation, types, sizes, masses, and still other little-understood physical properties (such as magnetic fields) of the galaxies involved. Frequently we observe ribbonlike filaments of matter streaming out of one or both interacting systems in the general direction of the other, perhaps forming a continuous "bridge" between them (Fig. 32-18). Often, but not always, a filament also emerges in the opposite direction, suggesting effects similar to a tide and countertide. Sometimes the outer arm of a spiral joins with a corresponding arm of a neighbor. Occasionally the interaction looks more like a repulsion than an attraction. When two systems intermingle to a large degree, the main bodies being in actual collision, vast antennalike streamers emerge from the confused central mass and reach out into space to very large distances—up to 40 or 50 kiloparsecs—much in excess of the diameters of isolated galaxies; an example is NGC 4038-39 shown in Fig. 32-19(a). There also occur cases of apparently isolated galaxies—for instance NGC 7135, shown in Fig. 32-19 (b)—that nevertheless possess extensive appendages or comet-like tails that cannot be attributed to interaction with a neighbor (unless it is nonluminous). In such circumstances we can only speculate that an internal explosion or ejection may be the cause of the phenomenon.

FIGURE   *Intergalactic "bridge" links two spiral galaxies.*
32-18   (*F. Zwicky, 200-inch telescope, Mount Palomar Observatory.*)

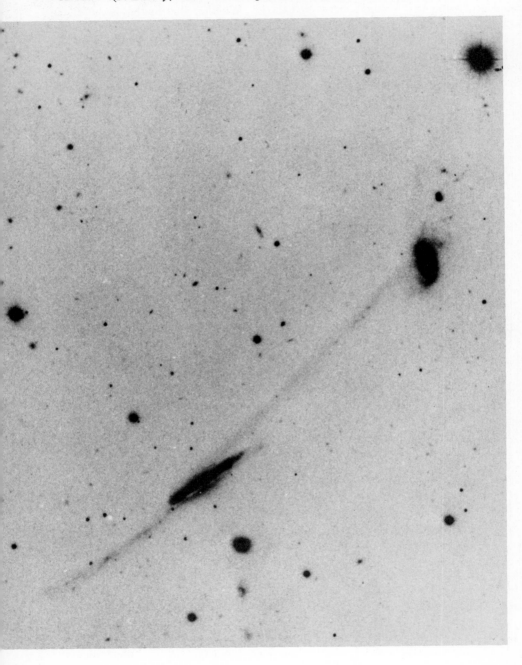

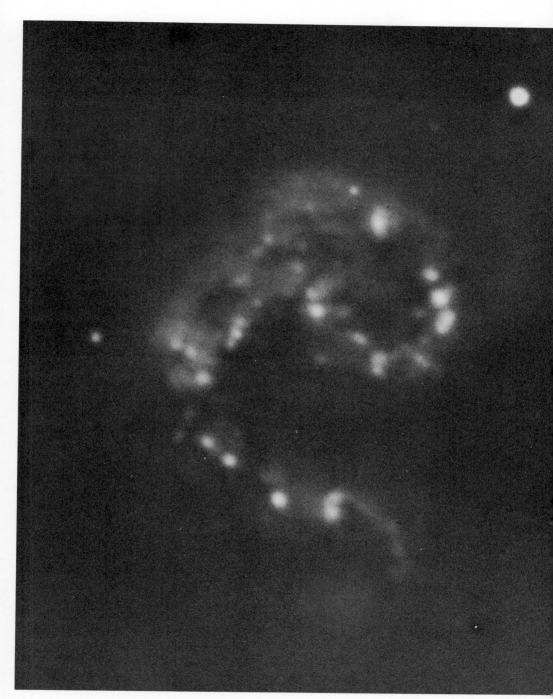

FIGURE
32-19
(a)

Interacting galaxies (NGC 4038–9).
(G. de Vaucouleurs, McDonald and Mount Stromlo Observatories.)

FIGURE   *Isolated galaxy (NGC 7135) with comet-like tail.*
32-19   *(G. de Vaucouleurs, McDonald and Mount Stromlo Observatories.)*
(b)

FIGURE
32-20 *Giant explosion in M 82 ejected glowing filaments of hydrogen far into space. (A. Sandage and W. Miller, Mount Wilson and Palomar Observations.)*

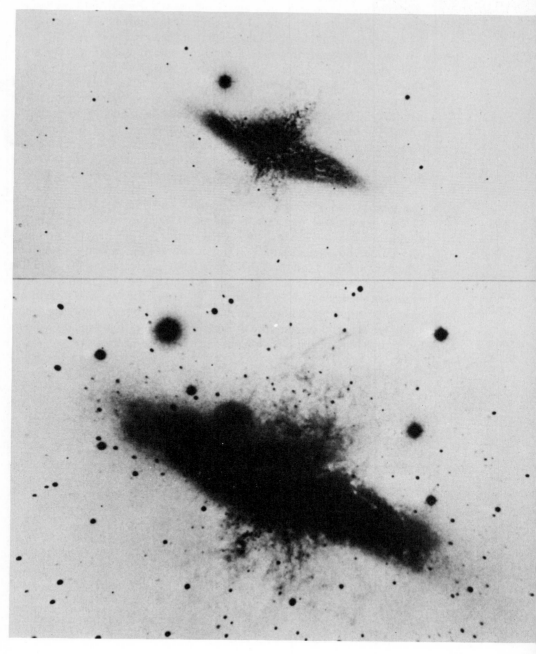

A beautiful example, investigated by Sandage at Mt. Palomar, is the irregular galaxy M 82, whose central regions must have been the seat of a giant explosion several million years ago; irregular filaments of glowing hydrogen gas fan out on both sides of the plane of the system, with velocities of the order of several hundred kilometers per second (Fig. 32-20).

In cases of strong interaction, an unusual state of excitation of the interstellar gas is indicated by the presence of emission lines in the spectra. These emission lines differ in several respects from the usual spectrum of H II regions or of the quiescent interstellar medium. When the interaction is mild, only the hydrogen lines appear in emission; when the interaction is strong, the forbidden lines of ionized oxygen, nitrogen, sulfur, and other elements are also present. Often the relative intensities of the lines indicate that atomic and electronic collisions rather than fluorescence have excited the luminescence of the gas. In some extremely violent collisions, lines of higher excitation, often broad or double, are visible, and the system is also observed as a strong emitter of radio waves (Chapter 33). In NGC 1275, a member of the Perseus cluster of galaxies, many lines appear double in a fraction of the area covered by this disrupted system and indicate a differential velocity of 5000 km/sec between the two emitting regions. Whether we should interpret this as the relative velocities of two colliding galaxies or (as in a nova or planetary nebula) as the expansion rate of an explosion shell is not clear. Finally, H. C. Arp of Mt. Wilson Observatory has called attention to lines or chains of galaxies, mainly ellipticals, which strongly suggest the possibility of explosive ejection of secondary galaxies by large parent galaxies by some as yet unknown mechanism.

32-10   **The Local Supercluster of Galaxies.**   The overall diameter of the Local Group is about 1 megaparsec. It is well isolated from the nearest of similar groups, the Sculptor Group in the southern sky and the M 81 group of Ursa Major in the northern sky, both at distances of the order of 2 or 2.5 Mpc. Some fifty groups and small clusters have been identified within a radius of 15 Mpc from the Local Group. The two richest concentrations of galaxies within this range are the Fornax cluster in the southern hemisphere and the Virgo cluster in the northern hemisphere. The distribution of these nearby groups over the celestial sphere, shown in Fig. 32-21, displays a definite concentration toward a great circle of the sphere. As first proposed by de Vaucouleurs in 1953, the explanation for this concentration toward a plane is the same that accounts for the concentration of stars toward the Milky Way, namely that our Galaxy and the Local Group are parts of a vast flattened supersystem of galaxies. The plane of maximum concentration of galaxies in this supercluster defines the "supergalactic" equator used as the horizontal axis in Fig.

32-21. Supergalactic longitudes L are measured from the intersection of the galactic and supergalactic planes in Cassiopeia. The gaps in the distribution near $L = 0°$ and $L = 180°$ are caused by obscuration by the Milky Way.

The flattening of the supercluster suggests that it is rotating, a prediction confirmed in 1957 by G. de Vaucouleurs through an analysis of the radial velocities of nearby galaxies. The velocities depend on supergalactic longitude in a manner similar to the Oort effect of galactic rotation in stellar velocities (see Fig. 31-9), but complicated by the effects of a general expansion of the system within the expanding universe (see Chapter 34). Our Galaxy and the Local Group are far out toward the edge of the supercluster and rotate around the distant center with a velocity of the order of 400 km/sec. The center is in or near the great Virgo cluster of galaxies at a distance of some 12 to 16 megaparsecs. By a new application of equation (32-2) we derive for the supercluster a total mass of the order of 2.10^{48} g or 10^{15} solar masses distributed among the tens of thousands of member galaxies and possibly also in the space between them.

The overall diameter of the Local Supercluster derived from counts of galaxies of successively fainter magnitudes is about 40 megaparsecs with a thickness of some 10 megaparsecs. Its volume is, then, of the order of 16,000 Mpc^3 or 5×10^{76} cm^3, and for a mass of 2×10^{48} g the average density is

$$\rho \simeq \frac{2 \times 10^{48}}{5 \times 10^{76}} \simeq 0.4 \times 10^{-28} \text{ g/cm}^3$$

As we shall see in the next section and in Chapter 34, this value is still a good deal higher than the average or "smoothed-out" density of the universe at large, confirming that we live in a condensed region of space.

32-11 The Large-Scale Distribution of Galaxies. The total number of galaxies brighter than the fifteenth magnitude is about 40,000 over the celestial sphere (corrected for galactic obscuration), or on the average 1 per square degree. A general catalogue of these brighter galaxies was prepared by F. Zwicky and collaborators from photographs taken with the 18-inch and 48-inch Schmidt telescopes. It is neither practical nor necessary to catalogue individual galaxies fainter than the fifteenth magnitude. On the other hand, the concentration in the Local Supercluster is perceptible in the northern galactic hemisphere down to the sixteenth or seventeenth magnitude. Clearly we must count galaxies to fainter magnitude limits if we are to obtain representative or "fair" samples of the universe at large.

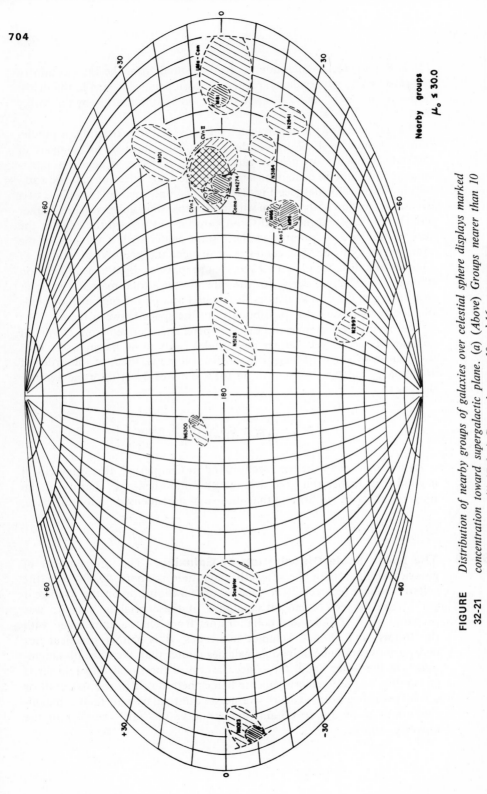

(a)

Nearby groups
$\mu_o \leq 30.0$

FIGURE
32-21
Distribution of nearby groups of galaxies over celestial sphere displays marked concentration toward supergalactic plane. (a) (Above) Groups nearer than 10 megaparsecs. (b) (Below) Groups between 10 and 16 megaparsecs.

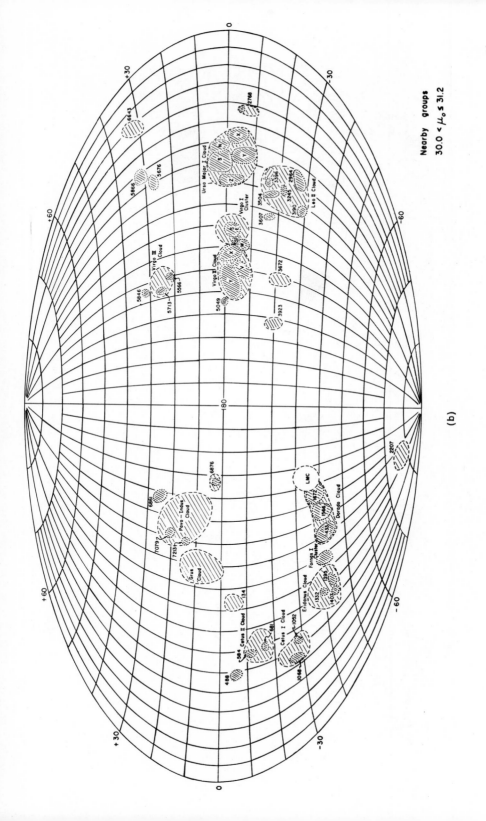

Nearby groups
$30.0 < \mu_o \leq 31.2$

(b)

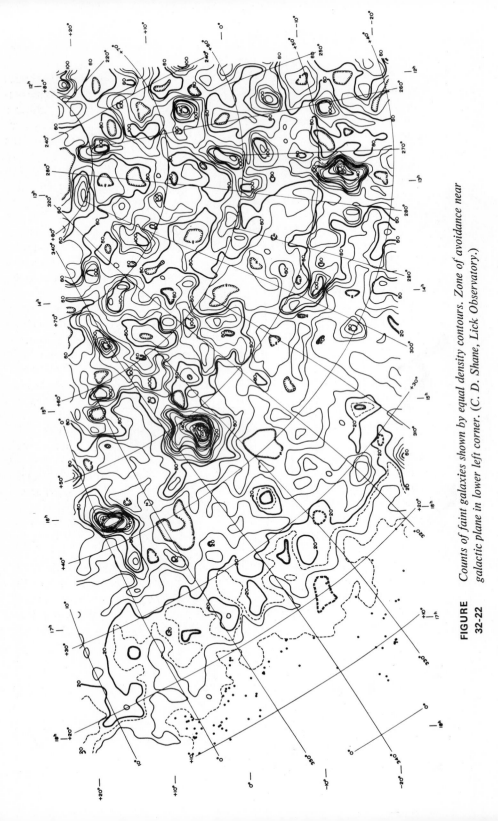

**FIGURE
32-22**
Counts of faint galaxies shown by equal density contours. Zone of avoidance near galactic plane in lower left corner. (C. D. Shane, Lick Observatory.)

The general distribution of fainter galaxies and clusters of galaxies has been studied by C. D. Shane and collaborators at the Lick Observatory. They counted all galaxies brighter than the nineteenth magnitude on the large homogeneous collection of plates taken with a 20-inch astrograph. From such counts contours of equal surface density, or isopleths (Fig. 32-22), were prepared for the three-quarters of the celestial sphere visible from California. The counts not only display in great detail the zone of avoidance caused by interstellar dust (cf. Sec. 32-2) but also disclose the great irregularity or "clumpiness" of the galaxy distribution. Galaxies are much more bunched together than stars. In fact most galaxies belong to groups, clusters, and superclusters. Several large superclusters, similar to the Local Supercluster, show up on the Lick charts. For example, the great cloud of galaxies centered around the double cluster in Hercules has an apparent diameter of 20°, corresponding to 40 megaparsecs at its estimated distance of 110 megaparsecs.

G. Abell, working with plates of the National Geographic-Mount Palomar Sky Survey taken with the 48-inch Schmidt telescope, reached similar conclusions from counts of clusters of galaxies. He found that the distribution of cluster centers on the celestial sphere is conspicuously "clumpy" (Fig. 32-22), and a statistical analysis reveals that the average size of the clumps is, again, of the order of 40 to 50 megaparsecs. This diameter appears to be an important characteristic length of the large-scale distribution of matter in the universe.

One may well ask whether still larger groupings exist. This question is difficult to answer, because available galaxy and cluster counts do not reach quite far enough to give sufficient perspective. Nevertheless, present indications are that the answer will turn out to be affirmative. Higher-order clusters of galaxies of the order of several hundred megaparsecs across (about one billion light years) are probably present.

This indefinitely clumpy distribution of galaxies has an important bearing on the cosmological theories of the structure of the universe on the largest scale (Chapter 34).

For the present we can estimate the average density of matter in the part of the universe probed by galaxy counts. Various estimates are in the range 10^{-30} to 10^{-31} g cm^{-3}, definitely lower than the value quoted earlier for the Local Supercluster. For want of better evidence we must assume as a working hypothesis that this low density is typical of the universe at large.

CHAPTER *33*

Radio and X-Ray Sources

33-1 **Discovery of Cosmic Radio Waves.** The discovery of radio waves from the heavens came largely by accident in 1931, although several physicists and astronomers, among them Thomas Edison and Sir Oliver Lodge, had speculated soon after the invention of wireless transmission that the sun might be a source of electric discharges and hertzian waves. But the crude radio equipment of the time was not sensitive enough, and early reception experiments in 1901 by C. Nordmann on Mt. Blanc in the Alps failed and were soon forgotten.

In 1931 a radio engineer at the Bell Research Laboratories, K. G. Jansky, was studying the sources of background noise of radio receivers and in particular the "atmospheric" component of the noise not caused by thermal agitation of electrons in the antenna, resistors, and tubes of the receiver itself. To find the direction of origin of the presumed atmospheric component he built a large directive aerial that could be rotated on a circular track. Within a year he came to the unexpected but definite conclusion that a sizable part of the "atmospheric" noise (at decameter wavelengths) had an extraterrestrial origin and came from fixed directions of the celestial sphere corresponding to the path of the Milky Way. The noise was strongest when the antenna beam pointed toward the galactic center in Sagittarius.

Strange to say, astronomers paid little attention to Jansky's results, which were first published in a technical journal in 1932–33, shortly before his death. Fortunately, his work was noticed and soon developed by another American radio scientist, G. Reber. As early as 1936 Reber built in his own backyard at Wheaton, Illinois, the first *radio telescope*—a 30-ft reflecting parabolic mirror at whose focus a cavity antenna and a radiometer received and amplified the weak extraterrestrial waves. The telescope could move up and down in the plane of the meridian to point at different declinations, while the rotation of the earth provided the motion in right ascension, each point of a zone of constant declination drifting across the antenna beam once every 24 sidereal hours. By piecing together many such "scans" at adjacent declinations, Reber produced in 1939 the first radio map of the Milky Way at a wavelength of 1.87 m (Fig. 33-1).

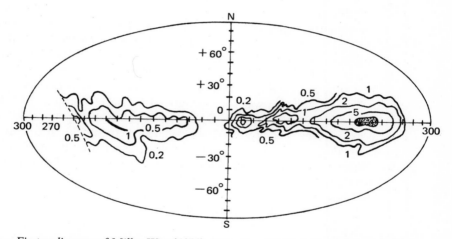

FIGURE 33-1 *First radio map of Milky Way (1939). Intensity of emission in galactic coordinates at wavelength 1.87 m. (G. Reber.)*

The radio isophotes agree generally with the optical isophotes, although later studies reaching fainter flux levels disclosed significant differences.

Reber's work finally attracted the attention of astronomers, who were completely unprepared for this revolutionary new method of probing the universe. Especially surprising was the fact that, while Reber's telescope readily detected radio waves from the Galaxy, it could not detect any radiation from the sun. (Reber did not succeed in detecting the sun until 1943–44.) At meter wavelengths the Milky Way is brighter than the sun!

Radio emission from the sun was in fact detected during World War II. On February 26, 1942, radar operators in Great Britain reported severe interference originating in the direction of the sun, moving across the sky with it, and disappearing at sunset. The effect reappeared the next day at sunrise and ceased only on the third day. At the same time, observatories noted a large sunspot and associated solar flares crossing the sun's central meridian. This first observation of a solar radio "outburst" was not made public until 1946, when the systematic radio patrol of the sun was organized in Great Britain and Australia and soon in many other countries. The main results of radio observations of the sun have been discussed in Chapter 9. Other applications to the study of meteors and the solar system have been mentioned in Chapters 12, 13, and 15.

33-2 Radio Stars. Radio stars were among many unexpected discoveries yielded by studies of cosmic radio waves in the early postwar years. As soon as radar techniques could be turned to peacetime uses, systematic surveys of the galactic radio emission were started in England and Australia, and later in other countries, to confirm and extend to various wavelengths the work of Jansky and Reber. During one of these surveys, in 1944–45, English observers J. S. Hey, Parsons, and Phillips of the Malvern Radar Establishment in England noticed that the "noise" coming from a small region in the constellation of Cygnus was not constant but fluctuated as if it originated in a small localized source, perhaps such as a star. This observation was soon confirmed by J. G. Bolton and G. J. Stanley, who in 1948 set up near Sydney, Australia, an ingenious interferometer experiment that produced interferences between the direct rays received by an aerial atop a high cliff and rays reflected by the surface of the sea (the radio equivalent of the optical experiment of Lloyd's mirror, Fig.

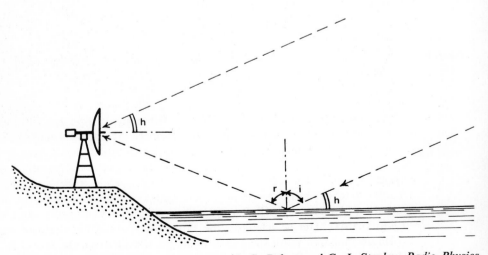

FIGURE 33-2 *Sea interferometer experiment. (J. G. Bolton and G. J. Stanley, Radio Physics Laboratory, Sydney.)*

33-2). The modulation of the signal as the Cygnus source rose above the sea horizon proved indeed the existence in this direction of a strong emitter of radio noise of small apparent diameter. The fluctuations were not in the source itself, but in the earth's ionosphere, an effect resulting from the small size of the source. Later more precise positions derived from other experiments led to the optical identification of the source Cygnus A with a distant peculiar galaxy, photographed in 1953 by W. Baade and R. L. Minkowski with the 200-inch telescope (Fig. 33-3).

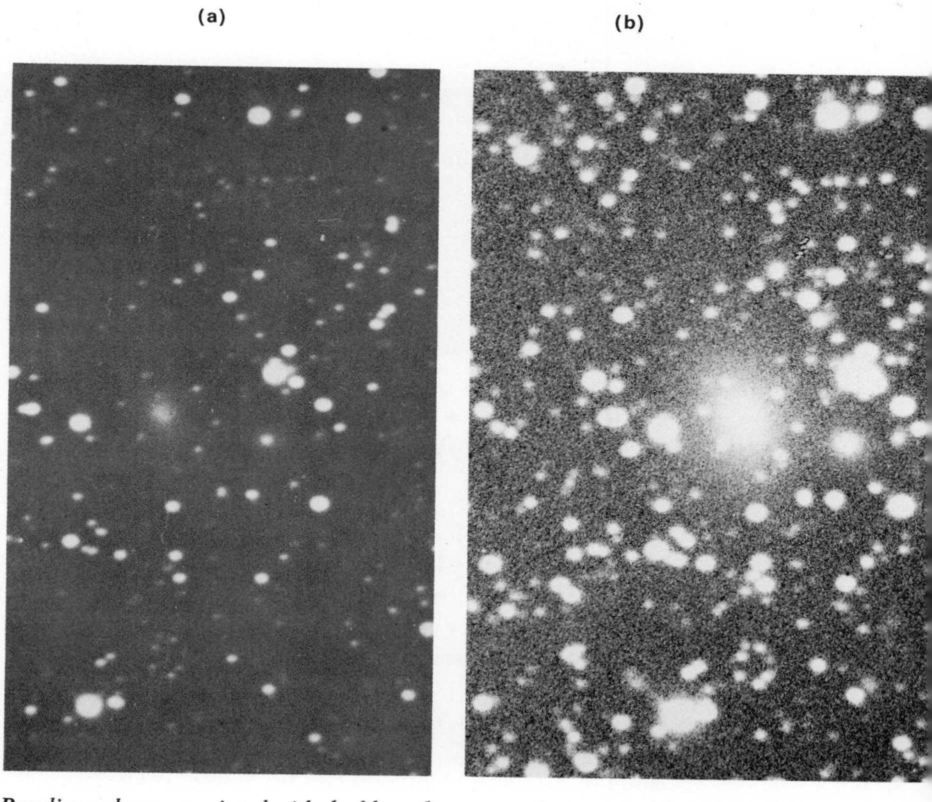

(a)   (b)

FIGURE 33-3   *Peculiar galaxy associated with double radio source Cygnus A. (a) Blue light. (b) Red light. Positions of radio sources outlined by ellipses. (H. J. Smith, McDonald Observatory.)*

In 1947 M. Ryle and F. G. Smith at the Cavendish Laboratory, Cambridge, England, discovered a second radio "star" in Cassiopeia. R. L. Minkowski later identified it optically with a faint peculiar galactic nebulosity, probably the gaseous remnant of a supernova.

FIGURE 33-4 *Peculiar galaxy M 87 associated with the radio source Virgo A. Gaseous jet is source of intense electron synchrotron emission. See Fig. 33-13. (G. de Vaucouleurs, McDonald Observatory.)*

The association of a class of radio sources with supernova remnants began to appear in 1949 when Bolton and Stanley detected a fairly strong source in Taurus at the position of the Crab nebula, the strange object (Fig. 27-9) previously identified by optical methods (Chapter 27) as the remnant of the supernova of 1054. Shortly afterward, the same radio astronomers first identified the radio sources Virgo A and Centaurus A with peculiar galaxies, respectively M 87 (Fig. 33-4) and NGC 5128 (Fig. 33-5).

Thus it became obvious that radio "stars" were not stars at all but a variety of extended galactic and extragalactic sources. Hence astronomers now employ the more general, noncommittal designation *radio sources*.

33-3 Surveys of Radio Sources. As soon as the great abundance and special significance of radio sources became evident, in the early 1950's, strenuous efforts were made in many countries, especially in England, Australia, and later in the United States, to build large radio telescopes (both in the form of a single large antenna and of arrays of smaller anten-

**FIGURE
33-5**   *Peculiar galaxy NGC 5128 associated with the radio source Centaurus A. (G. de
Vaucouleurs, Mount Stromlo Observatory.)*

nas) to collect more energy and increase the accuracy of the position
determinations. Here the long wavelengths of the radio waves require very
large instruments to obtain the desired resolution. The formula that
defines the diffraction-limited resolution of an optical telescope (Sec.
6-4) applies also to radio telescopes, since we are dealing in both cases
with electromagnetic waves. Thus, while a telescope of 14 cm aperture
will have a resolution limit of about 1 second of arc for the wavelength
$\lambda \simeq 0.55 \mu$ of yellow light, a really gigantic radio telescope would seem
necessary to achieve the same resolution, since the ratio of the apertures
must equal the ratio of the wavelengths. Radio waves of 0.55 meter are
10^6 times longer than those of 0.55 μ. Thus, a resolution of 1 second of arc
in the radio region requires a radio telescope of 14×10^6 cm or 140 kilometers across! No such telescope has been built, though since 1964 a

1000-ft fixed reflector has been in operation near Arecibo, Puerto Rico, but plans for very large arrays (VLA) over ten miles in length are being actively discussed, and certainly some such systems will be built in the near future.

Fortunately, one can obtain the desired resolution without building a solid, continuous reflector of such enormous size. A wire screen or arrays of spaced antennas can be used quite efficiently (Fig. 6-14). It may be even enough to *sample* the area of the full-size reflector, as in the technique of *aperture synthesis* (Sec. 6-9). Finally, interferometer techniques with widely spaced antennas of modest size can be applied to obtain precise positions of the stronger sources (Sec. 22-3).

By such techniques systematic surveys quickly led to the discovery of hundreds, then thousands of radio sources, distributed more or less at random over the whole celestial sphere.

FIGURE 33-6 *Radio map of Andromeda nebula. (J. D. Kraus and R. S. Dixon, Ohio State University, Columbus.)*

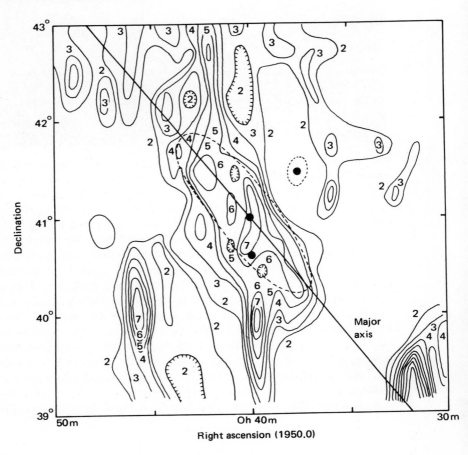

Precise positions, good to a fraction of a minute of arc, have been obtained for hundreds of sources, designated according to constellation and letter, such as Virgo A, Cygnus A, and so on, for the brighter ones, and according to catalogue number for the majority of faint sources. A famous source to be discussed later (Sec. 33-6) bears the designation 3C 273—that is, number 273 in the Third Cambridge Catalogue.

Precise positions often permit the identification of the source as being located in or associated with an object visible on direct photographs of the sky. Near the galactic plane most sources are extended, covering areas of several square minutes or even square degrees. Such sources are usually associated with emission nebulosities (H II regions), supernova remnants, or planetary nebulae.

Outside the galactic belt most sources are generally small and associated with peculiar galaxies. Some are supergiant galaxies, or galaxies with unusual nuclei; others are the so-called quasi-stellar objects (QSO's or Quasars) to be discussed later (Sec. 33-6). A few of the large nearby galaxies are weak, extended radio sources, in which respect they very much resemble our own galaxy. Some are large enough to permit detailed mapping of the intensity of radio emission (Fig. 33-6).

33-4   Radio-Source Structure, Spectrum, and Polarization.

The structure, spectrum, and polarization of radio sources give information on the origin, emission mechanism, and evolution of radio emission by cosmic bodies.

Detailed surveys of the distribution of galactic radio emission over a wide range of wavelengths were made during the 1940's and 1950's (Fig. 33-7). The emission shows a strong concentration to the galactic plane, which indicates an association in part with the flat or Population I component of the galactic disk. However, fainter emission from the whole sphere suggests the existence of a widespread, roughly spherical "corona" around the galactic system. In some regions tongues of enhanced continuous emission stretch far from the galactic plane, the strongest of which, known as "the spur," emerges north of the plane near longitude 30°. The nature and origin of this structure are still uncertain. In the galactic plane, strong sources associated with H II regions appear in emission at meter and shorter wavelengths, and in absorption in the decametric and longer waves where the background radiation is stronger.

A very strong complex source at the galactic center is identified with the galactic nucleus itself (Fig. 33-8). Radio waves easily penetrate the interstellar smog that blocks the optical view (Sec. 29-1).

Studies of the polarization of galactic radio emission have been made with receivers equipped with a rotatable receiving horn (or equivalent devices), the radio equivalent of polarizing filters. The observed complex

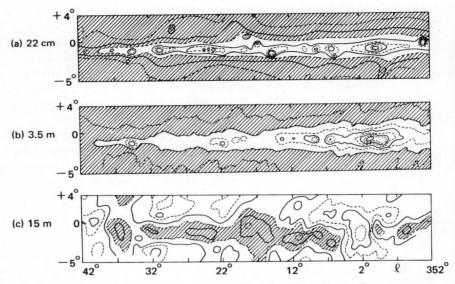

FIGURE 33-7 *Radio maps of our Galaxy at several wavelengths. The galactic plane appears in emission in the shorter waves (a, b) but in absorption in the longer waves (c).*

FIGURE 33-8 *Radio isophotes of galactic center at λ = 10 cm show strong source at nucleus. (Cooper and Price, CSIRO, Parkes.)*

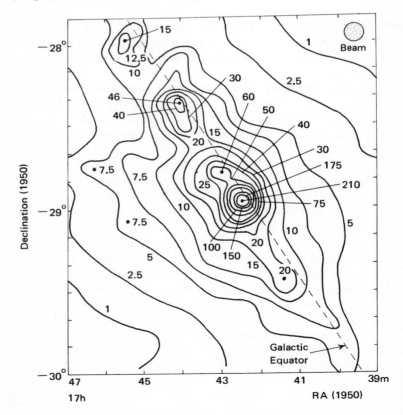

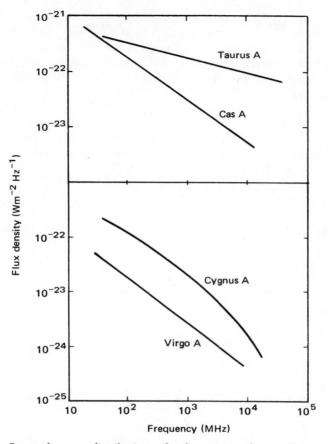

FIGURE   *Spectral energy distributions of radio emission from various*
33-9   *galactic and extragalactic sources.*

pattern of polarization indicates the presence of a magnetic field (Sec.
33-5), and the variation of polarization with wavelength (Faraday effect)
may give information on the structure and dispersion of the interstellar
medium at radio frequencies.

The spectral distribution of radio emission from the Galaxy and radio
sources provides fundamental information just as it does for starlight. If
(subject to possible modifications by interstellar absorption) the energy
distribution obeys Planck's formula (Sec. 8-7), we conclude that the radio
noise merely results from thermal emission of electrons in a gas whose
temperature we may then estimate. Thus H II regions are radio emitters
and absorbers corresponding to their electron kinetic temperature of
10,000° K.

But the spectra of the majority of cosmic radio sources depart strongly from the energy distribution of a black body. Hence other emission mechanisms, referred to as *nonthermal*, must be at work (Sec. 33-5). A number of typical spectral distributions are shown in Fig. 33-9 for our Galaxy (galactic poles and center), for normal and peculiar galaxies, and for a quasistellar source. Theories of cosmic radio sources must be able to explain these results.

FIGURE 33-10 *Intensity contours of radio emission from Centaurus A indicate enormous size of electron clouds ejected by central galaxy NGC 5128 (Fig. 33-5). (Cooper and Price, CSIRO, Parkes.)*

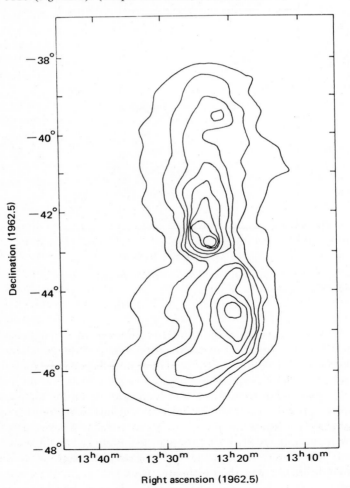

Right ascension (1962.5)

Another important factor is the total power emitted in the form of radio waves relative to the energy emitted as light. For our Galaxy and other nearby normal galaxies the power emitted in the radio region of the spectrum (which we may arbitrarily define as $\lambda > 1$ mm) is a minute fraction, about 10^{-6}, of the power emitted in the optical regions. The radio flux received from even the nearest galaxies is extremely small, of the order of 10^{-24} watt/m²/cps at a frequency of 100 MHz ($\lambda = 3$ m). This figure means that a 36-ft dish (area $\simeq 100$ m²) receiving in a rather broad 1-MHz bandwidth will collect

$$10^2 \times 10^6 \times 10^{-24} = 10^{-16} \text{ watt,}$$

or one billionth of an erg per second, of radio power. Such small signals can be detected and analyzed only because of the extreme sensitivity of radio receivers and techniques.

This ratio of 10^{-6} between radio and optical energy of a normal galaxy may also be compared with the corresponding ratio of 10^{-3} for the energy carried by the particles of cosmic radiation (Sec. 10-6). Thus, in general, light is the dominant form of energy in space.

There exist, however, unusual sources that emit a much larger fraction of the energy as radio waves. Thus the Crab nebula and the nearest radio galaxies such as Virgo A (M 87), Centaurus A (NGC 5128), and Perseus A (NGC 1275) possess radio fluxes from 10^{-4} to 10^{-3} of the optical energy. The ratio is about 1 for some of the more powerful and distant sources such as Cygnus A. Very special circumstances must be at work to cause such an outpouring of radio waves (Sec. 33-5).

Detailed analysis of the structure of nearby radio sources—that is, the distribution of emission over the area of an extended source—provides important information. Such a map (Fig. 33-10a) of contours of constant radio intensity resembles a graph of isophotes depicting the distribution of light emission from direct photographs of, say, a galaxy (Fig. 33-10b). A very surprising result of such studies was the discovery, about 1960, that many radio sources associated with galaxies are double, with components widely separated and symetrically placed with respect to the visible galaxy (Fig. 33-11). This observation quickly led to the concept that radio sources often consist of vast gas clouds ejected from galaxies, dissipating into the near vacuum of intergalactic space in a relatively brief time. The suggestion of a possible rapid evolution and short lifetime on a cosmic scale was directly confirmed when measurements of several galactic sources over even a few years disclosed a measurable fading of sources associated with supernova remnants. On a larger scale, the observation that the components of double radio sources tend to be larger and more diffuse as their separation from the parent galaxy increases furnishes indirect evidence for rapid evolution and expansion.

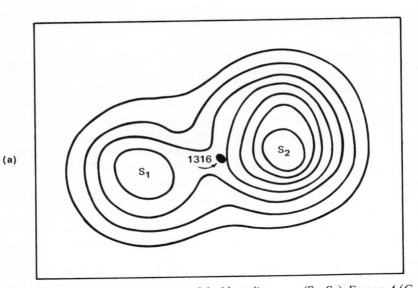

FIGURE 33-11   *(a) (Above) Intensity contours of double radio source* (S_1, S_2) *Fornax A (C. Wade, NRAO), associated with (b) (Below) peculiar galaxy NGC 1316 at center. (G. de Vaucouleurs, McDonald Observatory.)*

33-5   Physics of the Radio Sources. As we have already noted, radio sources can be classified as either thermal or nonthermal. Thermal emission, a well-known phenomenon, is discussed at length in Chapter 8 in connection with optical emission of radiation; it extends directly into the radio spectrum. Nonthermal emission can originate in a variety of phenomena such as electrical discharges (we can hear it on our radio sets during thunderstorms), plasma oscillations, and synchrotron emission. A plasma is an ionized gas cloud often contained by a magnetic field (as in the case of the Van Allen belts of the earth discussed in Chapter 10). It can be a powerful source of radio waves if the electrons oscillate in phase, as they do in the antenna of a radio transmitter. Such oscillations take place at times in the solar corona.

Synchrotron emission is so named because it was first observed to originate as a visible glow of the beam of atomic particles accelerated in a synchrotron. This phenomenon is not restricted to light emission but covers a wide range of wavelengths extending into the radio spectrum. The radiation occurs when electrons move in a magnetic field with velocities close to that of light. Such electrons are often dubbed "relativistic," because at such velocities their masses and other properties are described by the laws of special relativity (see Chapter 34). Physical theory shows that a relativistic electron moving in a circular or helical orbit around the lines of force of a magnetic field (see Chapter 10 and especially Fig. 10-11) emits an electromagnetic wave in a narrow cone around its direction of motion. It acts in some ways as a microscopic searchlight swung around at the end of a sling. The energy carried away by the radiation originates from the kinetic energy of the electrons. The wavelength of the radiation depends on the velocity of the electrons and the strength of the magnetic field.

The values of the velocity and field strength determine whether the radiation maximum occurs as ultraviolet, visible, or infrared light or as radio waves, each electron emitting over a broad range of frequencies (Fig. 33-12) around a characteristic or "critical" frequency f_c given (in megahertz) by the relation

$$f_c = 16HE^2 \sin \theta, \qquad (33\text{-}1)$$

where H is the magnetic field in gauss, E the electron energy in millions of electron volts, and θ the angle between the field lines of force and the velocity vector. A large astronomical source with a variety of electron velocities and magnetic-field intensities can produce a continuous spectrum extending all the way from the far ultraviolet to the radio domain. The electron cloud becomes visible both optically and as a radio source (and possibly an X-ray source). Sometimes different regions of the same overall source appear in the two spectral regions. Thus in the Virgo A radio source only a small, narrow, bright jet shows on photographs (Fig. 33-4), whereas radio emission originates in a much bigger volume more or less coextensive with the whole galaxy.

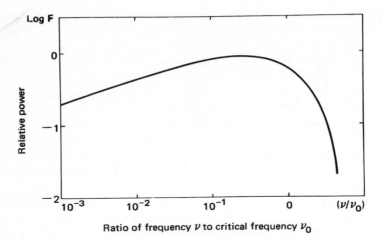

FIGURE
33-12 *Spectral energy distribution of synchrotron emission of a single electron.*

As a crucial test, this theory predicts that synchrotron radiation emanating from a volume occupied by a homogeneous magnetic field (so that the lines of force are parallel and the emitting electrons radiate in the same orientation) must be linearly polarized. The electric vibrations of the waves are all perpendicular to the direction of the magnetic field.

Both optical and radio observations confirm this prediction. The radiation is indeed strongly polarized. In consequence, the continuum light of the Crab nebula and other synchrotron sources displays spectacular differences in appearance, when photographed through different orientations of a polarizing filter (Fig. 27-10). Detailed measurements of such photographs permit the mapping of the magnetic field in the source. As a major complication, the field will generally not be homogeneous through the whole depth of the object. The overlapping of several emitting regions, all with different field orientations, tends to depolarize the radiation. The fact that significant polarization persists shows that homogeneous fields occupy large volumes of space in the source. Detailed consideration of the spectral energy distribution of the radiation enables one to calculate the strength of such fields. Sometimes segments observed in the radio and optical regions match smoothly as parts of a single energy-distribution curve (Fig. 33-13). This fact furnishes proof that a single mechanism operates for both the optical and the radio emission. Such analyses lead to typical estimated values of the order of 10^{-5} to 10^{-6} gauss, for a variety of sources, but field strengths as high as 10^{-3} or 10^{-4} gauss may occur in some concentrated emitting regions.

A comparison of the radio spectra of several sources reveals systematic differences from source to source (Fig. 33-9), which may arise from differ-

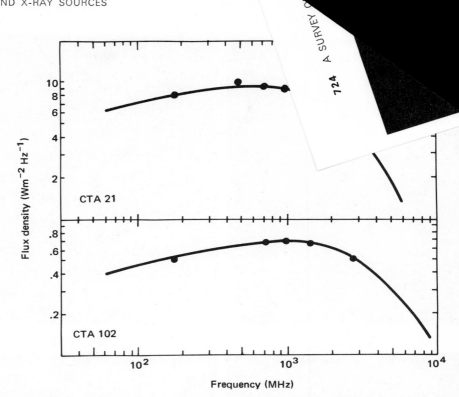

A SURVEY O

72⁴

FIGURE 33-13 *Spectral energy distribution of radiation from many radio sources can be explained by electron synchrotron emission. (K. I. Kellermann, Owens Valley Radio Observatory, California Institute of Technology.)*

ences in electron energy or in field strength or both, and which, in turn, may reflect different stages of evolution in a given class of source.

The classification of sources depends on the shape of the radio spectrum (linear or curved), its slope or spectral index, and the nature of the associated optical source—whether galactic or extragalactic, and in the latter class according to galaxy type. Some very powerful extragalactic sources are not associated with galaxies, peculiar or otherwise, but with very compact "quasi-stellar" optical objects whose nature is still the subject of much debate and controversy.

33-6   The Quasi-Stellar Objects and Radio Sources.   Among the radio sources identified with optical objects the most intriguing are those that appear as stars on direct photographs, occasionally accompanied by faint nebulous jets or appendages. The brightest of all, designated 3C 273, appears in Virgo as a thirteenth-magnitude star with a faint nebulous jet

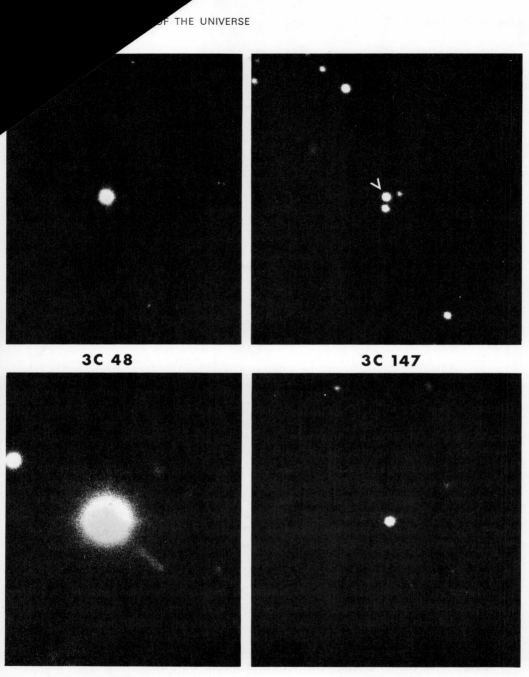

FIGURE   *Quasi-stellar radio sources. Note jet of 3C273. (Mount Wilson and Palomar*
33-14   *Observatories.)*

(Fig. 33-14). The peculiar nature of these objects became evident when astronomers discovered that they possessed large spectral redshifts, indicative of great distances (see Chapter 34) and hence very high luminosities, far brighter than anything previously known. To make the problem even more complex, the optical luminosity often varies in an erratic fashion, over intervals of a few months or years (Fig. 33-15). This fact leads to the conclusion that the sources must be very small—for a large object could scarcely show rapid variations because of the differences of light time from different areas of the source. Finally, the colors, and in particular the color-color relation, are unlike those of any other previously known object (Fig. 33-16). From a very few in 1962, the number of identified quasistellar objects or "quasars" had grown to over one hundred in 1967, almost one thousand in 1970 and we expect that several thousands will be found in the next few years.

The quasars are extremely distant objects. No proper motion can be detected even for 3C 273, the brightest and presumably nearest of them. The good correlation that exists between apparent magnitude and redshift (Fig. 34-10) indicates that quasars partake in the expansion of the universe (Chapter 34), but the redshifts are much larger than for ordinary galaxies. Thus the majority of quasars lie at distances greater than most observable galaxies, and many must be farther away than 1000 megaparsecs. We conclude that quasars have extraordinarily high luminosities and that many bring evidence of events in a very distant past, several billion years ago. Hence they are specially important in the study of cosmology (Chapters 34, 35).

For example, the nearest of all, 3C 273, is at an estimated distance of the order of 500 megaparsecs (some 10^3 times the distance of the nearest galaxies); the corresponding distance modulus is 38.5. Since the mean apparent magnitude is $m \simeq 13$, the absolute magnitude is about -25.5 and the intrinsic luminosity over fifty times greater than that of supergiant galaxies. Some quasars are several magnitudes brighter still. From such figures the energy output of a quasar turns out to be 10^{13} times the solar output or from 10^{46} to 10^{47} ergs per second, or again 100 to 1000 times the peak emission of a supernova. And this prodigious output of energy seems to last for many centuries!

Photographic records of 3C 273 in the Harvard Observatory plate collections extending back three-quarters of a century show no systematic change in the average luminosity of the object (Fig. 33-16). If the lifetime of a quasar were 10^3 years, the total energy radiated would exceed 10^{57} ergs and correspondingly higher values for longer periods. In particular the associated "jets," even if ejected by the central source at velocities close to the velocity of light, would take some 10^6 years to reach the observed positions.

We may be faced with the necessity of accounting for a total energy output in excess of 10^{60} ergs. Through the fundamental relation between

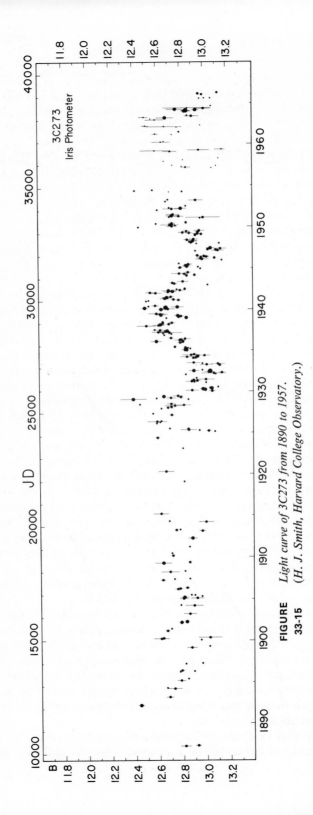

FIGURE *Light curve of 3C273 from 1890 to 1957.*
33-15 *(H. J. Smith, Harvard College Observatory.)*

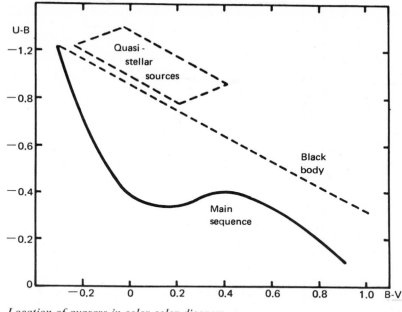

U-B

Quasi-
stellar
sources

Black
body

Main
sequence

B-V

Location of quasars in color-color diagram.

mass and energy (Sec. 24-3), this amount of energy would require the total dissipation of some $10^{60}/(3 \times 10^{10})^2 = 10^{39}$ grams or about 10^6 solar masses. But few natural conversion processes have an efficiency of more than 1 per cent and most offer much less. So we have to consider reactions involving masses in excess of 10^8 solar masses. Yet this large mass, equal to that of several hundred globular clusters or of a dwarf galaxy, must be compressed into a very small volume, perhaps not more than a few light-weeks or light months in diameter, as the time scale of the light variations suggests.

The difficulty of finding a suitable energy source is so severe that some astronomers and physicists have attempted to reduce it by arguing that quasars are not really as distant as their redshifts suggest. Perhaps the redshift of the spectral lines is caused, at least partly, by the relativistic effect of intense gravitation fields (see Sec. 34-5) on the frequencies of atomic vibrations. If so, the absolute luminosities would not be as high as we have estimated and the total power output not so extreme. However, this interpretation still lacks convincing proof, and at present most astronomers believe that the large distances implied by the redshifts must be accepted at face value.

Further evidence on the nature of quasars comes from a study of their optical and radio spectra. For most, if not all, other astronomical objects

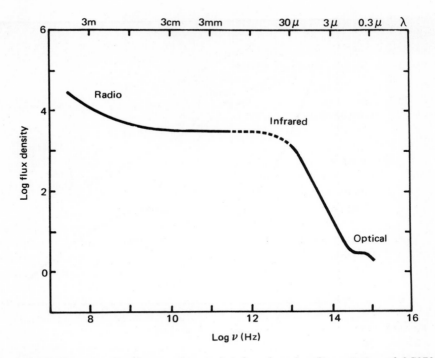

FIGURE
33-17   *Spectral energy distribution of optical, infrared, and radio emission of 3C273.*

previously known, the total energy in the radio spectrum is but a small fraction of the optical emission (Sec. 33-4). For quasars the opposite is true; large as the optical output may be, it is still dwarfed by the more powerful radio emission in the infrared and radio range (Fig. 33-17). In fact, the bulk of the energy lies in the infrared region most difficult to observe from the earth's surface. Part of this energy may result from synchrotron radiation, as in the more common radio sources (Sec. 33-5). and may originate mainly in the nebulous jets. But this mechanism is unlikely to explain most of the optical emission of the quasi-stellar component. Here spectroscopic studies reveal a rich, complicated array of emission and absorption lines (Fig. 33-18), which nevertheless are more easily identified than, for instance, the spectra of supernovae.

Among the strong emission lines, the Balmer series of hydrogen is conspicuous, but the lines are wide and diffuse as if emitted in a chaotic shell made up of many clouds with random velocities of several thousand kilometers per second. When the redshift is large enough, the Lyman-alpha line normally at $\lambda = 1216$ Å becomes visible in the near ultraviolet. Other emission and absorption lines attributed to various ionized ele-

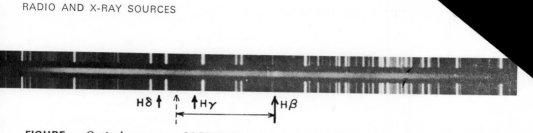

HδↆↆↆHγ ↆHβ

FIGURE 33-18
Optical spectrum of 3C273 showing large red shift of Hβ corresponds to recession velocity of 47,000 km/s. (McDonald Observatory.)

ments, including Mg II, C III, C IV, and He II, are visible against a continuum that varies in intensity. Such lines can form only in a low-density outer shell somewhat similar (on a much larger scale) to the emission shells of novae and supernovae.

The "hard core" of the quasars where the energy generation must take place remains hidden, and only tentative theoretical speculations have so far been offered to explain the mysterious objects. Among the many conflicting suggestions that have been discussed, we mention only the possible release of enormous quantities of energy through gravitational collapse of very large masses of cosmic gas or, alternatively, vast electron clouds pinched by rapidly shrinking magnetic fields (collapsing with the matter) with accompanying emission of radiation by the plasma or synchrotron mechanisms (Sec. 33-5). Many more facts now unknown or only dimly perceived will have to be collected and analyzed before we can correctly explain the physics, origin, and evolution of the quasi-stellar sources.

It is also probable that we shall discover many other objects of this general type that for some reason do not appear as radio sources (perhaps, for example, radio emission is only a relatively brief phase of evolution). Such "radio-quiet" quasi-stellar objects show up in the comparison of photographs taken in yellow and in ultraviolet light, because of the excess emission in the latter spectral range. Many such objects have been found in photographic surveys of the galactic polar caps. Some show the large redshifts characteristic of quasars, but without a detectable radio emission.

33-7 Cosmic X-Ray Sources. Among the strange objects detected in the heavens by modern techniques, we must mention here the sources of X-ray emission discovered in the early 1960's by means of special detectors flown outside the atmosphere in rockets or satellites (Sec. 16-5). The possibility that some celestial bodies, and even ordinary stars and the sun, might be emitting X rays had long been recognized on simple physical grounds. Indeed, at the high temperatures of the chromosphere and corona, and in solar flares, the emission of X rays is an immediate consequence of the laws of radiation (Secs. 8-1, 8-7).

is idea was first put to a direct test in 1962 at the suggestion of B. si of MIT by an industrial group headed by R. Giacconi and shortly rward by H. Friedmann of the Naval Research Laboratory. Photon nters receiving from small areas of the sky defined by suitable dia- ragms and collimating tubes, similar to a bunch of hypodermic needles, detect the X-ray emission. To separate the different spectral ranges, one changes the material of a thin window in front of the detector. For instance, windows of beryllium, aluminum, and mylar transmit X rays in the range $\lambda < 8$ Å, $8 < \lambda < 20$ Å, and $49 < \lambda < 60$ Å, respectively. Such receivers placed in the nose cone of a rocket will sweep small circles of the celestial sphere as the rocket spins and tumbles near the apex of its flight. The observational data, telemetered to the ground, register the X-ray count in various directions, as the rocket shifts position in a known manner. This procedure permits one to locate on the celestial sphere the directions from which the X-ray flux arrives.

One of the strongest celestial sources of X rays is the Crab nebula (Fig. 27-9). the remnant of the supernova of 1054 (Sec. 27-6). The radiation does not originate in a point source or star, but in the amorphous central region of the nebula seen on photographs taken in the continuum light (Fig. 27-10), not from the filaments. We have explained earlier that this continuum emission represents the synchrotron radiation of relativistic electrons in magnetic fields (Sec. 33-5). Thus it is possible that the X rays result from the same mechanism and constitute merely the "tail end" of the synchrotron emission spectrum. However, other mechanisms are possible—for instance, thermal emission by a very hot gas at tempera- tures as high as 100 million degrees. More detailed observations will be needed to definitely identify the physical mechanism responsible for X-ray emission.

Apart from nebulosities associated with supernova remnants, other sources have been observed in various parts of the sky. Some seem to be point sources, and a few have been identified with stellar objects, Thus the strongest X-ray source, denoted Scorpius XR-1, coincides with a thirteenth-magnitude variable star with unusual colors ($B - V = +0.3$, $U - B = -0.8$). Irregular variations of up to one magnitude occur from one day to the next. The spectrum shows broad emission lines of He II, C III, N III, and the Balmer lines of hydrogen. Such characteristics are common among old novae, and more discoveries of X-ray emission from nova and supernova remnants may be expected. Such sources will be strongly concentrated to the galactic plane. However, a number of X-ray sources have been discovered near the galactic poles, and some of them, at least, are extragalactic. Possible or definite identifications with clusters of galaxies, or, more plausibly, with peculiar galaxies already known to be strong radio sources (Virgo A) and quasars (3C 273), suggest that explora- tion of distant parts of the universe by X radiation will play an increasingly significant role in astronomy.

33-8 **Pulsars.** One of the most unexpected discoveries of recent years is that of variable radio stars having very short and extremely stable periods of pulsation. These objects, named pulsars, were first detected late in 1967 by Jocelyn Bell and A. Hewish, of the Mullard Radio Astronomy Observatory, Cambridge, England. A very large array of 2048 dipole antennas covering an area of 4.5 acres and operating at a wavelength of 3.7 meters was used for a survey of faint radio sources, and in particular of their intensity fluctuations. Most sources of small apparent diameter fluctuate in intensity, because the radio waves on their way to Earth cross the irregular plasma clouds of varying density spewed throughout the solar system by the sun (Chapter 9). This phenomenon, called "interplanetary scintillation," is very similar, except for its much larger scale, to the ionospheric scintillation that had originally led to the detection of radio sources (see Sec. 33-2) and again is the radio analogue of the optical scintillation of the stars caused by the density inhomogeneities of our atmosphere (Sec. 6-15). The interplanetary scintillation is normally lowest when sources are observed near midnight in a direction opposite to the sun, and it grows larger during the day when the incoming waves cross the inner, denser regions of the solar system.

Records taken during the low-scintillation period of the night in the summer of 1967 revealed a faint fluctuating source with a curiously regular pattern of repetition. Soon measurements of the repetition cycle disclosed an amazing stability of the period of the pulsation fixed at 1.33730113 second. The period is so stable that the object constitutes a highly precise clock with which some refined measurements of, say, the earth orbital velocity (see Sec. 7-6) could be made. This first pulsar was designated CP 1919 for Cambridge Pulsar at right ascension 19^h19^m (and $+21°$ declination). Although its position is precisely known, efforts to identify it with an optically visible and possibly variable star have not been successful.

TABLE 33-1 *Elements of ten pulsars*

Pulsar	Right Ascension	Declination		Period (seconds)
CP 0328	$3^h28^m52^s$	$+54°$	$23'$	0.714 518 563
NP 0532[a]	5 31 31	$+21$	59	0.033 093 47
CP 0808	8 08 50	$+74$	42	1.292 241 26
CP 0834	8 34 22	$+6$	07	1.273 764 2
CP 0950	9 50 29	$+8$	11	0.253 064 6
CP 1133	11 33 36	$+16$	08	1.187 911
HP 1506	15 07 50	$+55$	41	0.739 677 626
PSR 1749	17 49 49	-28	06	0.562 645 1
CP 1919	19 19 37	$+21$	47	1.337 301 13
PSR 2045	20 45 48	-16	28	1.961 663 3

[a] Crab-nebula pulsar identified with remnant of SN 1054.

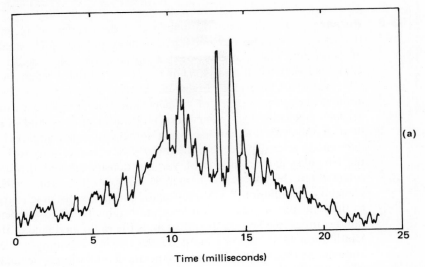

Time (milliseconds)

(*a*) (*Above*) *Fine structure of one pulse of CP-0950 observed at* λ = 1.55 m *with 1000-ft radio telescope. (J. M. Camella, H. D. Craft, and F. D. Drake, Arecibo.)*

**FIGURE
33-19
(a) (b)**

(*b*) (*Below*) *Large variations appear in the strength and structure of successive pulses from CP-0808 observed at* λ = 3.7 m *with the Cambridge antenna array. (A. Hewish.)*

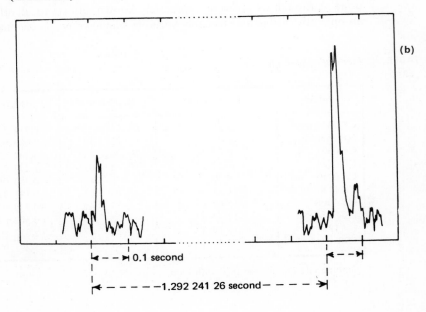

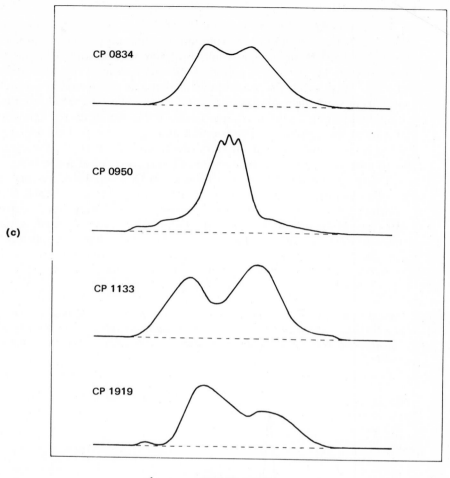

(c)

CP 0834

CP 0950

CP 1133

CP 1919

← — — 50 milliseconds — — →

FIGURE
33-19
(c)
Average pulse shape of four pulsars observed with 250-ft radio telescope (A. J. Lyne and B. J. Rickett, Jodrell Bank.)

In very short order several other pulsars were discovered at Cambridge and then at other radio observatories (Table 33-1). All have extremely short and stable periods ranging from a few hundredths to a few seconds. Most pulsars exhibit a double-peaked flash at each period; the total duration of the flash is very short, some 5 to 50 milliseconds, with a peak-to-peak time interval less than half this value (Fig. 33-19a). Apparently this fine structure varies with wavelength, but it is difficult to ascertain the precise shape of individual pulses because of the confusing effects of scintillation (Fig. 33-19b). By averaging many cycles one can determine a mean pulse shape, which is a distinctive characteristic of each object (Fig. 33-19c).

The extraordinary brevity and stability of the repetition period of pulsars suggests that it is related to the rotation or vibration period of extremely small and dense bodies such as white dwarfs or neutron stars (Chapter 27) or perhaps to the revolution period of a binary system of two such bodies in a very small orbit. The general idea is that a very fast rotator (or oscillator) is at work, helped perhaps by some focussing mechanism to produce a highly directional rotating beam of radio waves similar to the light cone of a revolving lighthouse. The rotation and oscillation periods of hyperdense bodies, such as neutron stars (density $\simeq 10^{13}$ g cm^{-3}, radius $\simeq 1$ to 10 km) theoretically calculated are of the right order of magnitude.

A very fast pulsar, NP 0532, discovered in October, 1968, has the unusually short period 0.033093 second; its location in the Crab nebula (Fig. 27-10) suggested that the supernova remnant, probably a collapsed neutron star (Sec. 27-7), was the source of the radio pulses. A seventeenth-magnitude star near the center of the nebula was positively identified optically as the source when in January, 1969, several groups of observers at the Steward, Kitt Peak, and McDonald Observatories detected fast light

FIGURE 33-20 *Optical light curve of the Crab nebula pulsar. (R. E. Nather, B. Warner, and M. MacFarlane, McDonald Observatory.) Over two thousand cycles of the 33-millisecond period were added in phase to produce the average light curve; note unequal peaks.*

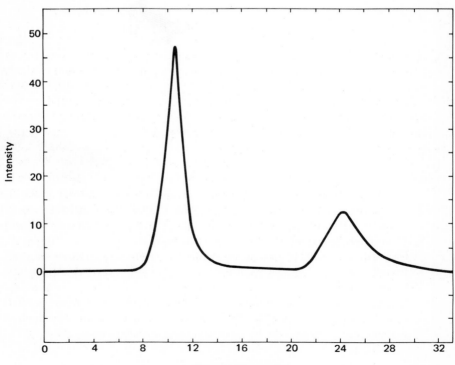

Time (milliseconds)

pulsations in exact synchronism with the radio periodicity (Fig. 33-20). Comparison of the apparent pulsation periods (corrected for the effect of the Earth motion) measured over a few months shows that the period increases (the pulsar is slowing down) at the rate of 39.10^{-9} second per day. This rate is consistent with theories of pulsar emission in which the rotational energy of a fast-spinning and oscillating neutron star is lost to plasma jets ejected by the star. The observed radio and optical pulses are emitted by the synchrotron mechanism (Sec. 35-5) in extremely strong magnetic fields.

33-9   Cosmic γ-Ray Sources.   As we push our exploration of the radiations from space to progressively shorter wavelengths the arbitrary boundary between "hard" X rays and γ rays (see Fig. 8-1) lies near $\lambda = 1$ Å. Observations are difficult because of the strong γ-ray emission by the Earth's atmosphere under the impact of cosmic radiation (Sec. 10-2) and are possible only from balloons flying at extreme altitudes or better still from artificial satellites. The instrumentation used, called a "γ-ray telescope," consists of a thick plate of plastic material which scintillates under the impact of γ rays and a spark chamber which records the negatron-positron pair formed by the materialization reaction of high-energy γ rays, $\gamma \rightarrow e^{+} + e^{-}$. Special devices are used to insure that only events resulting from γ rays (and not charged particles) are measured. By such methods, several groups of physicists succeeded in 1967–68 in detecting a weak γ-ray flux arriving from space near the top of the atmosphere. A large part of this flux is diffuse, i.e., it seems to come from all directions of space and may well be of extragalactic origin (see Sec. 35-6). A fraction comes from regions near the galactic plane and in particular the direction of the galactic center. A few discrete sources of γ radiation have also been detected, including galactic sources such as the Crab nebula and possible extragalactic sources such as the radio-galaxy Virgo A. Although the rate of arrival of γ-ray photons is very small (say, less than 1 cm^{-2} per minute), it is more than compensated for by the very great energy in each quantum (say, of the order of 100,000 electron volts) so that the total energy of the γ-ray and X-ray energy from Virgo A, for example, may greatly exceed the energy radiated by the same source at radio wavelengths. It is clear that much new information of fundamental significance will result in coming years from the rapid progress of these studies of the shortest known electromagnetic waves from the cosmos.

CHAPTER 34

Relativity and Cosmology

34-1 Euclidean Geometry. Euclid's geometry is more or less familiar to us all. This branch of mathematics describes the properties of lines, triangles, circles, and other plane figures. Or geometry may deal with more realistic forms: solid figures such as cubes or spheres. When studying plane or solid geometry we learned to prove all kinds of propositions, such as "two triangles are congruent if two sides and the included angle of one are equal to two sides and the included angle of the other." Remember how the proof of this proposition goes? "Pick up one triangle and lay it upon top of the other so that the equal legs coincide . . . ." Finally, the two triangles lie one exactly on top of the other so that their identity or "congruence" seems obvious—so obvious that one often forgets the basic axiom necessary for the proof: "A plane figure may be moved from one place to another without change of size or shape."

An axiom, incidentally, is a statement that seems to be self-evident but somehow or other evades proof. Consider the axiom more closely. If one moves a triangle drawn on a piece of paper, and happens to tear, twist, bend, or stretch the paper, the triangle will no longer be congruent to the original. If you object that distorting the paper was taking unfair advantage, then we must amend our original axiom to say that "a person may move a figure from one place to another without changing its size or shape, if he does not distort the figure during the motion." In practice, even the slightest motion may have some effect on the figure. The mere weight of the paper may distort it. Thus we suddenly become aware that the world of geometry is unrealistic, a world wherein certain imaginary and absurdly rigorous standards of length and shape exist—an idealized world, very different from the one we live in.

If, in fact, Euclidean geometry and the world of classical or Newtonian astronomy are nothing more than unrealistic idealizations, mere fictions, then other fictitious worlds may be equally self-consistent and equally deserving of our attention.

34-2   The Geometry of Minkowski and Einstein.   One difficulty with Euclidean geometry lies in the assumption that a straight line is self-evident. Alternatively, one sometimes defines a straight line as the "track of a point that does not change its direction." This statement, which seems to say so much, actually means nothing. Unless we have some standard of "straightness," we have no way of determining whether a point changes direction or not.

The problem is primarily one of standards. Unable to find a completely logical method for establishing an absolute standard, the nineteenth-century German mathematician, H. Minkowski, attempted to build up a geometry that did not have to rest on artificial and unprovable postulates. He found that he could do without straight lines, merely substituting for them circles of very large diameter. Over any given distance, a circle of large diameter can be made to come indefinitely close to any idealized straight line. Minkowski extended his geometry to describe worlds of more than three dimensions. In other words, he did not limit himself solely to length, breadth, and thickness—as did the old Euclidean geometry.

When Einstein tried to develop a mathematical theory of the universe, he found Minkowski's geometry extremely useful as a means of expressing relationships of physical interest. The physical world clearly requires more than ordinary geometry, because nothing happens in an empty world composed of only spatial elements. To study the real world, we have to add what at first sight may seem to be an entirely independent consideration: time. The world of physics is a world of events. To define an event we have to specify its coordinates in space and time. A certain meeting may occur on the third floor of a building at the corner of Fifth and Main Streets at *four o'clock in the afternoon.* Thus time enters the world of events on an equal footing with one of the spatial coordinates. In brief, an event is a point in the four-dimensional world of space-time. Further, events themselves have four-dimensional geometrical properties somewhat resembling those of points in three-dimensional geometry.

In Euclidean geometry the simplest congruence was the equality of two segments. One mentally picked up one of the segments, made one end coincide, then looked at the other end, if that, too, coincided, then the segments were of the same length. This idealized mental process is entirely impractical for the laboratory. Time must elapse between our making the segments coincide at one end and checking them at the other end. What right have we to assume that conditions remain exactly the same while we move from one end of the segment to the other?

Thus, if we are to use it for measurement in a physical world, even Euclidean geometry contains some hidden postulate concerning time. To follow Euclid exactly, what we really need is two observers, one for each end of the segment, who will *simultaneously* check on the coincidence between the two segments. This would seem to be a simple and satisfactory compromise, although we must now see what the word "simultaneous" implies.

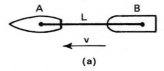

(a)

FIGURE 34-1 (*a*) *and* (*b*) *Simultaneity and the Barge experiment.*

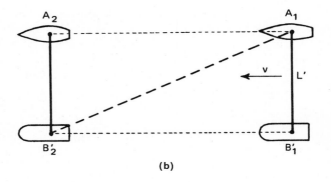

(b)

34-3 The Principle of Simultaneity. Simultaneity is a special feature that distinguishes the geometry of Minkowski and Einstein from that of Euclid and Newton. Consider the following practical problem. In Fig. 34-1(a), let *B* represent a barge being towed through a heavy fog by a ship *A*. The air is completely calm, so that any wind comes from the motion of the ship and barge. The man on the barge wishes to set his chronometer to agree with that of the ship. He listens for the sound of the ship's bell, striking midnight. He knows, from earlier measures, that the distance *AB* is 340 meters. He further knows that sound travels 340 meters a second. Hence, when he hears the chime, he decides that the time is one second after midnight. He sets his chronometer at that time and concludes that it is in synchronism with the ship's chronometer. He will be correct, however, only if the ship and barge are stationary. For, if the ship is moving, a wind will blow from A to B. The sound waves, helped along by the wind, will travel faster than in still air. In general, if c is the speed of sound in quiet air and v the velocity of the ship relative to the air, the effective speed of sound will be $c + v$. And if L is the distance between the ship and barge, the time T_1 for the sound to travel from A to B is simply

$$T_1 = \frac{L}{c + v}, \tag{34-1}$$

Thus in the previous example, for $L = 340$ meters, $c = 340$ meters per second and $v = 3.4$ meters per second, we have

$$T_1 = \frac{340}{340 + 3.4} = \frac{1}{1.01} \simeq 0.99 \text{ sec.} \tag{34-2}$$

If the man on the barge is unaware of the wind speed or cannot measure it, he will inadvertently set his clock $1.00 - 0.99 = 0.01$ second slower than the ship's clock. In other words, his chronometer will not be synchronous with that of the ship. He cannot, therefore, properly determine whether two events are simultaneous or not. Since successive chimes heard by the barge operator are similarly delayed, we conclude that he can at least adjust his clock to have the same rate as the ship's clock, even if he is uncertain about the exact time.

Now let us suppose that the ship and barge can communicate with one another. The operators of the two vessels decide to perform a simple experiment. At the precise moment when the barge operator hears the chime he strikes a bell that returns, like an echo, to the ship, where the ship's operator can measure the exact time required for the sound to execute the round trip from A to B and back to A. On the return trip, however, the sound is traveling against the wind, so that its effective speed is $c - v$. Hence the time T_2 for the return trip is

$$T_2 = \frac{L}{c - v} \tag{34-3}$$

and the time, T, for the round trip is

$$T = T_1 + T_2 = \frac{L}{c + v} + \frac{L}{c - v} = \frac{L(c - v) + L(c + v)}{(c + v)(c - v)}$$

$$= \frac{2Lc}{c^2 - v^2} = \frac{2L}{c} \cdot \frac{1}{1 - v^2/c^2} = \frac{2L}{c} \cdot \frac{1}{1 - \beta^2}, \quad \text{where } \beta = v/c. \tag{34-4}$$

For the foregoing example, with $L = 340$ meters and $c = 340$ meters per second, we should have $T = 2$ seconds, if the ship and barge are stationary ($v = 0$). But if $v = 3.4$ meters per second, so that $v/c = 0.01$, the elapsed time required for the signal to go from A to B and back again is

$$T = \frac{2}{1 - (0.01)^2} = \frac{2}{0.9999} \simeq 2 \times 1.0001 \text{ sec.} \qquad (34\text{-}5)$$

From the fact that the signal took 2.0002 seconds instead of 2.0000 seconds for the round trip, the man on shipboard may deduce the speed of the vessel with respect to the surrounding air.

The foregoing determination of the speed rests, however, on a very accurate knowledge of the distance, L, between the two clocks. The experimenters now decide to vary the procedure, so as to eliminate any uncertainties in L. First of all, they replace the tow rope by a rigid rod, AB (Fig. 34-1b). After carrying out the first measurement with the barge at B, they now send the barge to B′, at right angles to the ship's motion, keeping

$$AB' = L' = L = AB.$$

On this second measurement the sound travels at right angles to the motion, and hence its velocity is constant and equal to c for both the going and return trips. When the ship and barge are respectively at A_1 and B_1' the ship's bell strikes. During the interval T_1', while the sound is moving from the ship to the barge, the two vessels will have moved a distance

$$A_1 A_2 = B_1' B_2' = v T_1'$$

and

$$A_1 B_1' = A_2 B_2' = L'.$$

The sound, during the time T_1', will have moved along the diagonal $A_1 B_2'$, whose length is

$$A_1 B_2' = c T_1'.$$

By the familiar theorem of Pythagoras we must have the square of the hypotenuse equal to the sum of the squares of the other two sides. Thus

$$(A_1 B_2')^2 = (A_1 A_2)^2 + (A_1 B_1')^2$$

or

$$c^2 T_1'^2 = v^2 T_1'^2 + L'^2.$$

Hence

$$T_1' = \frac{L'}{c} \cdot \frac{1}{\sqrt{1 - v^2/c^2}} = \frac{L'}{c} \cdot \frac{1}{\sqrt{1 - \beta^2}}. \qquad (34\text{-}6)$$

The time for the round trip is

$$T' = 2T_1' = \frac{2L'}{c} \cdot \frac{1}{\sqrt{1 - \beta^2}} \cdot \qquad (34\text{-}7)$$

Thus with $L' = L$, we have two separate equations (34-4) and (34-7),

$$T = \frac{2L}{c} \cdot \frac{1}{1 - \beta^2}, \qquad T' = \frac{2L}{c} \cdot \frac{1}{\sqrt{1 - \beta^2}},$$

predicting the respective times for the signals to traverse the two paths ABA parallel and $AB'A$ perpendicular to the motion.

We have supposed that the man on shipboard can determine both T and T'. These two equations have just two unknowns, L and v, which can be determined. The results are:

$$L = \frac{cT'^2}{2T}, \qquad v = c \sqrt{1 - \left(\frac{T'}{T^2}\right)^2} \cdot \qquad (34\text{-}8)$$

These simple formulas are the correct ones for interpreting the experiments with ship and sound waves, however difficult they may be to perform.

But why use slow sound waves for the experiment, when fast light signals will greatly enhance the accuracy of time determination? After all, light travels about a million times faster than sound. For celestial experiments light is a necessity, because sound will not traverse the vacuum of interplanetary space. On the other hand, the distances and velocities of the heavenly bodies are correspondingly much higher.

The basic principles are similar, whether one deals with sound or light. The foregoing reasoning suggested a significant experiment for detecting the motion of bodies through space. We shall use the same formulae, except that c now represents the speed of light rather than sound.

34-4 The Michelson-Morley Experiment. The experiment of Michelson and Morley, performed at the University of Chicago in 1887, was designed to detect the motion of the earth through space by means of equation (34-8). Their technique differed somewhat from that discussed for sound waves, but the same theory applied. In effect, they measured the time T for a light beam to traverse the path ABA (Fig. 34-2), where A represents the source and B a mirror reflecting the beam back on itself. They then rotated the apparatus around A as a center, to measure the time T' over the identical path $AB'A$. If the earth and the equipment on it were moving parallel to AB, say, Michelson and Morley expected to measure the velocity in terms of the difference between T and T', as given in equations (34-4) and (34-7) to calculate L and v, by (34-8). To their surprise the experiment always gave $T = T'$, irrespective of the time of day or time of year.

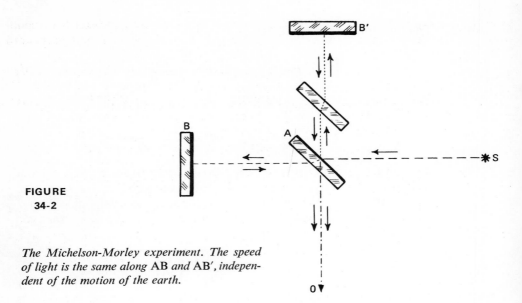

**FIGURE
34-2**

The Michelson-Morley experiment. The speed of light is the same along AB *and* AB′, *independent of the motion of the earth.*

The Irish physicist Fitzgerald made the following suggestion. Despite all precautions, perhaps L' is not equal to L. Then, since $T' = T$, we may equate (34-4) and (34-7) to find

$$L = L'\sqrt{1 - \beta^2} . \tag{34-9}$$

In brief, since L is a distance parallel to the motion, and L' the distance measured perpendicular to the motion, Fitzgerald was suggesting that a body automatically contracts in a direction parallel to its motion. The Dutch physicist H. A. Lorentz, following up this suggestion, showed that such a shortening was consistent with the electrical nature of matter and, in fact, could be predicted from the basic laws of electromagnetic theory. And so the phenomenon described by equation (34-9) is called the Lorentz-Fitzgerald contraction.

34-5   The Special Relativity Theory. Einstein's theory of relativity developed directly from these experiments. In principle a man on the moving barge could determine his velocity with respect to the stationary air with the aid of sound waves. Nineteenth-century scientists had speculated on the existence of a universal medium, something like air but infinitely more tenuous, that could transmit the light vibrations from one place to another. They called this postulated medium the "luminiferous aether" or simply the "aether." Michelson and Morley had designed their experiment to measure the speed of the earth with respect to this aether. And when the experiment failed, by virtue of the Lorentz-Fitzgerald contraction, "absolute motion" or the speed of a body in the aether no longer had significance. Indeed physicists shortly ceased to use the meaningless word "aether."

Einstein built his theory on two basic postulates. First, he denied all possibility of measuring absolute motion. An observer could measure his *relative* motion only. Hence the theory of *relativity*. Second, any observer, attempting to measure the speed of light in empty space, would always get the same value.

This assumption, apparently so simple, led to major changes in physical theory. We have already noted the difficulties encountered by the operator of the barge as he tried to set his clock in exact synchronism with the ship's bells. He could set his clock to run at the same rate, but he had to know his absolute speed in air in order to make the two times agree exactly. But if we cannot measure our absolute speed in space, we cannot set two clocks in precise synchronism. The correction, which increases with the distance, L, between the clocks, is equal to L/c.

These two clocks, by hypothesis, are moving in the same direction and at the same speed—that is, their relative positions do not change. Their relative velocity is zero. Apart from the difficulty with synchronism, two observers—one at each clock—would measure similar intervals of time. And they would use the same units of length for spatial measurement.

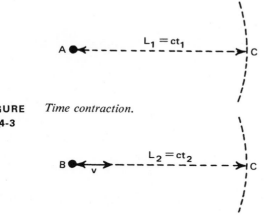

FIGURE 34-3   *Time contraction.*

But what happens if the two observers and their clocks are moving relative to one another? In Fig. 34-3 let A be the reference observer; let B be a second observer, moving toward the right with velocity v. At the moment of closest approach, each observer sends out a flash of light, which moves with velocity c toward the right. By hypothesis these two flashes must move together (for if one or the other got ahead, we could then distinguish between A and B). After a time t_1, according to observer A, the flash has moved a distance

$$AC = L_1 = ct_1.$$ (34-10)

Observer B, after time t_2, determines that the flash has moved a distance

$$BC = L_2 = ct_2.$$ (34-11)

We assume, with Einstein, that the speed of light is constant for both observers and equal to c.

Now, according to equation (34-9), the length L_2 in the moving system is contracted to

$$L_2 = L_1\sqrt{1 - \beta^2} = ct_2 = ct_1\sqrt{1 - \beta^2},$$ (34-12)

and it follows that

$$t_2 = t_1\sqrt{1 - \beta^2}.$$ (34-13)

The relative motion, therefore, leads to a contraction of time as well as of space intervals in the moving system. If $t_1 = 1$ year, t_2 will be less than a year; for example $t_2 = 0.99$ year if $\beta = v/c = 0.1$. Therefore, as viewed by observer A, the time of B advances more slowly than his own. Observer B ages less rapidly than observer A, and in fact all physical processes will slow down, including electromagnetic vibrations, whose wavelengths will be increased (that is "redshifted").

But this argument leads to a seeming paradox. We arbitrarily chose A as the stationary observer. We could equally well have chosen observer B, in which case we should have replaced (34-12) and (34-13) by

$$L_1 = L_2\sqrt{1 - \beta^2} = ct_1 = ct_2\sqrt{1 - \beta^2}$$ (34-14)

and

$$t_1 = t_2\sqrt{1 - \beta^2}.$$ (34-15)

Now it is observer A who ages less rapidly than observer B. How can both experiments be true? Which is correct? The special theory of relativity, on which the argument above is based, assumes that A and B are moving uniformly. Their paths cross only once. They cannot come back to check whether one or the other, if either, ages more rapidly.

34-6 The General Theory of Relativity. Einstein's general theory of relativity removes certain of the foregoing restrictions. It takes accelerated motion into account. If observer B, for example, wishes to return to point A, he can do so by use of rockets or some special device to reverse his motion. Now there is no question: it is observer B that has accelerated and not observer A. Thus, when A and B come together again, the fact that B is the younger should occasion no surprise.

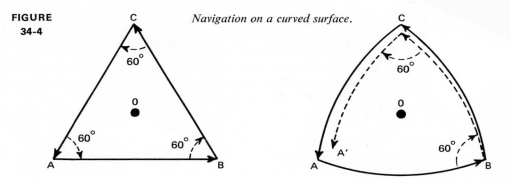

FIGURE
34-4

Navigation on a curved surface.

Einstein's general theory concerned measurements from an accelerated system, say from a planet moving in a gravitational field. This assumption made it possible for Einstein to derive a "natural" explanation for Newton's postulated law of gravitation. He explained gravitational effects as if they were caused by a curvature in the geometry rather than a true field of force. The significance of this statement appears evident in terms of a simple analogy. Suppose that a yacht is racing around a triangular course *ABC*, centered on an island *O* (Fig. 34-4). The day is foggy and it is necessary for the navigator to sail by "dead reckoning." He starts out and maintains a certain course *AB*. He then turns through an angle of 120 degrees and goes an equal distance *BC*, and finally swings through another angle of 120 degrees and comes back along *CA* toward his starting point. He would expect, by virtue of having traversed an equilateral triangle, *ABC*, with 60-degree angles at each corner, to come back to his starting point. He is somewhat surprised to find himself nearer the island than the starting point. Perplexed by the situation, he tries again, on a somewhat large course, and lands up with an even greater difference than he did before.

A "simple" explanation could occur to him. The island "exerts a gravitational pull" on the ship, so that if he wants to go in a strictly equilateral triangle or any other course he must always steer slightly away from the island to compensate for this mysterious force of attraction.

Any experienced navigator, of course, recognizes the real source of the difficulty. Actually the earth is curved, and an equilateral triangle on the surface of a sphere has angles that are bigger than 60 degrees. Hence, if he insists on turning at 120-degree angles he naturally swings in closer to the island on each leg, as shown in Fig. 34-4. What our navigator identified as a gravitational force pulling him toward the island was really a natural curvature in a space that he had mistakenly assumed to be flat.

According to Einstein's general theory of relativity, what Newton called gravitational force is really nothing more than a curvature of space associated with the presence of matter. Although the concept of gravitation ıl tool, which we may employ for calculation of the orbits of ₍venly bodies, we do not have to believe in it as a special force. ₍ad of adopting a Euclidean geometry, we may adopt a naturally ₍ved space in which the mysterious force disappears. Then the curved ₍aths of the planets and heavenly bodies arise as a natural consequence of the geometry of the medium in the neighborhood of a large mass.

Einstein, in developing this theory quantitatively, made three specific predictions concerning the motions of planets or of light rays, which differed from those calculated by pure Newtonian mechanics. First, the elliptical orbit of the planet Mercury around the sun should slowly rotate. The long axis of the ellipse should slowly advance at the very small regular rate of 43″ per century. Long before relativity, astronomers had detected and been puzzled by this departure from Newtonian theory (Sec. 12-1). The observed value agrees very closely with Einstein's theory.

A second spectacular prediction was that a ray of light from a distant star, just grazing the sun, would experience a deflection of 1″.74, concave towards the sun. Normally the sun is much too bright for us to make these observations. However, at the time of a total solar eclipse—especially when the sun happens to lie in some reasonably rich star field—we have a chance of determining this deflection experimentally. Numerous independent observations at different eclipses since 1919 generally accord with the theory of relativity, within the limit of error of these difficult measurements. Incidentally, the Newtonian theory of gravitation predicts a shift of 0″.87, exactly half that required by the Einstein theory.

Third, the Einstein theory of relativity predicts that light originating in a gravitational field (that is, a strongly curved space) will be shifted to the red, because light vibrations will be slower in such a field. Tests of this shift, for the sun, have led to general good agreement with theory—though the quantity measured is small and is partially masked by convection effects in the solar atmosphere.

The companion of Sirius, a white-dwarf star (Secs. 21-6, 22-5), possesses a much more intense gravitational field at the photospheric level where the visible light is emitted. Hence—even though our observations cannot be as accurate as for the sun—we can at least determine the gravitational shift, after making due allowance for the Doppler effect that arises from the motion of the star itself relative to the primary Sirius A.

In all these tests, the agreement of observation with theory is satisfactory and has generally been regarded as convincing evidence in favor of the Einstein theory. This theory, like all theories, must rest its case on observation.

One of the important corollaries of the Einstein theory was the principle of equivalence between mass and energy, which we have discussed earlier in connection with atomic energy (Sec. 24-4). The verification of this law by modern atomic physics must be regarded as perhaps the most important proof of relativity. For relativity requires that the mass as well as the dimensions of an object change with velocity. The relation, which has been precisely verified by measuring the change with velocity of the specific charge, e/m, of a fast-moving electron, is

$$m = \frac{m_0}{\sqrt{1 - \beta^2}}. \tag{34-16}$$

It shows that the mass increases indefinitely as β approaches unity—that is, as the velocity v approaches that of light.

Particles moving in spiral paths in atom-smashers such as cyclotrons, synchrotrons, and so on acquire velocities approaching that of light. The energy that an ordinary cyclotron can impart to a charged particle is limited because it does not allow for the variability of mass with velocity. The synchrotron, on the other hand, makes a correction for this variation. Indeed, the fact that we can measure the change of mass with velocity is another definite proof of relativity.

The many independent checks of the validity of Einstein's theories have encouraged physicists and astronomers to apply it to the universe as a whole.

34-7   Einstein's Static Universe.   Einstein himself made the first attempt to apply the laws of general relativity to the universe as a whole. First published in 1916, his description marks the beginning of all modern theories of cosmology. In the current terminology the word "cosmology" applies to the highly speculative branch of astronomy that attempts to describe the general properties of the universe in space and time, and the kinematics and dynamics of matter and radiation in it on the largest scale. The rules of the game allow the use of all known basic laws of physics and even permit consideration of possible variations of these laws in times past and future. Although problems of formation and evolution come into such discussions, the word "cosmogony" is reserved for the more restricted problem of the origin, formation, and evolution of the solar system, stars, and galaxies. Such divisions, convenient but arbitrary, may change again in the future.

In order to discuss the universe on the largest scale and "as a whole" (even though we do not know *a priori* whether such an expression has any definite meaning), some drastic simplifications must be made. Clearly it is impossible to follow the motion and evolution of every particle of matter and radiation in the universe. Since we are not concerned with the fate of each atom, star, or galaxy, but with the overall or average properties of the whole system, we can use an equivalent "smoothed-out" picture where only average quantities intervene, say the average density of matter in a "large enough" volume of space, or the average energy content of such a volume. Here we merely extend procedures that have proved eminently successful in the kinetic theory of gases in classical physics.

Thus, Einstein represented the universe by a fluid formed of innumerable "molecules" (stars or galaxies) with an average density ρ and an average kinetic and radiant energy corresponding to a certain pressure p. Because in 1916 no evidence existed to the contrary, he further assumed (1) that the geometric structure of space (the "spatial metric") is independent of time—in other words, that on the largest scale the universe is stable, static, and permanent—and (2) that space and the large-scale distribution of matter in it are homogeneous and isotropic (it is very much the same everywhere and in all directions).

Applying the principles of general relativity to these assumptions, Einstein derived a basic relation expressing the "distance" (in space *and* time) or "interval" between two points or particles in the universe. This relation resembled that of ordinary Euclidean geometry giving the distance between two nearby points on the surface of a sphere. Because of this analogy, the Einstein universe is said to be *spherical*, characterized by a *radius* of curvature R. This radius is constant, independent of time, so that it is a *static* universe. Its volume is finite and has a definite value given by $2\pi^2 R^3$ (not $\frac{4}{3}\pi R^3$ as in Euclidean geometry, because of the different geometry of relativity). In this volume a finite mass, the total mass of the universe, is present corresponding to the average density ρ. The higher the density, the greater the curvature of space (Sec. 34-6), and conversely the smaller the radius of curvature. Here we see that only an *empty* static universe ($\rho = 0$) could have an infinite radius, and only then could it be Euclidean. The presence of matter forces the light rays to be curved, even though, because the average density of matter is so low (see below), the curvature is very slight indeed over distances as large as a million light years.

As we noted earlier, the universe is characterized not only by the average density ρ of the idealized gas, but also by its pressure p (including radiation pressure). However, a simple calculation shows that the force exerted by this pressure is negligible compared with the pull of gravitation, and the latter depends only on ρ and the known value of the constant of gravitation. It follows that the curvature of space and the radius R of Einstein's universe are solely determined by the mean density ρ, and numerically, in cgs units (cm for R, g cm^{-3} for ρ),

$$R^2 = \frac{10^{27}}{\rho} \tag{34-17}$$

(the numerical coefficient, equal to $4\pi G/c^2$, is completely determined by the constant of gravitation G and the velocity of light c).

Observations of galaxies (Sec. 32-12) suggest that in the largest observable volume $\rho \simeq 10^{-30}$ g cm^{-3}, so that $R = \sqrt{10^{57}} = 10^{28.5}$ cm $\simeq$ 10,000 megaparsecs. This is much (but not very much) greater than the distance of the most distant galaxies that can be photographed. There are, however, serious difficulties with this static universe, as discussed in Sec. 34-8.

Before turning to more complicated world pic
to a curious property of this closed universe. Sin
curved paths, they must come back eventually to t
least not far from it, since the light source has it
time). For an analogy consider an airplane flying
pole; if it keeps going "straight ahead" following
tually return to the north pole after circling the e
holds true if the plane flies along a great circle from any point on the earth.
At the halfway point of its journey, the plane overflies the antipode of its
starting point. It is then at the greatest possible distance from its origin,
exactly half the circumference of the earth. Similarly, the photons emanating from a light (or radio) source can only go as far as the antipode of
the source in Einstein's universe. Beyond, they converge back toward their
starting point. Thus, the greatest distance between two points of this
universe is πR, or about 30,000 megaparsecs with the figure given above
for R.

This figure does not greatly exceed the distances that can be vaguely
estimated for some quasars (Sec. 33-6). This result brings up a very interesting question. If the real universe (even if it differs from Einstein's early
model) has this property of *antipodes*, if light rays can go "round and
round," it is at least theoretically possible that images of the same object
could be seen in two diametrically opposite directions of the celestial
sphere. If instead of an airplane we were to send out a giant atmospheric
disturbance (perhaps by exploding an H bomb) from the north pole, the
air wave would eventually reach the south pole from all directions after
travelling along the meridians. This analogy, of course, applies only in
principle, for an idealized picture of the earth and its atmosphere. A
similar picture holds for the Einstein universe (and other closed models
sharing this basic property). If we happen to be exactly at the antipode of a
powerful distant cosmic source (say, a quasar), it is possible, at least in
principle, that its radiation will reach us from all directions at the same
time. In other words, its "image" will cover the whole celestial sphere and
we should not see it as a recognizable object. However, if we are not
exactly at the antipode but only close to it, then we may be able to see
the same object in two nearly (but not exactly) opposite directions by the
light emanating from it at two different but close epochs. That this strange
situation may occur in the actual universe (as opposed to the idealized
fictional models of cosmology) is a fascinating possibility, under active
consideration by some astronomers.

Whatever the outcome of this search, it is practically certain that it will
not lead astronomers to accept the original static universe of Einstein as
a realistic model of the physical universe. We now proceed to review the
difficulties of this model and to describe some of the many other possibilities suggested by theory and observation.

The Cosmological Constant and the Cosmic Repulsion. In order to achieve a stable static model of the universe as seemed to be required by the astronomical observations then available, Einstein had to overcome a fundamental difficulty: a universe full of matter but with comparatively negligible internal pressure cannot be stable. The force of gravitational attraction between galaxies, stars, and atoms, however small it may be, will eventually cause matter to collapse "under its own weight" into a much smaller and denser universe, until the outward-directed pressure forces become sufficient to balance the inward-directed gravitational pull. We have already discussed this situation *a propos* the equilibrium of gaseous layers in a star (Sec. 25-2 and 28-1).

In order to prevent such a collapse, we must postulate a counterbalancing force, which will be much more effective at large distances than the negligible pressure forces. But it should not be significant on the scale of, say, the solar system, where no cosmic force other than gravitation can be detected. This *ad hoc* force is the cosmic repulsion represented in the relativistic equations by the so-called "cosmological constant" Λ (just as gravitation is present by its constant G). The introduction of this constant, which is not without mathematical justification (there is more to it than a mere "fudge factor"), has caused endless argument among theorists.

Einstein himself later repudiated it, but others thought it to be an essential logical requirement of the theory. Be that as it may, Λ, representing a weak cosmic repulsion significant only over great distances, provided the necessary force to exactly balance gravitation for the universe as a whole and so to permit Einstein's static model.

The mathematical development gave for the required value of the cosmological constant

$$\Lambda = \frac{1}{R^2} = \frac{4\pi G}{c^2}\,\rho, \tag{34-18}$$

where G is the constant of gravitation and c the velocity of light. (In cgs units $\Lambda \simeq 10^{-27}\,\rho$, whence $R^2 = 10^{27}/\rho$ in the static model.)

Even if we ignore the arguments over the introduction of Λ, another serious difficulty occurs, which later discussions brought to light. The static universe of Einstein *is not stable*. The static solution requires a rigorously exact balance between Λ and G, or between ρ and R, as we can see from equation (34-17). The slightest fluctuation in the density of matter or in the small but nonzero pressure forces will cause an instability that natural forces cannot counterbalance. The Einstein universe is in unstable equilibrium—and, as we know, no such situation can last very long in the real world. Even if we contrive to produce an exact balance between gravitation and cosmic repulsion in the initial condition of an artificial universe in the distant past, sooner or later one of the two forces will prevail in their tug of war and the universe will either give in to gravitation and collapse to a denser state, much denser than that presently observed, or it will succumb to the universal repulsive force and expand indefinitely, ever and ever faster into nothingness.

All the effort to stabilize the Einstein model came to nought. Other, more complex solutions were needed.

34-9 The De Sitter Universe.

Proposed by the Dutch astronomer in 1917, the De Sitter universe was another early and vain attempt to find a static solution of Einstein's equations without recourse to the objectionable introduction of an *ad hoc* cosmic repulsion force. Even in the idealized world of cosmological models it calls for a rather drastic solution to the problem of gravitational collapse. Since gravitation must be present if any matter exists in our universe, obviously gravitation will be absent and no collapse will occur if the universe contains no matter. More precisely, a static solution is possible if $p + \rho = 0$. One could possibly balance the necessarily positive density only with negative (inward) pressure forces, but it is difficult to find a plausible mechanism for such forces. Hence, we can only have both $\rho = 0$ and $p = 0$. If the De Sitter universe is indeed a stable, static solution of Einstein's equations, it is at an excessive price. It is completely devoid of matter and radiation! An empty universe, a pure vacuum—at first sight not a good model for the world we live in!

Nevertheless, it would be unfair to conclude that Einstein and De Sitter were indulging in idle doodling with empty mathematical symbols. Rather, we can regard their idealized models as extreme limiting cases of the real universe. After all, observation shows that the astronomical universe is indeed *almost* empty. As we estimated earlier, counts of galaxies indicate in our larger neighborhood (say, within one billion light years) an average density of only 10^{-30} gram per cubic centimeter (or one hydrogen atom per cubic meter). This is enormously more "empty" than our best laboratory "vacuum" and indeed than interstellar space within our own galaxy.

The metric of the De Sitter universe differs significantly from that of Einstein. In it, time progresses at different rates at different distances from the observer. We may say that time is "curved." In particular, the time interval between two events, say the period of an atomic vibration, appears greater to a distant observer than to one close to the source. Thus the spectral lines of a distant source will appear displaced to a greater wavelength than the corresponding lines of a similar nearby source. De Sitter wondered whether the displacement of spectral lines toward the red observed in the spectra of spiral nebulae could not be the result of this effect.

The relation became clearer when, within a few years, mathematical developments proved that the De Sitter universe is not really stable and observation established the generality of redshifts in galaxy spectra.

CLUSTER NEBULA IN	DISTANCE IN LIGHT-YEARS	RED-SHIFTS

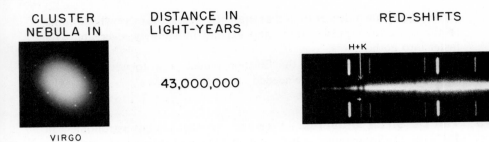

VIRGO

43,000,000

H+K

URSA MAJOR

560,000,000

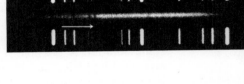

CORONA BOREALIS

728,000,000

BOOTES

1,290,000,000

HYDRA

1,960,000,000

FIGURE 34-5

Redshifted spectra of galaxies. (Mount Wilson Observatory.)

34-10 The Expansion of the Universe and the Law of Redshifts.

The "island-universe" concept, that spiral nebulae are external galaxies at enormous distances, was finally established between 1917 and 1924 (Sec. 32-1). During the same period the radial velocities of some forty galaxies were measured, mainly through the work of V. M. Slipher at Lowell Observatory. Most of these velocities were large (1000 km/sec and more) and positive—that is, *recession* velocities.

The idea of relating recession velocity V and distance Δ, encouraged by De Sitter's theory, was explored by several astronomers. In 1925, K. Lundmark of Sweden searched for a parabolic relation, with inconclusive results because the methods then available gave unreliable estimates of extragalactic distances. A definite, linear relation

$$V = H \cdot \Delta \qquad (34\text{-}19)$$

was finally established beyond doubt by E. Hubble at Mt. Wilson Observatory in 1929, thanks mainly to his improved distance estimates (cf. Sec. 32-4).   Modern estimates of the constant H are on the order of 100 kilometers per second per megaparsec.

This formula was soon confirmed by additional spectral observations of ever larger velocities, determined mainly by M. L. Humason at Mt. Wilson and later by N. U. Mayall at Lick Observatory. In 1935, the velocity-distance relation had been extended out to distances corresponding to recession velocities of 20,000 km/sec (Fig. 34-5). When the 200-inch telescope came into operation, fainter, more distant galaxies and much greater velocities could be observed, first up to 60,000 km sec^{-1}, then 120,000 km/sec, and eventually 150,000 km/sec, half the velocity of light (Fig. 34-6). Then the relative spectral displacement between the normal wavelength λ_0 and the observed wavelength λ is

$$z = \frac{\lambda - \lambda_0}{\lambda_0} = 0.5. \qquad (34\text{-}20)$$

The wavelength is increased by 50 per cent. For example, the H and K lines of ionized calcium, the most easily observed in galaxy spectra, are displaced from a mean $\lambda_0 \simeq 3950$ Å in the extreme violet to $1.5\lambda_0 \simeq 5900$ Å in the yellow part of the spectrum. Since all wavelengths are displaced in the same proportion, the spectrum as a whole suffers a shift toward the red, and the color of the galaxy turns redder. The color index of an elliptical galaxy, normally close to $+1.0$, increases to about $+1.4$ at $z = 0.1$ ($V = 30,000$ km/sec), $+1.8$ at $z = 0.2$, and so on. Most of the light originally emitted in the visible region is shifted into the infrared, while the normally unobservable ultraviolet appears in the blue and violet regions. This is the reason for the picturesque if not accurate description of the effect as a "redshift."

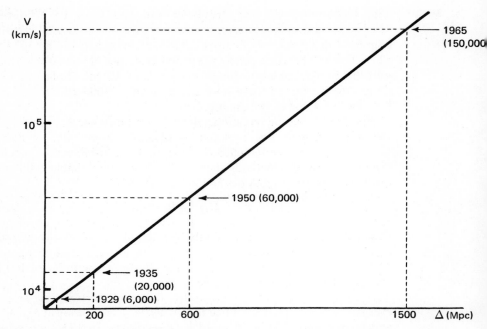

FIGURE
34-6 *Velocity-distance relation, showing largest velocity of galaxies known at various times.*

The expression is useful for at least two reasons. First, the simple Doppler relation (Sec. 8-14) used to convert spectral shift $\delta\lambda = \lambda - \lambda_0$ to velocity of recession V

$$z = \frac{\delta\lambda}{\lambda_0} = \frac{V}{c} \tag{34-21}$$

is valid only as a first approximation when the spectral shift is small. At large velocities a modified relativistic formula must be used (actually different formulas can be derived for different world models and different methods of measuring distances), and the intuitive interpretation of the velocity shift fails. In the theory of special relativity the relation is

$$1 + z = \sqrt{\frac{1 + V/c}{1 - V/c}}, \tag{34-22}$$

so that $V \to c$ when $z \to \infty$ while the elementary Doppler formula (34-21) would indicate $z = 1$ for $V = c$).

The second reason is that as newly measured recession velocities of still fainter and more distant galaxies kept increasing year after year, some scientists became uneasy about the very principle of interpreting the spectral shifts as caused by a Doppler effect. Are such enormous velocities really possible? Is it permissible to apply the Doppler formula (or its relativistic counterpart) so far beyond the familiar range of stellar velocities? Is it not possible that some new physical principle is at work over such great distances in space and time? Many suggestions were offered, but none has yet been able to replace the classical Doppler interpretation.

For instance, measurements of different spectral lines from the ultraviolet to the red in the spectra of galaxies with strong emission lines check precisely the condition that $\delta\lambda/\lambda_0 = $ constant, as required by the Doppler formula. Radio observations of redshifts of nearby galaxies by means of the 21-cm line of H I agree exactly with optical values and extend the verification with even greater precision over an enormous range of wavelengths.

Thus most astronomers agree that we have no choice but to accept the evidence at face value. The redshift is a velocity shift, and the galaxies are really moving away from us in all directions at the great velocities indicated, increasing in proportion to their distances. This general *expansion of the universe* is the single most important phenomenon in our modern picture of the world.

At first sight our view of the universe appears to be strangely geocentric and pre-Copernican. In all directions around us, wherever we see a galaxy in the sky, it is apparently receding from us as if the whole universe were exploding away with us at the center. A moment of reflection shows that this geocentric view is an illusion. Any other observer located in any other galaxy would observe the same phenomenon and be subject to the same illusion. It is space as a whole that is expanding everywhere and in all directions; there is no privileged center.

Consider the following classical analogy. Ants are scattered on the surface of a large rubber balloon, which is slowly and continuously inflated. The distance measured along the surface of the balloon, between any pair of ants A_1, A_2 increases slowly as the radius $R = R(t)$ of the balloon grows and in direct proportion to the distance $R(t)\chi$, if χ is the angular separation between A_1 and A_2 measured along the great circle through A_1, A_2. Apart from the effects of small individual wanderings ("proper motions"), any ant such as A_1 will observe that all other ants are moving away from it and at a speed increasing in direct proportion to distance. Our astronomical ant will observe a general expansion of its two-dimensional universe (the surface of the sphere) in our three-dimensional space. Looking on from our three-dimensional vantage point, we can readily see that ant A_1 is in no privileged position and is certainly not at the center of the expansion. Any other ant will also observe a general recession of its neighbors in the ant's "expanding universe." A similar situation obtains in the four-dimensional space-time continuum of general relativity. The galaxies are the three-dimensional "ants" in a four-dimensional universe, and the three-dimensional space accessible to our experience expands uniformly in all directions from all its points, none of which is more central than any other.

By analogy with the inflated sphere, the "radius" or scale factor of the universe $R(t)$ varies with time. Most of the recent studies have been concerned with three basic problems:

1. What is the present expansion rate in our neighborhood, as measured by the Hubble constant, $H = V/\Delta$, expressed usually in kilometers per second per megaparsec?

2. What is the exact form of the relation between V and Δ? Is the velocity strictly proportional to distance, or is the linear Hubble relation merely a first approximation valid over a relatively small range of distances?

3. What model of the universe agrees best with the observed velocity-distance relation and all other known properties of the extragalactic world?

The first question really amounts to measuring the distance Δ of galaxies or clusters of galaxies, since V is relatively easy to measure with precision on spectrograms. We have already discussed (Sec. 32-4) the methods used to measure extragalactic distances. As we know, a drastic change was introduced in the 1950's when the confusion between type I and type II Cepheids was cleared up. Accordingly the expansion constant, evaluated at $H = 550$ km/sec per Mpc by Hubble in 1936, was cut down first to 200, then to 100; finally, values as low as 50 km/sec/Mpc were even suggested. Currently most estimates fluctuate between 50 and 150 km/sec/Mpc. Many astronomers like to take $H = 100$ so that the distance of a galaxy in megaparsecs is roughly equal to 1 per cent of its velocity in kilometers per second. This, of course, is nothing more than a convenient, but rough rule of thumb, subject to further revision.

The value of the Hubble constant plays a central role in cosmology, because its inverse has the dimension of a time that is often associated with the "age" of the universe (Chapter 35). This relationship will appear more clearly in Sec. 34-11.

The second question concerns the possible departure from linearity of the Hubble relation. As early as 1925 Lundmark searched for a quadratic relation, but subsequent studies showed that the velocity-distance relation is very nearly linear over a very great range of distances. At small distances, say $\Delta < 50$ megaparsecs, a conspicuous departure from linearity in the northern galactic hemisphere, first pointed out by G. de Vaucouleurs in 1957 and further analyzed in 1968, results from the internal kinematics of the Local Supercluster of galaxies (Sec. 32-10). It possesses no cosmological significance (50 megaparsecs is an almost negligible distance in cosmology!), except insofar as it introduces an error or bias in all estimates of H based on the velocities and distances of groups of galaxies that are part of the Local Supercluster.

At very large distances the difficulty lies in the very definition of distance. In fact, distance can be defined precisely only within the context of a model universe and its particular metric. Since we are trying to determine which

model best fits the actual universe, we do not know *a p[*
measure the distance! At this point the third question become[

Consideration of the velocity-magnitude relation somewhat
difficulty. Since apparent magnitude is obviously related to d〔....〕ce, we
can use magnitude as a measure of distance. The advantage is that we can
directly measure magnitudes, but not distances. However, the observed
magnitudes require correction for various effects, which in turn vary
according to the model. We have no choice but to adopt each model in
turn as a *working hypothesis* and to pursue its consequences to the bitter
end to see whether the observed data interpreted in its framework agree or
disagree with expectations. We shall return to this problem of comparisons
of models with observations in Sec. 34-14, but let us first survey the wide
range of possible models to choose from.

34-11   Expanding-Models of the Universe.   The expanding universes of
Lemaitre and Eddington were the first models in which the law of redshifts
appears as a fundamental property of the universe. The first nonstatic
models were investigated theoretically by the Russian mathematician A.
Friedmann in 1922 and by the Belgian priest-astronomer G. Lemaitre in
1927, before the empirical discovery of the universal redshift. However,
only after Hubble's observation of the recession of galaxies did astronomers—in particular, A. S. Eddington in England—become actively interested in expanding models. Lemaitre, recognizing the inherent instability
of Einstein's universe, decided that the present universe, whether expanding or possibly contracting (theory gave no hint concerning which way the
instability would be resolved), may represent a transition phase in a continuous evolution, from an initial state similar to the Einstein universe (in
the distant past) to an ultimate state similar to the De Sitter universe (in
the indefinite future). We can illustrate this model, favored by Eddington,
by a plot of the function $R(t)$ giving the evolution of the "radius" of the
universe as a function of time (Fig. 34-7). The expansion rate is the rate
at which R increases per unit time interval and is proportional to the slope
of the tangent P_0P to the curve. The subtangent P_0P_1, which is inversely
proportional to the slope (that is, to the Hubble constant, is a time interval of great significance. With the currently preferred value $H =
100$ km sec^{-1} Mpc^{-1}, $P_0P = 1/H \simeq 10^{10}$ years. This time scale bears
some relation to the "age of the universe," the time since the Einstein
universe began its expansion. The precise relation depends on the details
of the model and on the ratio between the present radius R and the initial
R_0. As appears also in the figure, if $R(t)$ increases indefinitely, the volume
of space will also increase indefinitely and the density of matter will decrease indefinitely, approaching more and more closely the idealized empty
universe of De Sitter.

**FIGURE
34-7**

Variation of the radius of the universe with time in the Lemaitre-Eddington model. The tangent PP_0 intercepts a segment P_0P_1 measured by the reciprocal of the Hubble constant.

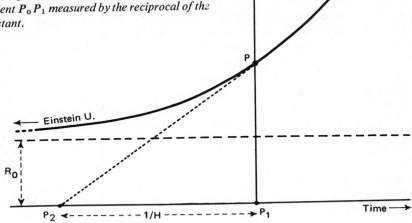

We shall see in our final chapter (Chapter 35) how this initial concept of Lemaitre and Eddington was soon modified by Lemaitre himself to include physical considerations, yielding a model of the origin and evolution of the universe known as the "primeval atom" or "big bang" theory.

For the present we merely point out the great variety of models that can be built around this general concept. These models vary according to the *signs* of the cosmological constant and the radius of curvature determined by the constant of gravitation and density of matter. Thus we can have contracting or expanding models, or even oscillating models (Table 34-1).

**TABLE
34-1**

Classification of homogeneous model universes of general relativity

Cosmological Constant	Curvature			Cosmic Force
	Negative	Zero	Positive	
Negative Zero Positive	Oscillating Expanding Expanding	Oscillating Expanding Expanding	Oscillating Oscillating Oscillating or expanding	Attraction Repulsion
Type of universe	Open, infinite	Euclidean, infinite	Closed, finite	

Some, having a negative or zero curvature, are open (including the Euclidean model of classical astronomy if the curvature is zero). Some with
positive curvature are closed and have a finite volume (Fig. 34-8). Of special
interest is a model, favored by Lemaitre, in which the universe begins at
some definite epoch with a very small radius and expands at first rapidly
(explodes would be a more appropriate description), then more slowly,
because in a small dense universe gravitation is dominant and slows down
the initial expansion.

**FIGURE
34-8**

*Families of models of the universe, including
the Lemaitre model (heavy line), with positive
constant, λ. The distance Ro represents the
distance of the Einstein universe.*

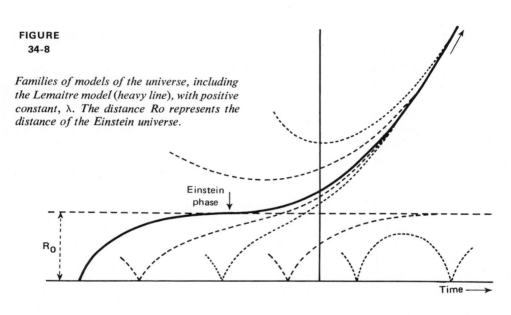

If Λ has the value Λ_0 appropriate to the Einstein universe, this model
will come almost to a standstill when its radius reaches the value R_0 of
the Einstein universe. How long the universe can remain in the precarious
balance between gravitation and cosmic repulsion is not specified by the
theory, since the decision depends on fluctuations in the density of matter
that are not considered in all idealized *homogeneous* models. Physical
intuition tells us that this transition phase cannot last very long, and observation shows that the force of cosmic repulsion won the contest (if not,
the universe would collapse back to minimum radius and possibly oscillate
indefinitely between almost 0 and R_0). Beyond this point, expansion takes
over and at an increasing rate, as in Fig. 34-8. As we shall see, physicists
like this model because the initial high densities offer a chance for nuclear
reactions to form chemical elements from hydrogen. Many astronomers like
it because the slow passage through the phase of a quasi-Einstein universe
gives a chance for fluctuations to form stars, galaxies, and clusters of
galaxies (Sec. 35-5).

However, a scientific question should not and will not be resolved by considerations of convenience or preference. We must explore all possibilities, all plausible views, and systematically check their predictions and properties against observations for possible points of agreement or disagreement.

Let us, then, continue our rapid survey of cosmological speculations to include some of the more recent views much debated during the past twenty years.

34-12 The Steady-State Universe of Bondi, Gold, and Hoyle. All attempts to build models of the universe are based on explicit or implicit assumptions or principles: homogeneity, isotropy, conservation of energy, and so on. Overriding is the so-called *cosmological principle*: "the universe looks very much the same from any location and in all directions." There is no privileged position in *space*. This principle, however, overlooks *time*; models based on it have different properties at different times. For example, the Lemaitre-Eddington models evolve from a dense past to a tenuous future. Cosmic evolution is a one-way street, and the present state is unique. A school of British theoreticians began in 1948 to question the validity of this hidden assumption of previous relativistic models.

If the universe is homogeneous and isotropic in space, why not also in time? And so they completed the cosmological principle by the added requirement ". . . and at all times." On the largest scale everything is very much the same anywhere, anytime. There is nothing unique in here and now. The universe was and will always be as we now see it. This assertion, which they dubbed the "perfect cosmological principle," is in a sense a logical extension of the Copernican view of the universe: neither our place nor our time is in any way remarkable.

This apparently innocent addition to the cosmological principle leads to some drastic conclusions; in particular, it leads to the rejection of another fundamental assumption, the principle of conservation of mass and energy. Observation shows that the universe is expanding. If the total mass and radiation are constant, its density must decrease with time as the volume of space increases. But then the future is not equivalent to the past. In other words, instead of assuming that mass (energy) is a constant, we may assume now that the density ρ is a constant. To maintain a constant density in an ever-expanding universe, matter must be created in space at a rate precisely compensating the rarefaction caused by the expansion.

The postulated spontaneous creation of matter *out of nothing* is the most disturbing feature of this theory of Bondi, Gold, and Hoyle. But it is a logical deduction from the hypothesis that the universe is simultaneously expanding and in a steady state. That this creation of matter violates the principle of conservation of mass and energy is not in itself a serious objection. Just as any other principle of physics, the conservation principle is justified by experience and with a precision that is quite limited. No one has verified it to, say, one part in 10^{12}. The creation rate required by the steady-state theory is much less than that. The universal expansion rate causes a given volume of space to increase each second by a fraction given by the Hubble constant and, according to Hoyle, the creation of just one atom of hydrogen per cubic foot every few *billion* years would be enough to keep ρ constant. No conceivable experiment could ever detect the stray brand new nucleus. Thus, we cannot discard the steady-state theory just because of this infinitesimal departure from the law of conservation.

The theory has some attractive features. Hoyle has shown that the metric of the steady-state universe is similar to that of the De Sitter universe, but, of course, it is not empty and it is infinite in space and time. The steady state model also leads to some definite predictions that may be compared with observations. In particular, distant galaxies should not be younger than nearby ones, as in the Lemaitre-Eddington models, and some very old galaxies should be present, much older than allowed by the Lemaitre model (Sec. 35.7). Recent observations of cosmic evolution and counts of radio sources tend to disagree with the predictions of the steady-state model. If it is finally disproved, it will at least have played a useful role in forcing scientists to question and investigate more deeply the basic principles of physics and cosmology.

34-13   Olbers' Paradox and the Redshift. Some of the most perplexing questions of astronomy are often raised by some of the simplest observations. For instance, one may ask, as did the German astronomer Olbers in 1826, "Why is the sky dark at night?" It is not a silly question if, as astronomers believed at the time, the universe is infinite and Euclidean. For then, Olbers reasoned, any line of sight, wherever an observer may direct his gaze, should eventually cross the surface of a star, and the heavens would uniformly have the brilliance of the sun. Absorbing matter in space will not help because, given enough time, it will reach an equilibrium with the radiation field and shine as brightly as the stellar photospheres themselves. Not only light, but gravitation, obeying the same inverse-square relation to distance, will raise a similar problem.

This paradox was eventually resolved by the discovery of the universal redshift, for as the wavelength λ received by an observer is increased, and the frequency reduced, the energy carried by the photon is also reduced and in the same proportion, by virtue of Planck's equation (Sec. 8-1). If

$$E_0 = h\nu_0 = \frac{hc}{\lambda_0} \tag{8-2}$$

is the energy emitted, the energy received from a source having a redshift $z = (\lambda - \lambda_0)/\lambda_0$, is

$$E = h\nu = E_0 \frac{\nu}{\nu_0} = E_0 \frac{\lambda_0}{\lambda} = \frac{E_0}{1 + z}, \tag{34-23}$$

and $E < E_0$, going to zero as z increases indefinitely. Relation (34-23) describes the "energy effect" of the redshift.

Not only is the energy content of each photon reduced by the universal redshift, but the expansion also reduces the space density of photons.

FIGURE
34-9
*The expansion of the universe reduces
the space density of photons.*

Consider a beam of light from a star in a distant galaxy. In one second it emits N_0 photons, which travel to a distance c from the star and fill a volume of length c (Fig. 34-9). Now if we observe this same star from earth, in the expanding universe, the distance between us and the source increases each second by $v = cz$, so that photons emitted in one second by the source are now spread over a length $c + v$ (Fig. 34-9). The number of photons per unit volume is inversely proportional to length, hence to $c + v$, and

$$\frac{N}{N_0} = \frac{c}{c + v} = \frac{1}{1 + z}. \tag{34-24}$$

Equation (34-24) describes the "number effect." Its expression is identical to equation (34-23).

The two effects reinforce each other. Reducing the density and energy of photons received from distant galaxies neatly resolves the Olbers paradox. Hence, some theorists have argued that the expansion of the universe could have been predicted from the mere fact that the sky is dark at night!

This, however, is not true, for in 1914 the Swedish astronomer Charlier had found a way of avoiding the Olbers paradox in the framework of Euclidean geometry and a static universe. He showed that the sum total of the luminosities and attractions of all celestial bodies will not increase indefinitely even in an infinite static universe, provided matter is clumped in a hierarchy of ever larger cosmic systems, star clusters, galaxies, clusters of galaxies, superclusters, and so on, and if a definite mathematical inequality relates the size or spacing and populations or masses of systems of successive orders. By such an arrangement an infinite number of cosmic masses could add up to a finite energy flux and attraction (it could even add up to negligible quantities, given enough empty space between systems).

The discovery of the redshift removed the incentive for such a solution of Olbers' paradox. This does not mean that Charlier's view of the universe is irrelevant. In fact, the abundant evidence for clustering of galaxies in an apparently open-ended series of ever greater scale and scope (Sec. 32-10) indicates that *clumpiness* is a basic property of the distribution of matter in the universe—one that theoretical cosmologists had generally neglected until recently. Even though a smoothed-out density was a justifiable mathematical simplification for a preliminary survey of cosmological models, the *fluctuations* of this density may well be the major factors in determining the past, present, and future behavior of matter and radiation—in other words, the evolution of the universe. This topic will be discussed in our final chapter.

34-14   Tests of Cosmological Models. A formidable array of possible worlds has been proposed by theoretical cosmologists (many even more fanciful than the ones we have mentioned). The problem of narrowing down the choice to a few models and, eventually, identifying the type of world we live in is not a simple one. Too often the predictions of the model are too vague or the observations are not yet good enough to distinguish between various predictions.

For the testing of cosmological models several tools are available.

(a) *The Redshift-Magnitude Relation for Galaxies in Clusters.* This is the most direct test, but it is not very critical. The test centers on the presence or absence of depatures from linearity at large values of z, on the graph of z vs. magnitude m (Fig. 34-10). Different models make slightly different predictions according to the presence or absence of an acceleration (or deceleration) in the expansion. The departures are only slight in the range accessible to observations (Fig. 34-10), and various uncertain corrections that must be applied to the measured magnitudes render the test inconclusive. These corrections are required to allow not only for extinction in our galaxy (Sec. 29-2) and for the energy and number effects (Sec. 34-13), but also for the fact that we cannot measure the total energy from a galaxy or its bolometric magnitude (as defined in Sec. 19-7) but only its photographic or photovisual magnitude. The correction from m_{pg} to m_{bol} depends on the spectral energy distribution curve (Sec. 8-7), of which only a small part is accessible to observation. Then instrumental effects (seeing, photometer aperture, and so on) complicate the problem further still. Finally the total effect or K-correction is highly uncertain.

As if this were not enough, we have to worry about problems of selection. Can we pick strictly comparable galaxies in nearby and distant clusters? Is, say, the brightest galaxy in a cluster a standard candle? And, finally, we encounter problems of evolution. When we observe very distant galaxies, say, 10^9 light years away, we see them as they were a billion years ago. Is it not possible that their stars were then brighter on the average than similar stars in nearby galaxies?

The problem becomes inextricable, and astronomers have turned their attention to other observational tests of cosmological models.

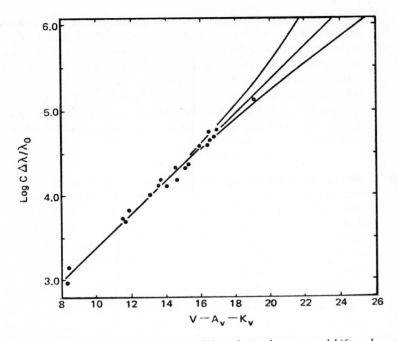

FIGURE 34-10 *Possible departures from linearity of the relation between redshift and magnitude, predicted by different models, are too small to be detected by present-day observations of clusters of galaxies. (A. R. Sandage, Mount Wilson and Palomar Observatories.)*

(b) *Galaxy Counts.* Consider, for a moment, galaxies scattered at random in a homogeneous, isotropic Euclidean universe with an average density of N_1 galaxies per cubic megaparsec. The number of galaxies counted on photographs, N, will increase in direct proportion to the volume explored (with Δ^3, if Δ is the average distance of the faintest galaxies photographed), so that $N \propto N_1 \Delta^3$ or $\log N = \log N_1 + 3 \log \Delta + \text{const}$. But the fundamental relation (19-6) between distance and apparent magnitude requires that $\log \Delta = 0.2m + \text{const.}$, so that finally

$$\log N = \log N_1 + 0.6m + \text{const.} \qquad (34\text{-}25)$$

If we count galaxies over a large enough area of the sky (near the galactic poles) to successively fainter magnitude limits, the slope of the graph of $\log N$ vs. m (Fig. 34-11) should be 0.6 in a homogeneous, static Euclidean universe.

If it is not 0.6, either the universe is not homogeneous, in our neighborhood, or we can try other models. In actuality, relation (34-25) is not obeyed by galaxies brighter than $m \simeq 15$ because of the density excess associated with the Local Supercluster (Sec. 32-10), but it is fairly well verified in the range $15 < m < 18$ according to galaxy counts made at Harvard Observatory in the 1930's and 1940's by H. Shapley and his colleagues. However, we need to go to fainter limits to reach distances at which differences between cosmological models may become significant. Except for the Lick Observatory Sky Survey to a limiting magnitude $m = 19$ (Sec. 32-10), no such counts have been made since World War II (a prewar survey by Hubble is no longer accepted because of uncertainties in the magnitude scale), and the value of galaxy counts for cosmological tests remains in doubt.

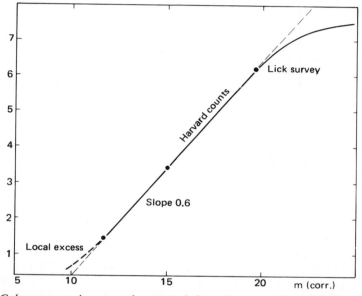

FIGURE
34-11   *Galaxy counts do not reach magnitude limits (corrected for redshift) faint enough to detect possible departures from the slope 0.6 appropriate for Euclidean space.*

(c) *Radio-Source Counts.* The main difficulty encountered in our quest for a critical test for model universes is that ordinary galaxies cannot be reliably observed out to distances large enough to cause significant differences between various theories. We need objects that can be detected and counted out to much larger distances. Such objects may well be found among radio sources, in particular among the fainter and presumably more distant sources associated with quasi-stellar objects (Sec. 33-6). Counts of radio sources to successive limits of flux density S have been secured since 1955 by English and Australian radio astronomers with initially

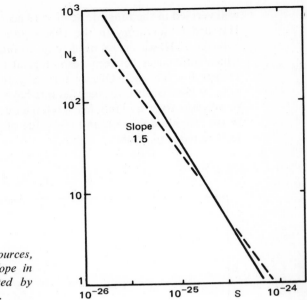

**FIGURE
34-12**

Relation between number of radio sources, N, *and flux density,* S, *has a slope in excess of the value 1.5 predicted by Euclidean and steady-state models.*

conflicting results. While the Australian data on the brighter sources were consistent with the Euclidean slope of 0.6 per magnitude (or 1.5 per unit in log S), the English data on the fainter sources gave larger values (up to 2 or more). Finally, after allowance for various statistical and instrumental effects, agreement was reached that the slope is about 1.8 (Fig. 34-12) over a large range of apparent intensities, but again this slope may vary with wavelength and, if so, which one do we pick?

This figure, of course, exceeds that (1.5) predicted from the simple Euclidean model and from the steady-state model. But how are we to interpret this result? Is it an effect of the metric (the geometry of the model), or of the redshift on the still poorly known energy curve of the sources? Or again, is it an index of evolution? If the most distant radio sources are intrinsically powerful quasars, they lie at distances of several billion light years and have values of z greater than 2. How can we be sure that their power output was the same billions of years ago? Is it not possible, then, that the excess of faint sources implied by a slope greater than 1.5 might simply reflect the fact that radio sources were more powerful when the universe was younger?

And so again we encounter the great problem of the evolution of the universe, of the past and future states of matter and radiation in it, of the birth and death of stars and galaxies. This problem is the topic of the next and final chapter of this book.

CHAPTER *35*

Cosmic Evolution
and Time Scales

35-1   Age and Evolution.   Throughout this rapid "survey of the universe" we have met many examples of variation and evolution in cosmic bodies. Comets, variable stars, galaxies, radio sources, and even the universe as a whole experience irreversible changes taking place over intervals that can serve to define cosmic time scales. Analogy with events in the realm of biology tempts us to assume that such periods somehow measure the *age* of the object or system. However, when we speak of "age" in the cosmic sense, we must take care that the word has an interpretable meaning. We shall assume, as we did in discussing the evolution of the solar system (Chapter 17), that the physical laws governing the interactions of matter and radiation have not changed over the time intervals we discuss. This assumption in itself immediately describes a certain state of ignorance. We do not, in fact, have much evidence beyond the pleochroic halos (Sec. 17-1) that these laws have not changed over the eons. We believe, but cannot prove, that the frequencies of spectral lines and atomic properties have remained constant for several billion years. At least the wavelengths and frequencies observed in the spectra of distant galaxies both in the long waves of the radio hydrogen line at 21 cm, and in the optical region, though shifted to longer wavelengths, are *relatively* unchanged. On the other hand, it is conceivable that the redshift of galaxies (Sec. 34-10) might result from changes in the laws or "constants" of physics over long periods of time or great distances, so that our time scales could be in error when applied to a distant past. If so, our "age" estimates may be wrong.

This reasoning warns us of the danger in using expressions such as "the age of the universe," a phrase that has meaning only in a restricted class of cosmological models and cosmogonical hypotheses. In some models— for example, the steady-state model—age of the universe is completely meaningless, because the steady-state concept implies an infinite age for the universe as a whole. Another common error is to estimate the time elapsed since the formation of a certain class of celestial bodies, say planets or globular clusters, and to describe the result as a determination of the "age of the universe." It is nothing of the sort.

Suppose an explorer from another world watches New York City and its denizens over a period of only one year, particularly recording its population, birth rate, death rate, and other pertinent facts. He may perhaps correctly conclude that the average age of the New Yorkers is thirty years, but of course he would be wrong if he concluded that the age of New York City was also thirty years. Or he may incorrectly use an idealized theoretical model and argue that since the excess of the birth rate over the death rate is currently, say, 1 per cent per year, the population of New York grew exponentially from just one couple ($N = 2$) to its present number of $N = 8 \times 10^6$ in (log 8.10^6 $-$ log 2)/log 1.01 = 1528 years, and again he would be grossly misled if he reported that calculation as an estimate of the age of New York City. The lesson is clear. Even if we can correctly evaluate the average ages of planets, of stars and star clusters, of our own and other nearby galaxies, and so on, the results may have little relevance to the question of the "age" (if any) of the whole universe. For all we know (or rather don't know), different parts of the universe may have different "ages" or no ages at all.

Since our view may strike the student as excessively skeptical, let him consider a second, nonfictional illustration. We shall merely record modern man's successive estimates of the "age" of the World (Fig. 35-1). Beginning with the biblical estimates of "over 4000 years" since the "creation" of the World (as late as the year 1658, Archbishop J. Ussher asserted that this interesting event took place on Sunday, 23rd October, 4004 B.C.), we note late nineteenth-century estimates of the "cooling time" of the earth (20 to 40 million years after Lord Kelvin) and the formation of the sun by gravitational contraction of Laplace's nebula (50 million years after Helmholtz) (Chapter 17). These estimates, which did not provide sufficient time even for the recorded geological and biological evolution, were revised upward several times after the discovery of radioactivity. Around 1930 radioactive dating of the oldest rocks had pretty well settled around a figure of 2 billion years, which, by some strange coincidence, happened to agree almost exactly with the reciprocal of the Hubble constant, then estimated at 1.8×10^9 years and widely (mis)interpreted as a measure of

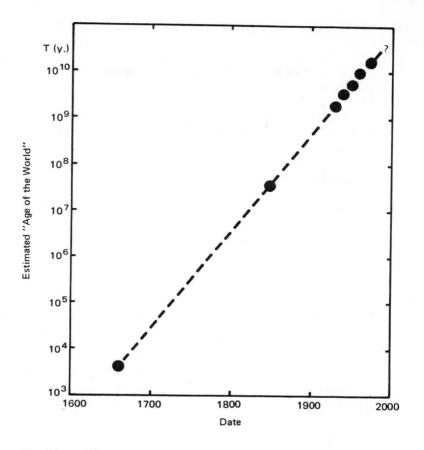

FIGURE
35-1 *The "Age of the World" has been frequently estimated during the past three centuries.*

the "age of the universe" (see Sec. 34-11). Modern estimates by improved methods of radioactive dating give 4.7×10^9 years for the age of the oldest rocks in the earth and meteorites (Sec. 35-2), while postwar revisions of the Hubble constant (Sec. 34-10) have conveniently upped its inverse first to 5×10^9 and now 1 to 2×10^{10} years (Fig. 35-1).

Is this, then, the correct and definitive value for the age of the universe? We doubt it. Estimates of the ages of the oldest stars in the Galaxy run currently around 20×10^9 years, and the Hubble constant is resisting all efforts to cut it down by another factor of two. We leave it to the reader to decide whether the graph in Fig. 35-1 will suddenly stop rising and be arrested at the level reached in A.D. 1970 or whether further changes will result from the unending pursuit of the (perhaps vain) question of the "age" of this great and mysterious universe.

TABLE 35-1 *Radioactive isotopes*

Element	Isotope	Half-Life	Emission Product
Uranium-238 Series			
Uranium	$_{92}U^{238}$	4.51×10^9 y	α
Thorium	$_{90}Th^{234}$	24.1 d	β
Protactinium	$_{91}Pa^{234}$	1.18 m	β
Uranium	$_{92}U^{234}$	2.48×10^5 y	α
Thorium	$_{90}Th^{230}$	8.00×10^4 y	α
Radium	$_{88}Ra^{226}$	1622 y	α
Radon	$_{86}Rn^{222}$	3.823 d	α
Polonium	$_{84}Po^{218}$	3.05 m	α
Lead	$_{82}Pb^{214}$	26.8 m	β
Bismuth	$_{83}Bi^{214}$	19.7 m	β
Polonium	$_{84}Po^{214}$	1.64×10^{-4} s	α
Lead	$_{82}Pb^{210}$	21 y	β
Bismuth	$_{83}Bi^{210}$	2.6×10^6 y	β
Polonium	$_{82}Po^{210}$	138.4 d	α
Lead	$_{82}Pb^{206}$	stable	—
Uranium-235 Series			
Uranium	$_{92}U^{235}$	7.13×10^8 y	α
Thorium	$_{90}Th^{231}$	25.6 h	β
Protactinium	$_{91}Pa^{231}$	3.43×10^4 y	α
Actinium	$_{89}Ac^{227}$	21.6 y	β
Thorium	$_{90}Th^{227}$	18.17 d	α
Radium	$_{88}Ra^{223}$	11.7 d	α
Radon	$_{86}Rn^{219}$	4.0 s	α
Polonium	$_{84}Po^{215}$	1.8×10^{-3} s	α
Lead	$_{82}Pb^{211}$	36.1 m	β
Bismuth	$_{83}Bi^{211}$	215 m	α
Thallium	$_{81}Tl^{207}$	4.78 m	β
Lead	$_{82}Pb^{207}$	stable	—
Miscellaneous Series			
Potassium	$_{19}K^{40}$	1.3×10^9 y	β
Calcium	$_{20}Ca^{40}$	stable	—
Rubidium	$_{37}Rb^{87}$	4.7×10^{10} y	β
Strontium	$_{38}Sr^{87}$	stable	—

After this necessary warning, we proceed with a review of modern estimates of the ages of various celestial bodies and current theories on the formation and evolution of the stars and galaxies.

35-2   Age of the Earth and Meteorites. For the earth and meteorites we can analyze actual samples in the laboratory with every resource of modern physics. We can derive intervals of time since major changes took place in the physical state of this material, or the time since nuclear changes occurred because of exposure to cosmic rays in space. The essence of various methods will be presented in this section, beginning with the most tangible and clear-cut deductions available from stony materials in our laboratories.

Table 35-1 lists a few of the radioactive isotopes useful in the astrophysical analysis of ages. Several major radioactive series of the heavy elements exist, that of long-lived thorium 232 as well as those of uranium 235 and 238, given in the table. The radioactive decay of potassium and rubidium also is important. The tables list the element, the isotope, the half-life in years (y), days (d), minutes (m), or seconds (s). The final column gives the major emission product, an alpha particle $_2He^4$, or a beta particle, an ordinary negative electron. A few nuclei may follow alternative routes. Instead of emitting α followed by β, they may do the reverse $\beta\alpha$, ending up with the same daughter product.

The isotopes with short lifetimes, of which many examples occur in nature, have all decayed since the early history of the solar system and are produced currently only by cosmic rays. Hence they are valuable for measuring the intervals since meteorites have been broken from larger masses so that cosmic rays could produce these rare isotopes. On the other hand, when a material containing the heavier isotopes listed, such as rubidium, potassium, and uranium, has once been frozen, the accumulation of the daughter isotopes makes possible an age measurement since solidification occurred.

Apart from the problem of making precise enough measurements, two major difficulties arise in determinations of ages by these methods. Some daughter isotopes, such as helium, tend to leak out of certain crystal structures, and the leakage is very sensitive to temperature. Thus careful attention must be given to the ability of the material to hold a light gas. Correspondingly, some substances can be interchanged or leached out by the action of water in earth rocks or meteorites after they fall. The second serious problem concerns the amount of the daughter isotope originally present in the mix. Usually no direct method exists for determining this amount, so we must employ indirect methods. For example, the present abundances of the isotopes U-235 and U-238 and their decay products, respectively Pb-207 and Pb-206, would tell us the length of time since the material froze, if we knew the original abundances of Pb-207 and Pb-206 that were present. Unfortunately, the amount of primordial lead is not directly known, and the method becomes quite complex.

Let us consider the best method, theoretically one of the simpler ones, of dating many rocks. Rubidium-87 decays to strontium-87 with a half-life of 4.7×10^{10} years. In a sample of rock or meteorite we can measure two ratios of atomic numbers present, Sr-87/Sr-86 and Rb-87/Sr-86. Because the Sr-86 atoms are stable, their abundance in the rock or meteorite has remained constant since solidification, while the decay of Rb-87 has increased the number of Sr-87 atoms at the expense of the Rb-87 atoms. One-half of the Rb-87 atoms decay in 4.7×10^{10} years, or some 2.5 percent in 10^9 years. If we knew the original ratio of Sr-87 to Sr-86 at solidification or the date of the solidification, we could calculate from our two measures the original ratio of Rb-87 to Sr-86. Suppose now that we find another sample that solidified at the same time as the first but in which the original ratio of Rb-87 to Sr-86 was different. Since chemical separation does not readily change the isotope ratios in a single element, we may safely assume that the ratio of Sr-87 to Sr-86 at solidification was the same for both samples. Now let us measure four quantities, the ratios Sr-87/Sr-86 and Rb-87/Sr-86 in both samples. We have four unknowns, the time since solidification, the original ratio of Sr-87/Sr-86, and two values of Rb-87/Sr-86, one for each sample. We can solve for all four unknowns, and thus obtain the desired age since solidification.

The method works beautifully for stony meteorites, giving for nine samples ages ranging from 4.3 to 5.0×10^9 years. The mean value is 4.7×10^9 years, with an estimated uncertainty of only 100 million years or 2 per cent. The solidification ages derived from K-40 → A-40 and (U, Th) → He^4 methods are also greater than 4.0×10^9 years for some meteorites, but are generally lower for most. Losses of the gases introduce a serious uncertainty. In two meteorites where the Pb-207, Pb-206 isotopes from U-235 and U-238 are unusually abundant with respect to Pb-204, the derived age is about 4.6×10^9 years. We may conclude that the material of the oldest meteorites (and perhaps all of them) accumulated about 4.6 to 4.7×10^9 years ago.

The oldest earth rocks, ranging up to about 3×10^9 years, are younger than the oldest meteorites, but the earth itself is older, the rocks having been frozen well after it formation. The measurement of lead, uranium, and thorium isotopes, coupled with a theory of their mixing and distribution, leads to an age for the earth of about 4.6×10^9 years, since formation. This value may well increase somewhat as measurements and theory progress, to keep pace with the better-determined solidification ages of the meteorites. The geologic ages over which geologic processes can be traced cover a much shorter period of time, barely 20 per cent of the earth's past.

TABLE 35-2   *Relative abundances of isotopes in terrestrial rocks and meteorites*

Isotope	Rocks	Meteorites
$_2He^4$	1	0.70
$_2He^3$	1.3×10^{-4}	0.30
$_{10}Ne^{20}$	0.909	0.33
$_{10}Ne^{21}$	0.0026	0.33
$_{10}Ne^{22}$	0.0882	0.33
$_{18}A^{36}$	0.00337	28
$_{18}A^{38}$	0.00063	44
$_{18}A^{40}$	0.996	28

In meteorites Table 35-2 clearly illustrates the effect of cosmic rays on nuclear abundance. It gives the approximate relative abundances of different isotopes for terrestrial rocks and for meteorites. Note that cosmic rays tend to produce comparable amounts of neighboring isotopes, where the terrestrial abundances of the naturally occurring, but stable, isotopes may differ markedly.

The Apollo Program has provided lunar rocks for which age determinations can be obtained.

35-3   Evolution of the Sun and Stars.   If the earth, planets, meteorites, —the solar system—were formed about 5 billion years ago, it is probably safe to conclude that the sun itself was already in existence at that time. But was it then only recently formed or was the planetary system a late addition? Most theories of the origin of the planets (Chapter 17) require that the solar system formed shortly after the sun itself or during its condensation from interstellar material.

Fortunately we are no longer limited to speculations about the history of the sun and planets. General theories of stellar evolution developed during the past quarter-century give a plausible history and time scale for the formation and development of a star in terms of its original mass and chemical composition. Observations of color-luminosity diagrams of galactic and globular clusters agree remarkably with the predictions of these theories.

The general picture is that a spiral galaxy, such as ours, which is rich in interstellar gas and dust clouds, will frequently develop conditions just right for the condensation of such clouds through the combined influence of gravitation in the cloud itself, radiation pressure from other stars on its surface, and quite possibly thermal and magnetic forces. Photographs

FIGURE 35-2   *Circle and crosses indicate locations of infrared sources in Orion nebula.*

of the Milky Way give ample evidence of the fragmentation of large cloud complexes into smaller clouds and "globules" (Sec. 29-1), which many astronomers regard as "proto-stars" in a very early stage of condensation.

From here on, the fate of the star is essentially determined by the mass of the condensing globule, since mixing in the galactic plane tends to keep the chemical composition fairly uniform. The cloud of gas and dust condenses at an accelerating pace, becoming progressively smaller, denser, and hotter with the release of potential gravitational energy and its conversion into kinetic energy as the atoms and particles fall freely inward. Eventually a stage is reached when the center of the proto-star becomes optically opaque to radiation. Thermal equilibrium then develops, and the star begins to radiate more or less like a black body at the still very low temperature of its opaque core. By Wien's law (Chapter 8) most or all of its radiation is emitted in the infrared spectrum; for example, at $T = 1000°$ K, maximum emission is at $\lambda_m = 2900/1000 = 2.9$ microns. In recent years surveys of the sky with sensitive infrared detectors have led to the discovery of a large number of such infrared stars that are almost or completely invisible on ordinary photographs (Fig. 35-2). At a later stage the star, still embedded in a dark cloud, falls in the class of the variable T Tauri stars (Sec. 26-13), and its evolution accelerates so much that it may unexpectedly appear as a "newborn" star (not a nova). A remarkable example is the star FU Orionis, which became visible suddenly in 1936 in the middle of a dark cloud. The phase of collapse from an infrared gas and dust cloud as large or larger than the whole solar system, to a visibly shining "new" star perhaps as large as the orbit of Mercury, takes place very rapidly, probably in only a very few years.

FIGURE 35-3 *Evolutionary tracks in the temperature-luminosity diagram of contracting stars of 0.5, 1, 3, and 15 solar masses from the end of rapid collapse to the main sequence.*

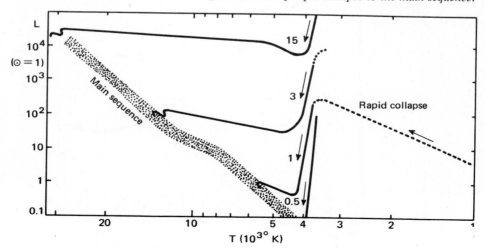

The relation between the energy output (the bolometric luminosity) and the effective temperature of the contracting star can be calculated step by step and described by an *evolutionary track*, showing the path of the star in the Hertzsprung-Russell diagram at various stages of the contraction (Fig. 35-3). The speed with which the star travels along this path varies greatly with the mass. Large masses collapse rapidly under their own extreme weight. Small masses shrink only very slowly in their weak gravitational field (Fig. 35-4). During the phase of rapid collapse, internal temperatures rise rapidly. The surface temperature soon reaches 4000, then 5000° K, and the star begins to shine visibly while continuing to contract more slowly. It moves to the left along the track of Fig. 35-3 until the temperature in its core reaches a value high enough, about $10^{6°}$ K, to

FIGURE 35-4 *Variation of sun's radius during past five billion years. Note rapid collapse in early phases of evolution, followed by T Tauri and hydrogen-burning stages. The scale at right shows radii of orbits of Pluto, Earth, and Mercury.*

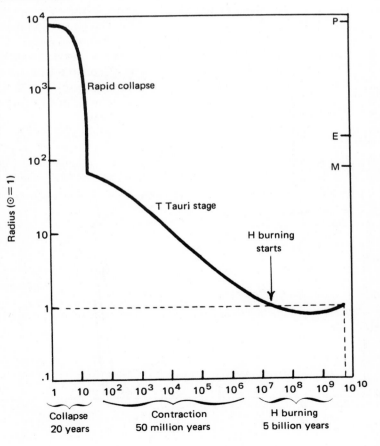

ignite the nuclear reactions converting hydrogen to helium through the p-p chain and C-N cycle, as explained in Chapter 25. The now mature star is stabilized on the main sequence, where it will remain for a long time until depletion of hydrogen in its core leads to the next phase of nuclear burning and initiates, much later, a third phase of stellar evolution.

The probable variation of the sun's radius during its collapsing and contracting phases and until it reached its present position on the main sequence is shown in Fig. 35-4.

Stars of highest mass and luminosity consume energy and therefore mass at a prodigious rate, up to a million times greater than that of the sun. For the brightest supergiants of absolute magnitude $M = -10$, we can predict that their lifetimes on the main sequence will be short, perhaps 10^5 years or less (See Fig. 35-5, right-hand scale). Stars of lower mass, such as the sun, are more parsimonious and use their reserves more slowly. Thus the sun, having already spent 5×10^9 years in the hydrogen-burning stage, has probably enough hydrogen left to last several more billion years. Dwarf stars of 0.1 solar mass will stay practically "forever" on the main sequence (over 10^{11} years).

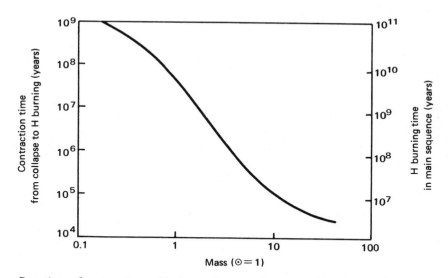

FIGURE 35-5   *Durations of contracting and hydrogen-burning phases as a function of stellar mass.*

When a substantial fraction of the hydrogen of the core has been converted into helium (perhaps a third to a half for stars of one solar mass), the core of the star contracts, its temperature increases, the outer regions expand, and the representative point in the H-R diagram moves off the main sequence along a path fixed by the mass and chemical composition.

The evolutionary tracks of stars originating in a small section of the main sequence are bunched in such a way that the H-R diagrams observed in open and globular clusters (Chapter 30) are readily explained and so can be related to a particular stage of stellar evolution or "age" beyond the initial or "zero-age" main sequence (Fig. 35-6).

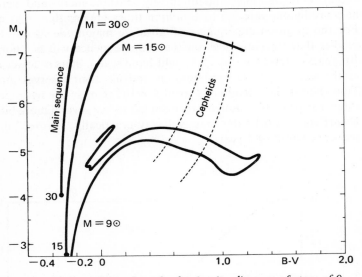

FIGURE 35-6 *Evolutionary tracks in the color-luminosity diagram of stars of 9, 15, and 30 solar masses during the hydrogen-burning phases. The vibrational instability strip of the Cepheid variables is crossed once or twice. Compare with Fig. 30-4.*

During these later stages of stellar evolution, the stellar core becomes so hot that even the very stable helium nucleus becomes subject to nuclear reactions (Sec. 25-9) and the helium-burning phase begins. The stellar atmosphere is enormously distended and the star becomes a red giant (Sec. 20-9). Later still, the star will move again toward the left in the H-R diagram and cross a zone of instability in the "Hertzsprung gap" between the red giant branch and the upper part of the main sequence. In this zone the star becomes unstable and subject to oscillations or pulsations; depending on its mass, it is then observed as a variable of the Cepheid, RR Lyrae, or β CMa types (Fig. 26-11).

As the various sources of nuclear energy become exhausted in turn (Chapter 25), the star finds itself eventually unable to sustain its prodigal output of energy. It has only one major energy source left, gravitational contraction, which reenters the picture in what may be called the "old age" in the life of a star. By rapidly shrinking its radius the star reduces its total light emission to space and so conserves its diminishing reserves. During the collapse, a rotating star sheds matter into space by virtue of the speed-up in its rotation and the increased centrifugal force resulting from the law of conservation of angular momentum. It may even go through a series of sudden releases of energy during unstable phases and appear as a nova (Chapter 27). A more drastic collapse will lead to a supernova explosion (Sec. 27-6). Eventually the star ends up as a white dwarf with a degenerate core of hyperdense matter (Sec. 21-7) or possibly as a still denser "neutron" star in which almost all the internal mass has reached nuclear densities and most of the emergent radiation appears as X rays. Some astronomers have speculatively considered this dead-end avenue of stellar evolution as a possible explanation for some of the cosmic X-ray sources (Sec. 33-7).

If an indefinite future is available for evolution, the ultimate stage of the residual matter of a star, after it has exhausted all its remaining energy sources, might be that of a hard, cold, hyperdense chunk of matter, perhaps not more than a few kilometers (for a neutron-star remnant) or thousands of kilometers (for a white-dwarf remnant) in diameter, but with masses still of the order of the solar mass, or a substantial fraction of it If so, and if the universe were old enough (say, 10^{11} years or more), a very large proportion of the total mass density in space might be locked up in such celestial ashes. Perhaps this is the reason for the puzzling discrepancy between the various methods of estimating galaxy masses (Sec. 32-7). Nevertheless, most astronomers currently favor cosmological models and cosmogonical theories that allow for a relatively short past history of our universe and, hence, a much smaller number of "dead bodies." We must now review the arguments for and against these speculations.

35-4   The Age of the Galaxy.   One of the strongest arguments for a short time scale of cosmic evolution comes from studies of star clusters in our galaxy. We have already discussed at some length (Chapter 30) the evidence for a fairly rapid evolution and dissipation of the galactic or open clusters. By comparison between the observed H-R diagram of a cluster and the theoretical evolutionary tracks of stars (Sec. 35-3), in particular the location of the "break-off" from the main sequence, we can fairly precisely date the time elapsed since the cluster stars reached the hydrogen-burning stage (Sec. 35-3). We also know that for any given star the contraction phases are much shorter than the time spent on the main sequence (Fig. 35-4). Hence with only a small correction the "break-off" point value gives the age of the cluster—the time elapsed since it condensed out of interstellar clouds. Many of the clusters are very young on a cosmic time scale, often less than 10^8 years, but some, such as M 67, register ages of several billion years, and the oldest dated clusters (and field stars near the sun) are some 10^{10} to 2×10^{10} years old. This figure approaches the range of ages, 1.5 to 2.5×10^{10} years, indicated by the break-off points of globular clusters (Sec. 30-7). Presumably the Galaxy must be at least as old as its oldest clusters.

The H-R diagrams of open and globular clusters show significant differences, however, and their spatial distributions are quite different, as we know (Secs. 30-3, and 30-6). Some factor other than age must be at work. Spectroscopic studies indicate that this second factor is chemical composition. We see that open clusters are still forming out of interstellar material spread in the flat or disk component of the galactic system (Sec. 31-2). Globular clusters, on the other hand, form a nearly spherical system and only briefly traverse the galactic plane in their orbital motion around the galactic center (Sec. 30-6). The conclusion is that globular clusters were formed, if not all at once, at least over a relatively short period (say, the first billion years) of the early history of the galactic system while it was itself in the process of condensing out of some primordial, intergalactic cloud (Sec. 35-7). The galaxy was then a roughly spherical cloud, and its chemical composition had not yet been appreciably modified by mixing with the products of nuclear reactions in stars (Chapter 25). Later, after the remaining gas had collapsed in a rapidly rotating disk (Sec. 35-7), the composition of the interstellar medium slowly altered as the result of a mixture of atoms ejected by evolving stars by various mechanisms: steady corpuscular radiation similar to the "solar wind" (Chapters 9, 10), or rapid ejection of their outer layers shedding excess angular momentum (Sec. 35-3), or in nova or supernova explosions.

In this fashion the products of nuclear reactions in stellar interiors (Chapter 25) became slowly dispersed in the interstellar medium, which steadily changed its chemical composition. Stars formed in the early history of the Galaxy out of the original medium are the population II (or "first-generation" stars found in globular clusters and the high-velocity stars sharing their spheroidal distribution (Sec. 32-3). Second- and later-generation stars and clusters that formed out of the modified medium in the galactic disk are the low-velocity population I stars found in spiral arms and open clusters. Their ages vary from "just born," such as FU Orionis (Sec. 35-3), to mature stars such as the sun, and finally to very old stars formed just after the initial collapse of matter in the galactic disk. On this theory the Galaxy must have been 15 to 20 billion years old when the sun was born 5 or 6 billion years ago.

In Sec. 35-7 we shall discuss further the formation of galaxies, but to do so we must operate within the framework of some specific class of cosmological models, since the large-scale evolution of the universe evidently plays a major role in determining the mechanisms at work on the galactic and supergalactic scales. It should be also obvious that from the above discussion of cluster ages, which serve to define the age of the Galaxy, the latter may have little or nothing to do with the "age of the universe," even if the latter expression can be defined.

It is only within a restricted class of models and theories that we can relate the two concepts and use the age of the Galaxy as a significant time scale for the universe at large. Let us, then, return briefly to the cosmogonical implications of some cosmological models.

35-5 **Lemaitre's Primeval Atom and the "Big Bang" Theory.** We explained in Sec. 34-11 how the evolution of the homogeneous model universes of general relativity can be represented by a graph (Fig. 34-7) of the radius of curvature $R(t)$ vs. time. Lemaitre favored a model in which $R(t)$ is mathematically zero (or physically very small) at a time t_0, the instant of "creation." He postulated that, no matter how this "creation" may have occurred (something he did not discuss), the universe started in this model as a hyperdense sphere of matter or "primeval atom," which physicists later described as a giant neutron ball.

We know that matter at nuclear densities is highly unstable beyond atomic number $\simeq 100$ or a mass of $\simeq 250$ nucleons. The original neutron in this picture had a mass equal to that of the whole universe, some 10^{78} to 10^{79} neutrons packed into a volume not much larger than that of the present solar system, an extremely unstable configuration. The superatom must have almost instantly exploded with a violence that defies description. The neutrons rapidly decayed into protons and electrons, forming deuterons by neutron capture, then helium nuclei, and so on, until an immense number of nuclear reactions built up the various atomic species of increasing atomic number.

The nuclear reactions in the primeval fireball have been investigated by many physicists. At the extremely high temperatures (billions of degrees during the first few minutes) and intense radiation and particle fluxes of the fireball a much greater variety of nuclear transmutations is possible than even in a hydrogen bomb. The complication of the chemistry of this initial cosmic brew, dubbed "ylem" by G. Gamow, defies a complete analysis. But a plausible scheme can be outlined through which nearly all the lighter elements up to iron can be formed in the first thirty minutes or so, and in about the correct observed cosmic abundances (Fig. 35-7). Because of this rapid explosion, proceeding initially at the velocity of light, the temperature of the fireball had to drop rapidly from perhaps $10^{12\,\circ}$ K within the first *microsecond*, to $10^{9\,\circ}$ K after five minutes, to $40 \times 10^{6\,\circ}$ K after one day, then more slowly to $6{,}000^\circ$ K after 300,000 years. According to Gamow, most atom-building reactions stopped after about thirty minutes. The heavier elements must have been produced much later in the secondary nuclear reactions that are still taking place in stellar interiors (Chapter 25).

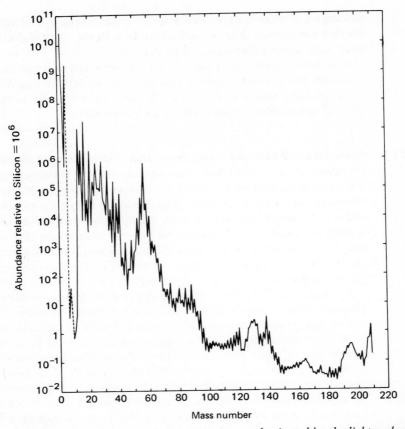

FIGURE
35-7 *The relative abundances of atomic species are dominated by the lighter elements, hydrogen and helium. The abundances of other isotopes may be explained by nuclear reactions in the primeval fireball and in stars.*

In the early phases of this explosion tremendous pressure forces were present, and the idealized cosmological models with zero pressure do not apply. The universe was a big ball of nuclear fire and radiation. Thus, in the era of the hydrogen bomb, the "primeval atom" hypothesis of Lemaitre has become popularly known as the "big bang" theory. After several hundred million years matter had dispersed and cooled enough and radiation been diluted to a point where the laws of ordinary macroscopic physics could take over, and more conventional cosmogonical mechanisms came into play. Most of these mechanisms require long periods of time to be effective. If it takes millions of years for interstellar matter to condense into stars (Sec. 35-3), the formation of galaxies and clusters of galaxies may have taken hundreds of millions of years or more.

In order to provide a long period of quasi-stable conditions during which small density and temperature fluctuations could initiate condensations on galactic and supergalactic scales, Lemaitre argued that the initial expansion was slowed down and finally almost brought to a standstill by the force of gravitational attraction, which was very strong in the initial dense phases of the universe. Thus the universe may have spent perhaps a billion years going through the quasi-stable Einstein phase (Sec. 34-11, Fig. 34-8). While the universe as a whole was almost in equilibrium, local fluctuations occurred—regions that were slightly hotter or colder, denser or more rarefied, than the average. Eddies of the cosmic fluid could react to the usual physical laws of gas and radiation and, according to Lemaitre, start the formation of clusters of galaxies. Observation suggests that galaxies form in groups, clusters, or superclusters (Sec. 32-10) and not as isolated individuals in space, just as stars form in galaxies and not (as far as we know) as isolated individuals in intergalactic space. Lemaitre visualized the great clusters of galaxies having mean densities of the order of 10^{-27} g cm^{-3} as remnants of the Einstein phase of the universe. Thus by equation (34-19) the radius of curvature of the universe while it passed slowly through the quasi-stable phase was $R_0 = \sqrt{10^{27}/\rho} \simeq 10^{27}$ cm $\simeq 300$ megaparsecs. (As we saw in Sec. 34-7, with the present density of 10^{-30} or so the radius of curvature of an Einstein universe is much larger.)

Eventually, as matter collected into vast clouds that later condensed further into galaxies (Sec. 35-7), the cosmic repulsion overcame gravitation, at least in the vast low-density intercluster spaces. The universe as a whole resumed its expansion, at first very slowly, then at an increasing rate following the $R(t)$ curve for the Lemaitre model (Fig. 34-8).

It is quite clear that the time $t - t_E$ elapsed since the passage through the Einstein phase, and _a fortiori_ the time $t - t_0$ since the primeval fireball, the true "age" of the universe in Lemaitre theory, have nothing to do with the time scale defined by the inverse $1/H$ of the Hubble constant, which is merely the present value $H(t)$ of the slope of the $R(t)$ curve. From current estimates quoted in Chapters 33 and 34, we surmise that $t - t_E \simeq 2H^{-1} \simeq 20 \times 10^9$ years, and $t - t_0$ should be greater still, although some astronomers maintain that $t - t_0 = H^{-1} = 13 \times 10^9$ years, in contradiction to estimated ages of star clusters.

35-6 The Background Radiation. Some recent observations of the faint continuous background radiation at centimeter wavelengths neatly fit the Lemaitre scheme and in particular the concept of an initial fireball. The intense radiation from the fireball will expand with the universe and be steadily degraded in density and energy, as explained in Sec. 34-13. Theory shows that it will appear now as a weak black-body radiation at a temperature $T_b = 3°$ K, arriving uniformly from all directions of space. Such a low level of radio noise is difficult to separate from other sources of noise in radio receivers. Nevertheless, comparison of the radio noise received by

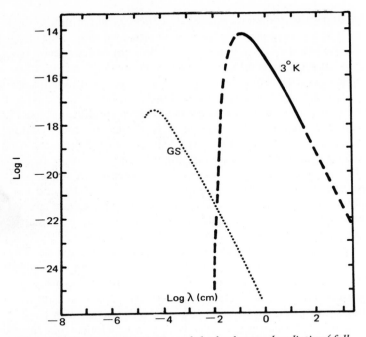

**FIGURE
35-8**   *The spectral energy distribution of the background radiation (full curve) differs greatly from that of stars in galaxies (dotted curve) and agrees with that of a black body at 3° K (dashed curve). It is consistent with the theory of the primeval fireball.*

giant skyward horn antennas with the computed theoretical noise of the receiver system reveals a residual excess of radiation that is not associated with the Milky Way or other identified sources of cosmic radio emission (Chapter 33). Observations indicate that it is very nearly isotropic, arriving uniformly, from all directions of space, and that its intensity measured at several wavelengths from 0.26 to 20.7 cm is consistent with the spectral energy distribution of black-body radiation at 3° K (Fig. 35-8).

Another kind of background radiation is the integrated optical flux of all galaxies, including those too faint to be photographed individually. This radiation must make a small contribution to the light of the night sky, but it has not yet been positively detected against the much stronger contributions from the airglow, the zodiacal light, and integrated starlight. An early estimate by de Vaucouleurs in 1948 based on the then available galaxy counts (Sec. 34-14b) indicated that in blue light all galaxies in the universe contributed at most 0.3 percent of the luminosity of the night sky. A recent revision of this estimate gives an even lower limit, about 0.1 percent, and an unsuccessful attempt by F. E. Roach to detect this faint "cosmic" component in observations of the light of the night sky is consistent with this limit. It is possible that future observations from satellites and ... scopes working against a darker sky background will eventually g... positive results and so provide another tool to probe the radiation from very distant parts of the universe.

A third type of background radiation is the diffuse isotropic emission of X rays (i.e., from all directions of the celestial sphere) in the energy range of 10^3 to 10^6 eV (Sec. 33-9). Proposed interpretations of the observed spectral energy distribution curve call on a variety of possible mechanisms, such as an interaction between cosmic-ray (i.e., high-energy) electrons with the low-energy photons of the 3°K radiation in intergalactic space or, perhaps, an interaction of lower-energy cosmic electrons with the low-density intergalactic plasma, or possibly an unresolved distribution of discrete extragalactic X-ray sources. While observations and theories will have to be refined before a definite interpretation emerges, it seems probable that we have here, at least potentially, a third clue to the large-scale properties and the distant past of the Universe.

35-7 **Evolution of Galaxies.** The formation and evolution of galaxies should be considered in the framework of either the big-bang or the steady-state theory. Actually, since in the first theory galaxies formed during the slow passage through the Einstein phase, conditions were not too drastic-ally different from those visualized by the steady-state concept, at least as far as time scales and average densities in intergalactic space are con-cerned. In both theories galaxies or clusters of galaxies form out of inter-galactic gas at densities of the order of 10^{-27} to 10^{-28} g cm^{-3}, but in the Lemaitre view this process occurred only once during a transitional phase of universal evolution, perhaps 20 billion years ago, while in the steady-state view this formation is taking place all the time.

What are the physical mechanisms available to start the condensation of gas masses of the order of 10^{10} solar masses (galaxies) or possibly 10^{14} solar masses (clusters of galaxies)? The oldest concept, first considered by Newton and developed in modern times by Sir James Jeans, is *gravitational instability*. In a gaseous medium density fluctuations take place all the time, as we know. On a cosmic scale these fluctuations can have a wide range of densities and linear sizes and, therefore, masses. Small fluctuations are unstable and will dissipate almost as soon as they form. However, some fluctuations may be large enough and possess sufficient mass to be stable. In other words, once accidentally formed, the aggregate gravitational force of the atoms in the gas blob will stabilize the system and give it enough cohesion to resist the dissipative forces. Then, Jeans conjectured, the self-gravitating cloud could slowly condense to become a protogalaxy. The process can recur on smaller and smaller scales inside the protogalaxy to lead eventually to the formation of stars and star clusters; various mechanisms have been discussed in recent years by F. Hoyle and others.

Gravitational instability, unfortunately, requires extremely long periods of time when the initial densities and density fluctuations are very low. More efficient mechanisms are needed to initiate the fragmentation and condensation of intergalactic matter. One such mechanism, which jet air travel has made familiar, is *hydrodynamic turbulence*. A fluid flowing at high or supersonic speed is generally subject to turbulence or eddy-formation. A whole hierarchy of eddies of different sizes and energy content is possible, and on a cosmic scale, with the help of gravitation, some of them may become stable (once formed, they will not disperse again). This mechanism, discussed by the German physicist C. von Weizsacker, could play an important role in promoting the initial condensation of intergalactic matter. The energy in the turbulent motions would be converted into heat and radiation and eventually escape from the condensing blob. Later, energy should be released by gravitational contraction while the condensing blob increases its rate of spin and develops into a flat system or thick disk. This evolution is a necessary consequence of the law of conservation of angular momentum (Sec. 17-4). If the centrifugal force becomes too large, the rotating disk expels excess angular momentum by shedding matter into space at the periphery. Von Weizsacker thought that various galaxy types might be formed by this general principle, and in particular that population II stars would form in the spheroidal phase of galactic evolution before population I stars could form in the disk.

This general picture of an early collapsing phase in the evolution of our Galaxy has received support through the work of O. Eggen and A. Sandage on the dynamical properties of various types of stellar population. It also fits in with the concepts of chemical evolution of the interstellar medium through slow mixing with the products of nuclear reaction resulting from stellar evolution, as outlined in Sec. 35-3.

The physics of the formation of subfragments, down to protostar size, in a condensing galaxy has been discussed by Hoyle in the framework of the concept of *gravitational turbulence*, which combines some basic ideas of the two previous hypotheses. He showed in particular that the gaseous protogalaxy must soon reach a sort of radiative equilibrium at a temperature close to $T = 10^{4\circ}$ K, at which ionization of hydrogen begins. The ionization-recombination equilibrium of hydrogen plays the role of a powerful thermostat, and the further contraction and successive fragmentation of the protogalaxy is isothermal, until stars begin to form and dust condenses into interstellar space. Dust, absorbing energy from starlight, acts as a powerful coolant by radiating energy in the infrared spectrum, and the temperature of the residual gas soon drops to the low values currently observed in the interstellar medium of our galaxy (Chapter 29).

Other mechanisms that could induce or assist the condensation of galaxies have been discussed, some in the framework of the steady-state

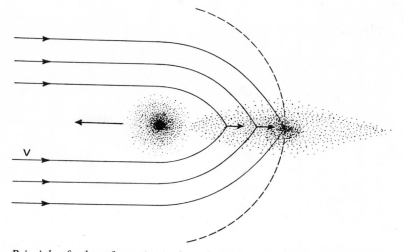

FIGURE 35-9   *Principle of galaxy formation in the wake of another galaxy, according to W. Sciama.*

concept, some without reference to a specific model. We shall only mention the attractive idea, developed by D. W. Sciama in England, that galaxy condensation nuclei could be formed by gravitational focussing of the intergalactic matter in the wake of a preexisting galaxy as it sails through the intergalactic medium (Fig. 35-9). Since this mechanism is specifically described in the context of the steady-state theory, in which past time is infinite (Sec. 34-12), it is assumed that the parent galaxy was formed earlier in the same manner and so on *ad infinitum*. It merely provides a mechanism to maintain automatically on the average a constant density of galaxies per (large) unit volume of space as required by the continuous-creation hypothesis. Some evidence relating to this general mechanism of galaxy formation in the wake of another was recently adduced by the British astronomers N. Woolf and K. Freeman, who noted curious similarities between some peculiar-looking galaxies and chains of galaxies and laboratory photographs of turbulent eddies in the wake of an obstacle in a supersonic flow (Fig. 35-10). Here, again, as in von Weizsacker's theory, turbulence is invoked to assist gravitation to initiate condensation.

Unrelated to any definite cosmological model or even to any known properties of matter, some ideas of the Armenian astronomer V. A. Ambartsumian may be mentioned here if only for their unorthodox value. There is now ample evidence that violent events occur within some galactic nuclei. We have previously discussed the explosion filaments in M 82 (Sec. 32-9), the jet in M 87 (Fig. 33-4), the peculiar spectra of some galaxies (Sec. 32-6), and other indices that extremely energetic events take place in the centers of some galaxies. From these facts and other arguments Ambartsumian concluded that galaxies may form explosively from some hyperdense nucleus in an unknown state of matter rather than slowly by condensation of low-density intergalactic gas. He does not attempt to

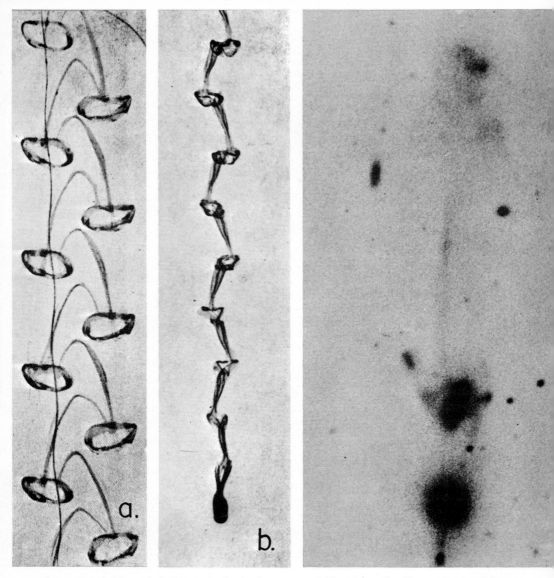

FIGURE 35-10 *(a through j) Strange galaxies have some similarities with eddy pattern in a laboratory flow. (After N. Woolf and K. Freeman.)*

explain the origin or the physics of this hyperdense nucleus. Accepting the concept as a working hypothesis, he shows how some properties of galaxies and groups of galaxies (Sec. 32-8) could result. Particularly surprising is his idea of a "fission" of galactic nuclei, which he sees as a way of accounting for some radio sources (Secs. 33-4, 33-5); nevertheless, observation demonstrates the existence in some systems of galaxies of a very bright, compact double nucleus, which might be an illustration of the phenomenon (Fig. 35-11). Ambartsumian's ideas have received some

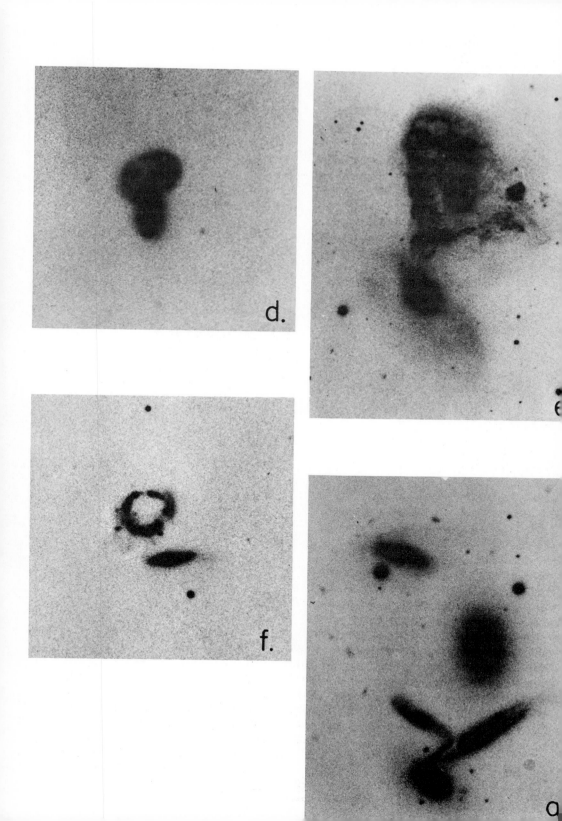

d.

e.

f.

a.

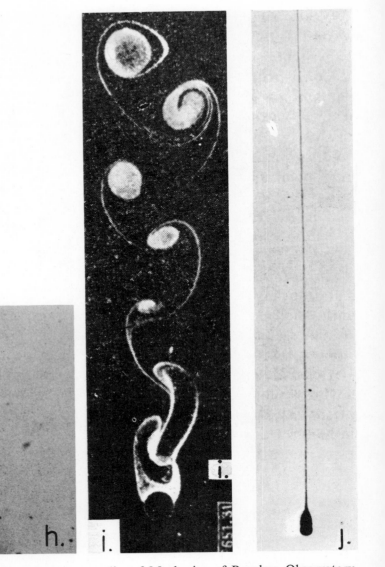

further support in recent studies of Markarian of Burakan Observatory, Armenia, and especially of H. C. Arp, of Mt. Wilson and Palomar Observatories. Both have discussed lines or chains of galaxies that could hardly be expected to be formed by chance. Arp speculates that these alignments are formed by the explosive ejection of whole galaxies or compact fragments by a giant "mother" galaxy, which is often a strong radio source as well; among the brightest examples are M 87, the Virgo radio source, and NGC 5128, the Centaurus A radio source (Fig. 33-5).

Strange as these speculations are and outside present-day physical concepts, we must always remember that so were novae before 1572, meteorites before 1803, supernovae before 1917, radio sources before 1946, quasars before 1960, pulsars before 1967, and so on.

Speculation has a place in science, but the mere fact that a conjecture seems to explain a given set of observations does not prove that the idea is correct. For the range of speculations is infinite. A hypothesis that does not contain some element of prediction is barren. The great theories, like Einstein's theory of relativity, are great because they lead to further contact with observation and to *precise* numerical verifications. Here we have mentioned these related but farfetched suggestions of Ambartsumian and Arp as illustrations of attempts to bridge by imagination great gaps of our factual knowledge.

FIGURE 35-11 *The peculiar multiple galaxy NGC 1741 with a highly luminous double nucleus illustrates Ambartsumian's concept of "nuclear fission" on a galactic scale. (G de Vaucouleurs, McDonald Observatory.)*

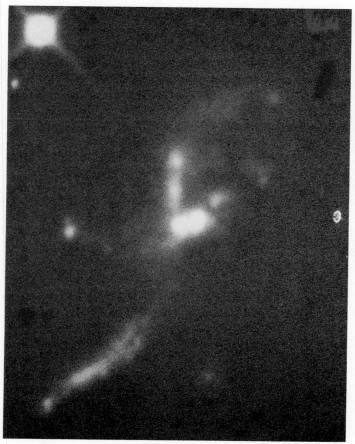

As we have tried to show throughout this book, progress in the understanding of the universe depends more on the successful welding of known physical laws and skillful observation than it does on the invention of entirely new and completely unsubstantiated physical concepts. The world is complex, and we should not expect to comprehend its full significance at a glance. Much hard work remains for future generations of astronomers in this old but always vigorous and evergrowing science.

APPENDIX

Bibliography

The general references listed first below are relevant to several or many of the separate chapters of this book. Certain of the books listed under specific chapter headings are relevant also to other, related chapters. The philosophical references throughout are usually of a general nature, relating astronomy to the other physical sciences.

Introduction to Astronomy

Abell, George, *Exploration of the Universe*. New York: Holt, Rinehart and Winston, 1964, 2nd ed., 1969.

Beet, E. Agar, *The Sky and its Mysteries*. London: G. Bell, 1962.

Dampier, William C., and Margoret Dampier, *Readings in the Literature of Science*. New York: Harper & Row, 1959.

de Vaucouleurs, G., and L. Rudaux, Larousse *Encyclopedia of Astronomy*. London: Batchworth Press, 1959.

Evans, D. S., *Observation in Modern Astronomy*. New York: Amer. Elsevier, 1958.

Martin, Martin Evans, and Donald H. Menzel, *The Friendly Stars*. New York: Dover, 1964.

Fanning, Anthony E., *Planets, Stars and Galaxies*. New York: Dover, 1966.

Hawkins, Gerald S., *Splendor in the Sky*. New York: Harper & Row, 1961.

Hoyle, Fred, *Astronomy*. London: Macdonald, 1962.

Mayall, R. Newton, and Margaret W. Mayall, *Sky Shooting*. New York: Dover, 1967.

Menzel, Donald H., *Field Guide to the Stars and Planets*. Boston: Houghton Mifflin, 1964.

Millman, Peter M., *This Universe of Space*. Cambridge, Mass.: Schenkman (Pitman Pub. Corp., N.Y.C.) 1961.

Norton, A. P., *A Star Atlas and Reference Handbook*. Cambridge, Mass: Sky Publ. Co., 14th ed., 1959.

Page, Thornton, and Lou Williams Page, *Neighbors of the Earth*. New York: Macmillan, 1965.

————, *Wanderers in the Sky,* 2 vols. New York: Macmillan, 1965.

Payne-Gaposchkin, C., *Introduction to Astronomy*. Englewood Cliffs, N.J.: Prentice-Hall, 1954, 2nd ed., 1959, 3d ed., 1970.

Sidgwick, J. B., *Introducing Astronomy*. New York: Macmillan, 1957.

Struve, Otto, and Velta Zebergs, *Astronomy of the Twentieth Century*. New York: Macmillan, 1962.

Struve, Otto, Beverly Lynds, and Helen Pillans, *Elementary Astronomy*. New York: Oxford, 1959.

Struve, Otto, *The Universe*. Cambridge, Mass.: M.I.T. Press, 1962.

Whitrow, G. J., *The Structure and Evolution of the Universe*. New York: Harper & Row, 1959.

Wood, H., *Unveiling the Universe*. Sydney: Angus and Robertson, 1967.

History and Philosophy of Science

Buchler, Justus, *Philosophical Writings of Peirce*. New York: Dover, 1955.

Bury, J. B., *The Idea of Progress*. New York: Dover, 1955.

Campbell, Norman, *What is Science?* New York: Dover, 1952.

Dampier, William C. and Margoret Dampier, *Readings in the Literature of Science,* New York: Harper and Row, 1959.

Heisenberg, W. *The Physicist's Conception of Nature*. London: Hutchinson, 1958.

Höffding, Harald, *History of Modern Philosophy,* 2 vols. New York: Dover, 1955.

Joad, C. E. M., *Guide to Philosophy*. New York: Dover, 1936.

Jevons, W. Stanley, *The Principles of Science*. New York: Dover, 1958.

Johnson, A. H., *Whitehead's Philosophy of Civilization*. New York: Dover, 1962.

————, *Whitehead's Theory of Reality*. New York: Dover, 1962.

Kline, Morris, *Mathematics and the Physical World*. New York: Dover, 1963.

Körner, S., *Observation and Interpretation in the Philosophy of Physics*. New York: Dover, 1962.

Planck, Max, *A Survey of Physical Theory*. New York, Dover, 1960.

Poincaré, Henri, *Science and Hypothesis*. New York: Dover, 1952.

————, *Science and Method*. New York: Dover, 1952.

————, *The Value of Science*. New York: Dover, 1958.

Windelband, W., *History of Ancient Philosophy*. New York: Dover, 1956.

von Weizsäcker, C. F., *The History of Nature*. Chicago: University of Chicago Press, 1949.

History of Astronomy

Abetti, G., *The History of Astronomy*. London: Sidgwick, 1954.

Berry, A., *A Short History of Astronomy*. New York: Dover, 1961 (original: London: J. Murry, 1898).

de Vaucouleurs, G., *Discovery of the Universe: An Outline of the History of Astronomy from the Origin to 1956*. New York: Macmillan, 1957.

Doig, P., *A Concise History of Astronomy*. London: Chapman and Hall, 1950.

Chapter 1   Astronomy and the Origins of Science

Cajori, Florian, *A History of Physics*. New York: Dover, 1962.

Cassirer, Ernst, *Language and Myth*. New York: Dover, 1953.

Cumont, Franz, *Astrology and Religion among the Greeks and Romans*. New York: Dover, 1960.

Graves, Robert, *The Greek Myths*, 2 vols. Baltimore: Penguin, 1955.

Harvey-Gibson, R. J., *Two Thousand Years of Science*. New York: Macmillan, 1929.

Hawkins, Gerald S., and John B. White, *Stonehenge Decoded*. New York: Doubleday, 1965.

Munitz, Milton K., *Theories of the Universe*. New York: Free Press, 1957.

Scully, Vincent, *The Earth, the Temple, and the Gods*. New Haven: Yale University Press, 1962.

Chapter 2   Ancient Astronomy

Allen, Richard Hinckley, *Star Names and Their Meanings*. New York: Dover, 1963.

Berry, Arthur, *A Short History of Astronomy*. New York: Dover, 1961.

Brown, N. O., *Hesiodus. Theogony*. New York: Liberal Arts Press, 1953.

Dreyer, John L., *A History of Astronomy from Thales to Kepler*. New York: Dover, 1953. (orig. Cambridge: Cambridge University Press, 1905).

Heath, T. L., ed. [1897], *The Works of Archimedes*. New York: Dover,

Heath, Sir Thomas, *A Manual of Greek Mathematics*. New York: Dover, 1963.

Neugebauer, O., *The Exact Sciences in Antiquity*. Princeton, N.J.: Princeton University Press, 1952.

Page, Thornton, and Lou Williams Page, *Wanderers in the Sky*. New York: Macmillan, 1965.

Pannekoek, A., *A History of Astronomy*. New York: Barnes & Noble, 1961.

Sarton, George. *Ancient Science and Modern Civilization*. New York: Harper & Row, 1959.

Chapter 3 The Copernican Revolution

Armitage, A., *The World of Copernicus*. New York: New American Library, 1951.

Baumgardt, C., *Johannes Kepler, Life and Letters*. New York: Philosophical Library, Inc., 1951.

Creu, Henry, and Alfonso de Salvio, *Galileo's Two New Sciences*. Evanston, Ill.: Northwestern University, 1946.

Dreyer, John L. E., *Tycho Brahe*. New York: Dover, 1890.

Fermi, Laura, and Gilberto Bernardini, *Galileo*. New York: Basic Books, 1961.

Gade, J. A., *The Life and Times of Tycho Brahe*. Princeton, N.J.: Princeton University Press, 1947.

Lubbock, Constance A., *The Herschel Chronicle*. New York: Macmillan, 1933.

Munitz, M. K., *Theories of the Universe*. New York: Free Press, 1957.

Reichen, Charles-Albert, *A History of Astronomy*. New York: Hawthorn, 1963.

Rosen, Edward, *Three Copernican Treatises*. New York: Dover, 1959.

Taylor, F. Sherwood, *Galileo and the Freedom of Thought*. London: Watts, 1938.

Chapter 4 The Law of Gravitation

Andrade, E. N. Da C., *Isaac Newton*. London: Max Parrish, 1950.

Bergmann, Peter G., *The Riddle of Gravitation*. New York: Scribner, 1968.

Brodetsky, S., *Sir Isaac Newton*. London: Methuen, 1927.

Cohen, I. Bernard, *The Birth of a New Physics*. New York: Doubleday, 1960.

Lodge, Sir Oliver, *The Pioneers of Science*. New York: Dover, 1960.

MacPike, E. F., *Hevelius, Flamsteed and Halley*. London: Taylor and Francis, 1937.

More, Louis Trenchard, *Isaac Newton*. New York: Dover, 1962.

Chapter 5 The Motions of the Earth

Cohen, I. B., *Roemer and the First Determination of the Velocity of Light*. New York: The Burndy Library Inc., 1944.

Macpike, E. F., *Hevelius*, Flamsteed and Halley. London: Taylor and Francis, 1937.

Murchie, Guy, *The Music of the Spheres*, 2 vols. New York: Dover, 1967.

Chapter 6 Tools of Astronomy

de Vaucouleurs, G., *Astronomical Photography*. New York: Macmillan, 1961.

Howard, N. E., *Standard Handbook of Telescopic Making*. New York: Crowell, 1959.

Hyde, F. W., *Radio Astronomy for Amateurs*. London: Lutterworth Press, 1962.

King, H. C., *The History of the Telescope*. Cambridge, Mass.: Sky Publishing Corp., 1955.

Kuiper, Gerard P., and Barbara M. Middlehurst, eds., *Telescopes*. Chicago: University of Chicago Press, 1960.

Miczaika, Gerhard R., and William M. Sinton, *Tools of the Astronomer*. Cambridge, Mass.: Harvard University Press, 1961.

Minnaert, M., *Light and Colour*. New York: Dover, 1954.

Moore, Patrick, *Telescopes and Observatories*. New York: John Day, 1962.

Neal, Harry Edward, *The Telescope*. New York: Messner, 1958.

Sidgwick, J. B., *Amateur Astronomers Handbook*. London: Faber, 1955.

Smith, F. Graham, *Radio Astronomy*. Baltimore: Penguin Books, 1960.

Texersau, J., *How to Make a Telescope*. New York: Interscience Publ., 1957; Nat. Hist. Libr., 1963.

Thackeray, A. D., *Astronomical Spectroscopy*. London: Eyre and Spottiswoode, 1961.

Woodbury, David O., *The Glass Giant of Palomar*. New York: Dodd, Mead, 1963.

Chapter 7 Measurements in the Solar System

Ryabov, Y., *An Elementary Survey of Celestial Mechanics*. New York: Dover, 1961.

Smart, W. M., *Spherical Astronomy*. Cambridge: Cambridge University Press, 5th ed., 1962.

Chapter 8 Radiation and Atomic Structure

Bragg, Sir William, *Concerning the Nature of Things*. New York: Dover, 1948.

Jones, G. O., J. Rotblat, and G. J. Whitrow. *Atoms and the Universe*. London: Eyre and Spottiswoode, 1956.

Lewis, Gilbert Norton, *Valence*. New York: Dover, 1966.

McCrea, W. H., *Physics of the Sun and Stars*. London: Hutchinson, 1950.

Nicholson, D. E., *The Universe, Matter and Life*. London: English Universities Press, 1962.

Weidner, R. T., and R. L. Sells. *Elementary Modern Physics*. Boston: Allyn and Bacon, 1960.

Whittaker, Sir Edmund, *From Euclid to Eddington*. New York: Dover, 1958.

Chapter 9 The Sun and Its Radiations

Abetti, Giorgio, *The Sun*. New York: Macmillan, 1957.

Daniels, Farrington, *Direct Use of the Sun's Energy*. New Haven: Yale University Press, 1964.

Gamow, George, *The Birth and Death of the Sun*. London: Pelican, 1945.

Hawkes, Jacquetta, *Man and the Sun*. New York: Random House, 1962.

Kuiper, Gerard P., *The Sun*. Chicago: University of Chicago Press, 1953.

Menzel, Donald H., *Our Sun*. Cambridge, Mass.: Harvard University Press, 1959.

Rau, Hans, *Solar Energy*. New York: Macmillan, 1964.

Chapter 10 The Earth as a Planet

Bates, D. R., ed., *The Planet Earth*. New York: Pergamon, 1957.

Gamow, George, *Matter, Earth, and Sky*. Englewood Cliffs, N.J.: Prentice-Hall, 1958.

————, *Planet Called Earth*. New York: Viking, 1963.

Geikie, Sir Andrew, *The Founders of Geology*. New York: Dover, 1962.

Gutenberg, B., *Internal Constitution of the Earth*. New York: Dover, 1951.

Kuiper, Gerard P., ed., *The Earth as a Planet* (*The Solar System*, vol. 2). Chicago: University of Chicago Press, 1954)

Leet, L. Don, *Earth Waves*. Cambridge, Mass.: Harvard; New York: Wiley, 1950.

Chapter 11 The Moon

Alter, Dinsmore, *Lunar Atlas*. New York: Dover, 1968.

Baldwin, Ralph B., *The Measure of the Moon*. Chicago: University of Chicago Press, 1963.

Fisher, Clyde, *The Story of the Moon*. New York: Doubleday, 1943.

Hess, Wilmot N., Donald H. Menzel, and John A. O'Keefe, eds., *The Nature of the Lunar Surface* (a Symposium). Baltimore: Johns Hopkins Press, 1966.

Whipple, Fred L., *Earth, Moon, and Planets*, 3rd ed. Cambridge, Mass.: Harvard University Press, 1968.

Wilkins, H. Percy, and Patrick Moore, *The Moon*. New York: Macmillan, 1955.

Chapter 12 The Terrestrial Planets

de Vaucouleurs, Gerard, *The Planet Mars*. London: Faber, 1949.

Firsoff, V. A., *The Interior Planets*. London: Oliver and Boyd, 1968.

Glasstone, S., *The Book of Mars*. Washington: N.A.S.A., 1968.

Moore, Patrick, *The Planet Venus*. New York: Macmillan, 1957.

Sagan, Carl, and Jonathan Norton Leonard, *Planets*. (Life Science Library). New York: Time, Inc., 1966.

Sander, W., *The Planet Mercury*. London: Faber, 1963.

Chapter 13   The Giant Planets

Alexander, A. F. O'D., *The Planet Saturn*. London: Faber, 1962.

Grooser, Morton, *The Discovery of Neptune*. Cambridge, Mass.: Harvard University Press, 1962.

Kuiper, Gerard P., and Barbara M. Middlehurst, eds., *Planets and Satellites* (*The Solar System*, vol. 3). Chicago: University of Chicago Press, 1961.

————— *Jupiter: Handbook of the Physical Properties*, Washington, D.C.: NASA SP-3031, 1967.

Peek, B. M., *The Planet Jupiter*. New York: Macmillan, 1958.

Whipple, Fred L., *Earth, Moon and Planets*, 3rd ed. Cambridge, Mass.: Harvard University Press, 1968.

Chapter 14   Asteroids and Comets

Middlehurst, Barbara M., and Gerard R. Kuiper, eds., *The Moon, Meteorites, and Comets*. (*The Solar System*, Vol. 4). Chicago: University of Chicago Press, 1963.

Peltier, Leslie, *Starlight Nights*. New York: Harper & Row, 1965.

Porter, J. B., *Comets and Meteor Streams*. New York: Wiley, 1952.

Richter, Niklaus B., *The Nature of Comets*, trans. and rev. by Arthur Beer. New York: Dover, 1963.

Watson, Fletcher G., *Between the Planets*, 2nd ed. Cambridge, Mass.: Harvard University Press, 1956.

Chapter 15   Interplanetary Debris

Hawkins, Gerald S., *Meteors and Comets*. New York: McGraw-Hill, 1964.

Heide, Fritz, *Meteorites*. Chicago: University of Chicago Press, 1964.

Krinov, E. L., *Giant Meteorites*. New York: Pergamon, 1966.

Lovell, A. C. Bernard, *Meteor Astronomy*. Oxford: Clarendon Press, 1954.

Mason, Brian, *Meteorites*. New York: Wiley, 1962.

Middlehurst, Barbara M., and Gerard P. Kuiper, *The Moon, Meteorites, and Comets*. (*The Solar System*, Vol. 4). Chicago: University of Chicago Press, 1963.

Nininger, H. H., *Out of the Sky*. Denver: University of Denver Press, 1952.

Watson, Fletcher G., *Between the Planets*, 2nd ed. Cambridge, Mass.: Harvard University Press, 1956.

Wood, John A., *Meteorites and the Origin of the Planets*. New York: McGraw-Hill, 1968.

Chapter 16 Space Exploration

Bates, D. R., *Space Research and Exploration*. London: Eyre and Spottiswoode, 1957.

Clarke, Arthur C., *The Exploration of Space*, New York: Harper & Row, 1959.

————, *Interplanetary Flight*, 2nd ed. Bath, England: Pitman Press, 1960.

————, *The Promise of Space*. New York: Harper & Row, 1968.

Doyle, R. O., ed., *A Long-Range Program in Space Astronomy*. Washington: N.A.S.A., 1969.

Hayes, E. Nelson, *Trackers of the Skies*. Cambridge, Mass.: Howard R. Doyle, 1968.

Fleisig, Ross, *Lunar Exploration and Spacecraft Systems*. Plenum, 1962.

Hirsch, Lester M., *Man and Space*. New York: Pitman, 1966.

Hubert, Lester F., and Paul E. Lehr, *Weather Satellites*. Waltham, Mass.: Blaisdell, 1967.

King-Hele, Desmond, *Observing Earth Satellites*. London: Macmillan, 1966.

Levitt, I. M., *A Space Traveller's Guide to Mars*. New York: Holt, Rinehart and Winston, 1956.

————, *Target for Tomorrow*. New York: Fleet, 1959.

Nayler, J. L., *A Dictionary of Astronautics*. N.Y.C.: Hart, 1964.

Newell, Homer, E., *Express to the Stars*. New York: McGraw-Hill, 1961.

Shelton, William, *American Space Exploration*. Boston: Little, Brown, 1967.

Von Braun, Wernher, and Frederick Ordway III, *History of Rocketry and Space Travel*. New York: Crowell, 1967.

Chapter 17 Origin and Evolution of the Solar System

Jastrow, R., and A. G. W. Cameron, eds., *Origin of the Solar System*. New York: Academic Press, 1963.

Russell, Henry Norris, *The Solar System and its Origin*. New York: Macmillan, 1935.

Page, Thornton, and Lou Williams Page, eds., *The Origin of the Solar System*. (Anthology). New York: Macmillan, 1966.

Smart, W. M., *The Origin of the Earth*. Cambridge: Cambridge University Press, 1951.

Urey, Harold C., *The Planets, Their Origin and Development*. New Haven: Yale University Press, 1952.

Chapter 18 Extraterrestrial Life

Cameron, A. G. W., ed., *Interstellar Communication*. New York: W. A. Benjamin, Inc., 1963.

Condon, Edward U., *Scientific Study of Unidentified Flying Objects*. New York: Bantam, 1969.

Jones, H. Spencer, *Life on Other Worlds*. New York: Macmillan, 1940.

Lewinsohn, Richard, *Animals, Men, and Myths*. New York: Harper & Row, 1954.

Menzel, Donald H., *Flying Saucers*. Cambridge, Mass.: Harvard University Press, 1953.

————, and Lyle G. Boyd, *The World of Flying Saucers*. New York: Doubleday, 1963.

Oparin, Aleksandr Ivanovich, *The Origin of Life*. New York: Dover, 1953.

————, ed., *The Origin of Life on Earth*. New York: Pergamon, (1957?)

Shklovskii, L. S., and Carl Sagan, *Intelligent Life in the Universe*. Holden-Day, 1966.

Sullivan, Walter, *We Are Not Alone*. New York: McGraw-Hill, 1964.

Weisskopf, Victor F., *Knowledge and Wonder*. New York: Doubleday, 1963.

Chapter 19   Stellar Distances and Luminosities

Armitage, Angus, *William Herschel*. New York: Doubleday, 1963.

Doig, P., *An Outline of Stellar Astronomy*. London: Hutchinson, 1948.

Chapter 20   The Spectra and Temperatures of Stars

Dufay, Jean, *Introduction to Astrophysics: The Stars*. New York: Dover, 1964.

Smart, W. M., *Some Famous Stars*. New York: McKay, 1950.

Struve, O., *Stellar Evolution*. Princeton: Princeton University Press, 1950.

Chapter 21   Double Stars and Stellar Masses

Aitken, Robert G., *The Binary Stars*. New York: Dover, 1964.

Binnendijk, L., *Properties of Double Stars*. Philadelphia: University of Pennsylvania Press, 1960.

Chapter 22   The Diameters and Densities of Stars

Dufay, J., *Introduction to Astrophysics: The Stars*. New York: Dover, 1964.

Inglis, Stuart J., *Planets, Stars, and Galaxies*. New York: Wiley, 1962.

Chapter 23   Stellar Atmospheres

Dufay, Jean, *Introduction to Astrophysics: The Stars*. New York: Dover, 1964.

Goldberg, L., and L. H. Aller. *Atorus, Stars and Nebulae*. Philadelphia: Blakiston, 1946.

Chapter 24   The Nucleus of the Atom

Alfvén, Hannes, *Worlds—Antiworlds*. San Francisco: Freeman, 1966.

Hoyle, Fred, *Galaxies, Nuclei, and Quasars*. New York: Harper & Row, 1965.

Chapter 25   Sources of Stellar Energy, Nuclear Reactions and Atomic
Evolution

Clayton, D. D., *Principles of Stellar Evolution and Nucleosynthesis*. New York:
McGraw-Hill, 1968.

Fowler, W. A., *Nuclear Astrophysics*. Philadelphia: Amer. Philosophical Soc.,
1967.

Johnson, Martin, *Astronomy of Stellar Energy and Decay*. New York: Dover,
1959.

Chapter 26   Variable Stars

Glasby, J. S., *Variable Stars*. Cambridge, Mass: Harvard University Press,
1969.

Merrill, Paul W., *The Nature of Variable Stars*. New York: Macmillan, 1938.

Payne-Gaposchkin, Cecilia, *Variable Stars and Galactic Structure*. London:
University of London Press, 1954.

Chapter 27   The Cataclysmic Variables

Glasby, J. S., *Variable Stars*. Cambridge, Mass: Harvard University Press,
1969.

Payne-Gaposchkin, Cecilia, *The Galactic Novae*. New York: Dover, 1964.

Chapter 28   Stars with Atmospheric Shells

Goldberg, L., and L. H. Aller, *Stars and Nebulae*. Philadelphia: Blakiston,
1946.

Sobolev, V. V., *Moving Envelopes of Stars*. Cambridge, Mass.: Harvard
University Press, 1960.

Chapter 29   Interstellar Matter

Dufay, Jean, *Galactic Nebulae and Interstellar Matter*. New York: Dover,
1968.

Middlehurst, B. M., and L. H. Aller, eds., *Nebulae and Interstellar Matter*.
Chicago: Chicago University Press, 1968.

Payne-Gaposchkin, Cecilia, *Stars in the Making*. Cambridge, Mass.: Harvard
University Press, 1952.

Chapter 30   Star Clusters and Associations

Blaauw, A., and M. Schmidt, eds., *Galactic Structure*. Chicago: University of
Chicago Press, 1965.

Bok, Bart J., and Priscilla F. Bok, *The Milky Way*. Cambridge, Mass.: Harvard University Press, 1957.

Chapter 31 The Milky Way

Blaauw, A., and M. Schmidt, eds., *Galactic Structure*. University of Chicago Press, 1965.

Bok, Bart J., and Priscilla F. Bok, *The Milky Way*. Cambridge, Mass.: Harvard University Press, 1957.

Chapter 32 Galaxies

Baade, W., *Evolution of Stars and Galaxies*. Cambridge, Mass: Harvard University Press, 1963.

Hodge, P. W., *Galaxies and Cosmology*. New York: McGraw-Hill, 1966.

Hubble, Edwin, *The Realm of the Nebulae*. New York: Dover, 1958.

Sandage, A., ed., *Galaxies*. Chicago: University of Chicago Press, 1970.

Shapley, Harlow, *Galaxies*. New York: Harvard University Press, 1961.

————, *The View from a Distant Star*. New York: Basic Books, 1963.

Zwicky, F., *Morphological Astronomy*. Berlin: Springer-Verlag, 1957.

Chapter 33 Radio and X-Ray Sources

Burbidge, G. R., and M. Burbidge, *Quasi Stellar Objects*. London: Freeman, 1967.

Butler, S. T., and H. Messel, *Atoms to Andromeda*. Sydney: Conpress Printing, 1966.

Davies, R. D., and H. P. Palmer, *Radio Studies of the Universe*. London: Routledge and Kegan, 1959.

Hanbury Brown, R., and A. C. B. Lovell, *The Exploration of Space by Radio*. London: Chapman and Hall, 1957.

Hyde, F. W., *Radio Astronomy for Amateurs*. London: Lutterworth Press, 1962.

Kahn, F. D., and H. P. Palmer, *Quasars*. Cambridge, Mass.: Harvard University Press, 1967.

Lovell, Sir Bernard, *The Story of Jodrell Bank*. New York: Harper & Row, 1968.

Maran, S. P., and A. G. W. Cameron, eds., *Physics of Nonthermal Radio Sources*. Washington: N.A.S.A., 1964.

Palmer, H. P., R. D. Davies, and M. I. Large, eds., *Radio Astronomy Today*. Cambridge, Mass.: Harvard University Press, 1963.

Smith, F. G., *Radio Astronomy*. Baltimore: Penguin Books, 1960.

Steinberg, Jean Louis, and James Lequeux, *Radio Astronomy*. New York: McGraw-Hill, 1963.

Chapter 34 Relativity and Cosmology

Bondi, H., *Cosmology*. Cambridge: Cambridge University Press, 1952.

Born, Max. *Einstein's Theory of Relativity*. New York: Dover, 1962.

Caiderc, P., *The Expansion of the Universe*. London: Faber, 1952.

Eddington, Arthur Stanley, *The Nature of the Physical World*. New York: Macmillan, 1940.

Einstein, Albert, *Sidelights on Relativity*. London: Methuen, 1922.

Jeans, Sir James, *Through Space and Time*. New York: Macmillan, 1934.

Lyttleton, R. A., ed., *Rival Theories of Cosmology*. London: Oxford University Press, 1960.

McVittie, G. C., *Fact and Theory in Cosmology*. London: Eyre and Spottiswoode, 1961.

Schilpp, Paul Arthur, *Albert Einstein*, 2 vols. New York: Harper & Row, 1949.

Schlegel, Richard, *Time and the Physical World*. New York: Dover, 1968.

Schlick, Moritz, *Space and Time in Contemporary Physics*. New York: Dover, 1963.

Sciama, D. W., *The Unity of the Universe*. London: Faber, 1959.

Singh, Jagit, *Great Ideas and Theories of Modern Cosmology*. New York: Dover, 1961.

Steinmetz, Charles Proteus, *Four Lectures on Relativity and Space*. New York: Dover, 1967.

Whitrow, G. J., *The Structure and Evolution of the Universe*. New York: Harper & Row, 1959.

Chapter 35 Cosmic Evolution and Time Scale

Gamov, G., *The Creation of the Universe*. New York: Viking Press, 1952.

Johnson, M., *Time, Knowledge and the Nebulae*. London: Faber, 1946.

Woltjer, L., ed., *Galaxies and the Universe*. New York: Columbia University Press, 1968.

Textbook Questions

Chapter 1

1. Give two classical examples of practical applications of astronomy.
2. Give two modern examples of practical applications of astronomy.
3. How were the principles of modern scientific methods of thought and investigation first developed?
4. What is the principle of authority? Is it acceptable in science?
5. What is the difference between astrology and astronomy?
6. Discuss two areas of human endeavor outside the physical sciences where the scientific method might well be applied.

Chapter 2

1. What is a constellation? Name three familiar ones.
2. Explain the correspondence between the names of the planets and the days of the week in English and Spanish (or French).
3. Name some of the greatest ancient Greek astronomers and identify their major discoveries.
4. Explain why Hipparchus has been widely acknowledged as the "father of astronomy."
5. What was the major contribution of Ptolemy?

Chapter 3

1. What was the major contribution of Copernicus? Why was it regarded as revolutionary?
2. What were the major contributions of Tycho Brahe? Why was his work so important to the progress of astronomy?
3. State Kepler's three laws of planetary motions.
4. What were the main astronomical discoveries of Galileo? What other major contributions did he make to science?
5. Where, when, and why was Galileo forced to "abjure" as "absurd and false" the heliocentric system of the world?

Chapter 4

1. What were the major contributions of Newton to physics and astronomy?
2. State Newton's law of universal gravitation.
3. State Newton's three laws of motion.
4. Define the constant of gravitation.
5. Explain precisely the difference between mass and weight.
6. The acceleration of gravity at the surface of the earth is $g = 980$ cm sec^{-2}. Compute the distances that a body in free fall will drop in 0.5, 1, and 2 seconds and its velocity v at the end of these time intervals.
7. Describe the Foucault pendulum experiment. What does it prove?
8. Explain how Neptune was discovered.

Chapter 5

1. What is the velocity of light? Who first measured it and how?
2. What is the aberration of light? Who discovered it? What is the value of the constant of aberration? How is it related to the velocity of light and the orbital velocity of the earth?
3. What is precession of the equinoxes? Who discovered it and how? What is the period of the precession? What causes it?
4. What is nutation? Give its period and amplitude. What causes it?
5. Explain in principle the mechanism of the tides. What causes the difference between spring tides and neap tides?
6. Explain in principle the mechanism of the seasons.
7. Define the sidereal year and give its length.
8. Define the tropical year and give its length. Why does it differ from the sidereal year?
9. Explain the Gregorian reform of the calendar. Why was it needed and when was it introduced.?
10. Explain the need for a future calendar reform and the principle of the proposed World Calendar.

11. Define precisely the celestial equator, poles, zenith and nadir, meridian, hour circle, and hour angle.

12. Explain the difference between true (or apparent) solar time and mean solar time. How are they measured?

13. What is the equation of time? What are its causes?

14. How are standard time zones defined? What is Universal Time? What is Daylight Saving Time?

15. What is Ephemeris Time? When was it introduced and why?

16. Define right ascension and declination.

17. What is the relation between sidereal time and right ascension?

Chapter 6

1. Describe the principle of the refracting telescope. Give the expression for the magnification of the objective-eyepiece combination.

2. Define and give the numerical value of the resolution limit of a refracting telescope of objective aperture A.

3. What is chromatic aberration and how is it reduced in refractors?

4. Describe the principle of the reflecting telescope. Designate the principal focal locations and mirror combinations.

5. Describe the principle of the Schmidt camera. What are its special advantages?

6. What is an equatorial mounting? Describe two common types.

7. What is a radio telescope? Describe several types.

8. What are the main advantages of photography over visual observation?

9. Define the "magnitude" of a star. What are the principal methods of measuring stellar magnitudes?

10. Describe the principal elements of a modern photoelectric photometer.

11. Describe the principal types of spectrographs.

12. What is atmospheric refraction? Give its value and describe some of its effects.

13. Why do stars twinkle?

Chapter 7

1. What are the values of the diameter and oblateness of the earth?

2. Explain the geometric principle used to measure the distance of the moon. What is the principle of another, more modern physical method of measuring the moon's distance?

3. Define the six geometric elements of an elliptical orbit.

4. What is the solar parallax? How is it related to the astronomical unit? Give their numerical values.

5. Describe how Cavendish measured the constant of gravitation.

6. Explain how the relative masses of the sun and of planets with satellites are calculated by means of Kepler's third law.

7. Explain how the ratio of the mass of the moon to the mass of the earth is derived from observations of asteroids.

8. Define the velocity of escape. Give its value for the earth. Why is it also called the parabolic velocity?

Chapter 8

1. Define the main types of electromagnetic radiation and give approximate wavelengths.

2. Give the numerical values of 1 micron and 1 angstrom in centimeters.

3. What is the Planck constant and how does it relate the energy of light quanta to the frequency of the associated wave?

4. Define plane, circular, and elliptical polarization of light waves.

5. List the main types of light detectors and their approximate ranges of sensitivity.

6. State the three Kirchoff-Draper laws of radiation.

7. Define a black body and its radiation.

8. Give the expression of Wien's law of black-body radiation.

9. Give the expression of the Stefan-Boltzmann law of black-body radiation.

10. Describe Bohr's planetary model of the hydrogen atom.

11. Sketch the energy levels of the hydrogen atom and mark the transitions corresponding to the first Lyman and Balmer lines.

12. What (in general principle) is the relation between electronic structure and periodic classification of the elements?

13. What are the main characteristics of molecular spectra?

14. Define kinetic temperature, excitation temperature, ionization temperature.

15. What is the Doppler-Fizeau effect? Give its expression.

Chapter 9

1. What is the solar constant? How is it determined?

2. Why is the sun's disk darker near the edge?

3. Name the main divisions of the outer layers of the sun.

4. What are the Fraunhofer lines? Where do they form?

5. What is the story of the discovery of helium?

6. What is a curve of growth? Sketch one.

7. Describe the sunspots and their main properties.

8. What is the Zeeman effect? How was it used to discover magnetic fields in sunspots?

9. Describe the principle of the spectroheliograph.

10. Describe the principle of the coronograph.

11. Describe the main types of prominences.

12. What are solar flares?

13. What is the chromosphere? Why the name?

14. Describe the solar corona and its variations.
15. How can we estimate the temperature of the corona at one million degrees Kelvin?
16. Describe the main types of radio emission by the sun.

Chapter 10

1. What is the composition of the earth's atmosphere?
2. Describe the main zones of the atmosphere (elevations, temperature).
3. Describe the properties of the magnetic field of the earth (intensity, variations).
4. What are the Van Allen belts?
5. What is the polar aurora? Describe its mechanism and variations.
6. What are cosmic rays? Where do they come from?
7. Explain how earthquakes are used to probe the internal structure of the earth.

Chapter 11

1. Explain the difference and give the relation between the sidereal and synodic months.
2. What is the mean distance of the moon?
3. What are the moon's librations?
4. Explain the mechanism and different types of solar eclipses.
5. What is the saros and its cause?
6. Give some basic facts and figures about the moon (diameter, density, atmosphere, surface structure).
7. Explain the differences between the craters of the moon and volcanic craters on earth.
8. Describe the main techniques used to measure the temperature of the moon.
9. Explain the mechanism and different types of lunar eclipses.
10. What is the structure of the moon's top layer according to direct measurements?

Chapter 12

1. Describe and compare the phases of the interior and exterior planets.
2. Give the main elements of the orbit and describe the rotation of Mercury.
3. Transits of Mercury in front of the Sun can take place only in what months of the year? Why?
4. What peculiarity of the motion of Mercury was explained by Einstein's theory of relativity?
5. What is the rotation period of Mercury? Explain its relationship to the period of revolution.
6. What is the surface temperature of Mercury at the subsolar point?
7. Give the main elements of the orbit and describe the rotation of Venus.

8. Explain why Venus moves from greatest eastern elongation in the evening sky to greatest western elongation in the morning sky in only about 5 months but requires some 14 months more to complete the circuit (use a diagram).

9. What is the rotation period of Venus? How was it determined?

10. What is known of the composition and structure of the atmosphere of Venus?

11. What is the surface temperature of Venus? How can it be so high?

12. Give the main elements of the orbit and describe the rotation of Mars.

13. What is the evidence for an atmosphere on Mars? What are its composition and surface pressure and how were they determined?

14. What is the evidence for water ice and for dry ice on Mars?

15. Describe the seasonal variations observed on the surface of Mars.

16. Is the presence of moonlike craters on Mars surprising? Explain your answer.

17. Give the main known facts about the rotation and orbit of Pluto.

Chapter 13

1. What molecules are responsible for the strong absorption bands in the spectra of the major planets?

2. Give the diameters and ellipticities of the globes of the major planets (in km or earth diameters).

3. Give the number of known satellites of each of the major planets.

4. Describe the structure of the rings of Saturn. What is their probable origin?

5. Describe a plausible model of the internal structure of the major planets.

6. What is the red spot of Jupiter?

7. Describe the main properties of the various types of radio emission from Jupiter and their probable mechanism.

Chapter 14

1. When was the first asteroid discovered? By whom? Was the discovery accidental? Why?

2. How many asteroids are known? How many more remain to be discovered? Explain your answer.

3. Where, in the solar system, does the asteroid belt fit? What has Bode's law to do with the question?

4. What are the diameters of the larger asteroids?

5. Name two famous periodic comets; give their periods of revolution.

6. Describe the three main parts of a comet and their composition.

7. What is the icy-conglomerate model of a cometary nucleus?

8. How long are cometary tails? How do they vary with distance to the sun, and why?

9. Describe the spectrum of a comet (a) far from the sun, (b) very close to the sun. Why the differences?

10. Compare the general orbital properties of planets and comets.

Chapter 15

1. What is the radiant of a meteor shower? What is the geometrical explanation of this phenomenon?
2. Explain precisely the different meanings of the words meteor, meteorite, meteoroid, bolide, shooting star.
3. What is the range of velocities at which meteors encounter the earth? What is the explanation for the limiting values?
4. What are the two major classes of meteorites and their chemical composition?
5. Explain the principle of the method used to observe meteors by radar.
6. Can we observe meteors during daytime? How?
7. Name the major nighttime and daytime meteor showers.
8. Are the orbits of meteor streams and of comets related? Why? Name some such associations.
9. Describe the zodiacal light. What is its source?

Chapter 16

1. Explain the principle of rocket propulsion.
2. What is the escape velocity? How does it compare with the circular velocity? Give the values of both velocities for the earth.
3. The mass of the earth is $6 \cdot 10^{27}$ g, its radius $6.4 \cdot 10^8$ cm; compute the period of revolution of an artificial satellite (near the earth's surface).
4. What are the main biological hazards of space travel (a) close to the earth (artificial satellites), (b) in interplanetary space?
5. What are the major scientific applications of space exploration?

Chapter 17

1. What are the main properties of the solar system that must be explained by cosmogonic theories?
2. What is the estimated age of the earth? On what evidence?
3. What is angular momentum. How is this concept relevant to the problem of the origin and evolution of the solar system?
4. Describe Laplace's nebular hypothesis.
5. Describe several hypotheses of the origin of the moon.

Chapter 18

1. Is there any evidence for life beyond the earth? If not, why discuss it?
2. Why do discussions of the origin of life pay so much attention to the chemistry or photochemistry of C, O, N, H mixtures and compounds?
3. Outline some modern views on the origin of life.
4. Why are we so sure there is no life on the moon, but remain in doubt for Mars?

Chapter 19

1. How could Herschel estimate relative star distances even though no stellar distances were then known?
2. Define stellar parallax and describe a method for measuring it.
3. The parallaxes of the following stars are: Sirius, 0.375; Procyon, 0.″287; Altair, 0.″210; Betelgeuse, 0.″012. Calculate their distances in parsecs and in light years. (Check answers with Tables 19.1 and 19.2.)
4. Sirius has a proper motion of 1.″32 per year. Calculate its tangential velocity. (Note question 3.)
5. The apparent visual magnitude of Sirius is −1.ᵐ43. Calculate its absolute magnitude and the ratio of its absolute visual luminosity to that of the sun. (Note question 3.)
6. Why does the color index increase from blue to red stars?

Chapter 20

1. Who were the first astronomers to study stellar spectra? How long ago?
2. Application of the spectroscope to the study of the stars caused the second revolution in astronomy; why?
3. Describe the two main methods of photographing stellar spectra. Which is best adapted to general surveys and which to detailed studies; why?
4. Describe the Henry Draper spectral classification system and the main characteristics of each spectral class.
5. What are the main spectral lines and elements visible in the spectra of B-type stars, A-type stars, F-type stars, G-type stars, K-type stars, M-type stars?
6. What principles and radiation laws are used to measure stellar temperatures?
7. Why is it possible to compare a star to a black body? If this designation is not entirely justified, what are the differences between the spectrum of a star and that of a black body?
8. Explain the principle of the luminosity classification of stars from an inspection of their spectra.
9. What is a "spectroscopic parallax"? What is the physical basis for this method?
10. Define color index and describe by a graph its relation to spectral class.
11. What is the Hertzsprung-Russell diagram? Describe by a graph the H-R diagram of nearby stars.

Chapter 21

1. Explain the distinction between optical and physical double stars.
2. What two quantities do we measure for a double star?
3. Describe the *apparent* relative orbit of a double star. Why does it differ (in general) from the *true* orbit?

4. Explain how the orbital motions of double stars are used to compute stellar masses.

5. What is a "spectroscopic binary"? How is it observed? Does it appear double in a telescope? If not, how do we know it is a physical pair?

6. What is an "eclipsing binary"? Name a famous example. Is it seen double in a telescope? If not, how do we know it is a pair in orbital motion?

7. Describe the main different types of light curves of eclipsing binaries and the principles of their explanation.

8. What is the mass-luminosity relation? Who discovered it? What does it mean?

Chapter 22

1. Can we directly see the disks of stars (other than the sun) through telescopes? If not, why?

2. What are the principal methods used to measure the angular diameters of stellar disks?

3. Explain how we can estimate the diameter of a star if we know its absolute luminosity and surface temperature.

4. What is an "interferometer"? Describe the principle of the Michelson interferometer. How is it used to measure (a) double stars, (b) stellar diameters?

5. Explain how occultations of stars by the moon can be used to measure stellar diameters.

6. What is the "intensity interferometer"? Can you explain its principle? To what stellar types does it apply the best?

7. Give typical mean densities of (a) red dwarf stars, (b) the sun, (c) red giant stars, (d) white dwarfs.

Chapter 23

1. Define and discuss what distinguishes the photosphere and the reversing layer in a stellar atmosphere.

2. Describe the effect of atmospheric pressure differences among stars of the same spectral class.

3. What are the Balmer lines of hydrogen? In what types of stellar spectra are they strongest?

4. What is the Lyman series of hydrogen lines? Can we observe it from the surface of the earth? If not why? From where and how can we observe this series?

5. What is the Stark effect? Is it observed in stellar spectra? If so, in what spectral classes is it most conspicuous?

6. In what spectral classes can we observe absorption lines of helium? Why are they strong in only one spectral class?

7. In what spectral classes are the absorption lines of metals particularly strong? What atoms are most commonly observed?

8. Name the atom or ion responsible for the following absorption lines: Fraunhofer C line (λ 6563), D lines (λ 5890, 5896), F line (λ 4861), H and K lines (λ 3968, 3933).

9. In what spectral classes are the absorption bands of molecules particularly strong? What molecules are most commonly observed?

10. Define the "equivalent width" of a spectral line. What is a "curve of growth"? Sketch one.

11. Name and describe briefly the three mechanisms by which heat can be exchanged. Which two are important in stellar atmospheres?

Chapter 24

1. Define atomic weight and atomic number.
2. Name the main fundamental particles of atomic nuclei.
3. State Einstein's principle of equivalence between mass and energy and give its expression.
4. How much energy (in ergs, calories, and kilowatt-hours) would be produced by the total conversion of 1 gram of matter into energy?
5. What is deuterium? heavy water? an alpha particle?
6. How much energy is released when 4 hydrogen atoms combine to form an atom of helium?
7. Define nuclear fission and nuclear fusion. Give examples of practical applications.
8. What is a chain reaction? Give an example.

Chapter 25

1. What three basic forces determine the equilibrium of a star? Which one is often negligible?
2. Why are we certain that the heat of the sun is not produced by ordinary combustion (that is, chemical reactions)?
3. Define exothermic and endothermic reactions. Give an example of each type.
4. Describe the formation of helium by the proton-proton chain. Why is it important for the production of stellar energy?
5. What is the carbon cycle? What is its net result? Why do we consider the carbon nucleus as playing the role of a catalyst?

Chapter 26

1. What is a variable star? How is it designated?
2. Define the main types of variable stars.
3. Describe the light curve and other properties of a long-period variable.
4. Describe the light curve and other properties of a cepheid variable.

5. What is the period-luminosity relation? Why is it so important?
6. Describe the light curve and other properties of a cluster variable.
7. What is a magnetic variable?
8. What is a flare star?

Chapter 27

1. What are the main types of cataclysmic variables? Why the name?
2. Describe the light curve and other properties of dwarf novae.
3. Describe the light curve and other properties of a fast and a slow nova. Give some examples.
4. What is a recurrent nova? Give some examples.
5. What are supernovae? Give some historical examples.
6. What is the Crab Nebula?

Chapter 28

1. What is symbiosis? What is a symbiotic star?
2. What are the chief characteristics of Wolf-Rayet stars?
3. What forces are probably responsible for the production of distended stellar envelopes?
4. What are planetary nebulae? Why are they so named? What are the temperatures of the nuclear stars?
5. What is the suspected relationship between planetary nebulae and novae?
6. What is the probable location of W stars in the spectral sequence?

Chapter 29

1. What proofs can we give for the presence of dust clouds in interstellar space?
2. Why are dust or smoke particles much more efficient light absorbers than planets or meteorites for the same total mass?
3. Describe the principle of the Wolf diagram and indicate how star counts are used to estimate the absorption by dust clouds.
4. What is interstellar reddening and its mechanism?
5. Define Rayleigh scattering and Mie scattering. Which is dominant in interstellar space?
6. Why is the daytime sky blue?
7. Why is the setting sun red?
8. What is interstellar polarization?
9. What is the evidence for the presence of isolated atoms and molecules in interstellar space?
10. How can we differentiate between stellar and interstellar lines in the spectrum of a star?

11. What are the main atoms and molecules detected in interstellar space?
12. What is a reflection nebula? Give an example.
13. What is an emission nebulosity? Give an example.
14. Explain the difference between H I and H II regions. What is the usual mechanism of formation of an H II region?
15. What is a Strömgren sphere?
16. What is the evidence for the presence of neutral hydrogen gas in interstellar space? What is its mean density?

Chapter 30

1. Define galactic (or open) clusters, globular clusters, and stellar associations. Give some examples of each type.
2. What types of star clusters are the following: the Pleiades, the Hyades, the Hercules cluster, Omega Centauri?
3. Explain how the distance of the Hyades can be measured by the method of the convergence.
4. Define the modulus relating absolute magnitude and distance. How is the modulus related to geometric distance?
5. Explain how the H-R or color-magnitude diagram can be used to estimate cluster distances.
6. What is a stellar association? What is the evidence that such associations are unstable and short-lived on a cosmic time scale?
7. How can we estimate the distances of globular clusters?
8. Describe the H-R diagram of a typical globular cluster and how it differs from those of open clusters.

Chapter 31

1. Define galactic equator and poles. Where is the north galactic pole? the galactic center?
2. What is the principle of Herschel's explanation of the Milky Way?
3. What are the approximate dimensions of the Milky Way system?
4. What is the solar apex? Explain how the motion of the sun in space can be derived from studies of the proper motions and/or radial velocities of stars.
5. What proofs do we have that the galactic system is rotating?
6. Explain the effects of differential rotation of the Galaxy on stellar radial velocities.
7. What is the approximate period of rotation of the sun around the galactic center? What is the mass of the Galaxy?
8. Explain how the mass of the Galaxy can be derived from the orbital velocity of the sun and its distance to the center.
9. What methods were used to discover the spiral structure of the Galaxy?

Chapter 32

1. How did the concept of "island universes" or galaxies beyond our own develop historically? When was it finally demonstrated and how?
2. Describe and sketch the main types of galaxies.
3. What are the two main types of stellar populations in galaxies?
4. Give the distances of the nearest and brightest galaxies.
5. What is the range of absolute magnitudes or luminosities of giant and dwarf galaxies?
6. What are the main emission lines in the spectra of galaxies? What causes them?
7. Give typical values of the rotation periods of galaxies of different types.
8. Describe the principle of the method used to estimate masses of rotating galaxies.
9. Describe some transient phenomena in galaxies.
10. Describe the evidence for clustering among galaxies.

Chapter 33

1. How were cosmic radio waves discovered?
2. Describe the first observations of radio emission from the Milky Way and from the sun.
3. Describe the first observations of small radio sources.
4. What are the main types of cosmic radio sources?
5. What are the two main mechanisms of radio emission? Give some examples.
6. What is the unit of radio flux?
7. What is synchrotron radiation?
8. What is a quasi-stellar radio source or "quasar"? What is the order of magnitude of the power emitted?
9. What techniques are used to detect cosmic sources of X rays?
10. What is a "pulsar"?

Chapter 34

1. What was the objective of the experiment of Michelson and Morley? What was its result?
2. Einstein built two distinct theories of "relativity." What are their fields of application?
3. What are the three classical astronomical proofs of the theory of general relativity?
4. What is the distinction between cosmology and cosmogony?
5. State the fundamental assumptions of classical cosmology.
6. What are the main differences between an Einstein universe and a Euclidian universe? Is either an acceptable model of the real world? Explain why.

7. What is the cosmological constant? Why was it introduced?
8. What is the empirical evidence for an expanding universe?
9. What is Hubble's law?
10. Describe the main properties of the Lemaitre-Eddington universes.
11. What is the basic idea of the steady-state model?
12. What is Olbers' paradox? Describe another way to resolve it. How is it avoided by the existence of the cosmological redshift?
13. What are the principal observational tests of cosmological theories?

Chapter 35

1. What is meant by "ages" of the earth, the Galaxy, the Universe?
2. Explain the principle of the radioactive method of estimating the ages of the earth, meteorites, and moon.
3. Describe, in general terms, the present concepts of the origin and evolution of the sun and stars.
4. Outline the general concepts of Lemaitre's primeval atom or "big bang" theory of the origin and evolution of the universe.
5. What is the so-called 3°K background radiation? Why is it related to the concept of an initial fireball?
6. Outline some unsolved problems of cosmology and cosmogony.

Mathematical Appendix

The authors have tried to keep the numerical and mathematical portions of this text as elementary as possible. Astronomy, however, is a physical science, and the development of certain concepts inevitably requires some mathematical analysis. To assist the student, we shall here briefly review some basic principles of several important branches of elementary mathematics.

C.1. Powers. In arithmetic one usually uses the symbol × to signify "times" or multiplication, as in $2 \times 6 = 12$. We may also use a raised dot to denote multiplication, as in $2 \cdot 6 = 12$. Or we may simply enclose the numbers in adjacent parentheses to indicate the process $(2)(6) = 12$. If we use letters in place of the numbers, we may omit the times symbol and say $a \times b = a \cdot b = (a)(b) = ab$.

Let a be any number whatever. The product of a by itself or $a \cdot a$, we term a square, written as a^2. The number, 2, called an exponent, denotes the number of times a appears in the product. We can, if we wish, write $a = a^1$. In similar fashion, we term the product $a \cdot a \cdot a$ as a cube, or a^3. Generalizing this notation, we say that the product of n factors, $a \cdot a \cdot \ldots \cdot a$, is a to the nth power or a^n.

If now we wish to multiply say, a^2 by a^3, the resulting product $(a \cdot a)$ $(a \cdot a \cdot a) = (a \cdot a \cdot a \cdot a \cdot a) = a^5$. The exponent $5 = 2 + 3$ indicates the number of times that a appears in the product. Thus we get the simple rule, basic in algebra, that to multiply one adds exponents. Thus $a^m \cdot a^n = a^{m+n}$.

A simple extension of the rule above includes division. Using the fraction line, we get $a^5/a^2 = a \cdot a \cdot a \cdot a \cdot a / a \cdot a = a^{5-2} = a^3$. Hence $a^m/a^n = a^{m-n}$. To divide, one subtracts the exponent of the divisor (that which divides) from that of the dividend (that which is to be divided). If n is greater than m, the difference is negative. For example, $a^2/a^5 = 1/a^3 = a^{-3}$. A negative exponent signifies the reciprocal of the expression with a positive exponent. In particular,

$$a \cdot a^{-1} = a/a = a^0 = 1.$$

Any quantity raised to the zero power is equal to 1 or "unity."

The exponents need not be integers. Suppose that we wish to multiply the square root of a by itself. The product must clearly be equal to a, or $\sqrt{a} \cdot \sqrt{a} = a$. Instead of the radical, $\sqrt{}$, we may use the notation $a^{1/2} \cdot a^{1/2} = a^1 = a$. The two exponents $\frac{1}{2} + \frac{1}{2} = 1$. Similarly we may take a cube root, or an nth root. Divide the exponent by the proper number: $a^{1/3}$ or $a^{1/n}$. For example, the nth root of a^m is $a^{m/n}$. We may generalize this concept further. An exponent can be any number, rational or irrational, positive or negative. Thus we may find $a^{1.7306\cdots}$ or $a^{-2.7143\cdots}$.

C.2.   **Numbers.**   In astronomy we frequently encounter numbers that are astronomically large or astronomically small. For example, the mass of the sun in grams is 1,989,000, ... followed by 27 more zeros. The mass of a hydrogen atom, in grams, is 0.000 ... 1673, where the intervening dots signify 20 additional zeros. The standard notation for numbers is clearly far too cumbersome for astronomical use.

Let us apply the exponential notation to our problem. A thousand, 1000, for example, is $10 \times 10 \times 10 = 10^3$. A million is 10^6. The exponent indicates the number of zeros following the 1. Thus 10^{33} is 1 followed by 33 zeros. The notation can be used for small numbers too. $0.001 = \frac{1}{1000} = 10^{-3}$; $0.000001 = \frac{1}{1000000} = 10^{-6}$, and so on. 10^{-24} signifies a 1 preceded by 23 zeros and a decimal point.

For numbers other than those of the form 10^n, where n is either a positive or negative integer, we may note the following examples:

$$200 = 2 \times 100 = 2 \times 10^2,$$
$$243 = 2.43 \times 100 = 2.43 \times 10^2,$$
$$0.2 = \tfrac{2}{10} = 2 \times 10^{-1},$$
$$0.00316 = \frac{3.16}{1000} = 3.16 \times 10^{-3}.$$

Hence we may express the masses of the sun and the hydrogen atom, previously noted, respectively as

$$1.989 \times 10^{33} \text{ g}, \qquad 1.673 \times 10^{-24} \text{ g}.$$

C.3.   Logarithms. As noted under (1), in the form 10^n, we are not necessarily limited to integral values of n. For example, the square root of 10 is

$$10^{1/2} = 10^{0.5} = 3.1623 \ldots .$$

Similarly

$$10^{1/3} = 10^{0.333 \cdots} = 2.154,$$
$$10^{2/3} = 10^{0.666 \cdots} = 4.641.$$

Thus, we could express the number

$$2154 = 2.154 \times 10^3 = 10^{0.333} \times 10^3 = 10^{3.333 \cdots}$$

by the rule of adding exponents. We can express any number as 10 to some special power. Table C.1 gives the exponents for the integers from 1 to 10.

TABLE C-1

Number	Exponent	Number	Exponent
1	0.0000	6	0.7782
2	0.3010	7	0.8451
3	0.4771	8	0.9031
4	0.6021	9	0.9542
5	0.6990	10	1.0000

Work out a problem in multiplication:

$$2 \times 3 = 10^{0.3010} \times 10^{0.4771} = 10^{0.7781} = 6.$$

Again we have added exponents. The small difference between 0.7781 and 0.7782 is the result of rounding-off errors. The 10, which is raised to the special power, is only incidental. The exponents we have used are also called *logarithms* to the base 10. The logarithm of a number uniquely specifies the number. For example, the masses in grams of the sun and of the hydrogen atom can be written, respectively as

$$10^{33.2987}$$

and

$$10^{-24} \times 10^{0.2235} = 10^{-24 + 0.2235} = 10^{-23.7765}.$$

We say that the logarithm of the solar mass is 33.2987 and the logarithm of the mass of a hydrogen atom is -23.7765. Sometimes we call the whole number to the left of the decimal the *characteristic* and the positive decimal part the *mantissa*. Thus, in the example above, -24 is the characteristic and 0.2235 the mantissa.

Logarithms are useful for multiplying or dividing, since addition or subtraction is simpler and less subject to error. However, since modern computers or even desk calculators have largely supplanted logarithms, we shall not pursue the question further in this text.

Our major use of logarithms here will be in diagrams. When a diagram embraces a wide range of powers of ten, logarithms are much handier to employ than the exponential form. Just remember that the logarithm is really the power to which 10 must be raised to equal the number.

C.4. **Algebra.** Only one chapter in this textbook makes any extensive use of algebra, that on relativity (Chapter 34), where we have rigorously derived the fundamental laws of special relativity in elementary fashion. Most advanced texts resort to higher mathematics, far beyond the scope of this book.

A full discussion of the principles of algebra would take us too far afield. An essential characteristic of algebra is that letters may take the place of numbers—we may substitute various sets of numbers for the letters. Use of letters permits us to derive general formulas that can be examined numerically if desired.

To assist the student in some of the more complicated equations, we have indicated separate steps in the derivation. For example, in equation (34.4), we start with the equation

$$T = \frac{L}{c + v} + \frac{L}{c - v}.$$

The right-hand side of this equation is the sum of two fractions. To carry out the indicated addition we must have the same denominators. Multiply the first fraction by $\frac{c - v}{c - v}$ and the second by $\frac{c + v}{c + v}$. The result is

$$T = \frac{L\,(c - v)}{(c + v)\,(c - v)} + \frac{L\,(c + v)}{(c + v)\,(c - v)}.$$

The product in the denominator is

$$(c + v)\,(c - v) = c^2 + cv - cv - v^2 = c^2 - v^2.$$

Now add the numerators.

$$T = \frac{L(c - v + c + v)}{c^2 - v^2} = \frac{2Lc}{c^2 - v^2}.$$

Divide the numerator and denominator by c^2 to get

$$T = \frac{2L}{c}\frac{1}{1 - v^2/c^2} = \frac{2L}{c}\frac{1}{(1 - \beta^2)},$$

with

$$\beta = \frac{v}{c}.$$

C.5.   Geometry. Astronomers use many of the results of geometry. For example, the circumferences C and area A of a circle of radius r are

$$C = 2\pi r, \qquad A = \pi r^2.$$

The area of a triangle is equal to one-half the product of the base and the altitude.

The area of a "thin" isosceles triangle whose apex angle is α (in degrees), and whose long side has length r, is

$$A = \frac{\pi\alpha}{360}r^2.$$

Such triangles often occur in astronomical problems involving determinations of parallax.

The Pythagorean theorem for a right triangle of sides a and b and hypotenuse c is

$$a^2 + b^2 = c^2.$$

C.6.   Plane Trigonometry. Certain relations concerning the sides and angles of right triangles are useful in astronomical problems. Figure C.1 shows a circle of radius r, center O, horizontal diameter $A'OA$, and vertical diameter BOB'. From the center, and within the sector $BAOB$, draw a radius to an arbitrary point M on the circumference. Let θ be the angle, measured counterclockwise from the line OA, between OA and OM. From M drop a perpendicular, MP, on the line OA. Then, by construction, the triangle OPM is a right triangle, with right angle at P. For simplicity, call the length of the three sides: $OP = a$, $PM = b$, and $OM = r = c$.

Define the quantities,

$$\text{sine } \theta \qquad = \sin \theta = \frac{b}{c},$$

$$\text{cosine } \theta \qquad = \cos \theta = \frac{a}{c},$$

$$\text{tangent } \theta \quad = \tan \theta = \frac{b}{a},$$

$$\text{cotangent } \theta = \cot \theta = \frac{a}{b}.$$

We call these quantities "trigonometric functions." No matter how large r may be, for a given value of θ, the numerical value of the ratios is fixed. They depend only on θ. A sample tabulation for a few selected values of θ appears in Table C.2. For θ between $0°$ and $45°$ use the upper designation, $\sin \theta$, $\tan \theta$, and so on. For θ between $45°$ and $90°$ use the lower designation. Tables of trigonometric functions are available, of course, for much finer divisions of angle. Note that

$$\frac{\sin \theta}{\cos \theta} = \tan \theta,$$

$$\cot \theta = \frac{1}{\tan \theta}.$$

Also, from the Pythagorean theorem

$$\frac{b^2}{c^2} + \frac{a^2}{c^2} = 1,$$

or

$$\sin^2 \theta + \cos^2 \theta = 1.$$

To use these functions in the calculation of triangles, note, for example, that

$$b = c \sin \theta$$

and

$$a = c \cos \theta.$$

TABLE C-2

θ	$\sin \theta$	$\tan \theta$	$\cot \theta$	$\cos \theta$	
$0°$	0	0.0000	∞	1.0000	$90°$
$5°$	0.0872	0.0875	11.430	0.9962	$85°$
$10°$	0.1736	0.1763	5.6713	0.9848	$80°$
$15°$	0.2588	0.2679	3.7321	0.9659	$75°$
$20°$	0.3420	0.3640	2.7475	0.9397	$70°$
$25°$	0.4226	0.4663	2.1445	0.9063	$65°$
$30°$	0.5000	0.5774	1.7321	0.8660	$60°$
$35°$	0.5736	0.7002	1.4281	0.8192	$55°$
$40°$	0.6428	0.8391	1.1918	0.7660	$50°$
$45°$	0.7071	1.0000	1.0000	0.7071	$45°$
	$\cos \theta$	$\cot \theta$	$\tan \theta$	$\sin \theta$	θ

Thus, if we know c and θ, we can calculate b and a. Suppose that $c = 20$ miles and $\theta = 15°$. From the table we find that

$$\sin 15° = 0.2588$$

and

$$\cos 15° = 0.9659.$$

Therefore

$$b = 5.176 \text{ miles}, \qquad a = 19.318 \text{ miles}.$$

If θ lies in the second quadrant $A'OB$ between 90° and 180°, one imagines folding the circle about BOB' until the new point appears in the first quadrant. The trigonometric functions are the same as before, except that the cos, tan, and cot are negative. Thus, if the angle were 165° instead of 15°, we should have

$$\sin 165° = 0.2588,$$
$$\cos 165° = -0.9659,$$
$$\tan 165° = -0.2679,$$
$$\cot 165° = -3.7321.$$

In the third quadrant, both sin and cos are negative and the tan and cot positive. In the fourth quadrant, the sin, tan, and cot are negative and the cos positive. These refinements are not necessary for this text, but students should know how to proceed for angles greater than 90°.

D

Units of Measurement and Conversion Factors

Length

1 angstrom unit	= 1 Å =	10^{-10} m =	10^{-8} cm =	10^{-4} μ
1 micron	= 1 μ =	10^{-6} m =	10^{-4} cm =	10^4 Å
1 centimeter	= 1 cm =	10^{-2} m =	0.39370 in. =	0.032808 ft
1 meter	= 1 m =	3.28084 ft =	10^2 cm =	1.09361 yd
1 kilometer	= 1 km =	10^3 m =	10^5 cm =	0.62137 mi
1 astronomical unit	= 1 A.U.=	1.49598×10^8 km =	1.49598×10^{13} cm =	9.2956×10^7 mi
1 parsec	= 1 pc =	3.08568×10^{18} cm =	206265 A.U. =	3.2616 l.y.
also				
1 inch	= 1 in. =	0.0254 m =	2.54 cm =	2.54×10^4 μ
1 foot	= 1 ft =	0.3048 m =	30.48 cm =	12 in.
1 yard	= 1 yd =	0.9144 m =	91.44 cm =	3 ft
1 statute mile	= 1 mi =	1609.344 m =	1.60934×10^5 cm =	5280 ft
1 nautical mile	= 1 n mi =	1852 m =	1.852×10^5 cm =	6076.1 ft
1 light year	= 1 l.y. =	9.46053×10^{17} cm =	63240 A.U. =	0.30659 pc

Mass

1 gram	= 1 g =	10^{-3} kg =	2.20462×10^{-3} lb
1 kilogram	= 1 kg =	10^3 gm =	2.20462 lb
1 ton, metric	= 1 ton =	10^3 kg = 10^6 g =	2204.62 lb
1 pound, avoir.	= 1 lb = 0.453592 kg =		453.592 g
1 ton, short	= — =	907.185 kg =	2000 lb

Time (for 1900.0)

1 second	= 1 sec	= 1/31,556,925.9747 of tropical year for 1900.0
1 second, sidereal	= 1 sid sec	= 0.997270 sec
1 hour	= 1 hr	= 3600 sec
1 day, mean solar	= 1 d	= 86,400 sec
1 day, sidereal	= 1 sid d	= 86,164 sec = 0.99727 d
1 month, synodic	= 1 mo	= 29.530588 d = 2.55144×10^6 sec
1 month, sidereal	= 1 sid mo	= 27.321661 d = 2.36059×10^6 sec
1 year, tropical	= 1 yr	= 365.242199 d = 3.15569×10^7 sec
1 year, sidereal	= 1 sid yr	= 365.256366 d = 3.15581×10^7 sec
1 year, anomalistic	= —	= 365.259641 d (perihelion to perihelion)
1 year, eclipse	= —	= 346.620031 d
1 year, Julian	= —	= 365.25 d
1 year, Gregorian	= —	= 365.2425 d

Frequency

1 cycle per second = 1 cps = 1 hertz = 1 hz

Velocity

1 km/sec	= 2236.9 mile/hr	1 A.U./yr	= 4.74 km/sec
1 mile/hr	= 44.704 cm.sec	1 pc/10^6 yr	= 0.978 km/sec

Energy

1 erg	= 1 g cm^{2}sec^{-2}	
1 electron volt	= 1 ev	= 1.60209×10^{-12} erg
1 Mev	= 10^6 ev	
1 Gev	= (1 Bev, old)	= 10^9 ev = 1.60209×10^{-3} erg
1 joule	= 1 j	= 1×10^7 erg
1 calorie	= 1 cal	= 4.1854×10^7 erg = 1 g calorie
1 calorie, kg	= 4.1854×10^{10} erg	
1 kilowatt-hour	= 1 kw-hr	= 3.6×10^{13} erg

Angle

1 second of arc	= 1 arc sec	= 1/206,265 rad	= 1/1,296,000 circle	
1 minute of arc	= 1 arc min	= 60 arc sec	= 1/3437.75 rad	= 1/21,600 circle
1 degree of arc	= 1 deg	= 60 arc min	= 1/57.2958 rad	= 1/360 circle
1 radian	= 1 rad	= $180°/\pi$	= 57.2958 deg	= 3437.75 arc min
				= 206,265 arc sec
1 complete circle	= 2π rad	= 360°	= 21,600 arc min	= 1,296,000 arc sec

Solid Angle

1 steradian	= 1 sterad	= 1/4π sphere	= 3282.81 sq deg
1 sphere	= 4π sterad	= 41,253.0 sq deg	

Temperature

1 degree centigrade $\quad= 1$ degree Celsius $= 1°C$
1 degree Kelvin $\qquad= 1$ degree absolute $= 1°K$
1 degree Fahrenheit $\quad= 1°F$
absolute zero $\qquad\qquad= 0°K = -273.16°C = -459.69°F$
water melts at $273.16°K = 0°C = 32°F$
water boils at $373.16°K = 100°C = 212°F$ $\Big\}$ at mean sea-level atm. pressure
(Rankine scale is $459.69 +$ temp. $°F$)
temperature in $°F = \frac{9}{5}(\text{temp. } °C) + 32° = \frac{9}{5}(\text{temp. } °K) - 459.69$
temperature in $°C = \frac{5}{9}(\text{temp. } °F - 32°) = (\text{temp. } °K) - 273.16$

Prefixes for multiplying factors

Multiplying Factor	Prefix	Symbol	Multiplying Factor	Prefix	Symbol
10^{12}	tera	T	10^{-2}	centi	c
10^9	giga	G	10^{-3}	milli	m
10^6	mega	M	10^{-6}	micro	μ
10^3	kilo	k	10^{-9}	nano	n
10^2	hecto	h	10^{-12}	pico	p
10	deka	da	10^{-15}	femto	f
10^{-1}	deci	d	10^{-18}	atto	a

EXAMPLE: 10^6 cycles per second $= 1$ Mhz; 0.01 m $= 1$ cm

Physical Constants

velocity of light in vacuum $= c \quad = 2.997925 \times 10^{10}$ cm/sec
constant of gravitation $\qquad= G \quad = 6.668 \times 10^{-8}$ cm³/(g sec²)
mass of hydrogen atom $\qquad= m_H \ = 1.673 \times 10^{-24}$ g
mass of electron $\qquad\qquad= m_e \ = 9.1084 \times 10^{-28}$ g

Astronomical Quantities

earth mass $\qquad\qquad\qquad 5.977 \ \times 10^{27}$ g
earth equatorial radius $\quad 6.3782 \times 10^8$ cm
sun mass $\qquad\qquad\qquad 1.989 \ \times 10^{33}$ g
sun radius $\qquad\qquad\qquad 6.960 \ \times 10^{10}$ cm
sun luminosity $\qquad\qquad 3.90 \ \ \times 10^{33}$ erg/sec
solar constant $\qquad\qquad 1.37 \ \ \times 10^6$ erg/(cm² sec)

Note that the British and Europeans use *millard* to identify a factor of 10^9, *billion* for 10^{12}, *trillion* for 10^{18}, and *quardrillion* for 10^{24}.

APPENDIX *E*

Constellation Charts

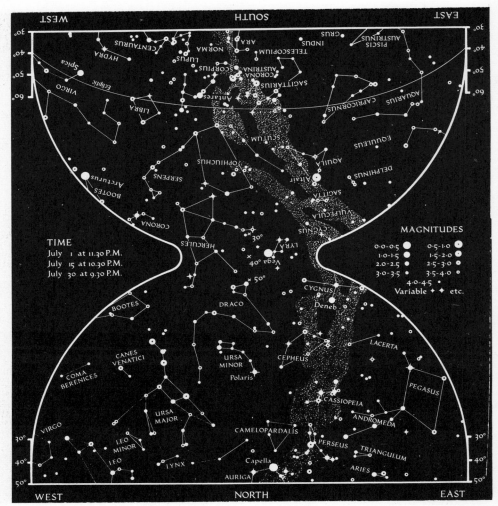

TIME
July 1 at 11.30 P.M.
July 15 at 10.30 P.M.
July 30 at 9.30 P.M.

MAGNITUDES

0.0-0.5	0.5-1.0
1.0-1.5	1.5-2.0
2.0-2.5	2.5-3.0
3.0-3.5	3.5-4.0
4.0-4.5	

Variable ✦ ✦ etc.

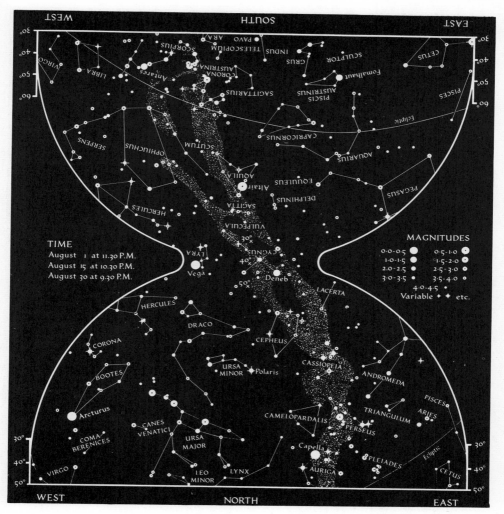

WEST · SOUTH · EAST

VIRGO · LIBRA · SCORPIUS · ARA · PAVO · TELESCOPIUM · INDUS · GRUS · SCULPTOR · CETUS · PISCES

SERPENS · OPHIUCHUS · SCUTUM · SAGITTARIUS · CORONA AUSTRINA · PISCIS AUSTRINUS · Fomalhaut · Ecliptic

HERCULES · AQUILA · Altair · SAGITTA · EQUULEUS · DELPHINUS · CAPRICORNUS · AQUARIUS · PEGASUS

VULPECULA · CYGNUS · Deneb · LACERTA

LYRA · Vega

TIME
August 1 at 11.30 P.M.
August 15 at 10.30 P.M.
August 30 at 9.30 P.M.

MAGNITUDES

0·0-0·5	●	0·5-1·0	●
1·0-1·5	●	1·5-2·0	●
2·0-2·5	●	2·5-3·0	○
3·0-3·5	○	3·5-4·0	○
	4·0-4·5		
Variable ✦ ✦ etc.			

HERCULES · DRACO · CEPHEUS

CORONA · CASSIOPEIA · ANDROMEDA

BOOTES · URSA MINOR · Polaris · PISCES · ARIES

Arcturus · CAMELOPARDALIS · TRIANGULUM · PERSEUS

CANES VENATICI · URSA MAJOR · Capella · PLEIADES · Ecliptic · CETUS

COMA BERENICES · VIRGO · LEO MINOR · LYNX · AURIGA

WEST · NORTH · EAST

30° · 40° · 50°

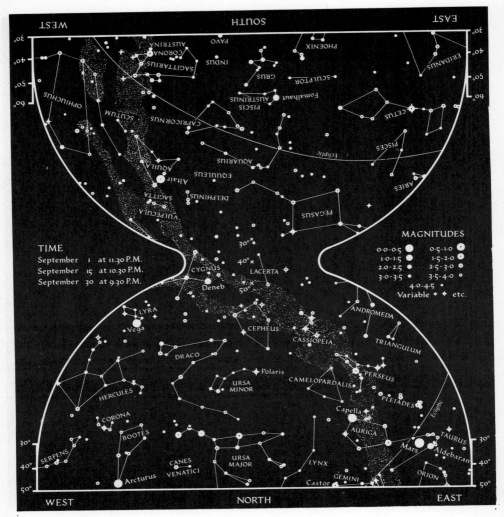

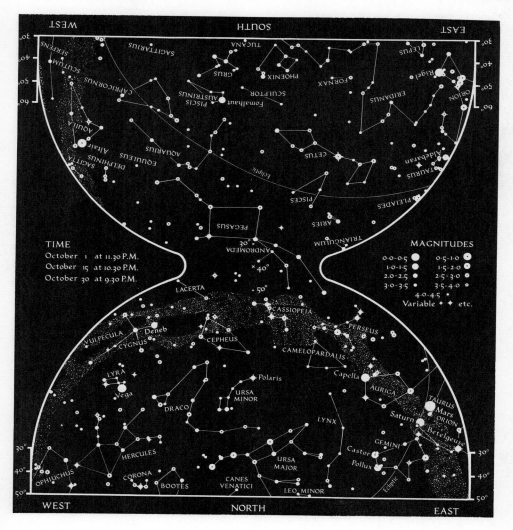

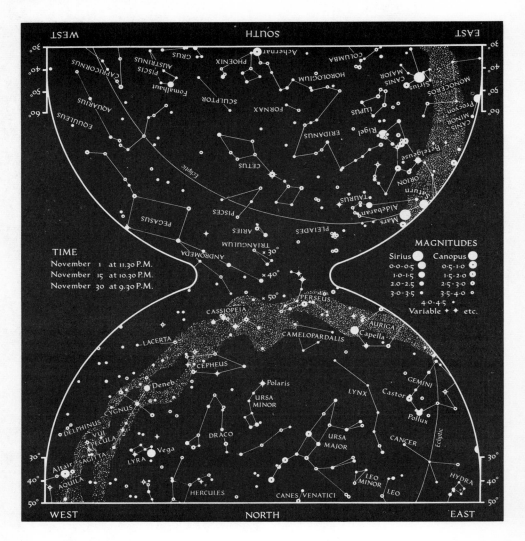

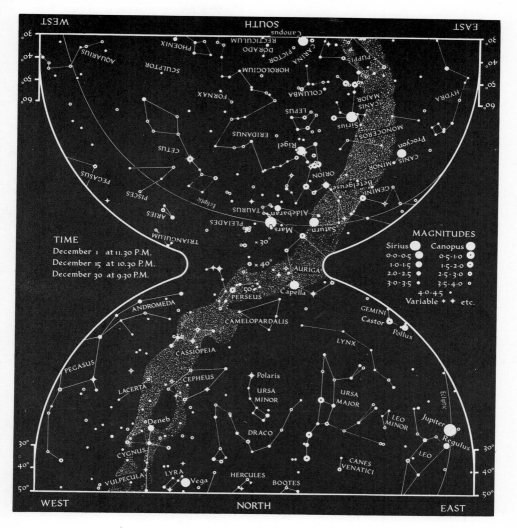

MAGNITUDES

Sirius   Canopus
0·0-0·5   0·5-1·0
1·0-1·5   1·5-2·0
2·0-2·5   2·5-3·0
3·0-3·5   3·5-4·0
4·0-4·5
Variable ✦   ✦ etc.

TIME

December 1 at 11.30 P.M.
December 15 at 10.30 P.M.
December 30 at 9.30 P.M.

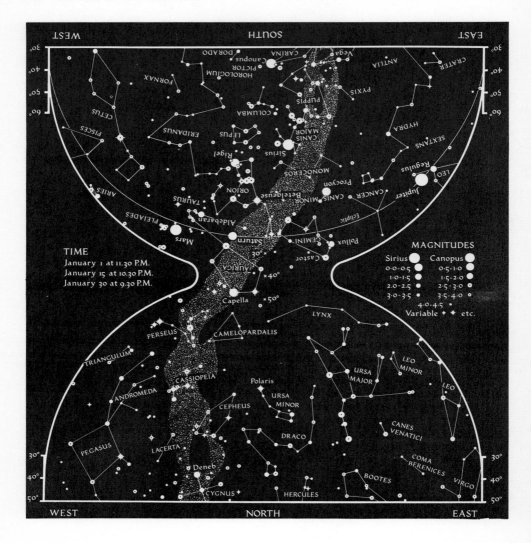

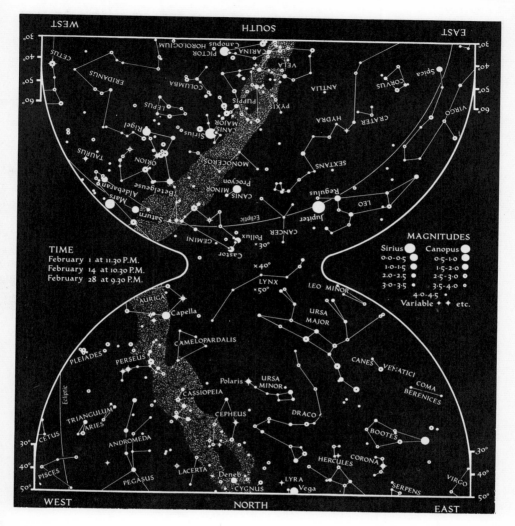

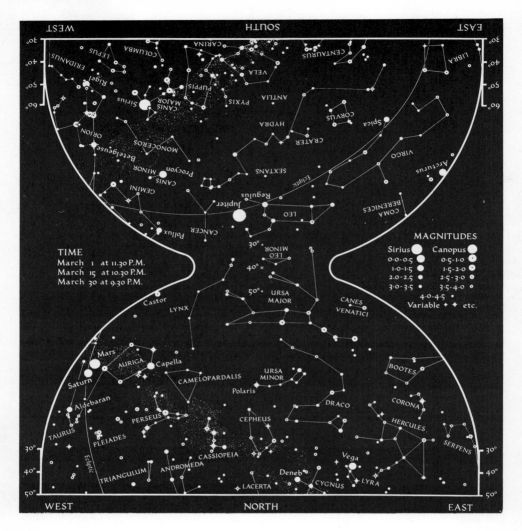

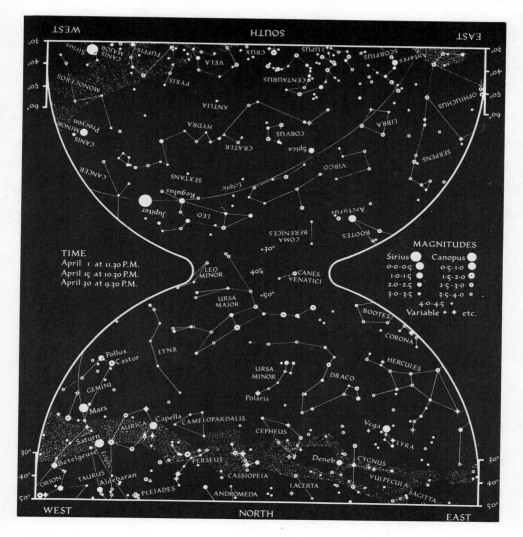

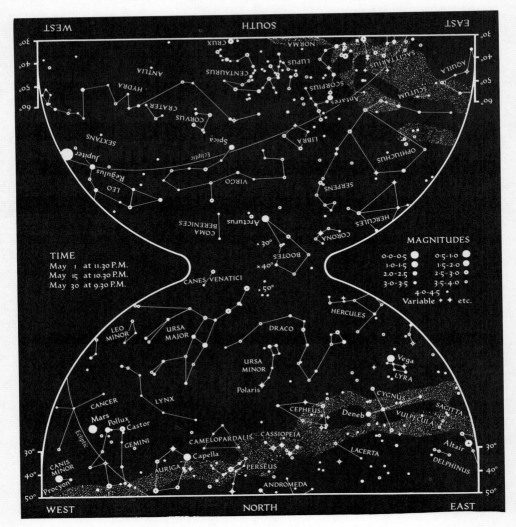

TIME
May 1 at 11.30 P.M.
May 15 at 10.30 P.M.
May 30 at 9.30 P.M.

MAGNITUDES

0.0-0.5	0.5-1.0
1.0-1.5	1.5-2.0
2.0-2.5	2.5-3.0
3.0-3.5	3.5-4.0

4.0-4.5
Variable ✦ ✦ etc.

WEST NORTH EAST

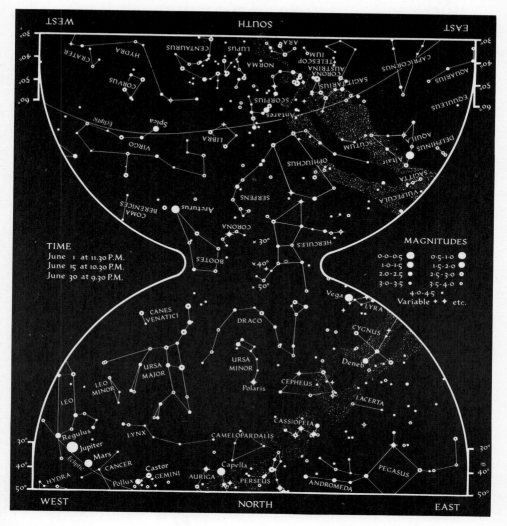

Name Index

Subject Index